Comprehensive Metaheuristics

Comprehensive Metaheuristics

Algorithms and Applications

Edited by

Seyedali Mirjalili
Center for Artificial Intelligence Research and Optimization,
Torrens University Australia, Brisbane, QLD, Australia
University Research and Innovation Center, Obuda University, Budapest, Hungary

Amir H. Gandomi
University of Technology Sydney, Sydney, Australia
University Research and Innovation Center, Obuda University, Budapest, Hungary

Academic Press is an imprint of Elsevier
125 London Wall, London EC2Y 5AS, United Kingdom
525 B Street, Suite 1650, San Diego, CA 92101, United States
50 Hampshire Street, 5th Floor, Cambridge, MA 02139, United States
The Boulevard, Langford Lane, Kidlington, Oxford OX5 1GB, United Kingdom

Notices

Knowledge and best practice in this field are constantly changing. As new research and experience broaden our understanding, changes in research methods, professional practices, or medical treatment may become necessary.

Practitioners and researchers must always rely on their own experience and knowledge in evaluating and using any information, methods, compounds, or experiments described herein. In using such information or methods they should be mindful of their own safety and the safety of others, including parties for whom they have a professional responsibility.

To the fullest extent of the law, neither the Publisher nor the authors, contributors, or editors, assume any liability for any injury and/or damage to persons or property as a matter of products liability, negligence or otherwise, or from any use or operation of any methods, products, instructions, or ideas contained in the material herein.

ISBN 978-0-323-91781-0

For information on all Academic Press publications
visit our website at https://www.elsevier.com/books-and-journals

Publisher: Mara E. Conner
Acquisitions Editor: Chris Katsaropoulos
Editorial Project Manager: Clark M. Espinosa
Production Project Manager: Sreejith Viswanathan
Cover Designer: Vicky Pearson Esser

Typeset by STRAIVE, India

Contents

Contributors

Benyamin Abdollahzadeh
Department of Computer Engineering, Urmia Branch, Islamic Azad University, Urmia, Iran

Silifat Adaramaja Abdulraheem
Department of Computer Science, Ahmadu Bello University, Zaria, Nigeria

Iyad Abu-Doush
Department of Computing, College of Engineering and Applied Sciences, American University of Kuwait, Salmiya, Kuwait; Computer Science Department, Yarmouk University, Irbid, Jordan

Mohammed Azmi Al-Betar
Artificial Intelligence Research Center (AIRC), College of Engineering and Information Technology, Ajman University, Ajman, United Arab Emirates; Department of Information Technology, Al-Huson University College, Al-Balqa Applied University, Al-Huson, Irbid, Jordan

Yusuf Sahabi Ali
Department of Computer Science, Ahmadu Bello University, Zaria, Nigeria

Khalifa Al-Jabri
Department of Civil and Architectural Engineering, College of Engineering, Sultan Qaboos University, Muscat, Oman

Ghazi Al-Rawas
Department of Civil and Architectural Engineering, College of Engineering, Sultan Qaboos University, Muscat, Oman

Ankush Anand
School of Mechanical Engineering, Shri Mata Vaishno Devi University, Katra, Jammu & Kashmir, India

Mohammed A. Awadallah
Department of Computer Science, Al-Aqsa University, Gaza, Palestine; Artificial Intelligence Research Center (AIRC), Ajman University, Ajman, United Arab Emirates

Parnian Hashempour Bakhtiari
Department of Civil and Architectural Engineering, College of Engineering, Sultan Qaboos University, Muscat, Oman

P. Shanthi Bala
Department of Computer Science, School of Engineering and Technology, Pondicherry University, Puducherry, India

Kusum Kumari Bharti
Department of Computer Science and Engineering, PDPM Indian Institute of Information Technology, Design and Manufacturing, Jabalpur, India

Gautam M. Borkar
Department of Information Technology, Ramrao Adik Institute of Technology, D Y Patil Deemed To Be University, Navi Mumbai, Maharashtra, India

Malik Shehadeh Braik
Department of Computer Science, Al-Balqa Applied University, Jordan

Özay Can
Department of Electronics and Automation, Technical Sciences Vocational School, Recep Tayyip Erdogan University, Rize, Turkey

Soumitri Chattopadhyay
Department of Information Technology, Jadavpur University, Kolkata, India

Aybike Özyüksel Çiftçioğlu
Department of Civil Engineering, Faculty of Engineering, Manisa Celal Bayar University, Manisa, Turkey

Ahmet Cevahir Cinar
Department of Computer Engineering, Faculty of Technology, Selçuk University, Konya, Turkey

Serdar Ekinci
Department of Computer Engineering, Batman University, Batman, Turkey

Hasan Eroğlu
Department of Electrical-Electronics Engineering, Faculty of Engineering and Architecture, Recep Tayyip Erdogan University, Rize, Turkey

Amir H. Gandomi
Faculty of Engineering and Information Technology, University of Technology Sydney, Ultimo, NSW, Australia; University Research and Innovation Center, Obuda University, Budapest, Hungary

Mohammadali Geranmehr
Department of Civil and Structural Engineering, University of Sheffield, Sheffield, United Kingdom

Farhad Soleimanian Gharehchopogh
Department of Computer Engineering, Urmia Branch, Islamic Azad University, Urmia, Iran

Ibrahim Hayatu Hassan
Institute for Agricultural Research; Department of Computer Science, Ahmadu Bello University, Zaria, Nigeria

Davut Izci
Department of Electronics & Automation, Batman University, Batman, Turkey

Sehej Jain
Department of Computer Science and Engineering, PDPM Indian Institute of Information Technology, Design and Manufacturing, Jabalpur, India

Isuwa Jeremiah
Department of Computer Science, Ahmadu Bello University, Zaria, Nigeria

Ersin Kaya
Department of Computer Engineering, Faculty of Engineering and Natural Sciences, Konya Technical University, Konya, Turkey

Muhammad Najeeb Khan
School of Mechanical Engineering, Shri Mata Vaishno Devi University, Katra, Jammu & Kashmir, India

Nima Khodadadi
Department of Civil and Environmental Engineering, Florida International University, Miami, FL, United States

Hisham M. Khudhur
Department of Mathematics, College of Computers Sciences and Mathematics, University of Mosul, Mosul, Iraq

Krishanu Kundu
Department of Electronics and Communication Engineering, G.L. Bajaj Institute of Technology & Management, Greater Noida, Uttar Pradesh, India

Aritra Marik
Department of Information Technology, Jadavpur University, Kolkata, India

Mansur Aliyu Masama
E-Library, Kebbi State University of Science and Technology, Aliero, Nigeria

Mwangi Mbuthia
Electrical and Information Engineering, University of Nairobi, Nairobi, Kenya

Seyedali Mirjalili
Centre for Artificial Intelligence Research and Optimisation, Torrens University Australia, Fortitude Valley, Brisbane, QLD, Australia; University Research and Innovation Center, Obuda University, Budapest, Hungary

Abdullahi Mohammed
Department of Computer Science, Ahmadu Bello University, Zaria, Nigeria

Rafaa Mraihi
Ecole Supérieure de Commerce de Tunis, Campus Universitaire de Manouba, Manouba, Tunisia

Mohammad Reza Nikoo
Department of Civil and Architectural Engineering, College of Engineering, Sultan Qaboos University, Muscat, Oman

Abraham Nyete
Electrical and Information Engineering, University of Nairobi, Nairobi, Kenya

Ali Öztürk
Department of Electrical-Electronics Engineering, Faculty of Engineering, Duzce University, Duzce, Turkey

Narendra Nath Pathak
Department of Electronics and Communication Engineering, Dr. B.C. Roy Engineering College, Durgapur, West Bengal, India

Anita R. Patil
Department of Information Technology, Ramrao Adik Institute of Technology, D Y Patil Deemed To Be University, Navi Mumbai, Maharashtra, India

Rishav Pramanik
Department of Computer Science and Engineering, Jadavpur University, Kolkata, India

Bochra Rabbouch
Higher Institute of Applied Sciences and Technology of Sousse, University of Sousse, Sousse, Tunisia

Hana Rabbouch
Higher Institute of Management of Tunis, University of Tunis, Tunis, Tunisia

Sajad Ahmad Rather
Department of Computer Science, School of Engineering and Technology, Pondicherry University, Puducherry, India

Foued Saâdaoui
Department of Statistics, Faculty of Sciences, King Abdulaziz University, Jeddah, Saudi Arabia

Davies Segera
Electrical and Information Engineering, University of Nairobi, Nairobi, Kenya

Sevil Sen
WISE Lab., Department of Computer Engineering, Hacettepe University, Ankara, Turkey

Amit Kumar Sinha
School of Mechanical Engineering, Shri Mata Vaishno Devi University, Katra, Jammu & Kashmir, India

Bahaeddin Turkoglu
Department of Computer Engineering, Faculty of Engineering and Natural Sciences, Konya Technical University, Konya, Turkey

Sait Ali Uymaz
Department of Computer Engineering, Faculty of Engineering and Natural Sciences, Konya Technical University, Konya, Turkey

Selim Yilmaz
WISE Lab., Department of Computer Engineering, Hacettepe University, Ankara; Department of Software Engineering, Muğla Sıtkı Koçman University, Muğla, Turkey

Ehsan Yousefi-Khoshqalb
Faculty of Civil, Water and Environmental Engineering, Shahid Beheshti University, Tehran, Iran

CHAPTER

Chaos theory in metaheuristics

1

Bahaeddin Turkoglu, Sait Ali Uymaz, and Ersin Kaya

Department of Computer Engineering, Faculty of Engineering and Natural Sciences, Konya Technical University, Konya, Turkey

1. Introduction

Optimization is the technique of finding the most suitable solution among the possible solutions for a particular problem. Many problems we encounter in the real world, such as timetabling, path planning, packing, traveling salesman, trajectory optimization, and engineering design problems, basically point to an optimization problem. There are several factors that affect the complexity of an optimization problem. Some of these include the size of the problem, the number of possible solutions, and the problem-specific constraints. Finding the ideal solution for high complexity optimization problems increases the cost (time, memory, etc.). The methods used in solving optimization problems are divided into two categories: deterministic and stochastic. Deterministic methods use gradient methods to reach the ideal solution. Deterministic methods achieve the same ideal solution when operated under the same initial conditions. Alternatively, stochastic methods try to reach the ideal solution using gradient-free techniques and contain randomness. Stochastic methods are examined in two classes as heuristic and metaheuristic [1].

Heuristic methods aim to improve the existing solution by trial and error according to the determined rule [2]. Metaheuristic methods aim to improve multiple agents until they reach the specified stopping criterion by operating multiple iterations with their own rules. As the final solution, the best solution obtained at the end of the search process is presented [3]. Although metaheuristic methods do not guarantee the optimum solution, they stand out with their ability to be applied easily to many problems and find solutions quickly [4,5]. Unlike heuristic methods, metaheuristic methods have a wider application area since they are not problem dependent [6,7].

Metaheuristic methods start the search process with random initial solutions. They use two basic search behaviors to find the ideal solution. These are exploration and exploitation. Exploration refers to the ability of a method to search the solution space, while exploitation refers to the capability to improve a solution. These are important concepts for metaheuristic methods and should be in an ideal balance. In the last few decades, nature-inspired metaheuristic methods have become popular. The main reason for this is that these methods provide effective performances in complex real-world problems. The most important factor behind their effective performance is the finding of structures inspired by natural phenomena that they use in searching the solution space and improving existing solutions. Many nature-inspired metaheuristic optimization algorithms have been proposed in the literature such

Comprehensive Metaheuristics. https://doi.org/10.1016/B978-0-323-91781-0.00001-6

as particle swarm optimization (PSO) [8], genetic algorithm (GA) [9], differential evolution (DE) [10], firefly algorithm (FA) [11], gravitational search algorithm (GSA) [12], harmony search (HS) [13], artificial algae algorithm (AAA) [14,15], artificial bee colony (ABC) [16], and gray wolf optimization (GWO) [17].

Chaos is a dynamic nonlinear system with random behavior that is sensitive to initial conditions. It appears random, but randomness is not necessarily required to provide chaotic behavior [8]. In recent years, chaotic maps have been used instead of random number generators used in metaheuristic methods. In particular, many successful applications have been presented that increase the performance of metaheuristic approaches by using chaotic systems in creating the initial population, increasing the convergence speed of the optimization technique to the global optimum, and searching the solution space. Examples include biogeography-based optimization with chaos [18], chaotic grasshopper optimization [1], chaotic krill herd algorithm [19], chaotic whale optimization algorithm [20], chaotic salp swarm algorithm [21], chaotic fruit fly optimization algorithm [22], and improved particle swarm optimization combined with chaos [23].

In this chapter, we first explain the dynamics and structure of the chaotic system. Then, we discuss the relationship between chaotic systems and metaheuristic algorithms and how chaotic approaches are applied in different components of metaheuristic algorithms. We also present metaheuristic algorithms with chaotic approaches that show successful performances.

The organization of the chapter is as follows. Section 2 explains chaos systems and examines 10 chaotic maps that are widely known in the literature. Section 3 discusses the relationship between chaotic systems and metaheuristic algorithms. Section 4 examines the chaotic version of the GSA, which has shown a successful performance, and explains the application results.

2. Chaos system and chaotic maps

Chaos theory refers to chaotic dynamic nonlinear systems. These systems show high sensitivity to initial parameters. In other words, small chaotic improvements in the initial parameters lead to big changes in the output and performance of the system. However, chaos systems are random. Although chaotic systems show random behavior, they do not need random parameters. Another feature of these systems is ergodicity. The ergodicity property of chaos can ensure chaotic variables traverse all states nonrepeatedly within a certain range according to its own laws. So, this can be used as a mechanism that avoids falling into the local minimum solution. Thanks to these features, chaotic maps inspired by chaos systems are frequently used to increase the performance of metaheuristic population-based optimization algorithms [24,25].

Chaotic maps are a stochastic, deterministic, and nonlinear strategy that is frequently used in generating long-term random numbers. The numbers in the sequences created by these maps have great advantages, such as not falling into repetitions, being spread over a wide spectrum instead of being stuck in a certain region, and low sequence production and storage costs. Because the randomly generated numbers in metaheuristic algorithms could be the same and within a certain range, these random numbers sometimes cause algorithms to be stuck into local minimums. For such reasons, chaotic sequences with a certain system, which can encompass the entire spectrum and do not fall into repetition, can be produced, and these sequences can be used instead of using randomly generated numbers. Although there are many different types of chaotic maps in the literature, such as Henon [26], Ikeda [27],

Zaslavskii [28], Intermittency [29], Liebovitch [30], and Tinkerbell [31], there are 10 chaotic maps that are frequently preferred in studies. These are Chebyshev [32], circle [33], Gaussian [20], iterative [34], logistic [35], piecewise [19], sine [18], singer [36], sinusoidal [37], and tent [38] chaotic maps. Chaotic sequences are created from these chaotic map equations with an initial parameter of the desired size. Instead of using random numbers in the required position, the next number is drawn from the generated chaotic sequence and used. The numbers in this sequence are unlikely to be the same and are spread over a wide spectrum [39].

2.1 Chebyshev map

Eq. (1) shows the Chebyshev chaotic map formulation. The number of k iterations in the equation, x_k k, represents the chaotic number.

$$x_{k+1} = \cos\left(k\cos^{-1}(x_k)\right) \tag{1}$$

Fig. 1 shows the spectrum of 100 numbers generated with the initial parameter $x_1 = 0.7$ using the Chebyshev map equation.

2.2 Circle map

Eq. (2) shows the formulation of the circle chaotic map. The number of k iterations in the equation, x_k k, represents the chaotic number, and a and b are the equation constants.

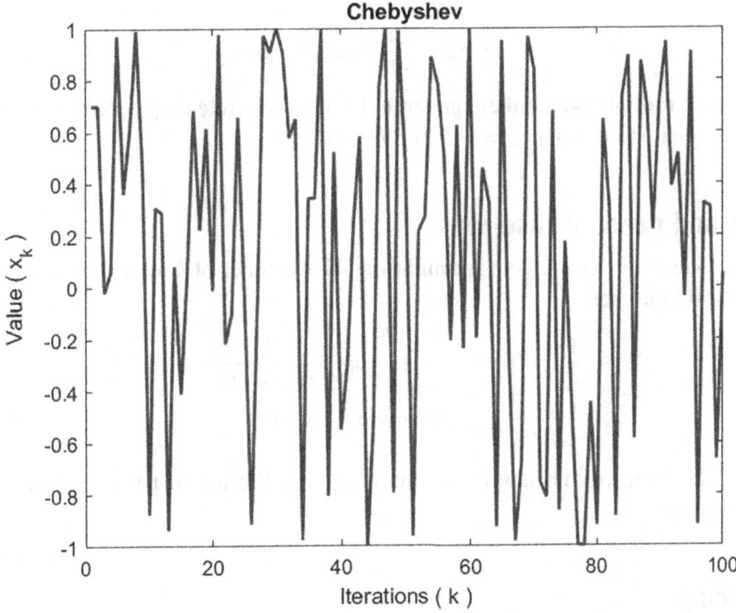

FIG. 1

Chebyshev map spectrum.

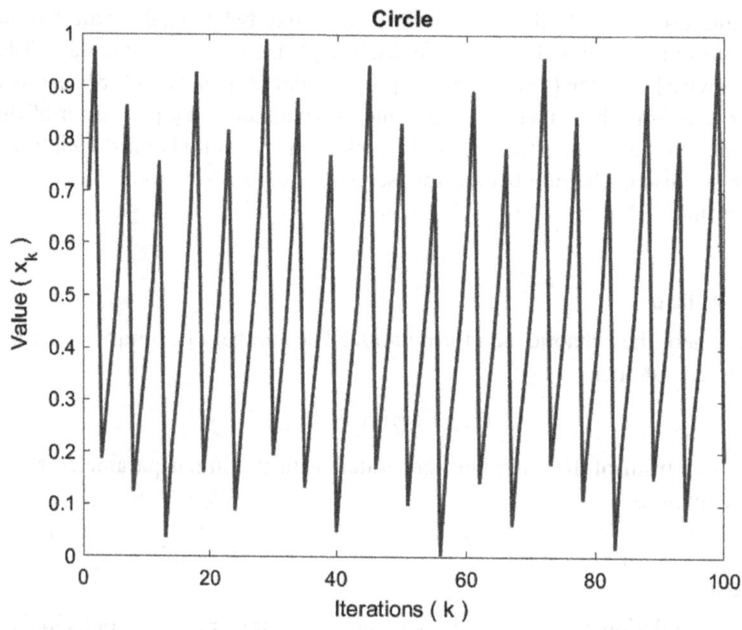

FIG. 2

Circle map spectrum.

$$x_{k+1} = x_k + b - \left(\frac{a}{2\pi}\right) \sin(2\pi k) \bmod (1) \tag{2}$$

Fig. 2 shows the spectrum of 100 numbers generated using the circle map equation with initial parameter $x_1 = 0.7$ and equation constants $a = 0.5$ and $b = 0.2$.

2.3 Gaussian and gauss-mouse map

Eq. (3) shows the Gaussian chaotic map formulation. The number of k iterations in the equation, x_k k, represents the chaotic number.

$$x_{k+1} = \begin{cases} 0 & x_k = 0, \quad \dfrac{1}{x_k \bmod (1)} = \dfrac{1}{x_k} - \left[\dfrac{1}{x_k}\right] \\ \dfrac{1}{x_k \bmod (1)} & \text{otherwise}, \quad x \geq 0 \end{cases} \tag{3}$$

Fig. 3 shows the spectrum of 100 numbers generated with the initial parameter $x_1 = 0.7$ using the Gaussian map equation.

2.4 Iterative map

Eq. (4) shows the formulation of the iterative chaotic map. The number of k iterations in the equation, x_k k, represents the chaotic number, and a is the equation constant.

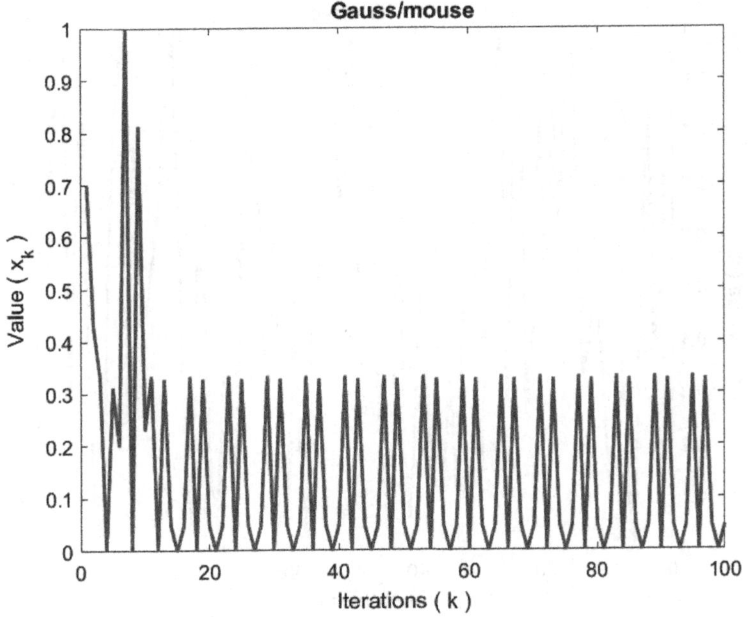

FIG. 3

Gaussian map spectrum.

$$x_{k+1} = \sin\left(\frac{a\pi}{x_k}\right), \quad a \in (0, 1) \tag{4}$$

The iterative map generates values between [−1, 1], and these values are normalized to [0, 1]. Fig. 4 shows the spectrum of 100 numbers generated using the iterative map equation with initial parameter $x_1 = 0.7$ and chaotic constant $a = 0.7$.

2.5 Logistic map

Eq. (5) shows the logistic chaotic map formulation. The number of k iterations in the equation, x_k k, represents the chaotic number, and a is the equation constant. The values that the constant a can take for the equation to generate chaotic numbers are in the range of [3.57, 4].

$$x_{k+1} = a x_k (1 - x_k) \tag{5}$$

Fig. 5 shows the spectrum of 100 numbers generated using the logistic map equation with initial parameter $x_1 = 0.7$ and chaotic constant $a = 4$.

2.6 Piecewise map

Eq. (6) shows the formulation of the piecewise chaotic map. The number of k iterations in the equations, x_k k, represents the chaotic number, and P is the equation constant.

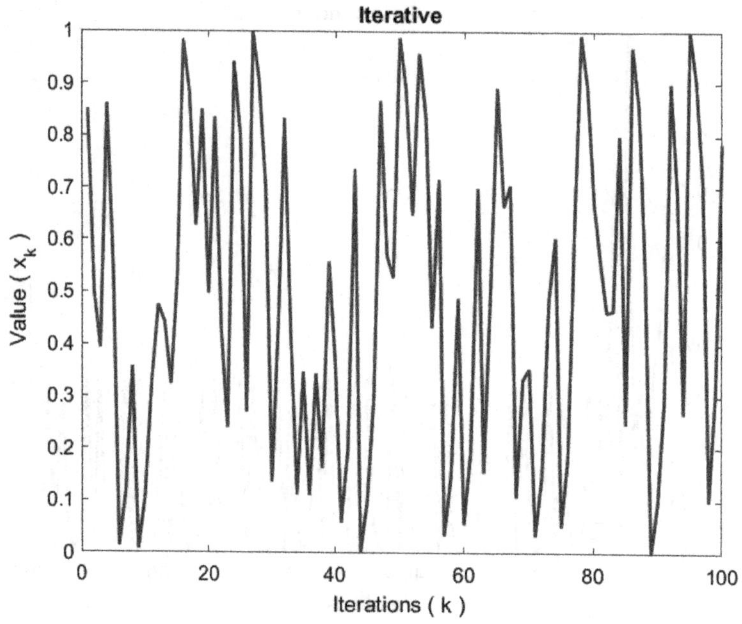

FIG. 4

Iterative map spectrum.

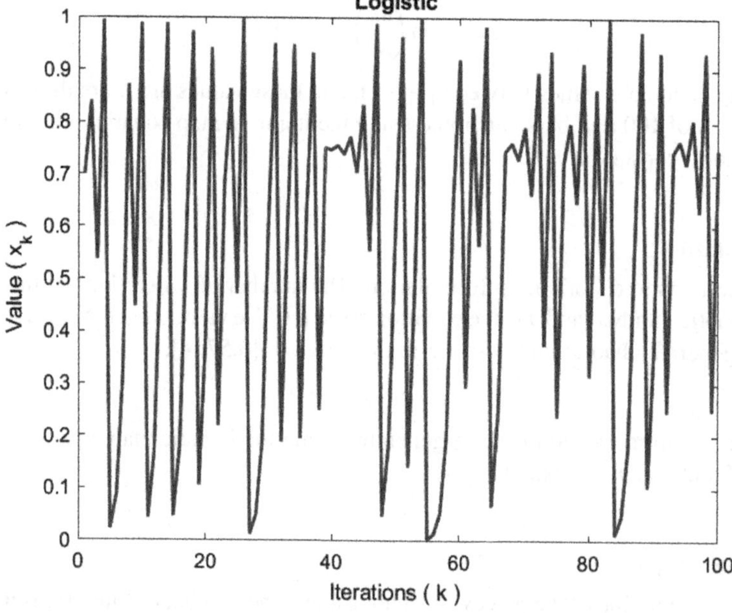

FIG. 5

Logistic map spectrum.

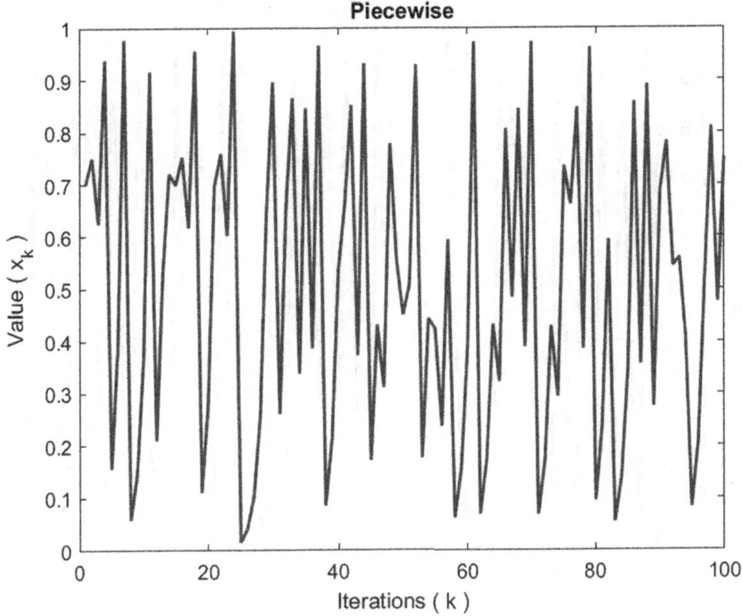

FIG. 6

Piecewise map spectrum.

$$x_{k+1} = \begin{cases} \frac{x_k}{P}, & 0 \leq x_k < P \\ \frac{x_k - P}{0.5 - P}, & P \leq x_k < \frac{1}{2} \\ \frac{1 - P - x_k}{0.5 - P}, & \frac{1}{2} \leq x_k < 1 - P \\ \frac{1 - x_k}{P}, & 1 - P \leq x_k < 1 \end{cases} \quad P = 0.4 \tag{6}$$

Fig. 6 shows the spectrum of 100 numbers generated with the initial parameter of the piecewise map equation $x_1 = 0.7$ and equation constant $P = 4$.

2.7 Sine map

Eq. (7) shows the sine chaotic map formulation. The number of k iterations in the equation, x_k k, represents the chaotic number and a is the equation constant.

$$x_{k+1} = \frac{a}{4} \sin(\pi x_k), \quad 0 \leq a < 4 \tag{7}$$

Fig. 7 shows the spectrum of 100 numbers generated using the sine map equation with initial parameter $x_1 = 0.7$ and equation constant $a = 4$.

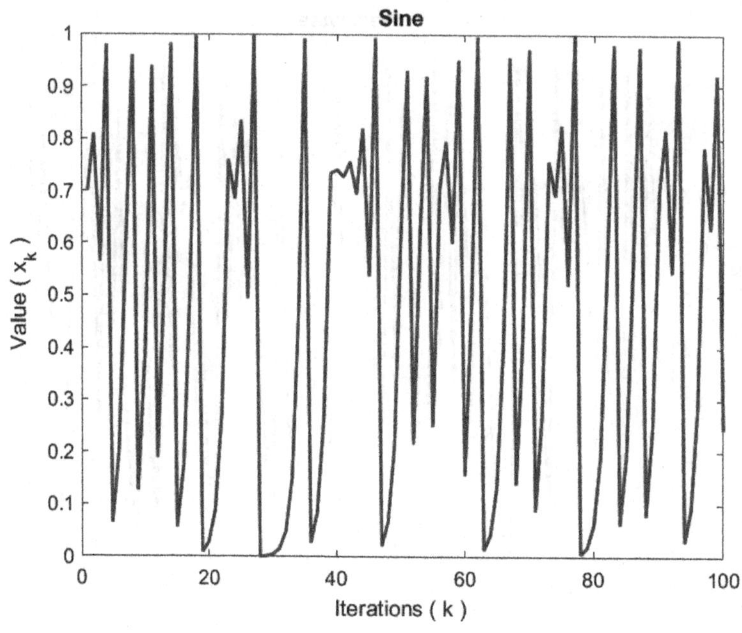

FIG. 7

Sine map spectrum.

2.8 Singer map

Eq. (8) shows the formulation of the singer chaotic map. The number of k iterations in the equation, x_k k, represents the chaotic number, and μ is the equation constant.

$$x_{k+1} = \mu\left(7.86x_k - 23.31x_k^2 + 28.75x_k^3 - 13.302875x_k^4\right) \tag{8}$$

Fig. 8 shows the spectrum of 100 numbers generated using the singer map equation with initial parameter $x_1 = 0.7$ and equation constant $\mu = 1.07$.

2.9 Sinusoidal map

Eq. (9) shows the sinusoidal chaotic map formulation. The number of k iterations in the equation, x_k k, represents the chaotic number, and a is the equation constant.

$$x_{k+1} = ax_k^2 \sin(\pi x_k) \tag{9}$$

Fig. 9 shows the spectrum of 100 numbers generated using the sinusoidal map equation with initial parameter $x_1 = 0.7$ and equation constant $a = 2.3$.

2.10 Tent map

Eq. (10) shows the formulation of the tent chaotic map. The number of k iterations in the equation, x_k k, represents the chaotic number.

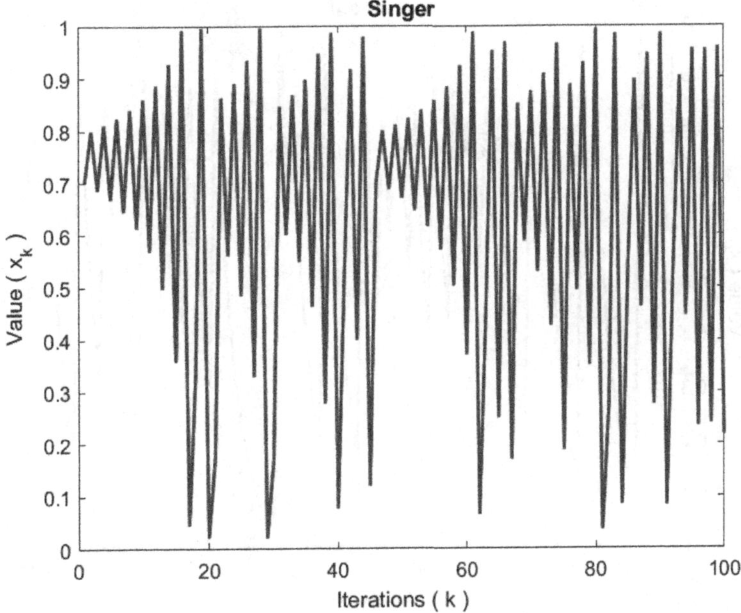

FIG. 8

Singer map spectrum.

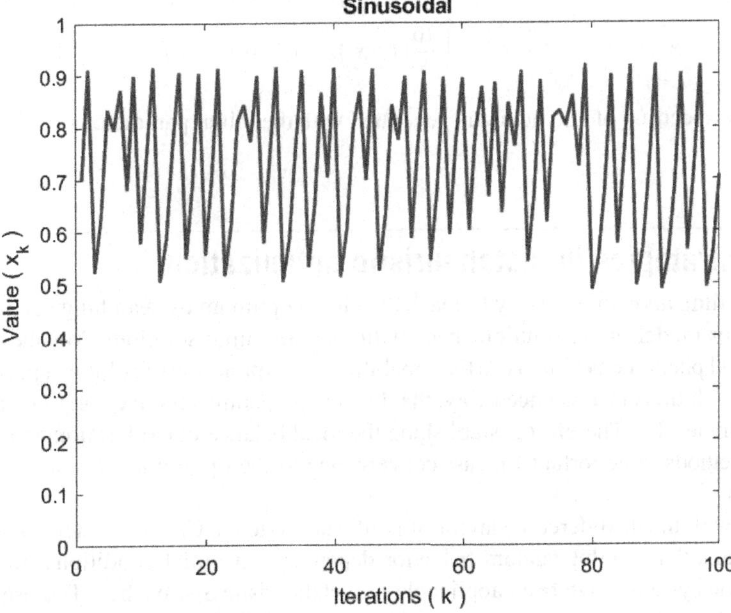

FIG. 9

Sinusoidal map spectrum.

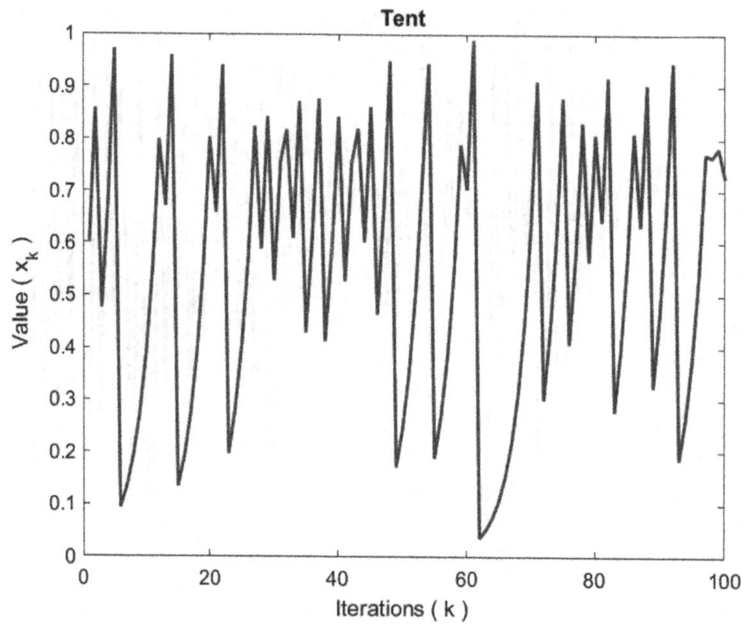

FIG. 10

Tent map spectrum.

$$x_{k+1} = \begin{cases} \dfrac{x_k}{0.7}, & x_k < 0.7 \\ \dfrac{10}{3}(1 - x_k), & x_k \geq 0.7 \end{cases} \tag{10}$$

Fig. 10 shows the spectrum of 100 numbers generated with the initial parameter $x_1 = 0.6$ using the tent map equation.

3. Chaotic strategies in metaheuristic optimization

Metaheuristic optimization methods try to reach the global optimum by searching according to the determined behavior model in the solution space with random initial solutions. Metaheuristic methods inspired by natural phenomena aim to use the capabilities of exploration of solution space and improvement of existing solutions in a balanced way, thanks to hyperparameters. Exploration and exploitation capabilities are trade-offs. Therefore, establishing the ideal balance of exploration and exploitation of metaheuristic methods is important for fast convergence to the optimum solution by avoiding local minimum points.

Chaos deals with the disordered behavior of nonlinear systems. Chaotic systems are nonlinear deterministic systems that exhibit random behavior dependent on initial conditions. Because of these properties, chaotic systems have been applicable in metaheuristic approaches. The use of chaotic approaches in metaheuristic methods emerges as the creation of the initial population, the realization of random movements of the candidate solutions, the determination of the hyperparameters of the methods, or the hybrid use of these models. There are many metaheuristic optimization methods with chaotic approaches in the literature.

3.1 Studies using different chaotic strategies in metaheuristics

Jothiprakash and Arunkumar presented the chaotic versions of GA and DE, which are evolutionary optimization algorithms, in their study in 2013. In the study, the initial populations of the GA and DE method were created using a logistic chaotic map. The proposed chaotic GA and DE approaches were used to solve the maximizing the hydropower production from a reservoir. Experimental study has shown that chaotic versions of GA and DE algorithms achieved more successful results than their original versions [40]. In 2010, Min-Yuan and Kuo-Yu proposed the k-means chaos genetic algorithm approach. In this approach, the initial population was generated with a chaotic approach and hybridized with classical genetic algorithm operators (selection, crossover, and mutation) and chaotic approaches [41]. Akhtar et al. used a chaotic version of the cuckoo search algorithm to solve substitution box design problems in their study in 2019. In this version, logistic and tent chaotic maps were used to create the initial population [42].

In a 2015 study, the crossover operator, which is an important process in the generation of new individuals in genetic algorithms, was combined with a chaotic approach, and a new chaotic crossover operator was proposed. Due to the new chaotic crossover operator, the exploration capability of the classical genetic algorithm increased [43]. In the differential evolution method, new individuals are produced within the framework of the strategy determined by using the individuals and their neighbors. In this approach, a new chaotic mutation operator has been introduced to generate new individuals. In this way, the ability of the proposed method to avoid local minima is increased. The logistic chaotic map is used in the new mutation operator proposed [44].

Liu et al. proposed two modifications to improve the performance of the classical PSO method. The first of these was the adaptive inertia weight factor modification, which enhanced the exploration capability of the method. Second, a chaotic local search modification was added to improve the fitness value of the best individual by using the logistic chaotic map. Chaotic local search modification increases the exploitation capacity of the method [23]. In the original GSA, a random number as an updating parameter is used to overcome early convergence and avoid local minima. An operator based on chaotic dynamics has been added to the standard GSA to balance between exploration and exploitation [45].

The biogeography-based optimization (BBO) method is an evolutionary algorithm inspired by biogeography. In the method, candidate solutions move in the solution space in line with the parameters of emigration, immigration, and mutation. This algorithm has problems with trapping local minima and slow convergence. Saremi and Mirjalili presented an approach in which emigration, immigration, and mutation probability parameters, which are important for the BBO, are determined with 10 different chaotic maps. The presented approach was tested in unimodal and multimodal numeric benchmark functions, and it was observed that the exploration and exploitation abilities of the method were increased [18].

Grasshopper Optimization Algorithm (GOA) [46] has two critical parameters: c1 and c2. The c1 parameter balances exploration and exploitation, while c2 reduces the attraction zone, comfort zone, and repulsion zone between grasshoppers. These are vital in deciding the rate of convergence and how to update the grasshoppers' position. In GOA, c1 and c2 are set to 1 and reduced linearly in the first step of the optimization process. In the chaotic GOA algorithm, the c1 and c2 parameters are adjusted using various chaotic maps [1].

The main parameters of the Krill Herd algorithm (KH) are inertia weights (ωn, ωf), which represent variations of global optimal gravity. These parameters are vital in deciding the rate of convergence and how to update the krill's position. In KH, the inertia weights (ωn, ωf) are initially set to 0.9 for scouting capability and are reduced linearly to 0.1 during the search process. Inertia weights (ωn, ωf) in chaotic KH were determined with chaotic maps [19]. In the Whale Optimization Algorithm (WOA), candidate solutions search for the optimum solution using the encircling prey, bubble-net attacking, and search for prey operators. In the bubble-net attacking and search for prey operators, the position of the individual is updated with a spiral movement and a randomly selected neighbor, respectively, while the position of the individual is updated using the best individual information in the population in the encircling prey stage. Updating of the candidate solutions depends on the p parameter. In the Chaotic Whale Optimization Algorithm (CWOA), the parameter p is created using 10 different chaotic maps. Thus, the convergence to optimum solution and solution space exploring capability of the classical WOA method has been increased [20].

Salp Swarm Algorithm (SSA) has three main parameters that affect its performance. These are $r1$, $r2$, and $r3$. While $r1$ decreases linearly over the iterations, $r3$ is responsible for determining whether the direction of the next position will be negative or positive. $r2$ and $r1$ are the two main parameters that affect the update position of a salp. Chaotic maps are used to adjust the $r2$ parameter of the chaotic SSA [21]. The Fruit Fly Optimization Algorithm (FOA) uses uniform distribution and random variables in the production of new food sources. Alternatively, chaotic FOA has presented a new parameter (alpha) whose value is determined by chaotic maps for the production of food sources [22].

Table 1 presents metaheuristic algorithms that are widely used in the literature and their chaotic versions.

Table 1 Metaheuristic algorithms and their chaotic versions.

Algorithm	Original paper	Chaotic version
Ant colony optimization	[47]	[48]
Artificial bee colony	[16]	[49–51]
Bat algorithm	[52]	[33,53]
Biogeograpy-based optimization	[54]	[18,55,56]
Butterfly optimization algorithm	[57]	[58]
Cuckoo search algorithm	[59]	[60–62]
Differential evolution	[63]	[64,65]
Dragonfly algorithm	[66]	[67]
Evolutionary algorithm	[68]	[69]
Firefly algorithm	[11]	[31,38,70,71]
Fruit fly algorithm	[72]	[22]
Grasshopper optimization algorithm	[46]	[1]
Gravitational search algorithm	[12]	[73–75]
Gray wolf optimizer	[17]	[76]
Krill Herd algorithm	[77]	[19]
Moth flame optimization	[78]	[79,80]
Multi verse optimization	[81]	[82]
Partcile swarm optimization	[8]	[23,24,83]
Salp swarm algorithm	[84]	[21,85]
Squirrel search algorithm	[86]	[87,88]
Whale optimization algorithm	[89]	[20,90–93]

4. An application with chaotic system

In this section, the study of "Chaotic gravitational constants for the gravitational search algorithm (CGSA)" [25] developed by Mirjalili and Gandomi is presented as a chaotic application.

The GSA was introduced to the literature by Rashedi et al. in 2009 [12]. This algorithm is a metaheuristic optimization algorithm inspired by Newton's laws of motion and gravity. Each agent in the search space is considered as a mass. According to the law of motion, these masses attract each other and exert a force on each other with the force of gravity. The masses exposed to these forces try to find the optimal solution moving in the search space. Each mass represents a candidate solution. The size of the mass means its fitness value. As the masses get heavier, they move slowly. This helps heavy masses with a good solution move slowly around the current solution to find a better solution. This process provides the exploitation ability of the algorithm. The lighter masses move faster, and heavy masses exert a force to attract lighter masses to themselves. This process also provides exploration capability. In this way, light masses enhance their solutions. At the end of the iterations, the heaviest mass denotes the best solution. The pseudo-code is shown in Algorithm 1.

In pseudo-code, x_i^d is the position of ith object in the dth dimension. $M_{ai}(t), M_{pi}(t), M_{ii}(t)$ are respectively the active, passive, and inertia mass, and $fobj_i(t)$ is the objective value of the agent i at the time t. The active, passive, and inertia mass of agent i is calculated according to its current objective function $fobj_i(t)$.

Within the scope of the publication, the gravitational constant (G), which is the key parameter of the GSA, is equipped with chaotic maps. This is the main parameter to balance exploration and exploitation. It is normalized by decreasing over the iterations. The previously mentioned 10 chaotic maps were used to determine the value of this parameter. The experimental study was evaluated on 12 benchmark function sets in Table 2, consisting of unimodal and multimodal functions.

Table 3 gives the rank results of the experimental study performed on benchmark functions using 10 chaotic maps. The GSA with sinusoidal chaotic maps performs better exploration according to these results. These results show that the CGSA can avoid local minima (explore the search space) better than GSA.

Algorithm 1 GSA pseudo-code.

Problem Definition, Objective function $f(x)$ *and Solution vector* $X_i=(x_i^1,\ldots,x_i^d,\ldots,x_i^m)$ $i=1,\ldots,N$

Initialize a population of n agents with random solutions

Define algorithm parameters

While *(t<Max Calculation or Stopping Criteria)*

    Evaluate each $(X_i : fobj_i)$

    Calculate each $(M_i : M_{ai}(t),\ M_{pi}(t),\ M_{ii}(t) \propto fobj_i(t))$

    Updating gravitational constant: $G(t)$

    Calculate Forces $F_i^d=\sum_{j\in kbest\, j\neq i} rand_i F_{ij}^d$

    Update acceleration: $a_i^d(t)=\frac{F_i^d(t)}{M_{ii}(t)}$

    Update Velocities: $v_i^d(t+1)=rand_i\, v_i^d(t)+a_i^d(t)$

    Update Positions: $x_i^d(t+1)=x_i^d(t)+v_i^d(t+1)$

end While

Table 2 Benchmark functions.

No.	Formula of functions	Range	Dim	Min	Type
F1	$F_1(x)=\sum_{i=1}^{\text{Dim}}(x_i+40)^2-80$	[−100, 100]	30	−80	*U*
F2	$F_2(x)=\sum_{i=1}^{\text{Dim}}\lvert x_i+7\rvert+\prod_{i=1}^{\text{Dim}}\lvert x_i+7\rvert-80$	[−10, 10]	30	−80	*U*
F3	$F_3(x)=\sum_{i=1}^{\text{Dim}}\left(\sum_{j=i}^{i}x_j+60\right)^2-80$	[−100, 100]	30	−80	*U*
F4	$F_4(x)=\max\{\lvert x_i+60\rvert, 1\le i\le \text{Dim}\}-80$	[−100, 100]	30	−80	*U*
F5	$F_5(x)=\sum_{i=1}^{\text{Dim}}\left(100\left((x_{i+1}+60)-(x_i+60)^2\right)^2+((x_i+60)-1)^2\right)-80$	[−30, 30]	30	−80	*U*
F6	$F_6(x)=\sum_{i=1}^{\text{Dim}}((x_i+60)+0.5)^2-80$	[−100, 100]	30	−80	*U*
F7	$F_7(x)=\sum_{i=1}^{\text{Dim}}-(x_i+300)\bullet\sin\left(\sqrt{\lvert(x_i+300)\rvert}\right)$	[−100, 100]	30	−418.98 × Dim	*M*
F8	$F_8(x)=\sum_{i=1}^{\text{Dim}}\left[(x_i+2)^2-10\cos(2\pi(x_i+2))+10\right]-80$	[−5.12, 5.12]	30	−80	*M*
F19	$F_9(x)=-20\bullet\exp\left(-0.2\sqrt{\frac{1}{\text{Dim}}\sum_{i=1}^{\text{Dim}}(x_i+20)^2}\right)+\exp\left(\frac{1}{\text{Dim}}\sum_{i=1}^{\text{Dim}}\cos(2\pi(x_i+20))\right)+20+\exp(1)-80$	[−32, 32]	30	−80	*M*
F10	$F_{10}(x)=\frac{1}{4000}\sum_{i=1}^{\text{Dim}}(x_i+400)^2-\prod_{i=1}^{\text{Dim}}\cos\left(\frac{(x_i+400)}{\sqrt{i}}\right)+1-80$	[−600, 600]	30	−80	*M*
F11	$F_{11}(x)=\frac{\pi}{\text{Dim}}\left\{10\sin(\pi y_1)+\sum_{i=1}^{\text{Dim}-1}(y_i-1)^2[1+10\sin^2(\pi y_{i+1})]+(y_{\text{Dim}-1})^2\right\}+\sum_{i=1}^{\text{Dim}}u((x_i+30),10,100,4)-80$ $y_i=1+\frac{((x_i+30)+1)}{4}$ $u_{y_i,a,k,m}=\begin{cases}k(x_i-a)^m, & x_i>a\\ 0, & -a\le x_i\le a\\ k(-x_i-a)^m, & x_i<a\end{cases}$	[−100, 100]	30	−80	*M*
F12	$F_{12}(x)=\frac{1}{10}\left\{\sin^2(3\pi(x_1+30))+\sum_{i=1}^{\text{Dim}-1}((x_i+30)-1)^2[1+\sin^2(3\pi(x_{i+1}+30)+1)]+((x_{\text{Dim}}+30)-1)^2(1+\sin^2(2\pi(x_{i+1}+30)))\right\}+\sum_{i=1}^{\text{Dim}}u((x_i+30),5,100,4)-80$	[−50, 50]	30	−80	*M*

Table 3 The rank of CGSA and GSA on the test functions.

Algorithms	Chaotic map	Benchmark test functions											
		F1	*F2*	*F3*	*F4*	*F5*	*F6*	*F7*	*F8*	*F9*	*F10*	*F11*	*F12*
Original GSA	–	11	11	11	11	11	11	11	3	11	10	11	11
CGSA1	Chebysev	3	6	4	3	4	3	3	5	4	3	4	7
CGSA2	Circle	4	5	8	7	9	9	7	8	6	7	9	8
CGSA3	Gaussian	10	10	10	10	6	10	10	1	10	11	10	10
CGSA4	Iterative	6	7	7	2	7	7	6	10	9	5	7	5
CGSA5	Logistic	9	9	3	9	5	6	9	2	7	6	3	6
CGSA6	Piecewise	4	3	5	5	8	5	8	6	8	4	8	4
CGSA7	Sine	8	4	9	6	10	8	5	4	3	9	5	9
CGSA8	Singer	1	2	2	8	2	2	2	7	1	2	1	2
CGSA9	Sinusodial	2	1	1	1	3	1	1	11	5	1	2	1
CGSA10	Tent	7	8	6	4	1	4	4	9	2	8	6	3

To prove the superiority of the proposed CGSA9 (as the best CGSA), compared with many of the most popular recent algorithms such as PSO, GA, FA, States of Matter Search (SMS), and Flower Pollination Algorithm (FPA). Table 4 shows the values of the hyperparameters of these algorithms.

All algorithms run the same fitness evaluation count with an equal number of search agents and 20 independent runs for a fair comparison. The benchmark set in Table 2 was used for comparison. Table 5 presents the mean and standard deviation results.

Table 4 Initial values for the controlling parameters of algorithms.

Algorithms	Hyperparameters	Values
PSO	Topology	Fully connected
	Cognitive and social constants	1.5, 1.5
	Inertial weight	Linearly decreases from 0.6 to 0.3
GA	Type	Real coded
	Selection	Roulette wheel
	Crossover	Single point (probability = 1)
	Mutation	Uniform (probability = 0.01)
FPA	Probability switch (ρ)	0.4
SMS	Beta	[0.9, 0.5, 0.1]
	Alpha	[0.3, 0.05, 0]
	H	[0.9, 0.2, 0]
	Phases	[0.5, 0.1, −0.1]
FA	α	0.1
	β_{min}	0.2
	γ	1

Table 5 Comparison of CGSA9 with well-known and recent algorithms.

Benchmark function		CGSA9	PSO	GA	FPA	SMS	FA
*F*1	Mean	**0.0000**	0.1169	1.0000	0.2079	0.8846	0.0002
	±std	**0.0000**	0.2290	1.0000	0.4158	0.0000	0.0001
*F*2	Mean	**0.0000**	0.0000	1.0000	0.0000	0.0022	0.0102
	±std	**0.0000**	0.0000	1.0000	0.0000	0.0000	0.0057
*F*3	Mean	**0.0000**	0.0250	0.0708	0.0021	1.0000	0.0014
	±std	**0.0027**	0.2239	1.0000	0.0595	0.0000	0.0273
*F*4	Mean	0.0278	0.2633	1.0000	0.3860	0.2803	**0.0000**
	±std	0.2632	0.4271	0.8132	1.0000	0.0000	**0.2411**
*F*5	Mean	**0.0000**	0.0012	1.0000	0.0569	0.3784	0.0000
	±std	**0.0000**	0.0042	1.0000	0.1271	0.0000	0.0000
*F*6	Mean	**0.0000**	0.2424	0.8355	0.1643	1.0000	0.0002
	±std	**0.0000**	0.5602	1.0000	0.7376	0.0000	0.0021
*F*7	Mean	**0.0000**	0.0267	1.0000	0.1112	0.9249	0.0029
	±std	**0.1125**	0.0757	1.0000	0.3260	0.0000	0.0359
*F*8	Mean	1.0000	1.0000	**0.0000**	1.0000	1.0000	1.0000
	±std	0.0125	0.0117	**1.0000**	0.0054	0.0000	0.0205
*F*9	Mean	**0.0000**	0.1889	0.5036	1.0000	0.1660	0.0631
	±std	**0.8809**	0.7409	1.0000	0.5614	0.0000	0.4355
*F*10	Mean	**0.0000**	0.9372	0.9604	0.9523	0.9545	1.0000
	±std	**1.0000**	0.2068	0.0894	0.0352	0.0000	0.0263
*F*11	Mean	0.6835	0.6795	0.8457	0.1424	1.0000	**0.0000**
	±std	0.2996	0.4168	1.0000	0.4209	0.0000	**0.0229**
*F*12	Mean	**0.0000**	0.0009	1.0000	0.0415	0.0852	0.0000
	±std	**0.0000**	0.0042	1.0000	0.1497	0.0000	0.0000

The results in Table 5 show that the CGSA9 algorithm outperforms other algorithms in most of the benchmark test functions. The best results in this table are indicated in bold style. The results in this chapter show that it is advantageous to integrate chaotic maps into GSA. The reason why the CGSA9 algorithm outperforms all other algorithms is that the sinusoidal chaotic map helps the algorithm to place a high emphasis on exploration in the initial stages of iterations, while the downward trend of gravitational constants (G) promotes exploration in the final stages of iterations.

References

[1] S. Arora, P. Anand, Chaotic grasshopper optimization algorithm for global optimization, Neural Comput. Applic. 31 (8) (2019) 4385–4405.

[2] A.R. Yildiz, An effective hybrid immune-hill climbing optimization approach for solving design and manufacturing optimization problems in industry, J. Mater. Process. Technol. 209 (6) (2009) 2773–2780.

[3] K. Hussain, et al., Metaheuristic research: a comprehensive survey, Artif. Intell. Rev. 52 (4) (2019) 2191–2233.
[4] J. Stork, A.E. Eiben, T. Bartz-Beielstein, A new taxonomy of global optimization algorithms, Nat. Comput. 1 (24) (2020) 219–242, https://doi.org/10.1007/s11047-020-09820-4.
[5] B. Turkoglu, E. Kaya, Training multi-layer perceptron with artificial algae algorithm, Eng. Sci. Technol. Int. J. 23 (6) (2020) 1342–1350, https://doi.org/10.1016/j.jestch.2020.07.001.
[6] M. Blocho, Heuristics, metaheuristics, and hyperheuristics for rich vehicle routing problems, in: Smart Delivery Systems: Solving Complex Vehicle Routing Problems, 2020, pp. 101–156.
[7] B Turkoglu, S.A. Uymaz, E. Kaya, Clustering analysis through artificial algae algorithm, Int. J. Mach. Learn. Cybern. 13 (4) (2022) 1179–1196, https://doi.org/10.1007/s13042-022-01518-6.
[8] J. Kennedy, R. Eberhart, Particle swarm optimization, in: Proceedings of ICNN'95-International Conference on Neural Networks, IEEE, 1995.
[9] S. Mirjalili, Genetic algorithm, in: Evolutionary Algorithms and Neural Networks, Springer, 2019, pp. 43–55.
[10] V. Feoktistov, Differential Evolution, Springer, 2006.
[11] X.-S. Yang, Firefly algorithms for multimodal optimization, in: International Symposium on Stochastic Algorithms, Springer, 2009.
[12] E. Rashedi, H. Nezamabadi-Pour, S. Saryazdi, GSA: a gravitational search algorithm, Inform. Sci. 179 (13) (2009) 2232–2248.
[13] S. Adeli, M.P. Aghababa, Metasearch engine result optimization using reformed genetic algorithm, in: 2019 9th International Conference on Computer and Knowledge Engineering (ICCKE 2019), 2019, pp. 18–25.
[14] S.A. Uymaz, G. Tezel, E. Yel, Artificial algae algorithm (AAA) for nonlinear global optimization, Appl. Soft Comput. 31 (2015) 153–171.
[15] B. Turkoglu, S.A. Uymaz, E. Kaya, Binary Artificial Algae Algorithm for feature selection, Appl. Soft Comput. 120 (108630) (2022), https://doi.org/10.1016/j.asoc.2022.108630.
[16] D. Karaboga, B. Basturk, A powerful and efficient algorithm for numerical function optimization: artificial bee colony (ABC) algorithm, J. Glob. Optim. 39 (3) (2007) 459–471.
[17] S. Mirjalili, S.M. Mirjalili, A. Lewis, Grey wolf optimizer, Adv. Eng. Softw. 69 (2014) 46–61.
[18] S. Saremi, S. Mirjalili, A. Lewis, Biogeography-based optimisation with chaos, Neural Comput. Applic. 25 (5) (2014) 1077–1097.
[19] G.-G. Wang, et al., Chaotic krill herd algorithm, Inform. Sci. 274 (2014) 17–34.
[20] G. Kaur, S. Arora, Chaotic whale optimization algorithm, J. Comput Des. Eng. 5 (3) (2018) 275–284.
[21] G.I. Sayed, G. Khoriba, M.H. Haggag, A novel chaotic salp swarm algorithm for global optimization and feature selection, Appl. Intell. 48 (10) (2018) 3462–3481.
[22] M. Mitić, et al., Chaotic fruit fly optimization algorithm, Knowl.-Based Syst. 89 (2015) 446–458.
[23] B. Liu, et al., Improved particle swarm optimization combined with chaos, Chaos, Solitons Fractals 25 (5) (2005) 1261–1271.
[24] B. Alatas, E. Akin, A.B. Ozer, Chaos embedded particle swarm optimization algorithms, Chaos, Solitons Fractals 40 (4) (2009) 1715–1734.
[25] S. Mirjalili, A.H. Gandomi, Chaotic gravitational constants for the gravitational search algorithm, Appl. Soft Comput. 53 (2017) 407–419.
[26] L. dos Santos Coelho, V.C. Mariani, A novel chaotic particle swarm optimization approach using Hénon map and implicit filtering local search for economic load dispatch, Chaos, Solitons Fractals 39 (2) (2009) 510–518.
[27] A. Ouannas, et al., On the dynamics, control and synchronization of fractional-order Ikeda map, Chaos, Solitons Fractals 123 (2019) 108–115.
[28] L. dos Santos Coelho, M.W. Pessôa, A tuning strategy for multivariable PI and PID controllers using differential evolution combined with chaotic Zaslavskii map, Expert Syst. Appl. 38 (11) (2011) 13694–13701.
[29] S. Elaskar, E. Del Río, New Advances on Chaotic Intermittency and Its Applications, Springer, 2017.

[30] J. Zhao, Z.-M. Gao, Liebovitch map enabled Sine Cosine algorithm, in: 2021 International Conference on Machine Learning and Intelligent Systems Engineering (MLISE), IEEE, 2021.
[31] L. dos Santos Coelho, V.C. Mariani, Firefly algorithm approach based on chaotic Tinkerbell map applied to multivariable PID controller tuning, Comput. Math. Appl. 64 (8) (2012) 2371–2382.
[32] R. Tang, S. Fong, N. Dey, Metaheuristics and chaos theory, in: Chaos Theory, IntechOpen, 2018, pp. 182–196.
[33] A.R. Jordehi, Chaotic bat swarm optimisation (CBSO), Appl. Soft Comput. 26 (2015) 523–530.
[34] S.A. Rather, P.S. Bala, Swarm-based chaotic gravitational search algorithm for solving mechanical engineering design problems, World J. Eng. 17 (1) (2020) 97–114, https://doi.org/10.1108/WJE-09-2019-0254.
[35] F.B. Demir, T. Tuncer, A.F. Kocamaz, A chaotic optimization method based on logistic-sine map for numerical function optimization, Neural Comput. Applic. 32 (17) (2020) 14227–14239.
[36] A.H. Gandomi, et al., Chaos-enhanced accelerated particle swarm optimization, Commun. Nonlinear Sci. Numer. Simul. 18 (2) (2013) 327–340.
[37] S. Talatahari, et al., Imperialist competitive algorithm combined with chaos for global optimization, Commun. Nonlinear Sci. Numer. Simul. 17 (3) (2012) 1312–1319.
[38] A.H. Gandomi, et al., Firefly algorithm with chaos, Commun. Nonlinear Sci. Numer. Simul. 18 (1) (2013) 89–98.
[39] R.M. May, Simple mathematical models with very complicated dynamics, in: The Theory of Chaotic Attractors, Springer, 2004, pp. 85–93.
[40] V. Jothiprakash, R. Arunkumar, Optimization of hydropower reservoir using evolutionary algorithms coupled with chaos, Water Resour. Manag. 27 (7) (2013) 1963–1979.
[41] M.-Y. Cheng, K.-Y. Huang, Genetic algorithm-based chaos clustering approach for nonlinear optimization, J. Mar. Sci. Technol. 18 (3) (2010) 15.
[42] T. Akhtar, N. Din, J. Uddin, Substitution box design based on chaotic maps and cuckoo search algorithm, in: 2019 International Conference on Advanced Communication Technologies and Networking (CommNet), IEEE, 2019.
[43] P. Snaselova, F. Zboril, Genetic algorithm using theory of chaos, Procedia Comput. Sci. 51 (2015) 316–325.
[44] G. Zhenyu, et al., Self-adaptive chaos differential evolution, in: International Conference on Natural Computation, Springer, 2006.
[45] X. Han, X. Chang, A chaotic digital secure communication based on a modified gravitational search algorithm filter, Inform. Sci. 208 (2012) 14–27.
[46] S. Saremi, S. Mirjalili, A. Lewis, Grasshopper optimisation algorithm: theory and application, Adv. Eng. Softw. 105 (2017) 30–47.
[47] M. Dorigo, M. Birattari, T. Stutzle, Ant colony optimization, IEEE Comput. Intell. Mag. 1 (4) (2006) 28–39.
[48] J. Cai, et al., Chaotic ant swarm optimization to economic dispatch, Electr. Pow. Syst. Res. 77 (10) (2007) 1373–1380.
[49] B. Alatas, Chaotic bee colony algorithms for global numerical optimization, Expert Syst. Appl. 37 (8) (2010) 5682–5687.
[50] C. Xu, H. Duan, F. Liu, Chaotic artificial bee colony approach to uninhabited combat air vehicle (UCAV) path planning, Aerosp. Sci. Technol. 14 (8) (2010) 535–541.
[51] W.-C. Hong, Electric load forecasting by seasonal recurrent SVR (support vector regression) with chaotic artificial bee colony algorithm, Energy 36 (9) (2011) 5568–5578.
[52] X.-S. Yang, A new metaheuristic bat-inspired algorithm, in: Nature Inspired Cooperative Strategies for Optimization (NICSO 2010), Springer, 2010, pp. 65–74.
[53] A.H. Gandomi, X.-S. Yang, Chaotic bat algorithm, J. Comput. Sci. 5 (2) (2014) 224–232.
[54] D. Simon, Biogeography-based optimization, IEEE Trans. Evol. Comput. 12 (6) (2008) 702–713.
[55] W. Zhu, H. Duan, Chaotic predator–prey biogeography-based optimization approach for UCAV path planning, Aerosp. Sci. Technol. 32 (1) (2014) 153–161.

[56] L. Wang, Y. Xu, An effective hybrid biogeography-based optimization algorithm for parameter estimation of chaotic systems, Expert Syst. Appl. 38 (12) (2011) 15103–15109.
[57] S. Arora, S. Singh, Butterfly optimization algorithm: a novel approach for global optimization, Soft. Comput. 23 (3) (2019) 715–734.
[58] S. Arora, S. Singh, An improved butterfly optimization algorithm with chaos, J. Intell. Fuzzy Syst. 32 (1) (2017) 1079–1088.
[59] X.-S. Yang, S. Deb, Cuckoo search via Lévy flights, in: 2009 World Congress on Nature & Biologically Inspired Computing (NaBIC), IEEE, 2009.
[60] L. Xiang-Tao, Y. Ming-Hao, Parameter estimation for chaotic systems using the cuckoo search algorithm with an orthogonal learning method, Chin. Phys. B 21 (5) (2012), 050507.
[61] P. Nasa-ngium, K. Sunat, S. Chiewchanwattana, in: Enhancing modified cuckoo search by using Mantegna Lévy flights and chaotic sequences, The 2013 10th International Joint Conference on Computer Science and Software Engineering (JCSSE), IEEE, 2013.
[62] G.-G. Wang, et al., Chaotic cuckoo search, Soft. Comput. 20 (9) (2016) 3349–3362.
[63] R. Storn, K. Price, Differential evolution—a simple and efficient heuristic for global optimization over continuous spaces, J. Glob. Optim. 11 (4) (1997) 341–359.
[64] L.S. Coelho, V.C. Mariani, Combining of chaotic differential evolution and quadratic programming for economic dispatch optimization with valve-point effect, IEEE Trans. Power Syst. 21 (2) (2006) 989–996.
[65] D. Jia, G. Zheng, M.K. Khan, An effective memetic differential evolution algorithm based on chaotic local search, Inform. Sci. 181 (15) (2011) 3175–3187.
[66] S. Mirjalili, Dragonfly algorithm: a new meta-heuristic optimization technique for solving single-objective, discrete, and multi-objective problems, Neural Comput. Applic. 27 (4) (2016) 1053–1073.
[67] G.I. Sayed, A. Tharwat, A.E. Hassanien, Chaotic dragonfly algorithm: an improved metaheuristic algorithm for feature selection, Appl. Intell. 49 (1) (2019) 188–205.
[68] X. Yao, Y. Xu, Recent advances in evolutionary computation, J. Comput. Sci. Technol. 21 (1) (2006) 1–18.
[69] R. Caponetto, et al., Chaotic sequences to improve the performance of evolutionary algorithms, IEEE Trans. Evol. Comput. 7 (3) (2003) 289–304.
[70] X.-S. Yang, Chaos-enhanced firefly algorithm with automatic parameter tuning, in: Recent Algorithms and Applications in Swarm Intelligence Research, IGI Global, 2013, pp. 125–136.
[71] L. dos Santos Coelho, D.L. de Andrade Bernert, V.C. Mariani, A chaotic firefly algorithm applied to reliability-redundancy optimization, in: 2011 IEEE Congress of Evolutionary Computation (CEC), IEEE, 2011.
[72] W.-T. Pan, A new fruit fly optimization algorithm: taking the financial distress model as an example, Knowl. Based Syst. 26 (2012) 69–74.
[73] C. Li, et al., Parameters identification of chaotic system by chaotic gravitational search algorithm, Chaos, Solitons Fractals 45 (4) (2012) 539–547.
[74] S. Gao, et al., Gravitational search algorithm combined with chaos for unconstrained numerical optimization, Appl. Math Comput. 231 (2014) 48–62.
[75] D. Shen, et al., Improved chaotic gravitational search algorithms for global optimization, in: 2015 IEEE Congress on Evolutionary Computation (CEC), IEEE, 2015.
[76] M. Kohli, S. Arora, Chaotic grey wolf optimization algorithm for constrained optimization problems, J. Comput Des. Eng. 5 (4) (2018) 458–472.
[77] A.H. Gandomi, A.H. Alavi, Krill herd: a new bio-inspired optimization algorithm, Commun. Nonlinear Sci. Numer. Simul. 17 (12) (2012) 4831–4845.
[78] S. Mirjalili, Moth-flame optimization algorithm: a novel nature-inspired heuristic paradigm, Knowl.-Based Syst. 89 (2015) 228–249.
[79] M. Wang, et al., Toward an optimal kernel extreme learning machine using a chaotic moth-flame optimization strategy with applications in medical diagnoses, Neurocomputing 267 (2017) 69–84.

[80] Y. Xu, et al., An efficient chaotic mutative moth-flame-inspired optimizer for global optimization tasks, Expert Syst. Appl. 129 (2019) 135–155.
[81] S. Mirjalili, S.M. Mirjalili, A. Hatamlou, Multi-verse optimizer: a nature-inspired algorithm for global optimization, Neural Comput. Applic. 27 (2) (2016) 495–513.
[82] A.A. Ewees, M. Abd El Aziz, A.E. Hassanien, Chaotic multi-verse optimizer-based feature selection, Neural Comput. Applic. 31 (4) (2019) 991–1006.
[83] J. Chuanwen, E. Bompard, A hybrid method of chaotic particle swarm optimization and linear interior for reactive power optimisation, Math. Comput. Simul. 68 (1) (2005) 57–65.
[84] S. Mirjalili, et al., Salp swarm algorithm: a bio-inspired optimizer for engineering design problems, Adv. Eng. Softw. 114 (2017) 163–191.
[85] M. Tubishat, et al., Dynamic salp swarm algorithm for feature selection, Expert Syst. Appl. 164 (2021), 113873.
[86] M. Jain, V. Singh, A. Rani, A novel nature-inspired algorithm for optimization: squirrel search algorithm, Swarm Evol. Comput. 44 (2019) 148–175.
[87] K.P. Kumar, C.V. Narasimhulu, K.S. Prasad, A new image restoration approach by combining empirical wavelet transform and total variation using chaotic squirrel search optimization, Int. J. Numer. Modell. Electron. Networks Devices Fields 34 (2) (2021), e2824.
[88] M. Sanaj, P.J. Prathap, Nature inspired chaotic squirrel search algorithm (CSSA) for multi objective task scheduling in an IAAS cloud computing atmosphere, Int. J. Eng. Sci. Technol. 23 (4) (2020) 891–902.
[89] S. Mirjalili, A. Lewis, The whale optimization algorithm, Adv. Eng. Softw. 95 (2016) 51–67.
[90] D. Oliva, M. Abd El Aziz, A.E. Hassanien, Parameter estimation of photovoltaic cells using an improved chaotic whale optimization algorithm, Appl. Energy 200 (2017) 141–154.
[91] G.I. Sayed, A. Darwish, A.E. Hassanien, A new chaotic whale optimization algorithm for features selection, J. Classif. 35 (2) (2018) 300–344.
[92] D. Yousri, D. Allam, M.B. Eteiba, Chaotic whale optimizer variants for parameters estimation of the chaotic behavior in permanent magnet synchronous motor, Appl. Soft Comput. 74 (2019) 479–503.
[93] R. Guha, et al., Embedded chaotic whale survival algorithm for filter–wrapper feature selection, Soft. Comput. 24 (17) (2020) 12821–12843.

CHAPTER

Metaheuristic approaches for solving multiobjective optimization problems 2

Selim Yilmaz[a,b] and Sevil Sen[a]

[a]*WISE Lab., Department of Computer Engineering, Hacettepe University, Ankara, Turkey,* [b]*Department of Software Engineering, Muğla Sıtkı Koçman University, Muğla, Turkey*

1. Introduction

Many real-life problems have at least two objectives that stay in conflict with each other. For example, many solutions consider not only the accuracy of the proposal but also its efficiency. Multiobjective optimization aims at optimizing these conflicting objectives simultaneously. Many metaheuristic approaches have been proposed in the literature to solve multiobjective optimization problems (MOOPs), and due to their powerful capability in dealing with these kinds of problems, new metaheuristics are emerging. These approaches are mainly classified into the following five classes: *aggregating*, *lexicographic*, *subpopulation*, *Pareto-based*, and *hybrid* approaches. Among them, the Pareto-based approaches are the most applied methods in the literature [1–3]. In Pareto-based approaches, the solution to MOOPs is generally not unique, but rather is a set of optimal, nondominated solutions called Pareto front. Hence, the Pareto front represents different trade-offs among conflicting objectives that decision-makers could choose the most appropriate solution according to their needs.

This chapter presents a novel metaheuristic algorithm called Multi Objective Electric Fish Optimization (MOEFO), which is based on Electric Fish Optimization (EFO) inspired by electric fish in nature [4]. Nocturnal electric fish have very poor eyesight and live in muddy, murky water, where the visual sense is very limited. Therefore, they rely on their species-specific ability called electrolocation to perceive their environment. The active and passive electrolocation capabilities of such fish are believed to be good candidates to perform local and global search, respectively. Hence it is proved in [4] that an algorithm based on such capabilities of electric fish is very successful for solving complex problems. The main motivation of this study is to enhance this promising algorithm for solving MOOPs.

In EFO, every individual carries two types of information called amplitude and frequency, which represent the degree of proximity of the fish (the candidate individual) to the best prey source (the global optimum). The frequency plays a key role in EFO to balance exploitation and exploration and is used to determine whether an individual will perform active or passive electrolocation. It enforces better individuals (active individuals), which are most likely to be in the vicinity of promising regions, to exploit their neighborhood, and it leads other individuals (passive individuals) to explore the search space so that they discover new regions. The amplitude, however, adjusts the length of search areas in which the active individuals perform local search. It also determines the likelihood of active-mode individuals to be sensed by passive-mode individuals, which promotes the better regions to be

Comprehensive Metaheuristics. https://doi.org/10.1016/B978-0-323-91781-0.00002-8

exploited more frequently. That's because active-mode individuals likely attract more individuals through their long-lasting higher amplitudes.

To adapt and improve EFO for solving MOOPs, some modifications have been made to the EFO algorithm. While the active and passive electrolocation phases of EFO are mainly kept the same in the proposed MOEFO algorithm, some adaptations are made to keep a set of nondominated solutions instead of a single solution by using a dominance-based selection. Here, not only is EFO adapted for solving MOOPs but also the complexity of the algorithm is reduced for efficiency such as by adapting a cellular-based approach [5,6], which places individuals in a static toroidal mesh topology at the initialization. As a result of all these modifications, an effective and efficient multiobjective approach is introduced and shared with the community [7].

The proposed MOEFO is compared with well-known metaheuristic approaches in the literature, namely Nondominated Sorting Genetic Algorithm-II (NSGA-II) [8], Strength Pareto Evolutionary Algorithm-II (SPEA2) [9], Indicator-based Evolutionary Algorithm (IBEA) [10], Multiobjective Evolutionary Algorithm based on Decomposition (MOEAD) [11], Multiobjective Cellular Genetic Algorithm (MOCell) [12], Generalized Differential Evolution 3 (GDE3) [13], and Optimized Multiobjective Particle Swarm Optimizer (OMOPSO) [14]. The rationale behind the selection of these metaheuristics as competitor algorithms is their popularity and high adoption in the studies carried out for performance assessment in the literature. Hypervolume (HV), SPREAD, EPSILON, and Inverted Generational Distance (IGD) metrics are used in evaluations. The experimental results show that MOEFO outperforms its competitors in all metrics other than SPREAD. It takes third place on evaluations that are made by using the SPREAD metric. The results show that the solutions found by the proposed MOEFO algorithm can converge better to the Pareto-fronts and that they are much more diverse than those found by the competitor algorithms. Moreover, it has a comparable running time with the competitor algorithms.

The proposed MOEFO algorithm is also explored on MaOPs, which are recently welcomed in the literature as more challenging tasks. Unlike the MOOPs, which have two to three objectives, MaOPs possess four or more objectives; hence they introduce some challenges such as diverseness and convergence for most of the existing metaheuristics. The experimental results show that the MOEFO algorithm is better than competitors on some MaOPs and is very competitive on other MaOPs according to the IGD metric.

The rest of this chapter is organized as follows. The following section provides background information and Section 2 summarizes related studies in the literature. Section 3 provides an overview of EFO and Section 4 introduces the MOEFO algorithm. Section 5 presents the experimental settings, then discusses experimental results with statistical findings. Finally, Section 6 concludes the chapter.

1.1 Definitions

In this section, we present some definitions used throughout the chapter. These definitions are the basic concepts of Pareto-based multiobjective optimization approaches. A general MOOP is expressed in the following form:

Definition 1 Multi Objective Optimization Problem: A type of problem where the objective is to find a solution $x^* = [x_1^*, x_2^*, \ldots, x_k^*]$ that minimizes a number of functions $f(x^*)$ satisfying the p inequality constraints $g(x^*)$ and q equality constraints $h(x^*)$. Considering x is a decision variable set, a function and constraint vectors can be written as follows (in case when the goal is maximization, *minimize* should be changed to *maximize*):

$$\begin{aligned} \text{minimize} f(x) &= [f_1(x), f_2(x), \ldots, f_n(x)], \\ \text{subject to}\, g(x) &= [g_1(x), g_2(x), \ldots, g_p(x)] \leq 0 \\ h(x) &= [h_1(x), h_2(x), \ldots, h_q(x)] = 0 \end{aligned}$$

The region constrained by g and h is *feasible region* (Ω) and any solution (x) within this region (i.e., $x \in \Omega$) is called a *feasible solution*. The following Pareto definitions are made with respect to the notations in the MOOP definition.

Definition 2 Pareto Dominance: A k-dimensional solution $u = [u_1, \ldots, u_k]$ is said to dominate another solution $v = [v_1, \ldots, v_k]$ (denoted as $u \prec v$); if $f(u)$ is partially less than $f(v)$, i.e., $\forall i: f_i(u) \leq f_i(v) \wedge \exists i: f_i(u) < f_i(v) \mid i \in \{1, \ldots, n\}$; $u, v \in \Omega$.

Definition 3 Pareto Optimal: A solution u is Pareto optimal if there is no other solution in Ω that dominates u, i.e., $\neg \exists \hat{u} \in \Omega : \hat{u} \prec u$.

Definition 4 Pareto Optimal Set: A set $\mathcal{P}^*$ that contains all the Pareto optimal solutions, i.e., $\mathcal{P}^* = \{u \in \Omega \mid \neg \exists \hat{u} \in \Omega : \hat{u} \prec u\}$.

Definition 5 Pareto Front: A set $\mathcal{PF}^*$ that contains all the objectives of Pareto optimal set ($\mathcal{P}^*$), i.e., $\mathcal{PF}^* = \{f(u) \mid u \in \mathcal{P}^*\}$.

2. Related works

The complex nature of search space of some objectives in MOOPs such as nondifferentiability and discontinuity makes traditional gradient-based techniques impossible or at least ineffective to apply. As such, researchers have proposed many population-based heuristic optimization methods that are not dependent on the characteristics of the search space. Such methods target to find as many Pareto-optimal solutions as possible. This chapter summarizes the well-known studies in this research area.

The first use of heuristic algorithms for the solution of MOOPs dates to the introduction of the Vector Evaluated Genetic Algorithm (VEGA) [15] in the literature. VEGA decomposes a problem with n objectives into n subproblems and splits the population into n subpopulations. Hence, each decomposed subproblem is addressed only by the corresponding subpopulation. The biggest drawback of VEGA is that it leads the entire population to converge to only one optimum due to using a selection mechanism that often relies on best solutions. The lack of proper selection scheme in VEGA led to the emergence of the Pareto concept [16]. MOEAD [11] is another method that decomposes the problem into smaller subproblems. The main goal is to minimize the maximum margin between each objective and its corresponding reference value. Like VEGA, it decomposes the problem into several subproblems and handles each of them simultaneously. The main drawback, however, is that its performance is directly proportional to the decomposition method employed.

Contrary to these approaches, IBEA [10] handles a MOOP in a single population. In the calculation of fitness value, it does not consider the objectives directly, but rather uses a particular performance metric obtained from the objectives such as HV and EPSILON and aims to increase the quality of solutions with respect to the chosen metric. While parent and offspring individuals are constructing the next population, the poor solutions regarding the chosen metric are eliminated to maintain the size of the population. However, the use of a single metric could lead to a poor performance with respect to other metrics. Like IBEA, Pareto-based SPEA2 [9], an efficiency-aware extension of SPEA [17], does not make direct use of the approximated Pareto front set for fitness evaluation either, but instead uses a fitness value calculated based on the Pareto domination count of a solution and its distance to other solutions. One of the main concerns about SPEA2 is that these two criteria used in fitness evaluation are treated evenly, which enables solutions that are completely dominated but farthest away from others to be kept in the population. NSGA-II [8] is the most popular Pareto-based algorithm in the literature. It is based on the Pareto dominance and distances of solutions in the objective space. Depending on the Pareto domination, NSGA-II splits the population into several fronts in which distance, called crowding distance, between the solutions is calculated. Different from SPEA2, nondominated solutions that are in better fronts are allowed in NSGA-II to survive. If the size of next front exceeds the size of the population, the distance plays a key role in determining individuals to remain in the population. NSGA-III [18], is an improved version of NSGA-II, which is proposed to primarily solve MaOPs. Even if NSGA-III follows the same search framework as that in NSGA-II, it slightly differs in the selection phase. Unlike to the crowding distance, NSGA-III relies on a set of well-spread reference points to keep the population as diverse as possible throughout the optimization. The multiobjective GDE3 [13] algorithm is a developed version of the multiobjective GDE [19] algorithm. The GDE algorithm replaces only the fitness-based greedy selection scheme with the Pareto dominance-based selection. GDE3 is a very similar algorithm to NSGA-II. It employs Pareto front and crowding distance-based survival schemes to keep a more diverse population in the objective space. MOCell [12] is also similar to NSGA-II in the selection of solutions to survive for the next generations. However, it applies the basic generic operators of GA to breed offspring in only a close neighborhood. This neighborhood is based on a predefined grid topology.

In addition to these well-known evolutionary-based algorithms, swarm-based heuristics have also been proposed to solve MOOPs. Such heuristics are generally proposed for solving single-objective problems, then their extensions are introduced for MOOPs. Particle Swarm Optimization (PSO) [20] and Artificial Bee Colony (ABC) [21] are the most popular swarm-based algorithms in the literature. Therefore, they have mostly been adapted for solving MOOPs. The initial attempt to adapt PSO for solving MOOPs is the MOPSO [22] algorithm. MOPSO is proposed as a Pareto-based approach, and it makes a slight modification on the search operators in the single-objective PSO algorithm. With this modification, the leader archive in MOPSO is determined according to the Pareto dominance. Another popular PSO-based multiobjective algorithm is OMOPSO [14]. OMOPSO, like MOPSO, considers the Pareto dominance of solutions for the construction of leader archive. Differently from MOPSO, crowding distance plays a role in OMOPSO when nondominated solutions exceed the predefined archive size. Decomposition-based PSO (MPSO/D) [23], Speed-constrained PSO (SMPSO) [24], Vector Evaluated PSO (VEPSO) [25], and Dynamic Neighborhood PSO (DN-PSO) [26] are among other popular PSO variants proposed for MOOPs.

The first modification to ABC for solving MOOPs is introduced as Vector Evaluated ABC (VEABC) [27], which is inspired by the VEGA and VEPSO algorithms. Therefore, like these algorithms, VEABC splits the whole population into subpopulations that are responsible for handling different objectives of the problem. Multi Objective ABC (MOABC) [2] is another ABC-based multiobjective optimization algorithm. The search framework of the single-objective ABC is generally preserved in MOABC except for the fitness proportionate selection scheme. It is replaced with a Pareto-dominance-based selection approach in MOABC. Asynchronous/synchronous (A/S), Pareto dominance/nondominated sorting (PD/NS), and MOABC algorithms (A-MOABC/PD, A-MOABC/NS, and S-MOABC/NS) are proposed as three different implementations of the single-objective ABC to extend the standard algorithm to handle MOOPs [28]. As in MOABC, the search framework employed in all variants is the same as the standard ABC algorithm. However, they show difference in their selection schemes both from each other and from the standard ABC algorithm. Nondominated Sorting-based ABC (NSABC) [29], Division-based Multi Objective ABC (dMOABC) [30], and Elitism-based Multi Objective ABC (eMOABC) [31] are other popular ABC-based multiobjective algorithms.

As discusses, PSO and ABC are maybe the most popular algorithms that have often been used to solve MOOPs in the literature due to their easy-to-implement structures and satisfying performances on different types of problems. Refer, respectively, to [1,32] for a detailed review of PSO- and ABC-based multiobjective optimization algorithms proposed in the literature. Apart from them, there are also multiobjective optimization algorithms that are built on other swarm-based single-objective optimization algorithms. These include Multi Objective Gravitational Search Algorithm (MOGSA) [33], Multi-Objective Gray Wolf Optimization (MOGWO) [34], Multi Objective Firefly Algorithm (MOFA) [35], and Multi Objective Teaching Learning-based Optimization (MOTLBO) [36].

3. An overview of electric fish optimization

The EFO algorithm is based on the following characteristics of electric fish: (i) active electrolocation, (ii) passive electrolocation, which is dependent on the activity of electric organ discharge (EOD), (iii) EOD frequency, and (iv) EOD amplitude behaviors. The local and global search is ensured in EFO through the modeled active and passive electrolocation, respectively. The balance between local and global search, however, is determined by EOD frequency. Because they are expected to further exploit the promising area in the vicinity of them, individuals with a higher frequency perform local search, while the other individuals perform global search. EOD amplitude is used to determine effective range in local search and the probability of neighboring individuals to be sensed in global search.

3.1 Initialization of population

The EFO algorithm starts by generating a population of individuals (N) randomly, which is common in most heuristic algorithms:

$$x_{ij} = x_{\min_j} + \phi\left(x\max_j - x_{\min_j}\right) \tag{1}$$

where x_{ij} represents the position of the ith individual in the population of size $|N|$ ($i = 1, 2, \ldots, |N|$) in the d-dimensional search space. $x_{\min j}$ and $x_{\max j}$ are the lower and upper boundaries for dimension $j|$ $j \in 1, 2, \ldots, d$, respectively. $\phi \in [0, 1]$ is a random value drawn from a uniform distribution. The algorithm then calculates the frequency (f) and amplitude (A) values of every individual with regard to Eqs. (2) and (3), respectively.

$$f_i^t = f_{\min} + \left(\frac{fit_{worst}^t - fit_i^t}{fit_{worst}^t - fit_{best}^t}\right)(f_{\max} - f_{\min}) \tag{2}$$

$$A_i^t = \alpha A_i^{t-1} + (1 - \alpha) f_i^t \tag{3}$$

where fit^t_{worst} and fit^t_{best} are, respectively, the worst and best fitness values, whereas fit^t_i is the fitness value of the ith individual at iteration t. $\alpha \mid \alpha \in [0, 1]$ is a constant value that determines the magnitude of the previous amplitude value. The initial amplitude value of the ith individual is set to its own initial frequency value f_i.

3.2 Active and passive electrolocation phases

After the initialization of individuals, the population is divided into two groups based on the frequency values of individuals in the population: active and passive. The individuals in these groups perform either active or passive electrolocation according to the group to which they are assigned. Hence, the search continues in parallel manner through passive (N_P) and active individuals (N_A) and creates a new population by updating individuals' frequency and amplitude values. These steps are iterated for each new population until the termination criterion is met.

Active individuals are allowed to search only their very vicinity as in nature. Hence, they fulfill the local search ability of the EFO algorithm. An individual in active mode first determines its active sensing range (r_i) depending on its amplitude value (Eq. 4), examines its neighborhood, and then evolves a new candidate solution through randomly selecting one of its neighbors (k) (Eq. 5) or through a random walk in case no neighbor exists in its vicinity (Eq. 6).

$$r_i = \left(x_{\max_j} - x_{\min_j}\right) A_i \tag{4}$$

$$x_{ij}^{cand} = x_{ij} + \varphi\left(x_{kj} - x_{ij}\right) \tag{5}$$

$$x_{ij}^{cand} = x_{ij} + \varphi r_i \tag{6}$$

where $\varphi \in [-1, 1]$ in Eqs. (5) and (6) is a random number generated from a uniform distribution and x_{ij}^{cand} represents newly evolved candidate solution.

Passive individuals, however, can easily exceed their vicinity as opposed to active individuals. They fulfill the global search ability of the EFO algorithm. They perceive their conspecific active individuals with a probability and then change their locations. The probability of active individuals being perceived by passive individuals is directly proportional to their own amplitude value and the distance to the target passive individual, as given in Eq. (7). Here, the EFO algorithm applies Euclidean distance formula to calculate the distance between two individuals in the space.

$$p_k = \frac{A_k / d_{ik}}{\sum_{j \in N_A} A_j / d_{ij}} \tag{7}$$

A reference location vector x_r is then built from perceived K individuals (Eq. 8) and then a new auxiliary solution (x^{new}) is generated from this reference vector (Eq. 9).

$$x_{rj} = \frac{\sum_{k=1}^{K} A_k x_{kj}}{\sum_{k=1}^{K} A_k} \tag{8}$$

$$x_{ij}^{new} = x_{ij} + \varphi\left(x_{rj} - x_{ij}\right) \tag{9}$$

EFO employs an acceptance condition (Eq. 10) on the generated solution to avoid a case in which a passive mode individual with higher frequency loses its promising location information completely.

$$x_{ij}^{cand} = \begin{cases} x_{ij}^{new} & \text{rand}_j(0,1) > f_i \\ x_{ij} & \text{otherwise} \end{cases} \tag{10}$$

Lastly, to further increase the diversity in a population, EFO employs the following equation that stochastically enables one parameter (j) of the newly evolved solution to be modified:

$$x_{ij}^{cand} = x_{\min_j} + \phi\left(x_{\max_j} - x_{\min_j}\right) \tag{11}$$

EFO has only two parameters (α and K) that need to be set. However, EFO is shown not to be sensitive to changes in parameters [4]. Distance calculation between individuals brings about an increase in the complexity of the algorithm, but leads EFO to show superior performance, particularly on a complex problem. The effective performance of EFO on unconstrained and constrained single-objective problems is statistically revealed in detail in [4]. Refer to [4] for pseudocode and further information on EFO.

4. Multiobjective electric fish optimization algorithm

In this study, EFO is extended for solving MOOPs. While the general framework based on local and global search is kept similar to as in EFO, some modifications have been made to the base algorithm in order to adapt it for MOOPs (1–2). Furthermore, by considering the more complex search space of such problems, some changes are applied to strengthen the exploration and exploitation search capabilities of the algorithm (3–4) and to reduce its complexity (5). The modifications to the EFO algorithm are:

i. The frequency updating of individuals (Eq. 2) has been modified. Hence the single fitness proportionate-based approach is replaced with a ranking- and crowding distance-based approach.
ii. The greedy selection has been replaced with a dominance-based selection and an external archive for storing Pareto optimal solutions is introduced.
iii. The conditional acceptance in passive search has been improved to support promising solutions to perform their search in the vicinity of a nondominated solution.
iv. Binary tournament selection has been employed in active and passive search to ensure a fine-tuned local and global search.
v. The distance calculation in EFO has been excluded here for the sake of efficiency of the proposed algorithm. Instead, a *cellular-based* approach has been adopted.

4.1 Population initialization

Here, the individuals are randomly generated as in EFO. Thus, Eq. (1) has been applied for the formation of the initial population. In EFO, an individual needs to measure its distance to the rest of the population to find neighboring individuals in the sensing/active range. However, this distance calculation at each iteration increases the computation complexity of the algorithm, which is also stated in [4]. Therefore, it is excluded in MOEFO; instead, the cellular-based approach is employed. Neighborhood

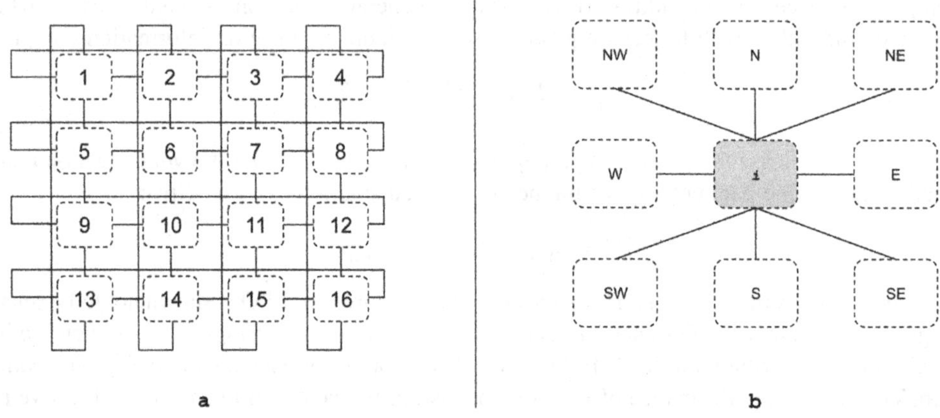

FIG. 1

(A) Toroidal mesh topology used in cGA. (B) A neighborhood demonstration of *i*th individual (N: North, S: South, W: West, E: East).

in this approach has been organized in a random manner and thus every individual has been directly connected with eights neighbors in this phase.

In the area of optimization, a cellular-based approach emerged from parallelization of a genetic algorithm [5,6]. This approach relies on the placement of every node, which correspond to individuals, in a toroidal mesh topology, and edges directly connected to nodes represent the one-hop neighbors of individuals.

Fig. 1A shows an exemplar mesh topology with 16 nodes. As a result of this topology, every individual intensively uses its nearby neighbors. In this study, a topology with eight neighbors is employed, as shown in Fig. 1B.

Even if the neighborhood in this approach has no relation to the geographical position of neighbor individuals, it still well balances the *exploration* and *exploitation*. While exploration is ensured by a slow diffusion of solution through the population, exploitation is ensured by application of search operators within the neighborhood [12]. It is worth stating here that not only distance calculation in EFO establishes the neighborhood, but it also maintains the "explore first exploit later" approach [4]. The cellular-based strategy employed here also contributes to the "explore first exploit later" approach, as the neighborhood of every individual has initially been organized in a random manner. This allows every individual to communicate with the randomly chosen individuals at the beginning of the iterations and to exploit them toward the end of the iterations. In addition to the cellular-based approach, an archive has been created in this phase to store Pareto-optimal solutions.

4.2 Frequency update

A substantial change has been applied to the frequency updating mechanism, as only single objective was used in EFO for the calculation of individuals' frequencies. This approach is not applicable in MOOPs where at least two conflicting objectives take part. As such, a modification based on Pareto ranking and crowding distance of every individual is proposed. While f_{min} and f_{max} represent the worst and the best, respectively, in EFO, they are evenly split into intervals that represent the total Pareto ranking of the population here. Individuals in the first ranking are placed into the first interval, those in the second ranking are placed in the second interval, and so on. Then crowding distance is applied separately on every subpopulation in the intervals such that the greater distance an individual has, the closer to the upper limit of that interval it has been placed. The calculation of the frequency of the ith individual (f_i) in the population is given in Eq. (12).

$$f_i = \left(1 - \left(\frac{front_i - 1}{front_{\max}}\right) - dist_i\right) \tag{12}$$

where $front_i$ and $front_{max}$ represent the front at which the ith individual belongs to and the total number of fronts, respectively, while $dist_i$ is a crowding distance measurement for the ith individual ($CrowDist_i$) normalized for $front_i$. The calculation of $dist_i$ is given as follows:

$$dist_i = \left(\left(1 - \frac{CrowDist_i - CrowDist_{\min}}{CrowDist_{\max} - CrowDist_{\min}}\right) / front_{\max}\right) \tag{13}$$

where $CrowDist_{\min}$ and $CrowDist_{\max}$ represents the value of minimum and maximum crowding distance measurements obtained from $front_i$, respectively. Fig. 2 demonstrates the frequency update mechanism in MOEFO. The figure is based on a scenario where the population is represented with four Pareto fronts, such that each front comprises a different number of solutions. The proposed frequency update approach assigns the maximum and minimum frequency values to every front. Hence, individuals in the first front have a frequency value between 1.00 and 0.75, individuals in the second front have a frequency value between 0.75 and 0.5, and so on. The distance of every individual is then calculated separately on every front with respect to the crowding distance measurement (refer to Eq. 13) and then normalized. Here, the individuals with *infinity* crowding distances represent the

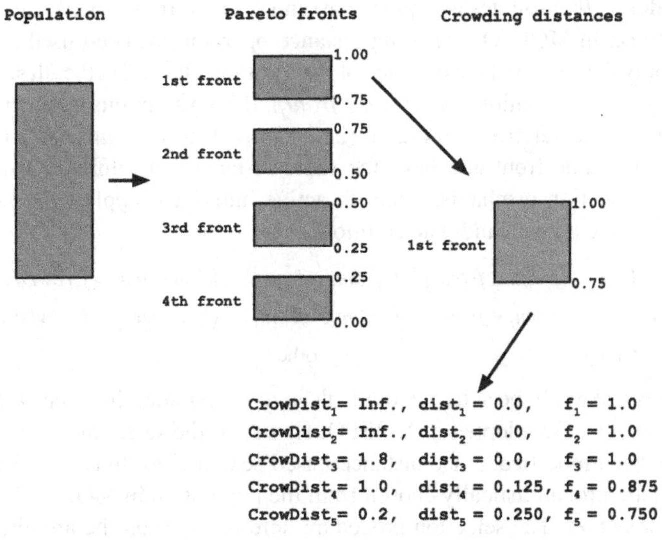

FIG. 2

A demonstration of the frequency calculation strategy in MOEFO.

solutions where their objective vectors are located in the extreme points of the front, which are known to be solutions that increase the diversity of the population. In MOEFO, these individuals are excluded in distance calculation (Eq. 13) and are assigned to the maximum frequency value of the front to which they belong. Hence, these individuals are promoted to perform local search and their probability of being selected by other individuals is increased.

As in EFO, this updating mechanism ensures better individuals to have higher frequency and thus enables them to perform local search, and vice versa. The update procedure for individuals' amplitudes remains the same as in EFO and thus it is calculated by using Eqs. (3) and (12).

4.3 Active and passive electrolocation phases

Only a few modifications have been made on the search performed by active and passive individuals. These are mostly limited to the selection scheme employed in active and passive electrolocation phases.

The main role of active individuals in EFO is to perform local search since they consider only the individuals in their very close neighborhood during the search. However, depending on the existence of neighbors that are evaluated using Euclidean distance formula, individuals could perform random walk or neighbor exploitation that could occasionally lead the algorithm to contribute to global and local search during the initial and final iterations, respectively. However, MOEFO does not rely on Euclidean distance between individuals, but instead it makes use of one-hop neighbors that are assigned at the initialization (see Section 4.1). The proposed scheme contributes also to global search at the initial iterations because the neighborhood is randomly established. Then, it contributes to local search as the neighbors continuously converge to the desired global minima. To further ensure a well-balanced search, *binary tournament selection* has been used in neighbor selection because it first picks two candidate solutions randomly (contributes to exploration) and then returns the better one among them (contributes to exploitation). In MOEFO, crowding distance operator has been used as comparator. This operator considers only the front and the distance of the solutions. It prefers the ith solution, if it belongs to the lower front than the jth solution (i.e., $front_i < front_j$). If both solutions belong to the same front, it chooses the solution with lesser crowded region (i.e., $front_i = front_j \wedge CrowDist_i > CrowDist_j$). If both solutions belong to the same front and have the same distances, it returns a random solution (see Eq. (14)). After this selection mechanism, the ith active individual applies the same formula as in EFO (see Eq. 5) to evolve a new candidate solution.

$$return \begin{cases} i & (front_i < front_j) \vee ((front_i = front_j) \wedge (CrowDist_i > CrowDist_j)) \\ j & (front_j < front_i) \vee ((front_i = front_j) \wedge (CrowDist_j > CrowDist_i)) \\ i \text{ or } j & \text{otherwise} \end{cases} \quad (14)$$

In passive search, EFO takes all the active individuals in the population into account to perform global search. This mechanism is also adopted in MOEFO. However, the selection approach applied to passive individuals has been modified, since distance-based search is excluded in MOEFO. As in EFO, K active individuals are probabilistically chosen from the population in MOEFO by applying K times *binary tournament selection*. The selection procedure here is based on the amplitude values that are implicitly determined by the crowding distance measurement (see Section 4.2). The reason for using amplitudes in the selection instead of crowding distance metric as in active search is that amplitude values of individuals do not change instantaneously. This adoption enables currently poor but formerly promising individuals to be selected, which further promotes the exploration capability of MOEFO.

The other modification has been made on the conditional acceptance condition given in Eq. (10) by replacing it with Eq. (15). This modification simply states that the probability of a promising solution with higher frequency being evolved through its neighbors is lowered, but that through a Pareto-optimal solution (a) from *archive* is considerably increased.

$$x_{ij}^{cand} = \begin{cases} x_{ij}^{new} & rand_j(0,1) > f_i \\ x_{ij} + \varphi\left(x_{aj} - x_{ij}\right) & \text{otherwise} \end{cases} \tag{15}$$

The final step in one iteration is on the replacement strategy of existing solutions with evolved candidate solutions. In EFO, *greedy selection* relying on only a single objective has been used. However, it is not suitable for MOOPs.

Therefore, a dominance-based comparator, which prefers a nondominated solution over a dominated solution, is used in this study. Thus, a candidate solution can be replaced with its parent solution only when it dominates, otherwise it can be replaced with its worst neighbor or discarded when it is worse than its worst neighbor from the point of domination view. Fig. 3 shows the pseudocode of the proposed MOEFO algorithm.

Algorithm : Pseudocode of MOEFO algorithm

```
Input: population size |N|, objective size |M|
Output: archive
population ← Initialize population;
Update frequency;
archive ← Update archive;
repeat
    for i = 1 to |N| do
        individual ← population[i];
        offspring ← {};
        if individual.frequency > rand then
            offspring ← active search(individual);
        else
            offspring ← passive search(individual);
        end
        if offspring ≺ individual then
            individual ← offspring;
        else if offspring ≺ individual.neighbor[worst] then
            individual.neighbor[worst] ← offspring;
        end
    end
    Update frequency;
    archive ← Update archive;
until termination criterion is met;
return archive;
```

FIG. 3

Pseudocode of MOEFO algorithm.

5. Experiments

The performance of the proposed MOEFO algorithm on MOOPs and MaOPs is evaluated by some metrics that are intrinsic for such problem types. The following sections introduce benchmark problems, performance metrics, algorithms used for comparison, and their settings. Finally, we present and discuss the experimental results.

5.1 Benchmark problems

In this study, we used 64 well-known unconstrained MOOPs with different characteristics. Among these, 30 problems are taken from DTLZ (Deb-Thiele-Laumanns-Zitzler) [37], LZ09 (Li-Zhang) [11], WFG (Walking FishGroup) [38], and ZDT (Zitzler-Deb-Thieler) [39]. Ten complex MOOPs are taken from the CEC09 algorithm contest [40], and 24 multimodal MOOP functions (MMF), which are proposed for the CEC2020 contest, are taken from the technical report given in [41]. MMFs differ from MOOPs in that there exists at least one local Pareto optimal solution, not dominated by any solution in the neighborhood, and at least two global optimal solutions, not dominated by any solution in the search space, corresponding to the same Pareto front in the objective space. Most of these problems have become the standard problems used in the literature to conduct a fair comparison. They differ from each other in search space, number of parameters to optimize (6–30), and number of objectives (2 and 3). Table 1 gives the basic characteristics of these problems. It is worth noting that the equations of functions *f*, *f_l*, or *f_a* in the MMF family are the same, but their reference data are different. For example, MMM15, MMM15_l, and MMF15_a have same equation to be optimized; however, they differ only in their reference data.

As for the performance evaluation on MaOPs, we use 10 different benchmark problems with different challenging features, which are proposed in [42]. Table 2 gives specific challenges of these problems. Refer to [42] for mathematical formulations and detailed discussions of these problems.

5.2 Performance metrics

The performance metrics applied on MOOPs differ from the metrics used for evaluating single-objective problems. Without the loss of generality, an algorithm is regarded as the best performing algorithm on a single-objective problem when its final solution is closest to the global optimum in comparison to others. However, for MOOPs, it depends on how well $\mathcal{PF}^*$ computed by nondominated solutions of an algorithm approximates to the true Pareto front ($\mathcal{TPF}^*$) of the problem. The following metrics reveal the convergence performance of an algorithm on MOOPs.

5.2.1 Hypervolume (HV)

Hypervolume (HV) measures the union of the volume between every member in $\mathcal{PF}^*$ and a reference point r, which represents a vector comprising the worst values of each objective. Therefore, the higher HV implies that $\mathcal{PF}^*$ well represents $\mathcal{TPF}^*$. HV is calculated by the following equation:

$$HV = \text{volume}\left(\bigcup_{i=1}^{|\mathcal{PF}^*|} v_i\right) \tag{16}$$

Table 1 The characteristics of MOOPs used in the experiments.

Family	Problem	O	V	Range
DTLZ	1	3	7	$0 \le x_1, x_2, \ldots, x_V \le 1$
	2–6		12	
	7		22	
LZ09	1, 7, 8	2	10	$0 \le x_1, x_2, \ldots, x_V \le 1$
	2–5, 9	2	30	
	6	3	10	
WFG	1–9	2	6	$0 \le x_k \le 2 \times k,\ k \in \{1,2,\ldots,V\}$
ZDT	1–3	2	30	$0 \le x_1, x_2, \ldots, x_V \le 1$
	4		10	$0 \le x_1 \le 1, -5 \le x_2, \ldots, x_V \le 5$
	6		10	$0 \le x_1, x_2, \ldots, x_V \le 1$
CEC09	1, 2, 5–7	2	30	$0 \le x_1 \le 1, -1 \le x_2, \ldots, x_V \le 1$
	3	2		$0 \le x_1, x_2, \ldots, x_V \le 1$
	4	2		$0 \le x_1 \le 1, -2 \le x_2, \ldots, x_V \le 2$
	8–10	3		$0 \le x_1, x_2 \le 1, -1 \le x_3, \ldots, x_V \le 1$
MMF	1	2	2	$1 \le x_1 \le 3, -1 \le x_2 \le 1$
	1_e	2	2	$1 \le x_1 \le 3,\ e^{-3} \le x_2 \le e^3$
	2	2	2	$0 \le x_1 \le 1, 0 \le x_2 \le 2$
	4	2	2	$-1 \le x_1 \le 1, 0 \le x_2 \le 2$
	5	2	2	$-1 \le x_1 \le 3, 1 \le x_2 \le 3$
	7	2	2	$1 \le x_1 \le 3, -1 \le x_2 \le 1$
	8	2	2	$-\pi \le x_1 \le \pi, 0 \le x_2 \le 9$
	10,10_l	2	2	$0.1 \le x_1, x_2 \le 1.1$
	11,11_l	2	2	$0.1 \le x_1, x_2 \le 1.1$
	12,12_l	2	2	$0 \le x_1, x_2 \le 1$
	13,13_l	2	3	$0 \le x_1, x_2, x_3 \le 1$
	14,14_a	3	3	$0 \le x_1, x_2, x_3 \le 1$
	15,15_a,15_l,15_a_l	3	3	$0 \le x_1, x_2, x_3 \le 1$
	16_l1,16_l2,16_l3	3	3	$0 \le x_1, x_2, x_3 \le 1$

O, *number of objectives;* V, *number of variables.*

5.2.2 SPREAD

SPREAD reveals the degree of spread of $\mathcal{PF}^*$ throughout the objective space through the following equation:

$$\Delta = \frac{\sum_{i=1}^{m} d(e_i, \mathcal{PF}^*) + \sum_{x \in \mathcal{PF}^*} |d(e_i, \mathcal{PF}^*) - \hat{d}|}{\sum_{i=1}^{m} d(e_i, \mathcal{PF}^*) + |\mathcal{PF}^*| \hat{d}} \tag{17}$$

where e_i stands for the ith extreme point in the mth objective in $\mathcal{TPF}^*$. $d(e_i, \mathcal{PF}^*)$ refers to the minimum distance between the extreme points in the set $\mathcal{PF}^*$ and $\mathcal{TPF}^*$, whereas $\hat{d}$ is the mean of the distances between every point in $\mathcal{PF}^*$ and in $\mathcal{TPF}^*$.

Table 2 The characteristics of MaOPs used in the experiments [21].

Problem	Characteristics	Range
MaOP1	Inverse of simplex, objective scales, multimodality	$0 \leq x_1, x_2, \ldots, x_V \leq 1$
MaOP2	Complicated PS	
MaOP3	Complicated PS, bias	
MaOP4	Complicated PS, bias	
MaOP5	Complicated PS, degeneracy	
MaOP6	Complicated PS, degeneracy	
MaOP7	Complicated PS, local degeneracy	
MaOP8	Complicated PS, local degeneracy	
MaOP9	Complicated PS, local degeneracy	
MaOP10	Complicated PS, local degeneracy	

5.2.3 EPSILON

EPSILON calculates the minimum factor (ϵ) that is necessary for at least one solution in $\mathcal{PF}^*$ to dominate all the vectors in $\mathcal{TPF}^*$. EPSILON is calculated according to the given equation:

$$E(\mathcal{PF}^*, \mathcal{TPF}^*) = \inf_{\epsilon \in \mathbb{R}^+} \{\forall \mathbf{p} \in \mathcal{TPF}^*, \exists \mathbf{v} \in \mathcal{PF}^* : \mathbf{v} \prec_\epsilon \mathbf{p}\} \tag{18}$$

here m refers to the number of objective functions, whereas ϵ represents a very small value used for a tolerance.

5.2.4 Inverted generational distance (IGD)

Inverted Generational Distance (IGD) is a good indicator that reveals how diverse the solutions in the objective space are and how well the solutions can converge. A small value of IGD suggests that the solution set obtained from the objective space is very close to $\mathcal{TPF}^*$. IGD is calculated as follows:

$$IGD(\mathcal{PF}^*, \mathcal{TPF}^*) = \frac{\sum_{v \in \mathcal{TPF}^*} d(v, \mathcal{PF}^*)}{|\mathcal{TPF}^*|} \tag{19}$$

here $d(v, \mathcal{PF}^*)$ gives the minimum Euclidean distance between the solution v and every point from the set $\mathcal{PF}^*$.

5.3 Competitor algorithms

For a comprehensive performance evaluation, well-known evolutionary- and swarm-based multiobjective optimization algorithms have been used as competitor methods against MOEFO. These include NSGA-II [8], NSGA-III [18], SPEA2 [9], IBEA [10], MOEAD [11], MOCell [12], GDE3 [13], and OMOPSO [14]. Most of the competitor algorithms are Pareto-based, relying mostly on Pareto-domination and crowding distance measurements for selection of solutions for the next generations. Among them, IBEA and MOEAD are not Pareto-based; they use a fitness-based greedy comparison

Table 3 The comparative features of competitor algorithms.

Algorithm	Category	Selection Pareto-domination (●) Greedy comparison (●) Crowding distance (○)	Search Procedure Crossover (■) Mutation (□)	Archive
NSGAII	Pareto-based	● ○	■ □	
NSGAIII	Pareto-based	● ○	■ □	
SPEA2	Pareto-based	●	■ □	
IBEA	Indicator-based	●	■ □	
MOEAD	Decomposition-based	●	■ □	✓
MOCell	Pareto-based	● ○	■ □	✓
GDE3	Pareto-based	● ○	■	
OMOPSO	Pareto-based	● ○	[a] □	✓

[a]*Crowding distance-based search procedure of the standard PSO [15].*

to select the individuals for the survival. While fitness calculation in IBEA depends on the indicator metric value (such as HV) of the individuals, it is maximum difference between individual's jth objective and ideal point for jth objective found by the population. Typical crossover and mutation operators are employed for breeding new individuals in these algorithms. Some of these algorithms (i.e., MOEAD, MOCell, and OMOPSO) possess an external memory (known as archive) to keep nondominated solutions until the end of the search, whereas the remaining algorithms output only the nonnominated solutions found at the end of the search. Among these competitor approaches, MOCell and MOEAD make use of a neighboring structure and every individual in these algorithms performs the search in its neighborhood.

These competitor algorithms are reviewed in Section 2 and their features are outlined in a comparative manner in Table 3. Metaheuristic Algorithms in Java (jMetal) [43,44], a Java-based framework for multiobjective optimization with metaheuristics, is used for the implementation of these competitor algorithms. jMetal provides some facilities such as state-of-the algorithms, most-popular benchmark problems, and the like. Refer to [43] for detailed description of this framework. The sections below detail experimental settings and performance comparisons for MOOP and MaOP benchmark sets.

5.4 Performance evaluation on MOOP benchmark sets

5.4.1 Parameter tuning

The parameter values of the competitor algorithms are taken from the default values of jMetal. As for the parameter set of the proposed MOEFO algorithm (i.e., α and K), the design of experiment (DoE) methodology has been conducted to determine their optimal values. DoE methodology first uses a second-order linear model to learn the relation between input (parameter values) and the output of the algorithm through a coefficient vector. It then uses quadratic programming with that coefficient vector to find the approximation to the optimal parameter.

As it would be too exhaustive to run DoE with every possible parameter setting, three-level full factorial design, in which each parameter is set to three levels (high, intermediate, and low), has been adopted here. To construct input and output data, MOEFO has been run ten times with each parameter setting for all the problems in Table 1 (i.e., $3^2 \times 10$ times per problem). The function evaluation number (FEN) and the population size have been set as 100,000 and 100, respectively. The performance metrics (HV, SPREAD, EPSILON, and IGD) are considered as the output of MOEFO for the given parameter setting. While the values of 0.01, 0.5, and 0.99 have been used for α, those of 1, 25, and 50 have been used for K in the DoE methodology. As a result, it is found that MOEFO performs better with respect to all the metrics overall when α and K are set to 0.5 and 37, respectively. Hence these values have been set for the experiments in this study.

In the experiments, every algorithm has been run 50 times with a population and generation size of 100 and 1000 (corresponding to 100,000 FENs), respectively. All the algorithms have been run in parallel using a client machine (Intel Core i7 CPU with four cores and eight threats, 16GB RAM). The MOEFO implementation in jMetal framework that is ready to replicate the same experimentation is shared with the community [7].

5.4.2 Results

This section examines the convergence and diversification performance of every algorithm on the employed benchmark problems by using HV, SPREAD, EPSILON, and IGD metrics. Fig. 4 shows the number of problems on which every algorithm has performed a satisfying performance (with ranking of 1st and 2nd algorithm).

As shown in the figure, MOEFO is superior than other algorithms by performing the best/second-best performance on the problems with respect to HV and EPSILON metrics. The results (the number

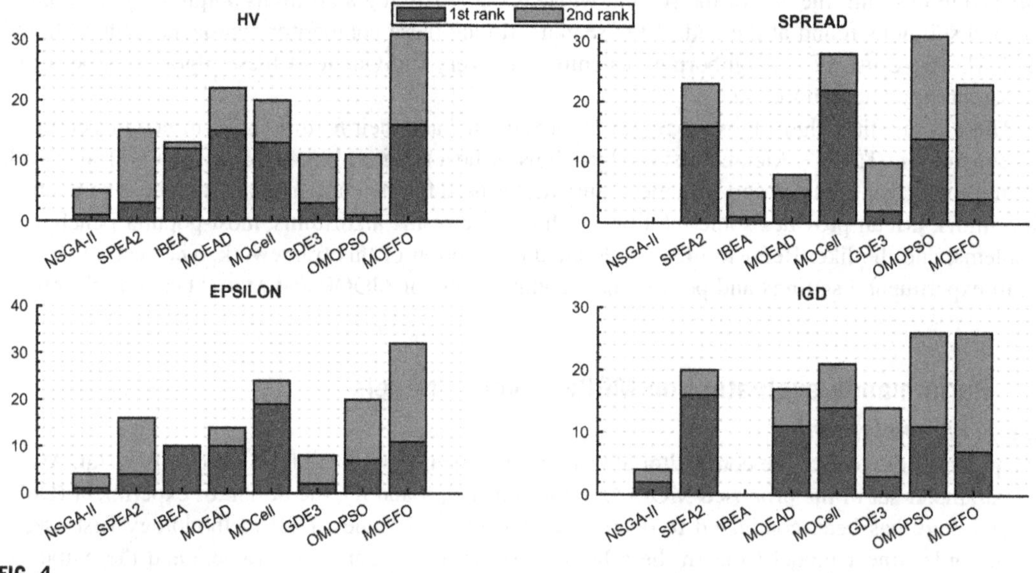

FIG. 4

Number of problems every algorithm has shown better performance.

of problems solved as the best algorithm, the number of problems solved as the second-best algorithm) on HV and EPSILON metrics are (16/15) and (11/21), respectively. OMOPSO, MOCell, and SPEA2 are the most competitive algorithms against MOEFO with respect to SPREAD and IGD metrics. In addition, it can be seen from the figure that NSGA-II, IBEA, and GDE3 showed poor performance overall.

These findings have also been shown separately for every problem family in Fig. 5 to reveal whether the performances of the algorithms are problem dependent. The figure shows that NSGA-II has the poorest performance on all problem families with no exception. SPEA2 showed the best performance on DTLZ and MMF problem families, whereas IBEA showed the best

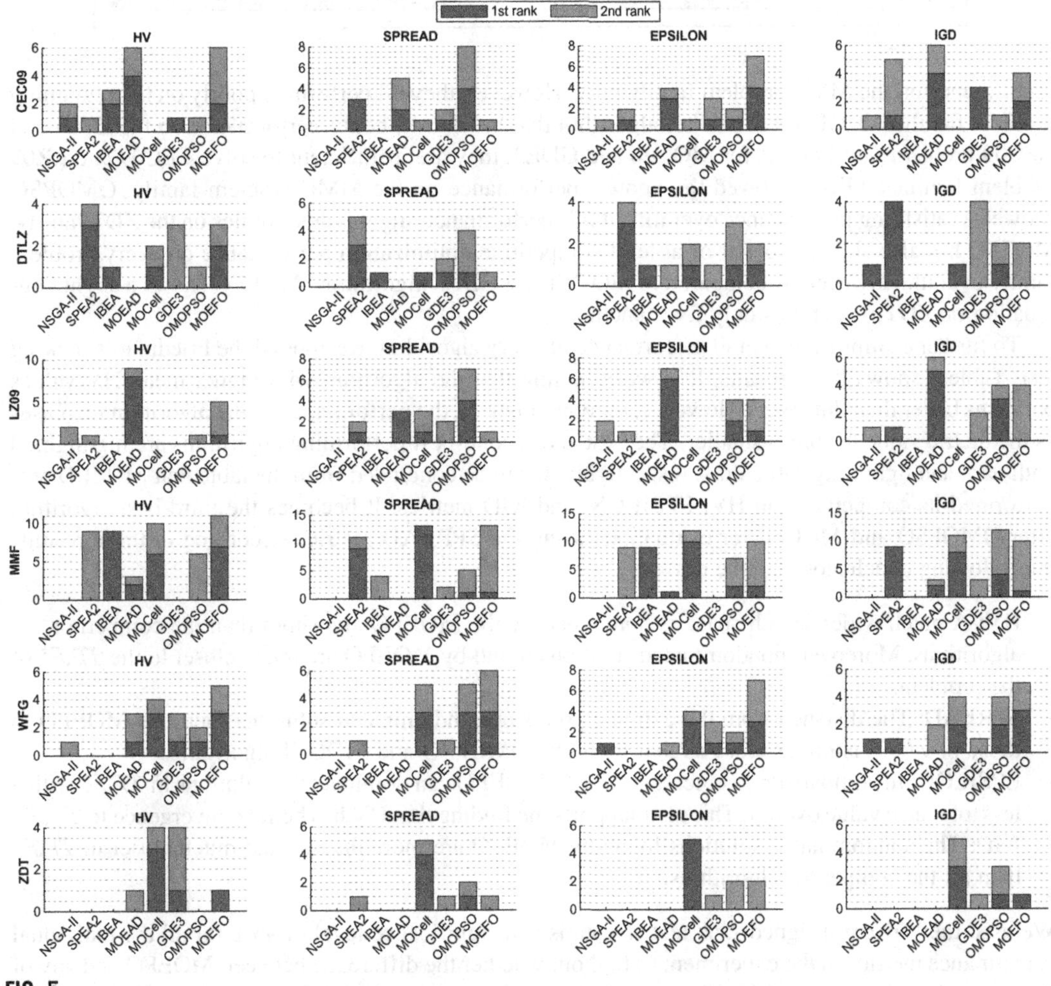

FIG. 5

Number of problems every algorithm has shown better performance for every problem family.

Table 4 Friedman's ranking values calculated separately on evaluation metrics for MOOPs.

Algorithms	Ranking			
	HV	SPREAD	EPSILON	IGD
NSGA-II	3.531	6.266	5.406	5.234
SPEA2	4.922	3.703	4.469	3.813
IBEA	3.469	6.438	5.969	7.406
MOEAD	4.453	5.844	4.703	4.578
MOCell	4.719	3.172	4.359	4.281
GDE3	4.898	3.984	3.891	3.766
OMOPSO	4.094	3.016	4.047	3.688
MOEFO	5.914	3.578	3.156	3.324

performance on the MMF problem family only. However, they all performed poorly on the remaining problem families. MOEAD is another algorithm that has shown better performance on the LZ09 and CEC09 problem families. As for MOCell and GDE3, they are insufficient to solve CEC09 and LZ09 problem families. GDE3 showed the poorest performance on the MMF problem family. OMOPSO showed a satisfying performance overall, but this performance slightly deteriorates on the ZDT family. MOEFO has also shown a better or at least competitive optimization performance on every problem family overall. However, as compared to MOCell and GDE3 algorithms, MOEFO shows a slight degradation on ZDT, which has only five functions.

To further examine the overall performance of every algorithm, we applied the Friedman's ranking test. Table 4 shows these results. It is worth noting that the algorithm with higher mean HV values performs better than the others. However, for other metrics, the performance of the algorithm increases as the metric value becomes smaller. The best and second-best performing algorithms are highlighted with dark and light gray cell colors, respectively. It can be concluded from the table that MOEFO outperformed its competitors on HV, EPSILON, and IGD metrics. It becomes the third-best algorithm after OMOPSO and MOCell for evaluations using the SPREAD metric. According to these results, the following conclusions can be drawn:

- HV: A much wider search space is dominated by the MOEFO algorithm than the competitor algorithms. Moreover, nondominated solutions found by MOEFO are much closer to the $\mathcal{TPF}^*$ of the problem.
- SPREAD: The distance from the extreme points of nondominated solutions found by MOEFO to those of $\mathcal{TPF}^*$ is not as close as that found by OMOPSO and MOCell algorithms.
- EPSILON: In comparison to other algorithms, MOEFO can dominate the solutions in $\mathcal{TPF}^*$ with a less tolerance value overall. This also supports the finding that HV has better convergence to $\mathcal{TPF}^*$.
- IGD: The nondominated solutions found by MOEFO are more diverse and much closer to $\mathcal{TPF}^*$ than all the competitor algorithms.

We used the Wilcoxon signed-rank test at a statistical significance level of 95% for all the individual performance metrics in the experiment to find out whether the difference between MOEFO and any of the competitor algorithms is significant. In this study, the null hypothesis is that there is no statistically significant difference between the median of the results produced by MOEFO and its competitor algorithms. To speculate that EFO performs much better on a given problem, the null hypothesis should

Table 5 Comparative number of problems where proposed MOEFO algorithm performs, in order of representation, "superior(+)/equal(=)/inferior(−)" than the competitors based on the measurement through Wilcoxon signed rank test (superior cases are highlighted with gray background color).

Metric	Problem family	MOEFO vs.						
		NSGA-II	SPEA2	IBEA	MOEAD	MOCell	GDE3	OMOPSO
HV	CEC09	5/3/2	4/3/3	5/3/2	4/1/5	8/2/0	9/0/1	9/0/1
	DTLZ	4/1/2	2/1/4	5/1/1	5/0/2	2/1/4	3/0/4	7/0/0
	LZ09	6/1/2	6/0/3	7/1/1	0/2/7	8/1/0	8/1/0	8/0/1
	MMF	20/3/1	13/1/10	15/0/9	17/1/6	12/2/10	13/6/5	18/1/5
	WFG	6/2/1	6/2/1	6/3/0	6/0/3	5/0/4	4/1/4	5/1/3
	ZDT	4/0/1	4/0/1	5/0/0	4/0/1	1/0/4	1/0/4	3/0/2
	Sum	45/10/9	35/7/22	43/8/13	36/4/24	36/6/22	38/8/18	50/2/12
SPREAD	CEC09	4/4/2	4/1/5	6/3/1	2/1/7	2/2/6	3/1/6	1/1/8
	DTLZ	7/0/0	2/0/5	5/1/1	7/0/0	3/3/1	2/3/2	1/3/3
	LZ09	4/2/3	3/3/3	5/2/2	3/0/6	2/2/5	1/2/6	1/0/8
	MMF	22/1/1	13/0/11	17/5/2	23/0/1	9/2/13	15/5/4	11/9/4
	WFG	9/0/0	7/0/2	9/0/0	8/1/0	3/2/4	6/1/2	4/3/2
	ZDT	5/0/0	3/1/1	5/0/0	4/1/0	0/0/5	3/0/2	2/0/3
	Sum	51/7/6	32/5/27	47/11/6	47/3/14	19/11/34	30/12/22	20/16/28
EPSILON	CEC09	6/1/3	6/3/1	8/1/1	5/1/4	9/0/1	6/2/2	7/1/2
	DTLZ	2/3/2	2/1/4	5/0/2	5/0/2	2/2/3	2/2/3	4/2/1
	LZ09	5/1/3	5/1/3	7/2/0	1/1/7	7/2/0	6/1/2	4/2/3
	MMF	20/2/2	14/0/10	15/0/9	19/2/3	3/14/7	14/6/4	9/8/7
	WFG	7/2/0	8/1/0	8/1/0	7/1/1	4/3/2	7/0/2	7/1/1
	ZDT	4/0/1	4/0/1	5/0/0	4/0/1	0/0/5	4/0/1	2/1/2
	Sum	44/9/11	39/6/19	48/4/12	41/5/18	25/21/18	39/11/14	33/15/16
IGD	CEC09	5/2/3	4/3/3	10/0/0	4/1/5	9/1/0	7/2/1	9/0/1
	DTLZ	3/2/2	2/1/4	5/1/1	4/1/2	2/2/3	2/2/3	5/2/0
	LZ09	5/1/3	5/0/4	7/2/0	0/2/7	7/1/1	5/1/3	3/2/4
	MMF	20/2/2	13/1/10	24/0/0	19/1/4	8/6/10	15/4/5	5/9/10
	WFG	6/2/1	6/1/2	7/1/1	7/0/2	4/3/2	6/1/2	3/4/2
	ZDT	4/0/1	4/0/1	5/0/0	3/1/1	1/1/3	4/0/1	2/1/2
	Sum	43/9/12	34/6/24	58/4/2	37/6/21	31/14/19	39/10/15	27/18/19

be rejected, or an alternative hypothesis should be accepted. The *P*-values in Wilcoxon's sign test are considered for acceptance of the null hypothesis. Thus, the null hypothesis can only be rejected if the *P*-value is less than 0.05 (due to the adopted significance level of 95%) and vice versa. Based on the significance comparison, the number of cases (i) where null hypothesis is rejected and MOEFO exhibited superior (+) or inferior (−) performance, and (ii) where null hypothesis is accepted and the performance between MOEFO and competitor algorithm is very similar (=) is found. Table 5 lists these '+/=/−' cases with respect to every performance metric for each problem family. The last rows of every problem metric (*Sum*) show the total count of the cases. The gray cell colors in the table indicate that MOEFO has shown better performance than the competitor algorithms with respect to the evaluation metric. For the sake of better readability, the *P*-values as well as cumulative positive and negative rank values obtained from Wilcoxon's test have been excluded from the table and they can be found in Ref. [7].

Like the Friedman's ranking evaluation, the Wilcoxon's statistical results also affirm better solutions by MOEFO on HV, EPSILON, and IGD metrics, while its performance becomes competitive with respect to the SPREAD metric, where OMOPSO and MOCell showed superior performances. From the cumulative performance of the Wilcoxon's test, MOEFO showed better performance than all competitor algorithms with respect to the HV, EPSILON, and IGD metrics. It is better than NSGA-II, SPEA2, IBEA, MOEAD, and GDE3 with respect to the SPREAD metric. Only OMOPSO and MOCell algorithms have shown better performance than the proposed MOEFO on only the SPREAD metric, according to the Wilcoxon's sign test. To conclude, both Friedman's ranking evaluation and Wilcoxon's sign test results reveal that NSGA-II, IBEA, and MOEAD are the poorest algorithms with respect to these metrics, whereas MOCell and OMOPSO are the most compelling algorithms to MOEFO overall.

Figs. 6 and 7 show the true Pareto front of problems and the Pareto front of algorithms to reveal the convergence performance of every algorithm on the objective space. Fig. 8 gives the corresponding

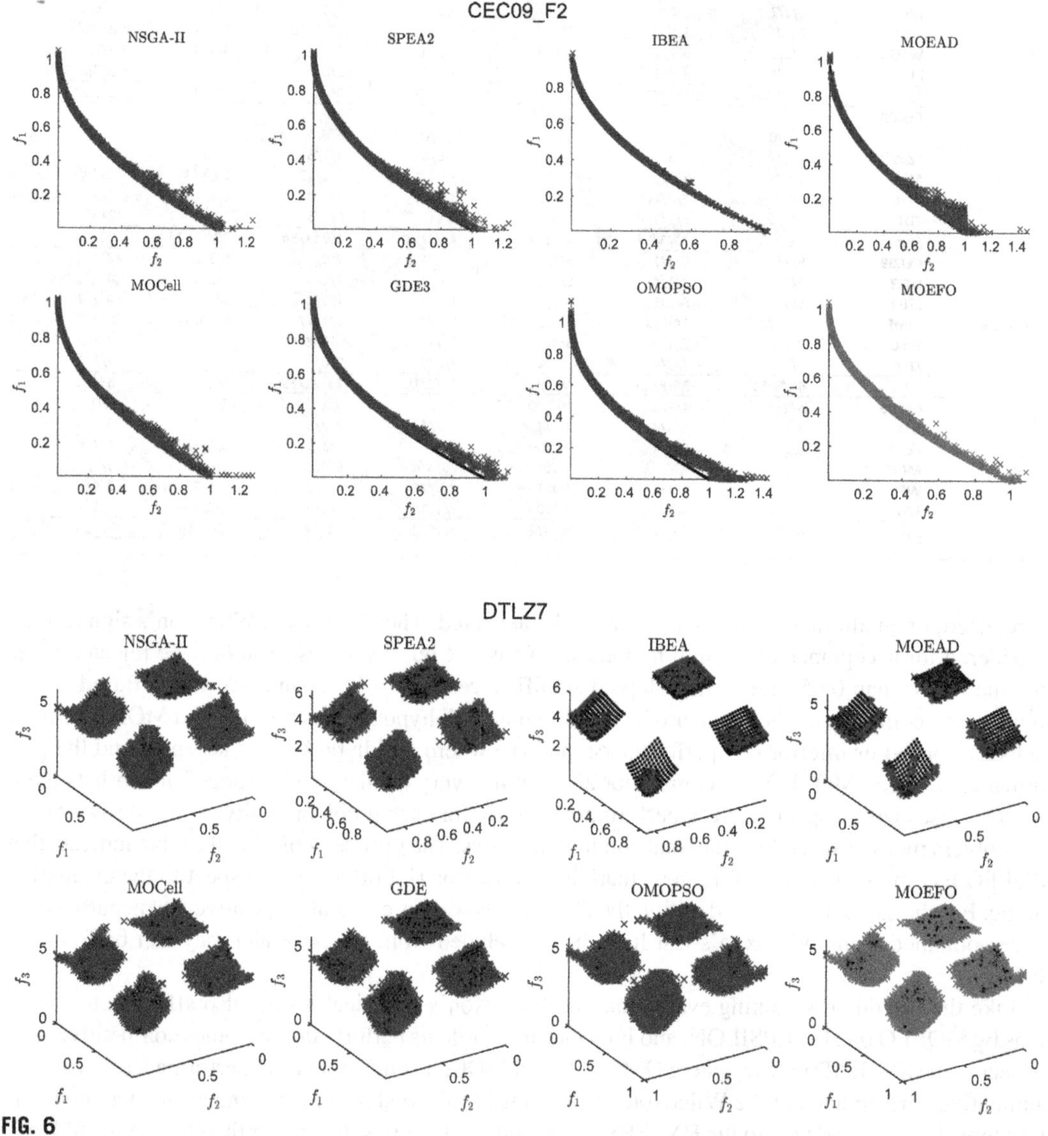

FIG. 6

The comparative convergence performance of the algorithms on CEC09_F2, DTLZ7, and LZ09_F9 problems.

(Continued)

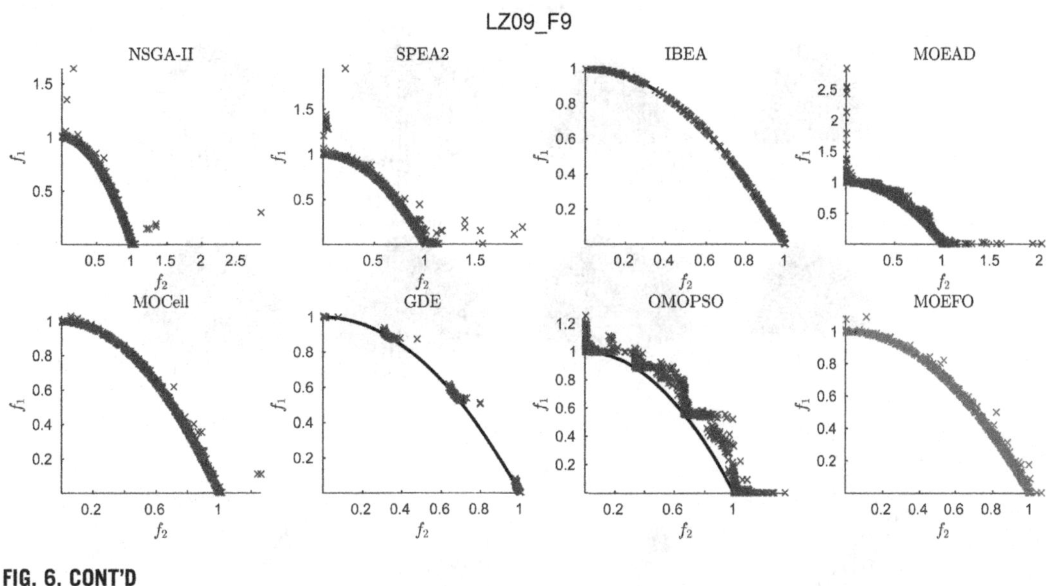

FIG. 6, CONT'D

box plots that show statistical best, worst, and mean performances. Note that only a subset of problems, in which the proposed MOEFO has become the first algorithm, the second algorithm, and worse than the second algorithm with respect to the HV metric, has been given here. Refer to [7] for all the convergence and box plots that reveal the performance of every algorithm on each problem by using each metric separately. The approximations to the true Pareto fronts show that MOEFO could competitively represent the true Pareto front of the problems.

Table 6 comparatively shows the average run times of the algorithms for each problem family. The results show that the CEC09 problem family requires the longest running time due to its greater dimensions (i.e., 30) to be optimized. In addition, OMOPSO requires the least running time, while IBEA, MOEAD, and MOCell require the longest running time overall. The proposed MOEFO algorithm, however, requires an admissible running time for these problem sets overall. It showed better performance than the competitors, except for OMOPSO, on the CEC09 problem family having higher dimensions, which shows that MOEFO can handle high-dimensional problems efficiently in terms of run time without sacrificing its optimization performance.

5.5 Performance evaluation on MaOP benchmark set

5.5.1 Parameter tuning

MOEFO is also explored on MaOPs, whose characteristics are summarized in Table 2. The experimental findings obtained in [45] are taken for comparison in this study. Thus, the parameter settings used in [45] are adopted for this experimental task. Therefore, MOEFO is run 20 times with 500,000 FENs.

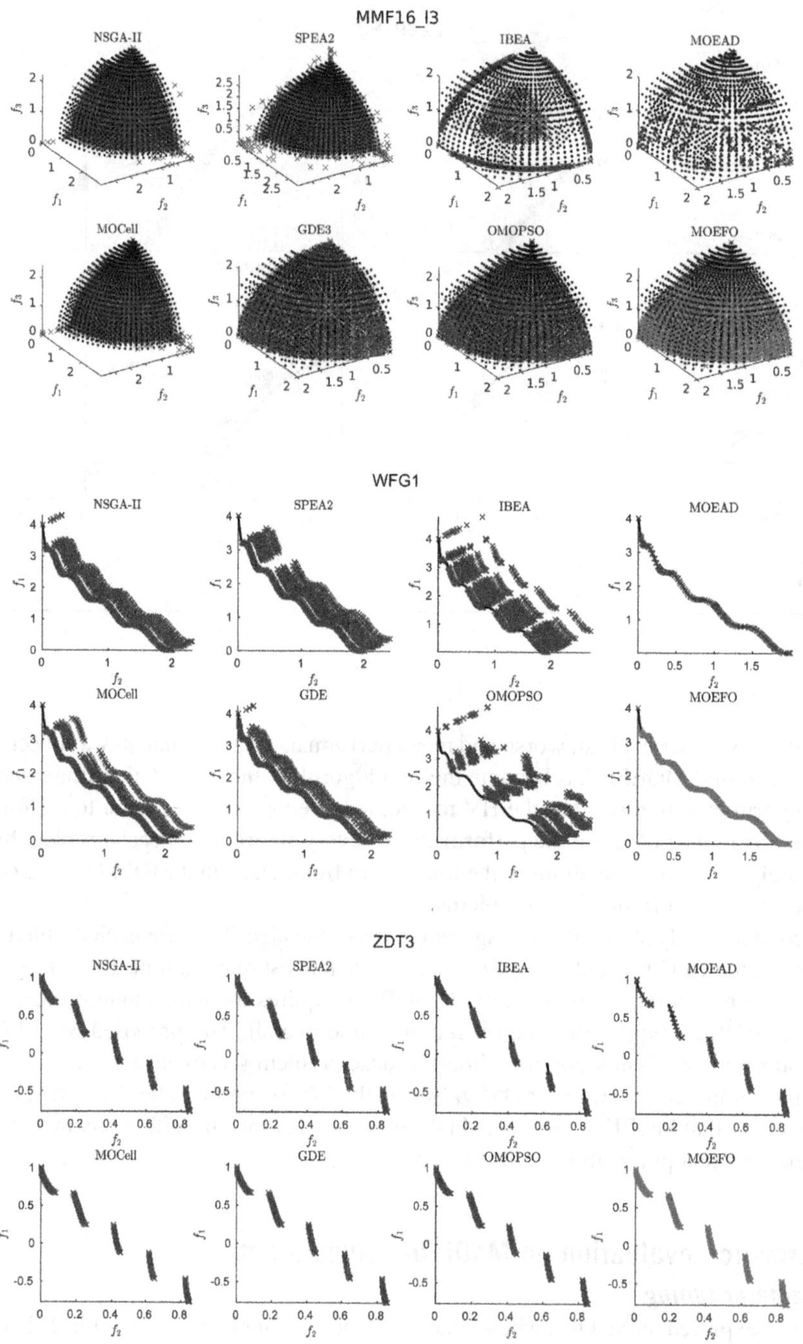

FIG. 7

The comparative convergence performance of algorithms on MMF16_I3, WFG1, and ZDT3 problems.

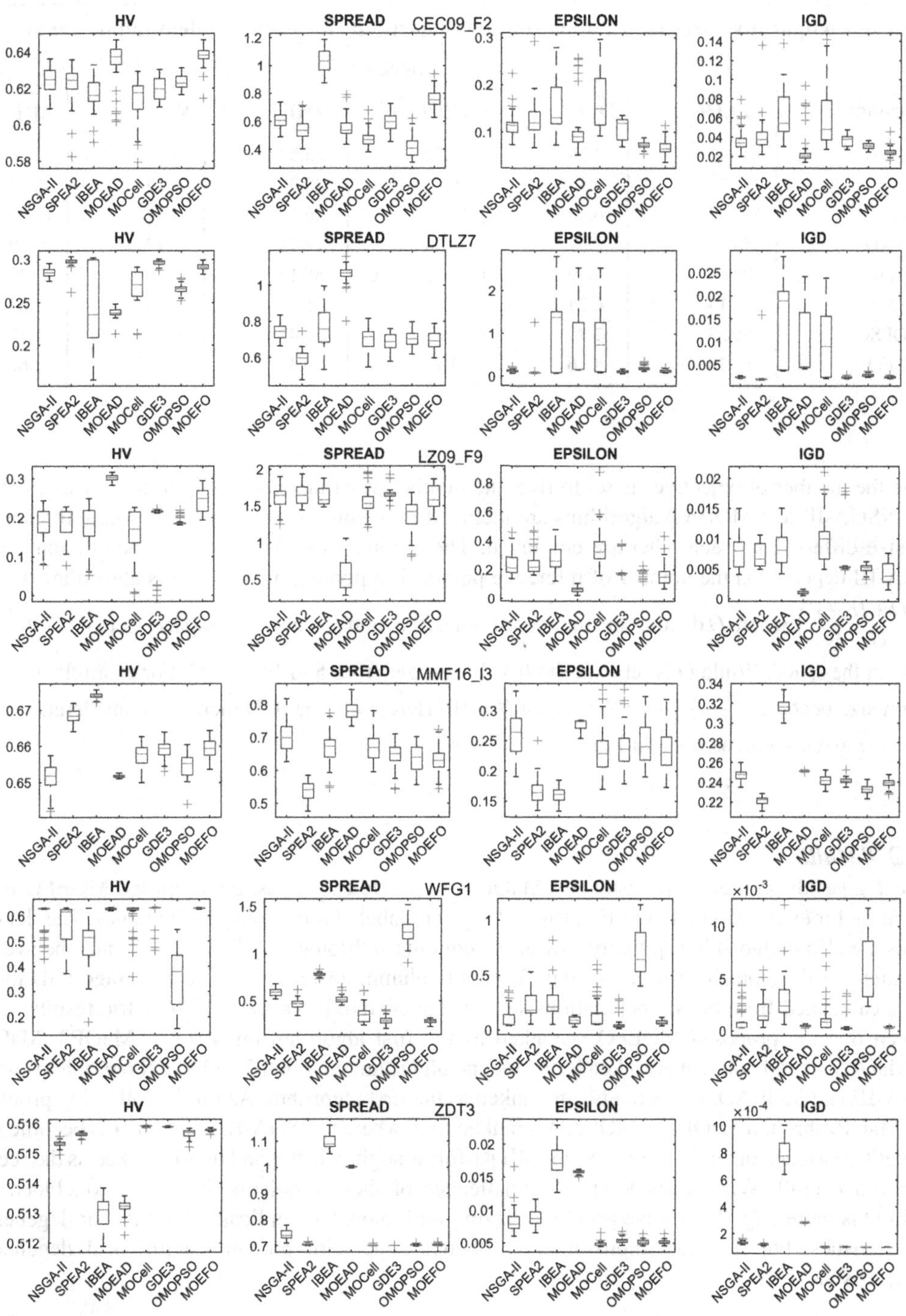

FIG. 8

The comparative box plots on CEC09_F2, DTLZ7, LZ09_F9, MMF16_L3, WFG1, and ZDT3 problems.

Table 6 Comparative average running times of algorithms for each problem family (in s).

	Problem family					
Algorithm	**CEC09**	**DTLZ**	**LZ09**	**MMF**	**WFG**	**ZDT**
NSGA-II	167.41	2.28	3.72	25.33	1.64	2.36
SPEA2	274.77	9.57	6.24	48.20	8.17	8.03
IBEA	278.03	16.43	16.36	56.37	14.91	15.02
MOEAD	258.31	2.37	6.94	112.21	1.12	1.29
MOCell	309.64	5.39	6.53	88.13	4.00	7.20
GDE3	118.14	4.24	1.45	25.14	2.33	2.63
OMOPSO	55.33	2.49	0.89	21.43	0.97	1.18
MOEFO	85.20	4.16	3.76	69.45	3.19	3.62

While the number of objectives is set to five, the number of variables to be optimized is set to 10. In [45], NSGA-III and MOEAD algorithms are included for comparison. Six settings where the population size differs from each other are employed. The rationale for this is that the population size in NSGA-III depends on the number of reference points. The population size in this algorithm is equal to $\binom{O+H-1}{O-1}$ where O denotes the number of objectives and H denotes the spaces between reference points in the space. While O is set to 5, the H value ranges from 5 to 10 in [45]. For example, the population size becomes $\frac{(5+10-1)!}{(10)!(5-1)!}=1001$, when $H=10$. Here, only the IGD metric is considered for performance assessment, as done in [45].

5.5.2 Results

Table 7 gives the comparative results on MaOPs. As previously states, the proposed MOEFO is run separately for every population setting (the settings are labeled a through f in the table). The detailed results are also included in [7]. Every row of the columns a through f in Table 7 indicates the average IGD metric values obtained from 20 runs. The end column, however, gives the average IGD metric values calculated from the six populations settings conducted. From these IGD metric results, it can be seen that the proposed MOEFO is ranked as the first algorithm on MaOP1, MaOP3, MaOP4, and MaOP5 overall, while it is ranked as the second algorithm on MaOP2 and MaOP6 problems where NSGA-III and MOEAD, respectively, are ranked as the first algorithm. As for the following problems (i.e., MaOP7 through MaOP10), MOEAD is ranked first, whereas NSGA-III and MOEFO become very competitive to each other. However, NSGA-III performs slightly better and thus is ranked as the second algorithm overall. When considering the challenges of these problems, it can be concluded that MOEFO is more eligible to solve problems having multimodal, complicated PS, bias, or degeneracy characteristics, but it needs improvement to better cope with problems with local degeneracy characteristics.

Table 7 Comparative average IGD values of MOEFO, MOEAD, and NSGA-III algorithms under six different population size settings (a = 126, b = 220, c = 330, d = 495, e = 715, and f = 1001) for MaOP problems.

Problems	Algorithms	Population settings a	b	c	d	e	f	Average
MaOP1	MOEFO	17.98	17.33	17.01	16.82	16.72	16.70	17.09
	MOEAD	18.46	18.56	18.25	18.34	18.34	18.41	18.39
	NSGA-III	19.72	17.82	17.46	17.84	17.84	17.40	17.93
MaOP2	MOEFO	0.18	0.14	0.12	0.11	0.09	0.08	0.12
	MOEAD	0.17	0.16	0.16	0.15	0.14	0.14	0.15
	NSGA-III	0.11	0.09	0.09	0.08	0.08	0.08	0.09
MaOP3	MOEFO	0.29	0.25	0.23	0.21	1.12	2.21	0.72
	MOEAD	6.22	7.39	11.65	13.94	16.10	17.93	12.21
	NSGA-III	10.82	13.62	16.27	17.94	19.06	20.65	16.39
MaOP4	MOEFO	0.39	0.38	0.38	0.37	0.37	0.36	0.38
	MOEAD	0.47	0.46	0.45	0.43	0.41	0.40	0.44
	NSGA-III	0.39	0.38	0.38	0.37	0.37	0.36	0.38
MaOP5	MOEFO	0.26	0.25	0.25	0.25	0.25	0.25	0.26
	MOEAD	0.41	0.35	0.31	0.28	0.26	0.23	0.31
	NSGA-III	0.47	0.46	0.46	0.45	0.45	0.43	0.45
MaOP6	MOEFO	0.58	0.55	0.55	0.49	0.46	0.43	0.51
	MOEAD	0.31	0.26	0.20	0.17	0.15	0.12	0.20
	NSGA-III	1.19	0.69	0.71	0.74	0.60	0.51	0.74
MaOP7	MOEFO	0.54	0.48	0.44	0.40	0.37	0.32	0.42
	MOEAD	0.27	0.25	0.21	0.19	0.17	0.15	0.21
	NSGA-III	0.50	0.42	0.31	0.28	0.24	0.21	0.33
MaOP8	MOEFO	0.44	0.40	0.34	0.31	0.28	0.26	0.34
	MOEAD	0.27	0.23	0.19	0.17	0.15	0.14	0.19
	NSGA-III	0.46	0.37	0.34	0.27	0.25	0.20	0.32
MaOP9	MOEFO	0.45	0.36	0.31	0.27	0.24	0.22	0.31
	MOEAD	0.31	0.27	0.22	0.21	0.19	0.18	0.23
	NSGA-III	0.43	0.36	0.29	0.25	0.21	0.19	0.29
MaOP10	MOEFO	0.43	0.36	0.32	0.29	0.25	0.23	0.31
	MOEAD	0.28	0.26	0.22	0.20	0.18	0.17	0.22
	NSGA-III	0.38	0.30	0.26	0.23	0.20	0.17	0.26

6. Conclusion

In this chapter, we modified the EFO algorithm for solving MOOPs. We adopted a Pareto-based approach and introduced the new MOEFO algorithm. We adapted EFO to solve MOOPs and to give a set of nondominated solutions as its output as well as reduced the complexity of the algorithm for efficiency. We explored the effectiveness of the proposed algorithm on both MOOPs and MaOPs. While NSGA-II [8], SPEA2 [9], IBEA [10], MOEAD [11], MO-Cell [12], GDE3 [13], and OMOPSO [14] are used for MOOPs comparison, NSGA-III [18] and MOEAD [11] are used for MaOPs comparison. We used well-known unconstrained MOOPs and MaOPs for fair comparison. We compared the algorithms by using four metrics for performance evaluation on MOOPs: HV, SPREAD, EPSILON, and IGD.

We applied Friedman's ranking and Wilcoxon signed rank tests for every metric to statistically show overall and problem-wise performance, respectively. The statistical results show that MOEFO outperformed its competitors on HV, EPSILON, and IGD metrics. It is the third best algorithm after OMOPSO and MOCell in comparisons based on the SPREAD metric. As for performance evaluation on MaOPs, only the IGD metric was taken into consideration. The results show that the proposed MOEFO has an outstanding optimization performance on most of the problems included. However, it needs to be further improved for problems having local-degeneracy characteristics.

References

[1] M. Reyes-Sierra, C.A.C. Coello, et al., Multi-objective particle swarm optimizers: a survey of the state-of-the-art, Int. J. Comput. Intell. Res. 2 (3) (2006) 287–308.

[2] R. Akbari, R. Hedayatzadeh, K. Ziarati, B. Hassanizadeh, A multi-objective artificial bee colony algorithm, Swarm Evol. Comput. 2210-6502, 2 (2012) 39–52, https://doi.org/10.1016/j.swevo.2011.08.001.

[3] R. Akbari, K. Ziarati, Multi-objective bee swarm optimization, Int. J. Innov. Comput. Inf. Control 8 (1B) (2012) 715–726.

[4] S. Yilmaz, S. Sen, Electric fish optimization: a new heuristic algorithm inspired by electrolocation, Neural Comput. Applic. 1433-3058, (2019), https://doi.org/10.1007/s00521-019-04641-8.

[5] H. Mühlenbein, M. Schomisch, J. Born, The parallel genetic algorithm as function optimizer, Parallel Comput. 0167-8191, 17 (6) (1991) 619–632.

[6] L. Darrell Whitley, Cellular genetic algorithms, in: Proceedings of the 5th International Conference on Genetic Algorithms, Morgan Kaufmann Publishers Inc., San Francisco, CA, USA, 1993, p. 658, ISBN: 1-55860-299-2.

[7] WISE-LAB, MOEFO Source Code and Experimental Results, 2020. https://wise.cs.hacettepe.edu.tr/projects/moefo. (Online; accessed 29 December 2021).

[8] K. Deb, A. Pratap, S. Agarwal, T.A.M.T. Meyarivan, A fast and elitist multi-objective genetic algorithm: NSGA-II, IEEE Trans. Evol. Comput. 6 (2) (2002) 182–197.

[9] E. Zitzler, M. Laumanns, L. Thiele, SPEA2: improving the strength Pareto evolutionary algorithm, TIK-report, 103, 2001.

[10] E. Zitzler, S. Künzli, Indicator-based selection in multi-objective search, in: X. Yao, E.K. Burke, J.A. Lozano, J. Smith, J.J. Merelo-Guervós, J.A. Bullinaria, H.-P. Schwefel (Eds.), Parallel Problem Solving from Nature-PPSN VIII, Springer Berlin Heidelberg, Berlin, Heidelberg, 2004, pp. 832–842, ISBN: 978-3-540-30217-9.

[11] H. Li, Q. Zhang, Multi-objective optimization problems with complicated pareto sets, MOEA/D and NSGA-II, IEEE Trans. Evol. Comput. 13 (2) (2009) 284–302, https://doi.org/10.1109/TEVC.2008.925798.

[12] A.J. Nebro, J.J. Durillo, F. Luna, B. Dorronsoro, E. Alba, Design issues in a multiobjective cellular genetic algorithm, in: International Conference on Evolutionary Multi-Criterion Optimization, Springer, 2007, pp. 126–140.

[13] S. Kukkonen, J. Lampinen, GDE3: the third evolution step of generalized differential evolution, in: 2005 IEEE Congress on Evolutionary Computation, vol. 1, IEEE, 2005, pp. 443–450.

[14] M.R. Sierra, C.A. Coello Coello, Improving PSO-based multi-objective optimization using crowding, mutation and e-dominance, in: C.A. Coello Coello, A.H. Aguirre, E. Zitzler (Eds.), Evolutionary Multi-Criterion Optimization, Springer Berlin Heidelberg, Berlin, Heidelberg, 2005, pp. 505–519.

[15] J.D. Schaffer, Some Experiments in Machine Learning Using Vector Evaluated Genetic Algorithms (Artificial Intelligence, Optimization, Adaptation, Pattern Recognition) (PhD thesis, Nashville, TN, USA), 1984. AAI8522492.

[16] D.E. Goldberg, Genetic Algorithms in Search, Optimization and Machine Learning, first ed., Addison-Wesley Longman Publishing Co., Inc., Boston, MA, USA, 1989, ISBN: 0201157675.

[17] E. Zitzler, Evolutionary Algorithms for Multi-Objective Optimization: Methods and Applications, Swiss Federal Institute of Technology, Zurich, 1999.

[18] K. Deb, H. Jain, An evolutionary many-objective optimization algorithm using reference-point-based non-dominated sorting approach, part I: solving problems with box constraints, IEEE Trans. Evol. Comput. 18 (4) (2014) 577–601.

[19] J. Lampinen, et al., DE's Selection Rule for Multi-Objective Optimization, Lappeenranta University of Technology, Department of Information Technology, 2001, pp. 03–04. Tech. Rep.

[20] J. Kennedy, R. Eberhart, Particle swarm optimization, in: Proceedings., IEEE International Conference on Neural Networks, 1995, vol. 4, November 1995, pp. 1942–1948.

[21] D. Karaboga, B. Basturk, A powerful and efficient algorithm for numerical function optimization: artificial bee colony (ABC) algorithm, J. Glob. Optim. 1573-2916, 39 (3) (2007) 459–471, https://doi.org/10.1007/s10898-007-9149-x.

[22] C.A. Coello Coello, M.S. Lechuga, Mopso: a proposal for multiple objective particle swarm optimization, in: Proceedings of the 2002 Congress on Evolutionary Computation. CEC'02 (Cat. No.02TH8600), vol. 2, May 2002, pp. 1051–1056, https://doi.org/10.1109/CEC.2002.1004388.

[23] C. Dai, Y. Wang, M. Ye, A new multi-objective particle swarm optimization algorithm based on decomposition, Inform. Sci. 0020-0255, 325 (2015) 541–557, https://doi.org/10.1016/j.ins.2015.07.018.

[24] A.J. Nebro, J.J. Durillo, J. Garcia-Nieto, C.A.C. Coello, F. Luna, E. Alba, SMPSO: a new PSO-based metaheuristic for multi-objective optimization, in: 2009 IEEE Symposium on Computational Intelligence in Multi-Criteria Decision-Making (MCDM), IEEE, 2009, pp. 66–73.

[25] K.E. Parsopoulos, D.K. Tasoulis, M.N. Vrahatis, Multiobjective optimization using parallel vector evaluated particle swarm optimization, in: Proceedings of the IASTED International Conference on Artificial Intelligence and Applications (AIA 2004), ACTA Press, 2004, pp. 823–828. Key Words.

[26] X. Hu, R. Eberhart, Multiobjective optimization using dynamic neighborhood particle swarm optimization, in: Proceedings of the 2002 Congress on Evolutionary Computation. CEC'02 (Cat. No.02TH8600), vol. 2, May 2002, pp. 1677–1681, https://doi.org/10.1109/CEC.2002.1004494.

[27] S.N. Omkar, J. Senthilnath, R. Khandelwal, G.N. Naik, S. Gopalakrishnan, Artificial bee colony (ABC) for multi-objective design optimization of composite structures, Appl. Soft Comput. 1568-4946, 11 (1) (2011) 489–499, https://doi.org/10.1016/j.asoc.2009.12.008.

[28] B. Akay, Synchronous and asynchronous pareto-based multi-objective artificial bee colony algorithms, J. Glob. Optim. 1573-2916, 57 (2) (2013) 415–445, https://doi.org/10.1007/s10898-012-9993-1.

[29] A. Kishor, P.K. Singh, J. Prakash, Nsabc: non-dominated sorting based multi-objective artificial bee colony algorithm and its application in data clustering, Neurocomputing 0925-2312, 216 (2016) 514–533, https://doi.org/10.1016/j.neucom.2016.08.003.

[30] Y. Xiang, Y. Zhou, A dynamic multi-colony artificial bee colony algorithm for multi-objective optimization, Appl. Soft Comput. 1568-4946, 35 (2015) 766–785, https://doi.org/10.1016/j.asoc.2015.06.033.

[31] Y. Xiang, Y. Zhou, H. Liu, An elitism based multi-objective artificial bee colony algorithm, Eur. J. Oper. Res. 0377-2217, 245 (1) (2015) 168–193, https://doi.org/10.1016/j.ejor.2015.03.005.

[32] Y. Liu, L. Ma, G. Yang, A survey of artificial bee colony algorithm, in: 2017 IEEE 7th Annual International Conference on CYBER Technology in Automation, Control, and Intelligent Systems (CYBER), July 2017, pp. 1510–1515, https://doi.org/10.1109/CYBER.2017.8446301.

[33] H.R. Hassanzadeh, M. Rouhani, A multi-objective gravitational search algorithm, in: 2010 2nd International Conference on Computational Intelligence, Communication Systems and Networks, July 2010, pp. 7–12,-https://doi.org/10.1109/CICSyN.2010.32.

[34] S. Mirjalili, S. Saremi, S.M. Mirjalili, L.d.S. Coelho, Multi-objective grey wolf optimizer: a novel algorithm for multi-criterion optimization, Expert Syst. Appl. 0957-4174, 47 (2016) 106–119, https://doi.org/10.1016/j.eswa.2015.10.039.

[35] X.-S. Yang, Multi-objective firefly algorithm for continuous optimization, Eng. Comput. 29 (2) (2013) 175–184.

[36] F. Zou, L. Wang, X. Hei, D. Chen, B. Wang, Multi-objective optimization using teaching-learning-based optimization algorithm, Eng. Appl. Artif. Intel. 0952-1976, 26 (4) (2013) 1291–1300, https://doi.org/10.1016/j.engappai.2012.11.006.

[37] K. Deb, L. Thiele, M. Laumanns, E. Zitzler, Scalable test problems for evolutionary multiobjective optimization, in: A. Abraham, L. Jain, R. Goldberg (Eds.), Evolutionary Multiobjective Optimization, Springer, 2005, pp. 105–145.

[38] S. Huband, P. Hingston, L. Barone, L. While, A review of multiobjective test problems and a scalable test problem toolkit, IEEE Trans. Evol. Comput. 10 (5) (2006) 477–506.

[39] E. Zitzler, K. Deb, L. Thiele, Comparison of multiobjective evolutionary algorithms: empirical results, Evol. Comput. 8 (2) (2000) 173–195.

[40] Q. Zhang, A. Zhou, S. Zhao, P.N. Suganthan, W. Liu, S. Tiwari, Multiobjective optimization test instances for the CEC 2009 special session and competition, Technical report, 2008.

[41] J. Liang, P. Suganthan, B. Qu, D. Gong, C. Yue, Problem definitions and evaluation criteria for the CEC 2020 special session on multimodal multi-objective optimization, Technical report, 2019.

[42] H. Li, K. Deb, Q. Zhang, P.N. Suganthan, Challenging novel many and multi-objective bound constrained benchmark problems, Technical report, 2017.

[43] J.J. Durillo, A.J. Nebro, jMetal: a java framework for multi-objective optimization, Adv. Eng. Softw. 0965-9978, 42 (10) (2011) 760–771, https://doi.org/10.1016/j.advengsoft.2011.05.014.

[44] J.J. Durillo, A.J. Nebro, E. Alba, The jMetal framework for multi-objective optimization: design and architecture, in: CEC 2010, July 2010, pp. 4138–4325. Barcelona, Spain.

[45] H. Li, K. Deb, Q. Zhang, P.N. Suganthan, L. Chen, Comparison between MOEA/D and NSGA-III on a set of novel many and multi-objective benchmark problems with challenging difficulties, Swarm Evol. Comput. 2210-6502, 46 (2019) 104–117, https://doi.org/10.1016/j.swevo.2019.02.003.

CHAPTER

A brief overview of physics-inspired metaheuristics 3

Soumitri Chattopadhyay[a], Aritra Marik[a], and Rishav Pramanik[b]

[a]*Department of Information Technology, Jadavpur University, Kolkata, India,* [b]*Department of Computer Science and Engineering, Jadavpur University, Kolkata, India*

1. Introduction

Computationally challenging optimization problems have always been of special interest to researchers around the globe. This is primarily due to these problems often having a very high dimensional search space or having highly complex and nonlinear objective functions at their core, which classical gradient-based methods fail to tackle efficiently. This has been the main reason for the development of metaheuristic algorithms that take inspiration from our surroundings (nature, swarms, and physical processes) and can provide a computationally cheap yet robust optimization procedure for hard problems. Parallelly, researchers have also noticed the undeniable success of modeling physics' processes to study highly complex phenomena, both in the real world and computer science. For instance, the resource allocation problem has been well tackled by statistical mechanics models [1], while certain aspects of statistical thermodynamics have been employed to explain microevolution of species [2], and so on. As a result, in spite of swarm-inspired algorithms [3, 4] being in the forefront as robust optimizers, researchers have shown keen interest in adapting principles and theories of physics and applying them to solve real-world optimization problems.

Recently, the world of metaheuristics has seen the advent of several novel search mechanisms based on various nonlinear physics processes. The novelty of these approaches lies in the fact that the nonlinear physical phenomena are leveraged as backbones to be modeled upon to formulate efficient search algorithms, whose mechanism is quite different from the conventional swarm and evolutionary algorithms. Such physics-inspired algorithms have shown great promise and robustness as global optimization strategies.

In this chapter, we discuss some of the most popular optimization algorithms derived from nonlinear physics. As such, physics has several categorical subdomains of study, including classical mechanics [5, 6], thermodynamics [7, 8], and optics [9], to name a few. Accordingly, we have grouped the existing metaheuristics into such classes based on the physical phenomena they were modeled upon. In our discussion, we explain the intuitions behind the algorithms, drawing parallels between the algorithmic steps and the original physical processes. Our discussion includes widely used as well as recently proposed algorithms to give readers a holistic overview regarding physics-based metaheuristics. Specifically, we cover 21 algorithms across six subdomains of physics in this chapter.

Comprehensive Metaheuristics. https://doi.org/10.1016/B978-0-323-91781-0.00003-X

2. Classical mechanics-based metaheuristics

The following section presents several optimization techniques based on classical mechanics, gravitation, and kinematics. They are central force optimizer (CFO), gravitational search algorithm (GSA), colliding bodies algorithm, and equilibrium optimizer (EO).

2.1 Central force optimization

Formato et al. [5] introduced CFO, a deterministic algorithm based on gravitational kinematics. For a deterministic, we provide the positive and negative curvature direction, whereas, for a stochastic-based optimizer, we do not give these conditions in advance.

The CFO algorithm uses probes as its population. These probes with iterations move closer to the fittest probes, searching the search space effectively. CFO defines each probe as having a position vector, an acceleration vector, and a fitness value. The position vector is a representation of the probe's current coordinates regarding each dimension of the search space. Under minimum storage requirements, both the previous and current positions should be stored.

CFO is an eight-step algorithm. The steps include initialization of position and acceleration, calculating initial probe position, calculating initial fitness values, updating probe positions, retrieving errant probes, calculating fitness values of the probes, computing new accelerations, and like any other algorithm, checking if the stopping criteria are met. The performance of the algorithm largely depends on the initial probe distribution.

2.2 Gravitational search algorithm

Inspired by the dynamics of gravity and mass interaction, Rashedi et al. [6] proposed the GSA, which is formulated on the laws of gravity and Newton's second law of motion. In GSA, each of the particles accelerates its velocity inversely proportional to R, which is contrary to the law of gravitation, which says acceleration is inversely proportional to R^2. The authors found experimentally that use of either yields better results.

The concept and formulation of GSA is similar to CFO, except that GSA is stochastic unlike the latter, which is deterministic. The particles (masses) are influenced by the position of all the other masses. Masses have four specifications: position, inertial, active, and passive gravitational mass. The position of the mass corresponds to a solution to the problem, and its gravitational and inertial masses are determined using a predefined fitness function. Fig. 1 shows the influence of particles.

The process of GSA comprises eight principal steps of initialization, fitness evaluation, calculation of gravitational constant, updating inertial and gravitational masses, updating total force, computing the acceleration and velocities, updating particle positions, and checking if the termination criteria are achieved or not.

2.3 Colliding bodies optimization

Kaveh and Mahdavi [10] proposed colliding bodies optimization (CBO), which is inspired by the phenomenon of collision between two bodies in one dimension. Unlike other algorithms, CBO does not use any memory or any internal parameter to save its best-obtained solutions. One object collides with

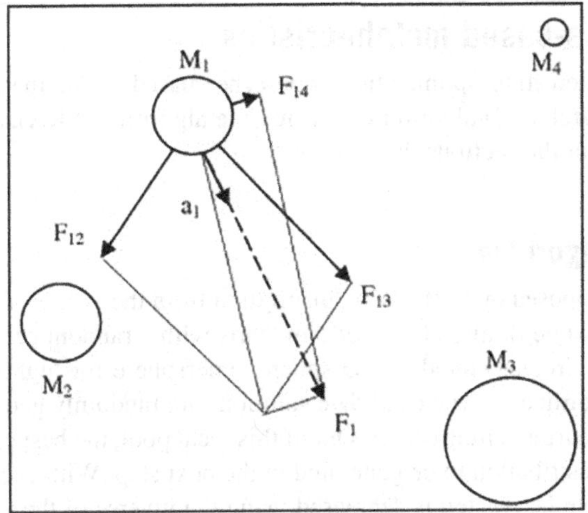

FIG. 1

The influence of particle on other particles.

Reproduced with permission from E. Rashedi, H. Nezamabadi-Pour, S. Saryazdi, GSA: a gravitational search algorithm, Inf. Sci. 179 (13) (2009) 2232–2248, Original figure.

another object to attain minimum energy. The authors originally proposed this algorithm to solve engineering structure-based problems.

We refer to each agent in CBO as a colliding body. Each of these colliding bodies has a mass and a velocity. CBO essentially comprises six principal steps of initialization, calculating the mass of each of the colliding bodies, then sorting the entire population and dividing it into two equal groups. One group starts from the best and the other group starts from the middle. We consider the first group to be stationary and the other group to be mobile and move toward the best-obtained solutions. Then, the new velocities are calculated after the collisions are done. The final positions are updated and, finally, the termination criteria are checked.

2.4 Equilibrium optimizer

EO was proposed by Faramarzi et al. [11]. The main inspiration of EO is dynamic mass balance in physics. The authors used a mass balance equation to mathematically model the mass balance phenomenon. The search agents consist of concentrations that are used to find the optimal solution in the search space.

Just like any other metaheuristic optimization algorithm, in order to explore the maximum search space we randomly initialize the search space. An equilibrium pool is created out of the four best-performing solutions and the average of the best four solutions. Next, to update the concentrations, a balance between exploration and exploitation is maintained by setting a parameter that is a function of several iterations. This is done to maximize the local search ability in the later stage and enhance the exploration during the initial stages. The generation rate is set following the first-order decay process to enhance the exploitable nature of the solution. Finally, the stopping criteria is checked.

3. Fluid mechanics-based metaheuristics

There have been some interesting optimization approaches based on the theories of fluid mechanics. We discuss the vortex search (VS) algorithm, flow regime algorithm (FRA), and Archimedes optimization algorithm (AOA) in the sections that follow.

3.1 Vortex search algorithm

The VS algorithm was proposed by [12], taking inspiration from the vortex-like phenomena occurring in irrotational, incompressible fluids. The algorithm starts with a random central point as the starting solution and generates a circular bounded (3D sphere/hypersphere for higher dimensions) Gaussian distribution around its coordinates. The candidate solutions are randomly generated from the distribution neighborhood of the current circle center. Out of this local pool, the best candidate solution is chosen as the center for the distribution to be generated in the next step. With each iteration, the radius of the spherical distribution to be created is decreased using the inverse of the incomplete gamma function. This radius decrement operation is the key to changing the focus of the search mechanism. Thus, the intuition of the algorithm is to find vortex-like search patterns that gradually converge to an optimal point, as shown in Fig. 2.

It is important to note that contrary to other physics-based metaheuristics, which are mostly population-guided, VS is a single solution-based search procedure. As a result, although the algorithm seeks to find the global optimum, the search mechanism essentially follows a local exploitation paradigm for the same.

3.2 Flow regime algorithm

Proposed in [13], the FRA derives from principles of hydrodynamics for its optimization metaphor. Specifically, it models the turbulent and laminar flow regimes of a viscous fluid to define its exploration and exploitation mechanisms. While a turbulent flow regime is characterized by chaotic property changes including rapid variation of flow velocity and pressure, in laminar flow, the motion of the particles of the fluid is very orderly as the fluid layers flow in parallel without intermixing. Intuitively, a turbulent regime represents a global exploration of the search space, while a laminar regime indicates

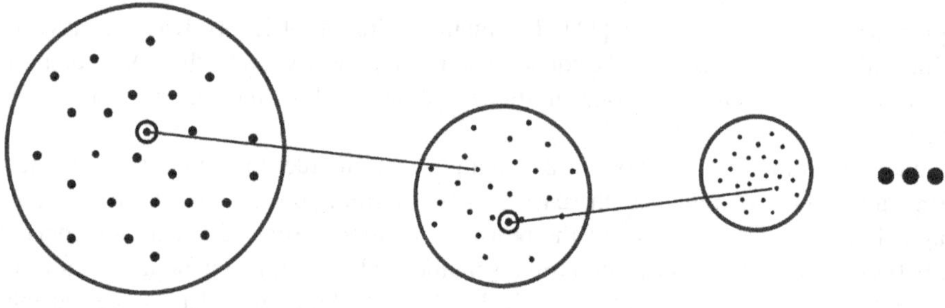

FIG. 2

Intuitive behavior of the vortex search algorithm.

local search or exploitation. FRA uses a parameter to balance between its search phases—a direct parallel drawn from Reynold's number in fluid dynamics (i.e., ratio of inertial to viscous forces acting on a flowing fluid), which is the threshold factor to determine the flow regime of the fluid. Further, the search processes are augmented by the use of arbitrary constants sampled from Levy and Gaussian distributions to increase their randomness.

3.3 Archimedes optimization algorithm

AOA is a recently proposed metaheuristic algorithm by Hashim et al. [14] that has its roots in the very famous Archimedes' principle of hydrostatics, which states that the upward buoyant force exerted on a body immersed in a fluid (fully or partially) equals the weight of the fluid displaced by the body. For a body immersed in a fluid, the forces at play are the gravitational and buoyant forces, which are in opposing directions. Accordingly, the body will be at equilibrium if and only if it is acted upon by forces that are balanced, that is, the weight of the body is exactly compensated by the hydrostatic buoyant force. Following Newton's laws of classical mechanics, the net acceleration of such a body on which balanced forces are acted upon will be zero. This constitutes the optimization paradigm AOA is modeled upon. Fig. 3 shows a schematic diagram of the scenario.

In particular, AOA defines a population as consisting of candidate agents that represent objects submerged in a fluid of given density. The candidate solutions (i.e., objects) have density, acceleration, and volume as their known attributes since these are the quantities on which the buoyant force would depend. The objects are also initialized with random positions that are updated at each step. The acceleration of an object is computed based on the condition of its collision with another object of the population. A higher value of acceleration would imply the search is in the exploration phase (collisions taking place), while a lower value would indicate exploitation (no collision). If a collision takes place, an arbitrary object is taken as a reference for updating the current values, while in case of no collision, the candidate solution follows the global best candidate. AOA uses several parameters to ensure balance between exploration and exploitation is ably maintained.

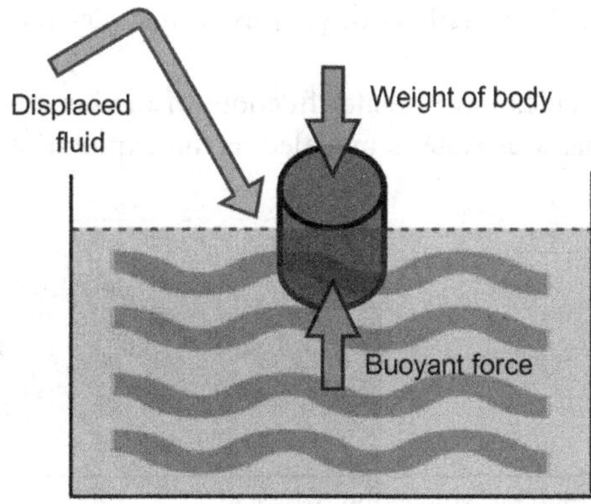

FIG. 3

Schematic diagram to represent Archimedes' principle.

4. Thermodynamics-based metaheuristics

Thermodynamics deals with the relationship between heat and other forms of energy of the system. There have been several interesting metaheuristic algorithms developed based on the four laws of thermodynamics. This section focuses on thermal exchange optimization (TEO), states of matter search (SMS), kinetic gas molecules optimization (KGMO), and Henry gas solubility optimizer (HGSO).

4.1 Thermal exchange optimization

Proposed by Kaveh et al. [8], TEO is modeled on the thermodynamic process of heat transfer. Specifically, it follows Newton's law of cooling, which states that the rate of heat loss of a body is proportional to the difference in temperatures between the body and its surroundings (Fig. 4).

In TEO, the position of each candidate agent is characterized by its temperature. Each agent is considered as a cooling object that performs thermal energy exchange with another candidate taken as surrounding fluid. It is worth noting that the choosing of cooling objects and the surrounding environment is similar to the divisive grouping of bodies in CBO (Section 2.3). Newton's law of cooling comes into play in order to update the temperatures. Furthermore, TEO also uses two mechanisms for escaping local minima, as well as utilizes memory caching to save a fixed number of recent best solution vectors and their corresponding cost function values, which can be used to enhance the performance of the algorithm without incurring much of a computational cost.

4.2 States of matter search

SMS is a metaheuristic algorithm that bases itself on the simulation of interchangeability among the physical states of matter. Classically speaking, matter exists in three states: gaseous, liquid, and solid. The states differ in the extent of their intermolecular forces, that is, solids have the highest intermolecular force of attraction, followed by liquids where the attractive forces are slightly weaker, and then gaseous state where the said forces are very weak in comparison. Accordingly, the packing of the constituent molecules decreases from solids to gases, resulting in variable extents of molecular interactions in the respective physical states. The authors of [15] used this very theoretical backdrop to develop SMS (Fig. 5).

It is intuitive that gas molecules have greater freedom of randomized movement as compared to liquids and solids. Thus, the gaseous state is modeled as the exploration phase. For solids, since the

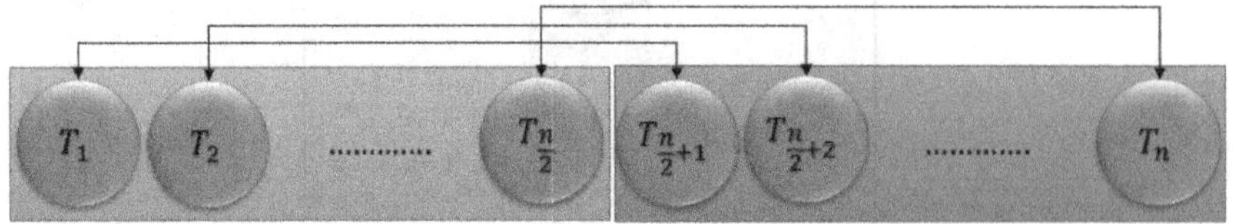

FIG. 4

A pictorial description of heat distribution among the particles.

Reproduced with permission from A. Kaveh, A. Dadras, A novel meta-heuristic optimization algorithm: thermal exchange optimization, Adv. Eng. Softw. 110 (2017) 69–84, Original figure.

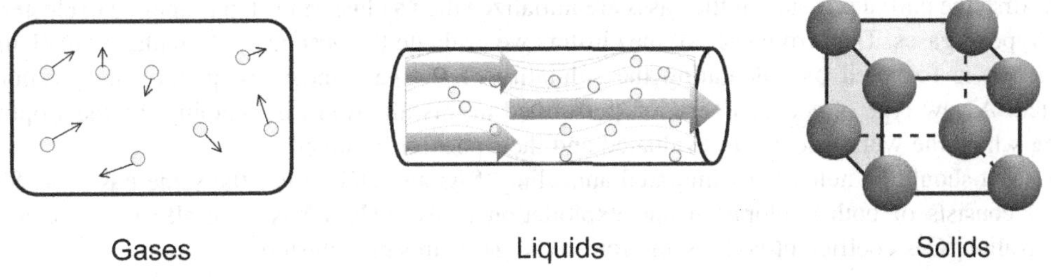

FIG. 5

Schematic representation of the packing extents across various states of matter.

interactions are very less allowed, they are modeled as the exploitation phase. As for liquids, the molecules exhibit moderate movement and thus represent both exploration and exploitation. SMS follows an evolutionary paradigm where the agents (i.e., molecules) start from the gas state (pure exploration) and the algorithm modifies the exploration and exploitation intensities until the solid state (pure exploitation) is reached. The algorithm incorporates collision and direction vector operators to ensure the balance between exploration and exploitation, while the solution vector updates are performed using a random position operator.

4.3 Kinetic gas molecules optimization

KGMO is a swarm-based metaheuristic algorithm based on the kinetic energy of gas molecules. Proposed by Moein et al. [7], KGMO uses gas molecules as agents in the search space and the performance of each agent is based on their kinetic energy. Each gas molecule in the search space has four specifications: kinetic energy, velocity, position, and mass. The velocity and position of the gas molecules are determined using their respective kinetic energies.

In the algorithm, after exploration of the entire search space, the gas molecules reach the point of lowest temperature and kinetic energy because of the attraction between gas molecules originated due to the weak Van der Waal's force between the molecules. KGMO is based on the principle of Boltzmann distribution, which implies that the velocity of the gas molecules is proportional to the exponential of the molecules' kinetic energy. After the initialization of each molecule's position and velocity in the search space, the kinetic energy determines the velocity of each gas molecule, which eventually updates the position of each gas molecule.

It is noteworthy that both KGMO and Gases Brownian Motion (GBM) optimization algorithms [16] are novel models with regard to evolution of swarm particles. They are completely novel because the evolution of swarm particles was based on newer exploration techniques that were inspired by the nonlinear equations that describe thermodynamic systems.

4.4 Henry gas solubility optimizer

Proposed by Hashim et al. [17], the HGSO algorithm imitates the huddling behavior of gas to balance exploitation and exploration. The algorithm is modeled on Henry's law, which states that the amount of dissolved gas in a liquid is proportional to its partial pressure above the liquid. Accordingly, the search agents in HGSO are individual gases, each having a partial pressure associated with it.

At first, the partial pressure of the gases are initialized then a clustering step is applied to cluster the same type of gases. Then from each of the clusters we evaluate the best gases. Calculation of Henry's coefficient is followed by calculating the solubility of the gas, and subsequently the position is updated. A new type of escape operation performed in this algorithm is escaping the local optima, during which the worst agents are evaluated and their position changed.

Also, it should be noted that simulated annealing (SA) and HGSO use the same gas laws. Since HGSO consists of both exploration and exploitation phases, HGSO is generally used for global optimization. Gas coefficient is the same for all the gases in same clusters.

5. Electromagnetism-based metaheuristics

The properties shown by charged particles in electric and magnetic fields provide very interesting insights into global optimization paradigms. Specifically, in this section we discuss some of the optimization techniques based on the principles of classical theories of electrostatics and electromagnetism. These include artificial electric field algorithm (AEFA), magnetic-inspired optimization (MIO), electromagnetic field algorithm (EFA), and ions motion optimization (IMO).

5.1 Artificial electric field algorithm

AEFA, proposed by Yadav et al. [18], is a population-based metaheuristic algorithm inspired by Coulomb's laws of electrostatic force. In AEFA, we charge the agents' particles, and the only force defining the movement of the particles in the search space is the attractive electrostatic force. AEFA proposes an isolated system of particles obeying Coulomb's law of electrostatic force and laws of motion.

After initializing the position of the particles, we assign the charges of the particles using the fitness functions. On the calculation of Coulomb's constant, and evaluation of the particles based on the fitness values, the force acting on one particle, because of the rest of the particles in the search space, is calculated. The velocity of the particle, which specifies the updated position of each particle in the search space, is based on the acceleration of each particle. The acceleration, caused by the electrostatic force of attraction acting on the particle, is calculated using the second law of motion.

For enhancing the exploration capability of the algorithm, Coulomb's constant is initialized to a higher value and is decreased with each iteration. AEFA uses both global optima history and local optima history. For updating the position of each particle in AEFA, the particles are attracted toward the best particle in the search space and the force value specifies the movement of the particle.

5.2 Magnetic-inspired optimization

MIO algorithm is a metaheuristic algorithm proposed by Tayarani-N and Akbarzadeh-T [19] inspired by magnetic field theory. In MIO, a magnetic particle scattered in the search space represents each solution. MIO is a cellular-based algorithm with the particles interacting in a lattice-like interactive population, as shown in Fig. 6.

The basic principle behind MIO is the magnetic force acting between the charged particles, which specifies the movement of the particles in the search space. We assign the fitness of each particle in the initialized lattice to the respective magnetic field of the particle. We calculate the mass of each particle

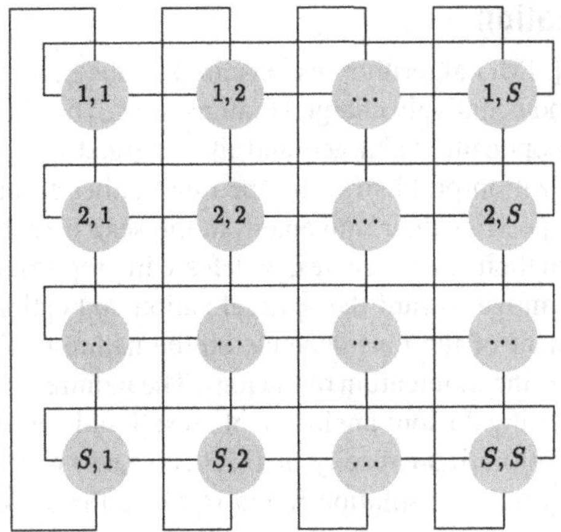

FIG. 6

Example of lattice structure in MIO algorithm. Lines between S charged particles show the connection between neighbor nodes.

based on the normalized magnetic field value. We assign fitter particles with higher mass values, which result in more attractive forces between the neighboring particles in the lattice. Each particle's velocity is calculated based on the magnitude of the force and the mass of the particle, which signifies the updated position of the respective particle. After several iterations, the particles around the inferior optimum will start moving toward the superior optimum because of the larger mass of the particles surrounding it.

The particles in MIO move toward the particle with greater magnetism (fitness) because of the long-range attractive forces. To overcome the retention of particles around the local optima because of the forces of attraction, MIO also proposes four novel operators: short-range repulsion, explosion, hybrid explosion-repulsion, and crossover interaction, which enhance the performance of the proposed algorithm.

5.3 Electromagnetic field algorithm

The authors of [20] presented EFA, a population-based metaheuristic approach inspired by the behavior of electromagnets. An electromagnet, unlike a permanent magnet, produces a magnetic field because of the direction of the electric current and has single polarity (positive/negative). EFA is inspired by the forces of attraction/repulsion acting between electromagnets of different polarities.

In EFA, each solution represents an electromagnetic particle formed by the electromagnets. The electromagnets are determined by the variables of the optimization problem. We divide the population into three fields: positive, negative, and neutral. The attraction-repulsion forces among electromagnets of these three fields specify the movement of the particles toward global minima. A nature-inspired ratio called the golden ratio, which determines the ratio between attractive/repulsive forces, enhances the algorithm. A new electromagnetic particle is formed for every electromagnet (variables) whose fitness is then compared with global minima of the search space, which, if worse, is substituted by the newly formed particle.

5.4 Ions motion optimization

Proposed by Javidy et al. [21], IMO algorithm was inspired by the nature of the force acting between positively charged (cations) and negatively charged (anions) ions. The algorithm imitates the attractive and repulsive forces between oppositely charged and like-charged ions, respectively. The candidate solution to a particular optimization problem is grouped into either anions or cations, and the nature of the force between the ion specifies their movement in the search space.

We evaluated ions based on their fitness values, which are in proportion with the objective function value of the ions. The anions move toward the optimal cation and cations move toward the optimal anion. The amount of movement of the ions depends on the nature of the force between them. The magnitude of the force specifies the momentum of the ions. The requirement of the ions for diversifying and intensifying is met by assuming the ions are in two phases: liquid phase (diversification) and crystal phase (intensification). The exploration capacity of IMO is weaker compared with other approaches like PSO, which are guided by the best solution found so far in the search.

6. Optics-based metaheuristics

Optics deals with the behavior and properties of light and its interaction with matter. The algorithms discussed in this section are based on the properties of the systems in optics. The two approaches are ray optimization and optics-inspired optimization.

6.1 Ray optimization

This is the first work in the world of metaheuristics derived from ray optics, proposed by Kaveh et al. [9]. The algorithm is inspired by Snell's law of refraction of light, which explains why and how the direction of a light ray changes when it faces a change in medium (i.e., while passing from one medium to another). RO specifically uses ray tracing as the backbone of its search mechanism using basic geometrical operations and is formulated in three steps: (1) scattering, (2) ray movement and motion refinement, and (3) ray convergence.

The initial population pool of RO comprises rays of light represented as candidate vectors. A candidate solution in RO follows the path traced by a ray to move about in the search space. Since ray-tracing is feasible only in two-dimensional (2D) and three-dimensional (3D) spaces, for bigger dimensions the solution vector is divided into groups comprising two or three members, which are then moved to new positions following the exploration and exploitation procedures described by the algorithm. Thus, each update of the candidate solution consists of position updates of its constituent groups in 2D or 3D spaces. The best positions of the local vectors along with the best global solution vector are stored in each iteration until the termination condition is reached.

6.2 Optics-inspired optimization

Proposed by Kashan [22], OIO was developed based on reflection of light by spherical mirrors. These mirrors have a very interesting property to converge and diverge images of the specific objects. In OIO, the search field of the interested problem is to be optimized as a wavy mirror in which the concave mirror is represented as a valley and the convex mirror is represented as a peak. In other words, the surface of the objective function is considered a spherical mirror that can either be converging (i.e., concave) or diverging (i.e., convex), and each candidate solution represents an artificial light point from which the new solutions are generated based on the final produced image upon reflection. Fig. 7 presents a detailed description of the approach.

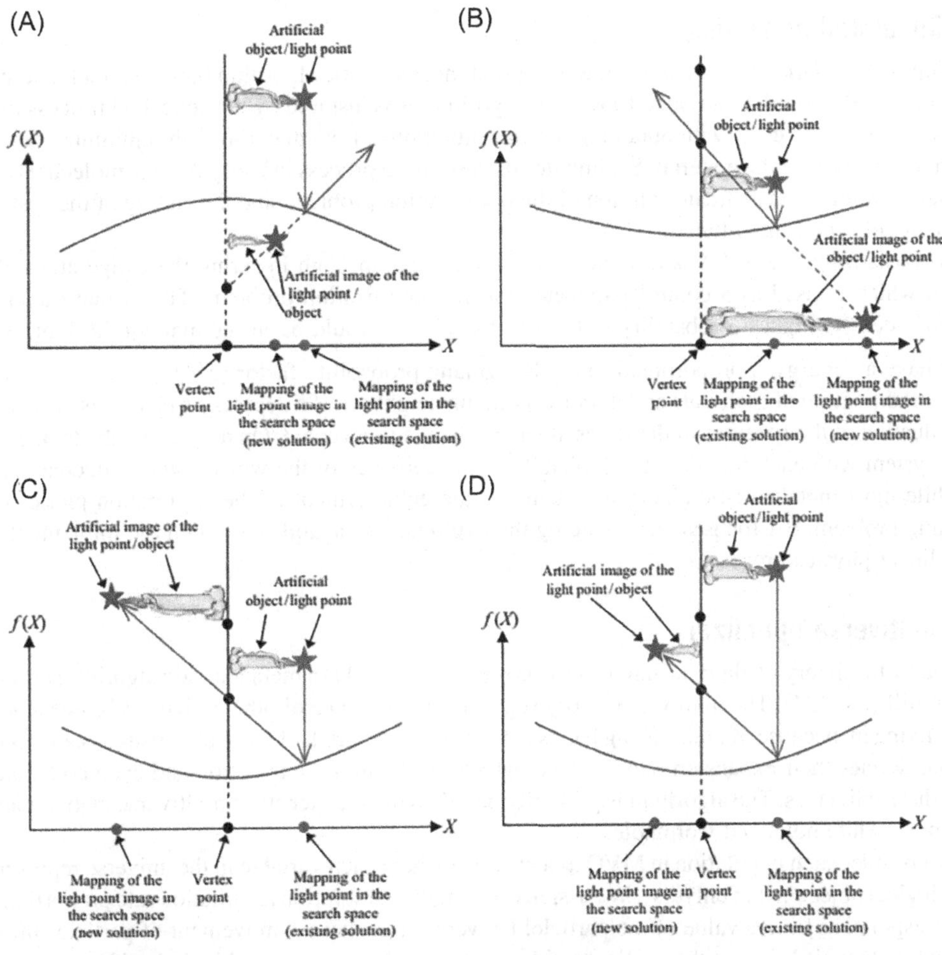

FIG. 7

The processes involved in OIO: (A) function as a convex surface; (B) function as concave surface and the artificial object/light point is between artificial focal point and the function surface; (C) function as concave surface and the artificial object/light point is between artificial focus and center of curvature; and (D) function as a concave surface and the artificial object/light point is beyond center of curvature.

Reproduced with permission from A.H. Kashan, A new metaheuristic for optimization: optics inspired optimization (OIO), Comput. Oper. Res. 55 (2015) 99–125, Original figure.

7. Other physics-based metaheuristic algorithms

In this section, we extend our discussion to some of the more unique physics-inspired optimization techniques that are too diverse to be categorized under the domains explored in this chapter. The discussed algorithms are SA, multiverse optimizer (MVO), atom search optimization (ASO), and nuclear reaction optimization (NRO).

7.1 Simulated annealing

SA, proposed by Kirkpatrick et al. [23], was the first metaheuristic algorithm based on nonlinear physics processes. Proposed in the early 1980s, this algorithm was inspired by the annealing process in material science, which deals with obtaining the configurations of the material with minimum energy of molecules or atoms of the material. SA imitates the annealing process in a way that the molecules of the material represent the candidate solution of the optimization problem, and the energy of the system is used to signify the fitness function.

The basic principle of SA is to obtain the optimal state through lowering the temperature of the system, which is used as a control parameter for the optimization problem. The primary approach is to consider the highest probability state of a system that would be in thermal equilibrium, which would have an energy ϵ, proportional to the Boltzmann probability factor ($e^{\frac{-\epsilon_i}{kT}}$).

On modification of each state and its corresponding neighbors, if a decreasing change is observed in the evaluation of the fitnesses of the states, then the change is accepted. On decreasing the temperature of the system with each iteration, the probability of acceptance of the worse solution decreases.

While most metaheuristic algorithms focus on the enhancement of the exploration phase of the searching problem, SA focuses on enhancing the exploitation capability through the implementation of nonlinear physics processes.

7.2 Multiverse optimizer

Inspired by the theory of alternate universes in cosmology, MVO is a metaheuristic algorithm proposed by Mirjalili et al. [24]. The multiverse theory suggests the existence of other universes besides the one we are living in, because of multiple big bangs in the beginning. Each of these universes has an inflation rate that defines their expansion rate in space, which results in the interaction and even collision between these universes. The algorithm is primarily based on three concepts of multiverse theory, namely black hole, white hole, and wormhole.

We consider each population in MVO as a universe where each variable in the universe represents a cosmological object in that universe. We assign each candidate solution an inflation rate proportional to the corresponding fitness value of the particle. Universes may have a movement of particles through black/white tunnels between them. We consider a universe having more white holes if it has a higher inflation rate, and black holes if it has a lower inflation rate. Objects move toward lower inflation rates from higher inflation rates, which enhances the average inflation rate.

After sorting the universes according to their inflation rates, the assignment of a white hole to a particular universe is done through a roulette wheel mechanism, based on the normalized inflation rates. To enhance exploitation performance, a wormhole tunnel enables the exchange of objects between a universe and the best universe formed so far.

7.3 Atom search optimization

ASO, proposed by Zhao et al. [25], is a metaheuristic optimization algorithm inspired by basic molecular dynamics. ASO is a population-based metaheuristic algorithm that mimics the atomic motion determined by interaction and constraint forces.

In ASO, we represent each solution by the position of the atom in the search space, whose value is determined by the mass of the atom. The heavier atoms seek to attract the other atoms of the search space toward them. The lower value of acceleration in heavier atoms ensures a better search for the local optima, while the higher value of acceleration in the case of lighter atoms enables the search for better promising regions in the entire search space. The acceleration of atoms comes from two parts, (i) interaction force caused by the L-J potential and (ii) the constraint force caused by bond-length potential, which is the weighted position difference between each atom and the best atom. Fig. 8 is a pictorial illustration of ASO.

In the initial stages of the algorithm, the atoms interact with each other through attractive and repulsive forces depending on the distance between them, which results in the lighter atoms being attracted toward the heavier atoms. The repulsive force denies the over-concentration of atoms around a local optimum, ensuring the exploration capability of the algorithm throughout the search space. With each iteration, the repulsive force grows weaker and the attractive forces grow stronger, ensuring the exploitation capabilities of the algorithm. With each iteration, the velocities and position of each particle are updated until the stopping criterion is met.

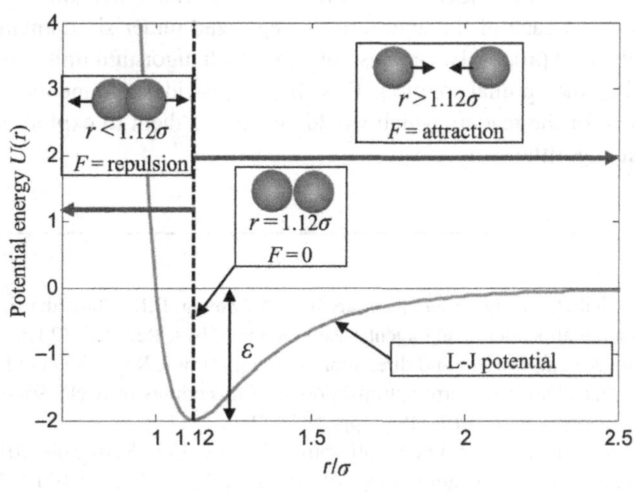

FIG. 8

A pictorial illustration to how ASO is modeled upon.

Reproduced with permission from W. Zhao, L. Wang, Z. Zhang, Atom search optimization and its application to solve a hydrogeologic parameter estimation problem, Knowl.-Based Syst. 163 (2019) 283–304, Original figure.

7.4 Nuclear reaction optimization

NRO, proposed by Wei et al. [26], is a metaheuristic algorithm based on the phenomena of nuclear reaction. It comprises two phases: the nuclear fusion (NFu) phase and the nuclear fission (NFi) phase. NFi inhibits the exploitation characteristics to find a better solution near the local optima, while NFu exhibits exploration capabilities for obtaining a global solution.

The NRO algorithm assumes that the whole search space is in a sealed container where all nuclei interact with each other, with the nuclei representing the variables of the optimization problem. In the NFi phase, NRO adopts the Gaussian walk method to generate new nuclei after fission. The Gaussian walk is primarily implemented to imitate the fission fragments with different states for enhanced exploitation of a search space after which we update the position of the nuclei using the fitness function.

The NFu phase is divided into two phases: ionization and fusion. The nuclei are evaluated based on their fitnesses and the fitter nuclei are retained for the enhanced exploitation of the NFu phase, while the less fit nuclei are used to improve exploration. We implement the fusion stage movement to change the state of an ion combined with information of other ions in the nuclear reaction population. Similar to the ionization phase, this phase uses the variants of difference operators for enhanced exploration. In the fusion phase, Levy flight is implemented to assist the current solution in escaping from the local optimal space.

8. Conclusion

In this chapter, we discussed the most recent and widely used metaheuristic algorithms based on nonlinear physics. The algorithms are categorized according to the branch of physics they draw inspiration from, such as classical mechanics, thermodynamics, and electromagnetism, among others. We described the intuition behind each of the algorithms categorized under six domains that include both recent as well as widely used proposals, emphasizing how each algorithm imitates the nonlinear physical processes for finding the optimal solution. This chapter provides a foundation of physics-inspired optimization paradigms for the readers which would encourage them to explore further into the vast world of metaheuristic algorithms.

References

[1] A. Chakraborti, D. Challet, A. Chatterjee, M. Marsili, Y.-C. Zhang, B.K. Chakrabarti, Statistical mechanics of competitive resource allocation using agent-based models, Phys. Rep. 552 (2015) 1–25.
[2] L.A. Demetrius, Boltzmann, Darwin and directionality theory, Phys. Rep. 530 (1) (2013) 1–85.
[3] J. Kennedy, R. Eberhart, Particle swarm optimization, in: Proceedings of ICNN'95—International Conference on Neural Networks, vol. 4, IEEE, 1995, pp. 1942–1948.
[4] S. Mirjalili, S.M. Mirjalili, A. Lewis, Grey wolf optimizer, Adv. Eng. Softw. 69 (2014) 46–61.
[5] R.A. Formato, Central force optimization, Prog. Electromagn. Res. 77 (1) (2007) 425–491.
[6] E. Rashedi, H. Nezamabadi-Pour, S. Saryazdi, GSA: a gravitational search algorithm, Inf. Sci. 179 (13) (2009) 2232–2248.
[7] S. Moein, R. Logeswaran, KGMO: a swarm optimization algorithm based on the kinetic energy of gas molecules, Inf. Sci. 275 (2014) 127–144.

[8] A. Kaveh, A. Dadras, A novel meta-heuristic optimization algorithm: thermal exchange optimization, Adv. Eng. Softw. 110 (2017) 69–84.
[9] A. Kaveh, M. Khayatazad, A new meta-heuristic method: ray optimization, Comput. Struct. 112 (2012) 283–294.
[10] A. Kaveh, V.R. Mahdavi, Colliding bodies optimization: a novel meta-heuristic method, Comput. Struct. 139 (2014) 18–27.
[11] A. Faramarzi, M. Heidarinejad, B. Stephens, S. Mirjalili, Equilibrium optimizer: a novel optimization algorithm, Knowl.-Based Syst. 191 (2020) 105190.
[12] B. Doğan, T. Ölmez, A new metaheuristic for numerical function optimization: vortex search algorithm, Inf. Sci. 293 (2015) 125–145.
[13] M. Tahani, N. Babayan, Flow regime algorithm (FRA): a physics-based meta-heuristics algorithm, Knowl. Inf. Syst. 60 (2) (2019) 1001–1038.
[14] F.A. Hashim, K. Hussain, E.H. Houssein, M.S. Mabrouk, W. Al-Atabany, Archimedes optimization algorithm: a new metaheuristic algorithm for solving optimization problems, Appl. Intell. 51 (3) (2021) 1531–1551.
[15] E. Cuevas, A. Echavarría, M.A. Ramírez-Ortegón, An optimization algorithm inspired by the States of Matter that improves the balance between exploration and exploitation, Appl. Intell. 40 (2) (2014) 256–272.
[16] M. Abdechiri, M.R. Meybodi, H. Bahrami, Gases Brownian motion optimization: an algorithm for optimization (GBMO), Appl. Soft Comput. 13 (5) (2013) 2932–2946.
[17] F.A. Hashim, E.H. Houssein, M.S. Mabrouk, W. Al-Atabany, S. Mirjalili, Henry gas solubility optimization: a novel physics-based algorithm, Futur. Gener. Comput. Syst. 101 (2019) 646–667.
[18] Anita, A. Yadav, AEFA: artificial electric field algorithm for global optimization, Swarm Evol. Comput. 48 (2019) 93–108.
[19] M.-H. Tayarani-N, M.-R. Akbarzadeh-T, Magnetic-inspired optimization algorithms: operators and structures, Swarm Evol. Comput. 19 (2014) 82–101.
[20] H. Abedinpourshotorban, S.M. Shamsuddin, Z. Beheshti, D.N.A. Jawawi, Electromagnetic field optimization: a physics-inspired metaheuristic optimization algorithm, Swarm Evol. Comput. 26 (2016) 8–22.
[21] B. Javidy, A. Hatamlou, S. Mirjalili, Ions motion algorithm for solving optimization problems, Appl. Soft Comput. 32 (2015) 72–79.
[22] A.H. Kashan, A new metaheuristic for optimization: optics inspired optimization (OIO), Comput. Oper. Res. 55 (2015) 99–125.
[23] S. Kirkpatrick, C.D. Gelatt, M.P. Vecchi, Optimization by simulated annealing, Science 220 (4598) (1983) 671–680.
[24] S. Mirjalili, S.M. Mirjalili, A. Hatamlou, Multi-verse optimizer: a nature-inspired algorithm for global optimization, Neural Comput. Appl. 27 (2) (2016) 495–513.
[25] W. Zhao, L. Wang, Z. Zhang, Atom search optimization and its application to solve a hydrogeologic parameter estimation problem, Knowl.-Based Syst. 163 (2019) 283–304.
[26] Z. Wei, C. Huang, X. Wang, T. Han, Y. Li, Nuclear reaction optimization: a novel and powerful physics-based algorithm for global optimization, IEEE Access 7 (2019) 66084–66109.

CHAPTER

4

Evolutionary computation techniques for optimal response actions against water distribution networks contamination

Parnian Hashempour Bakhtiari[a], Mohammad Reza Nikoo[a], and Amir H. Gandomi[b,c]

[a]*Department of Civil and Architectural Engineering, College of Engineering, Sultan Qaboos University, Muscat, Oman,*
[b]*Faculty of Engineering and Information Technology, University of Technology Sydney, Ultimo, NSW, Australia,*
[c]*University Research and Innovation Center, Obuda University, Budapest, Hungary*

1. Introduction

Water distribution systems (WDSs) are critical municipal infrastructures, the duty of which is to deliver safe and clean water to end consumers. As a significant part of the water supply network, WDSs are exceedingly complex, consisting of various storage tanks, pipes, and valves. This complexity and extensiveness make WDSs extremely vulnerable to safety threats like natural damages by disasters and accidents or deliberate contamination due to physical and cyber-attacks. The soluble nature of water makes it a potential source of transporting highly toxic chemical, biological, or radiological contaminants with minimal decay in the shortest possible time.

Among all possible reasons for anomalies in WDSs, deliberate contaminant intrusion has attracted interest in recent years because of the increasing rate of major attacks [1–4]. Various water quality monitoring systems and emergency management plans have been proposed in the literature, e.g., [5–10], and adopted by municipalities to improve the reliability and security of WDSs and, ultimately, reduce their unserviceability. A major part of emergency management plans consists of sudden response actions against contaminant intrusion. After detecting the contaminant intrusion into the WDS, proper contaminant containment and pollution discharge strategies should be taken to return the WDS condition to a safe state and protect public health in the shortest time [11–15].

Development of efficient contamination response actions requires the assessment of different social, political, and economic aspects. These aspectsare closely related to factors like contaminant type, WDS size, and detection facilities [16,17]. Hence, as a critical complex issue, numerus studies have been conducted over the decades to provide applicable response actions [18]. Most studies have adopted optimization approaches and stated their success in achieving reliable results considering single or multiple objectives (e.g., [9,19–21]). Substantial reduction of WDS recovery time, controlled consumption of contaminated water, and balanced operational costs are some outcomes of utilizing optimal response actions [11,22–25]. However, the computational burden of the traditional optimization models has always led to simplifications that are incompatible with the complex nature of the decision-making process [26,27].

Comprehensive Metaheuristics. https://doi.org/10.1016/B978-0-323-91781-0.00004-1

Over the last two decades, evolutionary algorithms have been widely used as suitable replacements for conventional optimization algorithms/models to provide optimal solutions to complex practical obstacles [28–30]. Inspired by Darwinian evolution, the evolutionary algorithm (EA) is proposed to be a robust, flexible, and conceptually simple heuristic search technique and optimization method [31–34]. Researches have shown that the EAs increase the probability of finding an optimal/near-optimal solution in the early stages while addressing the issues related to existing uncertainties, stochastic characteristics of decision variables, or noisy fitness functions [19,22]. Generally, EAs can be successfully utilized to solve both single- and multiobjective optimization problems. The EAs are representation independent; hence, any type of decision variables (e.g., discrete, integer, or discontinuous) and objective functions (e.g., continuous or differentiable) can be analyzed by these algorithms [32]. Furthermore, the EAs restrict the search space by utilizing prior information, which leads to computational burden reductions and, consequently, improving their applicability in solving large-scale problems [32,33].

The EA has been presented as a subset of evolutionary computation that is a term defined in the field of computational intelligence [35,36]. In fact, computational intelligence (CI) consists of a set of computational techniques inspired by the nature laws such as biologically inspired algorithms [37,38]. The CI concept has been widely used in situations where mathematical logic becomes unproductive due to the complexities, existing uncertainties, or stochastic nature of the problem [39]. Accordingly, evolutionary computation (EC) features a robust behavior and flexible nature that can effectively develop global optimal solutions for acute problems.

Among the well-known EAs, different variations of genetic algorithms have been widely used by researchers to achieve optimal response actions against contaminant intrusion into water distribution networks, as thoroughly discussed in the following.

2. Evolutionary computation

Since the latter half of the 20th century, with the development of computer systems, there has been an increasing demand for automation of decision-making processes [36,37]. In this regard, the concepts of CI have been stated by the literature as an effective solution to solve existing complex problems [33]. As a subbranch of artificial intelligence, CI facilitates the assessment and modeling of an intelligent behavior by learning from and adapting to complex changing environments, and provides a successful solution for real-world problems. The main CI paradigms consist of artificial neural networks, evolutionary computation, swarm intelligence, artificial immune systems, and fuzzy systems [38,39].

EC employs the principles of Darwinian natural evolution and develops robust algorithms, the EAs, which provide impressive performance [32,35]. As the title implies, EAs model genetic evolution, where the genotypes are representatives of individual characteristics. In general, the EA attempts to achieve the desired candidate of solutions and reach the objective goals. In this regard, EA develops a solution candidate as a set of genes called a "chromosome." Then, through an iterative process, the algorithm attempts to improve the features of provided candidates known as "parents" and produce new candidates labeled as "children" considering the desired objective (fitness function). This procedure is called evolution and continues until the sufficient candidate is achieved [31,32].

The EAs have been classified into seven categories: co-evolution, cultural evolution, differential evolution, evolutionary programming, evolution strategies, genetic algorithms, and genetic programming [33,36,40]. Despite the differences of mentioned paradigms, all EAs follow similar procedures.

2.1 Development and evaluation of the solution candidate

2.1.1 Representation schemes

Referring to the underlying concept of EC, the set of input individuals (e.g., the optimization decision variables) is referred to as chromosome, while each variable is labeled as "gene." The defined chromosomes provide the required information of the EAs. Hence, an appropriate representation scheme of chromosomes should be provided to ensure the accuracy of the obtained results.

Various representation schemes have been utilized by the existing EA paradigms, including the finite-state, integer, real-valued, mixed-integer, permutations, and tree format (specifically used by genetic programming). In the original form, the EA transforms the input individuals to the binary vectors of fixed length. For instance, in the case of an optimization problem with N_D decision variables, the established chromosome becomes a set of N_D bit strings following this procedure:

1. Decision variables of binary type

Here, the length of chromosomes equals the number of decision variables. In this regard, a chromosome of N_D bits would be obtained containing a value of 0 or 1.

2. Discrete-valued decision variables

Usually, the value of decision variable should be specified among the predefined set of numerical values. In such cases, the decision variable is encoded as a bit array of N_{Dis} dimension, where N_{Dis} equals the required number of bits representing the value of decision variable in binary form (as depicted in Fig. 1).

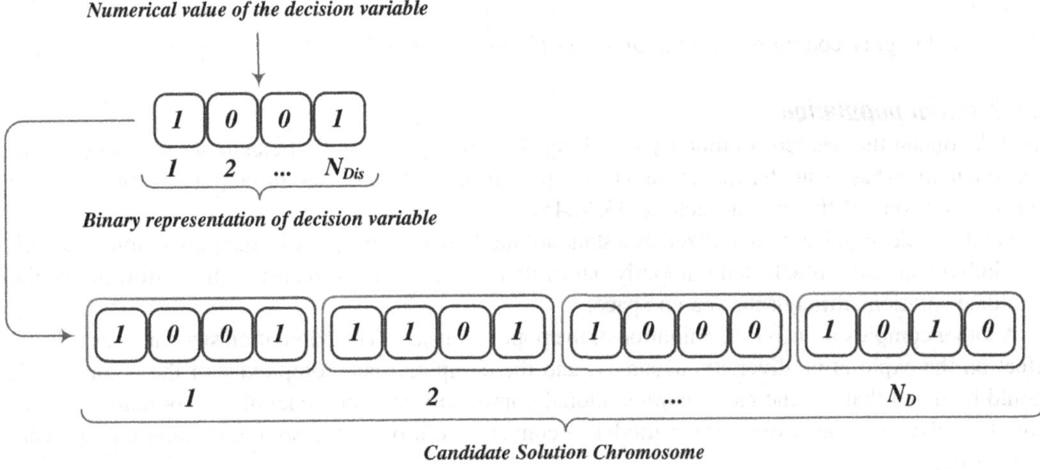

FIG. 1

Schematic representation of developing candidate solutions with discrete-valued decision variables.

3. Continuous-valued decision variables

If the decision variable is specified to be continuous-valued, the EA first maps the continuous search space to a discrete domain and then transforms it to a binary string (Fig. 1), as illustrated in the following:

- Determine the upper and lower bounds of variables: $DV_i \in [DV^i_{min}, DV^i_{max}]$
- Encode the decision variable as a N_{Cts}-dimensional bit array, where the N_{Cts} equals the required number of bits: $DV_i \rightarrow b_i = (b^i_1, b^i_2, \ldots, b^i_{N_{Cts}})$
- Map the N_D search space into a binary array:

$$DV = (DV_1, DV_2, \ldots, DV_i, \ldots, DV_{N_D}) \rightarrow b = (b_1, b_2, \ldots, b_i, \ldots, b_{N_D})$$

Hence, the resulting chromosome is comprised of $N_{Cts} \times N_D$ bits.

By using this mapping technique, the continuous search space is replaced by a discrete one, which clearly decreases the accuracy of the provided optimal solution. In fact, the developed EA is able to achieve the maximum accuracy of $\frac{DV^i_{max} - DV^i_{min}}{2^{N_{Cts}} - 1}$ for each decision variable, where $2^{N_{Cts}}$ represents all possible bit array values corresponding to the desired decision variable.

Despite all benefits of binary coding, studies have pointed out its weaknesses in providing solutions for optimization problems [41–43]. Hence, "gray coding" is suggested as a replacement to ensure algorithm efficiency. The binary coding can be easily converted to gray coding using the following equation:

$$g^i_j = \left[b^i_{j-1}\left(1 - b^i_j\right) + \left(1 - b^i_{j-1}\right)b^i_j\right] \tag{1}$$

Subjected to:

$$g^i_1 = b^i_1 \tag{2}$$

where g^i_j is the gray coding representation of bit b^i_j, which depends on b^i_j and its previous bit, b^i_{j-1}.

2.1.2 Initial population

The EAs obtain the desired solution by searching through a population of candidate solutions due to their population-based stochastic nature. Hence, proper generation of the initial population should be considered as one of the critical factors [33,44,45].

Random selection has been utilized as a standard method to develop the initial population, in which each individual value is selected randomly out of its relevant domain to ensure the uniformity of the candidate solutions through the search space.

Another complex issue is determination of the population size. The population size imposes a direct effect on the exploration diversity, accuracy, and the computational complexity of the algorithm. It should be noted that the increased computational complexity in each generation may reduce the required number of generations for the model to converge to a desirable solution, while the opposite is also true.

Hence, the initial population size is suggested to be determined through a trial-and-error procedure to achieve a balance among the population size, required number of generations, and the resulting computational burden. The provided balance ensures the efficiency of the developed EAs by reaching the global optimal solution in the shortest possible time [46,47].

2.1.3 Fitness function

In order to assess how good the developed chromosomes are, a mathematical function labeled as "fitness function" is provided within the EA framework. In the majority of EA paradigms, the fitness function equals the applied objective function (e.g., the objective function of an optimization problem). The mathematical structure of the fitness function, first, decodes the binary chromosomes (using Eqs. 3 and 4 for binary and gray coding, respectively) to determine the real value of candidate solutions [33]. The candidate solutions are then applied to calculate the value of objective function. At the final stage, the goodness of candidate solutions is evaluated and ranked by comparing the resulted objective values. Based on Darwin's law of natural selection, the candidate solution with the best objective value will be suggested as the selected solution by the developed EA.

$$RV_i = DV^i_{\min} + \frac{DV^i_{\max} - DV^i_{\min}}{2^{N_{Cts}} - 1}\left(\sum_{j=1}^{(N_{Cts}-1)} b^i_{(N_{Cts}-j)} 2^j\right) \tag{3}$$

$$RV_i = DV^i_{\min} + \frac{DV^i_{\max} - DV^i_{\min}}{2^{N_{Cts}} - 1}\left[\sum_{j=1}^{(N_{Cts}-1)} \mathrm{mod}\left(\sum_{k=1}^{(N_{Cts}-j)} b^{i-1}_{(N_{Cts}+k)} 2\right) 2^j\right] \tag{4}$$

where RV_i represents the real value of the assumed genes (e.g., the decision variables in optimization problems).

Different types of objective function exist that EAs should become compatible with. The most popular forms of objective functions are unconstrained, constrained, and multiobjective. The unconstrained objective function is the simplest form in which the fitness and objective functions are equal. To solve the problems with constraint objective functions, the EAs modify the fitness function containing two "original" and "constraint penalty" objective functions. Finally, the multiobjective problems are dealt with considering two scenarios of defining the fitness function as a weighted sum of all objective functions or utilizing a Pareto-based algorithm.

2.2 Applying the reproduction operators

Through a reproduction process, two or more offspring are produced from their determined parents utilizing one or both of the reproduction operators: mutation and crossover.

The crossover operator attempts to create new genes by combining the randomly selected corresponding genes from two or several parents. Alternatively, the mutation operator creates new candidate solutions by randomly changing the gene values, which consequently increases the population diversity. Mutation rate is controlled in EAs by defining a mutation probability. High mutation probability forces the EA to explore in the search space more extensively, especially for a small population. However, high probabilities can also distort the most-fit solution candidates during the final generations. As a result, two approaches have been proposed for determining the mutation probability:

- assignment of a constant low value to the mutation probability
- initializing the mutation probability to a large value and gradual reduction in a timely manner to take the extensive exploration into account

Considering the variety of these operators among the EA paradigms, we discuss the different implementations of mutation and crossover operators in the following sections.

2.3 Applying the selection operator

Selection is the major EA operator that formulates the evolution concept within the algorithm.

Following the reproduction procedure, most-fit candidate solutions should be passed on to the next generation, using the selection operator. New candidates can be selected both from the "Parent" and "Offspring" groups, or only the "Offspring."

Various selection operators have been proposed in the literature. Frequently used operators include Boltzmann selection, elitism, proportional selection, rank-based selection, random selection, and tournament selection. An important issue in efficiency of the selection operators is their relevant selective pressure, which is defined as the convergence speed of the EA only if the selection operator is being utilized. The high selective pressure of a selection operator will diminish the exploration ability and decrease the diversity of the EA population compared to operators with a low selective pressure.

2.4 Identifying the termination criterion

In an ideal condition, an EA stops after reaching its predefined optimal fitness level. However, the stochastic search-based nature of EAs prevents them from achieving the desired optimal solution. Hence, a termination criterion should be estimated in advance to stop the EA iterations. As one of the termination criterion definitions, the maximum number of generations is limited to a proper number. The generation termination number should be high enough to allow the EA to extensively explore the search space.

We discuss the genetic algorithm in the following section, considering its widespread application in providing emergency response actions after contamination of water distribution networks.

2.5 Genetic algorithm

The genetic algorithm (GA) was popularized by John H. Holland and has become the most well-known type of EA over the last decades [33,48,49]. The GA has been widely used in the field of machine learning, pattern recognition, and, specifically, optimization [50–52].

Like typical EAs, GA takes advantage of the main operators identified as representation scheme, selection, crossover, and mutation, the definitions of which differ among the existing GA variations [53,54]. We discuss the fundamental characteristics of GA selection, crossover, and mutation operators in the sections that follow.

Selection

The standard form of GA utilizes the proportional selection operator to select the best candidate solutions at the end of each generation. The proportional selection directs the selection toward the best developed genetic information (e.g., the decision variables) [55].

For instance, the most-fit values of the developed decision variables within an optimization problem are determined by assessing their selection probability proportional to the predetermined fitness (objective) value:

$$p_s[DV_i(g)] = \frac{f_{sc}[DV_i(g)]}{\sum_{j=1}^{N_P} f_{sc}[DV_j(g)]} \tag{5}$$

where $p_s[DV_i(g)]$ is selection probability of decision variable i at g^{th} generation, $f_{sc}[DV_i(g)]$ is the decision variable's corresponding scaling function, and N_P equals the population size. The scaling function is determined using Eq. (6):

$$f_{sc}[DV_i(g)] = f_{\max} - f[DV_i(g)] \tag{6}$$

where $f_{\max}$ equals the maximum possible fitness function, while $f[DV_i(g)]$ is the calculated value of objective function of decision variable i at g^{th} generation. Clearly, determination of maximum fitness function is not always a straight-forward process. Hence, the maximum obtained fitness function up to generation g ($f_{\max}(g)$) can be provided as a practical alternative.

The value of scaling function can also be provided in a range of (0, 1] using Eq. (7) in the case of minimizing an objective value:

$$f_{sc}[DV_i(g)] = \frac{1}{1 + f[DV_i(g)] - f_{\min}(g)} \tag{7}$$

where $f_{\min}(g)$ is the minimum obtained fitness function up to generation g.

After calculation of the relevant selection probabilities, two famous "roulette wheel" and "stochastic universal sampling" techniques are utilized by the proportional selection operator to rank and select the population of the next generation [56,57]. The roulette wheel technique determines the probability distribution of decision variables calculating their corresponding normalized selection property, which is determined considering their relevant normalized fitness value (the normalized fitness value is calculated by dividing the fitness functions by minimum (or maximum in some cases) obtained fitness function). The obtained probability distributions are then compared to select the best value of each decision variable. The mathematical structure of roulette wheel technique results a wide variety of offspring, many of them may not be the most-fit candidates. The stochastic universal sampling technique was then introduced to address this issue by setting a limit on the number of produced offspring [58,59].

It should be mentioned that the direct selection of candidates proportional to the fitness value may limit the extensive exploration of search space by domination of the strong candidate solutions. Hence, the proportional selection can be considered as operator with high selective pressure.

Crossover

In addition to direct selection of candidate solutions for the next generation, the GA reproduces some solutions using the crossover operator [33]. The crossover operator attempts to generate more fit candidate solutions considering the desired fitness function. Defining a predetermined crossover probability, the operator selects one or more solutions as parents, produces one or more solutions as offspring, and replaces them with the worst existing candidate solutions.

Various crossover operators have been presented by far, which are categorized based on the number of parents included and the representation schemes. In the case of binary representation scheme, the crossover operators usually generate one or two offspring from two parents. Fig. 2 shows the three types of procedures used by the crossover operator to generate binary-specific offspring. Among the crossover operators presented in Fig. 2, the one-point crossover is the most common that selects a random cut-off point and switches the parents' bitstrings from that point to produce offspring.

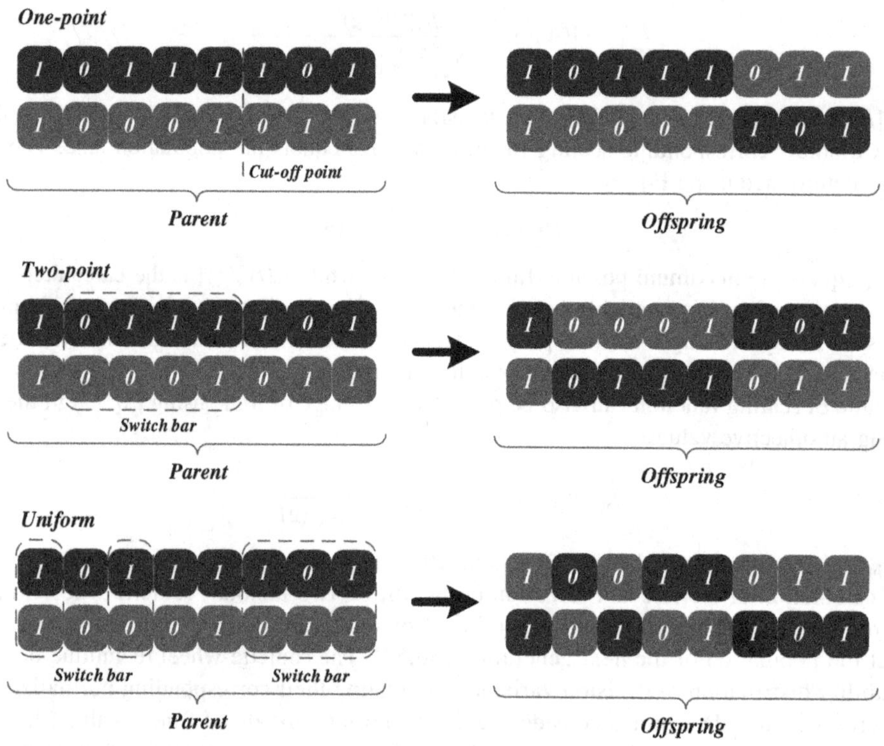

FIG. 2

Binary-specific crossover operators.

Mutation

In contrast to crossover, the mutation operator adds candidate solutions with new information to the populations to increase the diversity of provided solutions in each iteration [55]. High mutation rates may disfigure the more fit solutions, especially in final iterations. Hence, small values of mutation probability are more common giving convincing performances.

Considering binary representations, three mutation operators have been provided labeled as "uniform mutation," "inorder mutation," and "Gaussian mutation."

In general, the uniform mutation is applied within the GAs in which random bits are selected and their values are changed, as shown in Fig. 3.

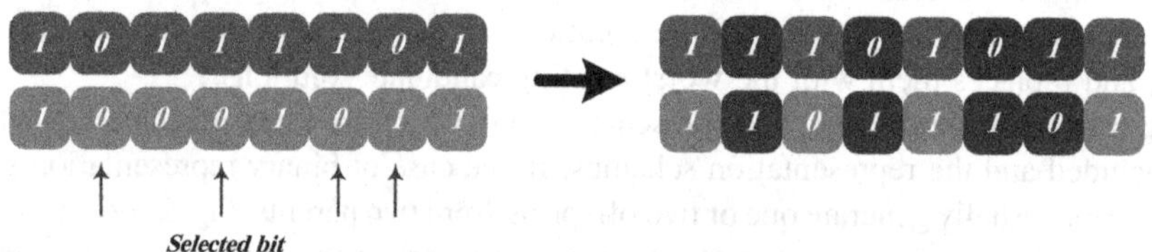

FIG. 3

Random mutation for binary representation schemes.

2.5.1 Multiobjective optimization

The GAs can be effectively utilized to solve multiobjective problems using three main approaches: of weighted aggregation, population-based non-Pareto techniques, and Pareto-based techniques [60]. The weighted aggregation and Pareto-based approaches have been widely used in the literature and we discuss these in the following sections.

Weighted aggregation

The weighted aggregation technique provides a weighted summation of multiple objectives and aggregates them in a single objective that is analyzed by the GA (as illustrated by Eq. (8)) [61,62].

$$OBJ = \sum_{s=1}^{N_{Sub}} w_s(Obj_s) \tag{8}$$

Subjected to:

$$\sum_{s=1}^{N_{Sub}} w_s = 1 \tag{9}$$

where OBJ is the aggregated objective function that equals the sum of subobjective functions (Obj_s) multiplied by the weights (w_s). Considering N_{Sub} objective functions, the summation assumed weights should not exceed one.

It is noteworthy that the assumed weights in this technique exert a direct effect on the accuracy of the provided solutions, while their determination is surrounded with considerable uncertainties. Moreover, this approach provides only one solution, and consecutive execution of optimization model is required to obtain different objective values [56].

Nondominated sorting genetic algorithm

The Pareto-based techniques have been widely used in the literature as trustable methods to solve multiobjective problems by providing Pareto-optimal solutions. The nondominated sorting genetic algorithm (NSGA) is one of the popular Pareto-based techniques [63,64]. Despite all its advantages, the NSGA technique is associated with extreme computational complexities. First, the nondominated sorting algorithm in NSGA sorts the candidate solutions by individually comparing their obtained objective values for each objective function in every generation, which becomes computationally impractical by increasing the identified population size. Moreover, the NSGA controls the diversity of provided solution by defining the "sharing" concept. The sharing parameter should be identified by the user and may exert significant impact on the optimization model convergence, while substantially increasing the computational efforts.

To address such problems, Deb et al. [65] proposed the NSGA-II as an improved version of NSGA. The NSGA-II takes advantage of fast nondominated sorting and crowded distance estimation procedures, and crowded comparison operator to reduce the computation complexities. Consequently, it can solve truly complex multiobjective optimization problems. The developed fast nondominated sorting procedure applies the nondomination concept to rank the candidate solutions and divide them into several nondominated sets. The first set includes the best nondominated solutions following by the second, third, and the worst obtained solutions. Using the elitism operator, the sorting procedure creates a group of former and current candidate solutions with a size twice the population and selects the best solutions considering the ordered sets until reaching the population size. If the length of a set exceeds the population size, the NSGA-II applies the crowded comparison operator to select the required number of solutions. The crowded comparison operator also controls the diversity of provided solutions by choosing the solutions that are located in less-crowded Pareto regions. Further improvements have also been made by researchers to reduce the computational runtime and required storage space, which makes NSGA-II an efficient multiobjective optimization algorithm [66,67].

3. Methodology development and applications

After detection of contamination in WDSs, the contaminant plume should be controlled and flushed out of the system to protect public health. Techniques such as proper valves closure to isolate the contaminated water, hydrant flushing to discharge the contaminated water, injection of chemical cleaning agents, or a reasonable combination of these have been suggested as effective methods to restore the normal condition of WDSs [68,69]. Studies have stated that the efficiency of these techniques is directly affected by the time elapsed after the contaminant intrusion into the systems [14,70]. Sudden decision-making about the pattern of hydrants opening and/or valve closures is quite challenging considering the complex nature of WDSs, which have hundreds of valves and numerous hydrant nodes [71,72].

In this regard, various simulation-optimization frameworks have been proposed by researchers to develop sudden response strategies for use after contamination of WDSs. Assessing different optimization models, EAs have shown a superior capability in problem analyzing and determining optimal solutions. For instance, Bashi-Azghadi et al. [22] compared the performance of dynamic programming and GA, and concluded that despite the superiority of the dynamic programming in producing optimal solutions, the GA acts much better by increasing the number of decision variables.

Different paradigms of CI (e.g., ant-colony, particle swarm optimization, evolutionary strategy-based search, and improved harmony search) have been utilized in the literature, among which the GA is the most widely used [9,12,19,73]. The GA has been utilized within the developed single- or multiobjective optimization models to obtain optimal contamination response actions. The regular type of GA is applied to solve single-objective problems; however, various modifications have been proposed by the literature to make the GA appropriate for solving multiobjective optimization problems [74,75]. Using various combinatorial approaches, the multiple objectives can be aggregated into one objective by applying different mathematical rules such as the weighted summation or distance measurement of objectives [20,76]. To address the doubts about accuracy of obtained mathematical rules, the NSGA was developed, which determines several optimal solutions labeled as "Pareto-optimal." Considering the conflicting nature of assumed objectives, each Pareto-optimal solution sacrifices one or more objectives to improve the other obtained objective value [67,77].

In the following, we discuss the assumptions and modifications required in GAs to obtain optimal contamination response actions.

3.1 Development of response actions

Optimal contamination response actions are developed by assessing the performance of hydrant opening and/or valve closure through a simulation-optimization process. Hence, the EA-based optimization model selects a population of individuals in each iteration and evaluates the value of fitness functions through a heuristic procedure to reach the optimal response actions. This process requires proper identification of important objectives and influential decision variables.

3.1.1 Response action objectives

Cleaning contaminated WDSs is a multidimensional issue considering various political, social and economic aspects. Some studies developed optimization models considering the single objective of minimizing the amount of consumed contaminated water, e.g., [12,14,78]. However, other objectives such as minimizing the required operations (i.e., closing the valves and opening the hydrants) or minimizing the unserviceability of the WDS, have to be analyzed simultaneously to obtain applicable contamination response actions [11,22]. Hence, the NSGA-II has been widely used to determine Pareto-optimal contamination response actions.

The defined objectives in a multiobjective problem are rationally conflicting and, consequently, the provided set of Pareto-optimal solutions (such as that depicted in Fig. 4) allows authorities to select the solution that is compatible with their goals and desires.

3.1.2 Defined decision variables

The developed response actions consist of a set of hydrants to be opened while closing several valves. Some studies have focused only on opening of hydrants, e.g., [12,17,24]. Thereby, a hydrant strategy is defined as the decision variable, which includes the hydrants that should be opened with their opening time and durations.

However, the procedure of controlling the contamination by valves closure and then discharge of the contaminated water by opening of hydrants is proved to be a more efficient method [9,21,23,79]. As an example, Rasekh and Brumbelow [14] proved that application of valves closure in conjunction

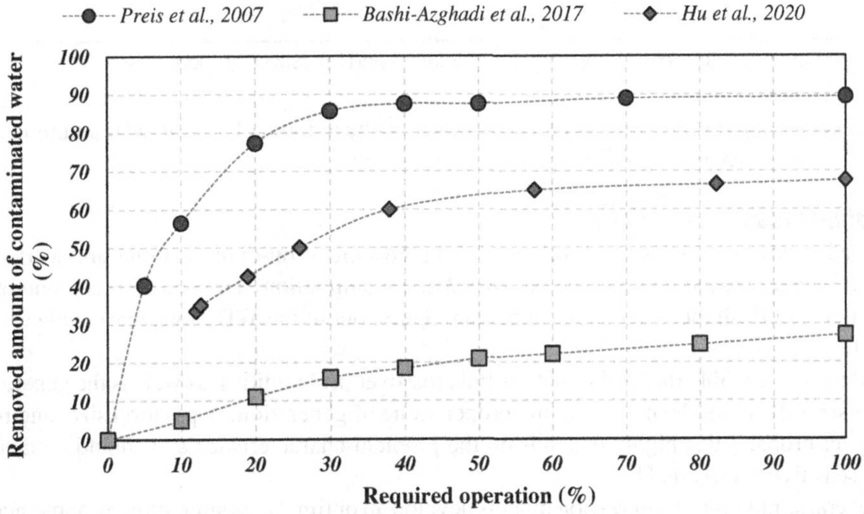

FIG. 4

Examples of developed Pareto-optimal contamination response actions using the NSGA-II optimization algorithm.

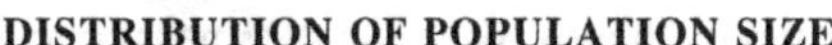

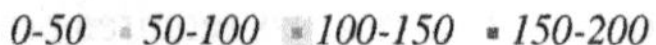

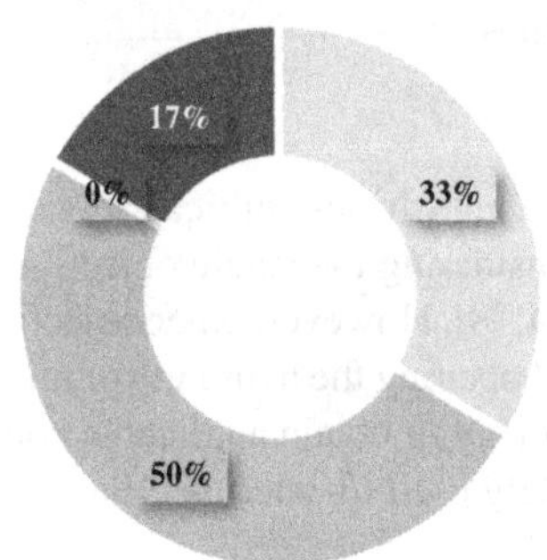

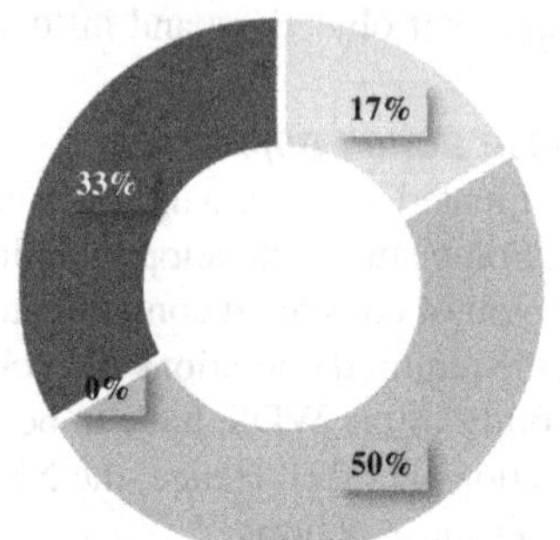

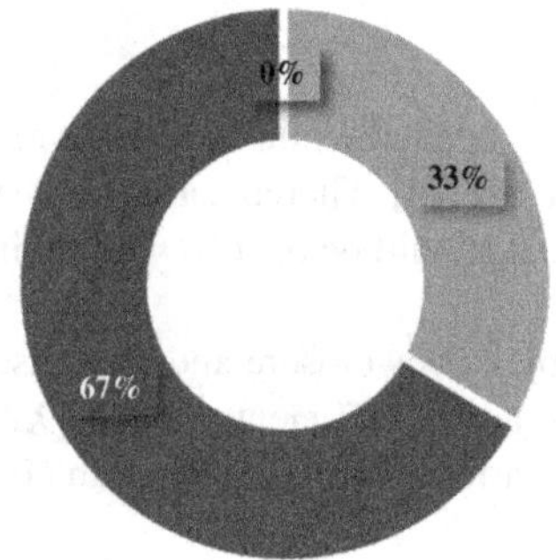

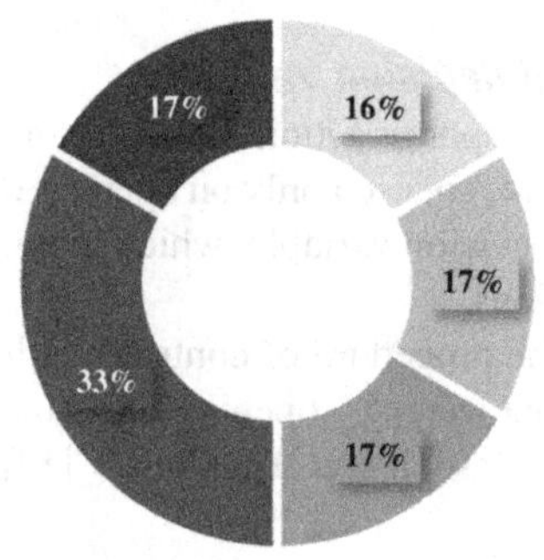

FIG. 5

The distribution of generation, population size, and mutation and crossover probabilities.

with hydrants opening resulted in about 80% reduction in the amount of consumed contaminated water compared to a 40% reduction by just opening the hydrants.

3.1.3 Optimal response actions

Previous studies have stated that the developed EA-based simulation-optimization frameworks are able to successfully analyze the complex nature of WDS contamination. The obtained contamination response actions can flush more than 80% of contamination out of the WDS by a reasonable carried operation [11,21,22].

Fig. 5 depicts the viable range of mutation and crossover probabilities, as well as the generation and population sizes. It should be noted that the proper value of generation, population size, and mutation and crossover probabilities highly depends on the problem characteristics and should be determined through a sensitivity analysis [14].

The general simulation-optimization models develop an optimal contamination response action that are fixed throughout the WDS cleaning period. However, studies have shown that during the hydrants' operation, the hydraulic condition of the WDS changes and the initial optimal solution may not be practical anymore [12,22]. To assess this claim, researchers divided the cleaning period into several parts,

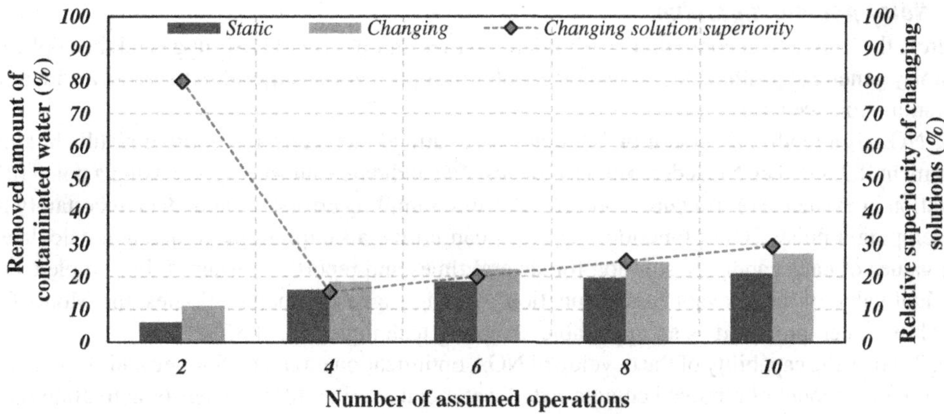

FIG. 6

Comparison of static and changing optimal response actions in removing the contaminated water, considering different number of assumed operations (number of hydrants opening and/or valves closure).

Adopted from Bashi-Azghadi, S.N., Afshar, A., Afshar, M.H., 2018. Multi-period response management to contaminated water distribution networks dynamic programming versus genetic algorithms. Eng. Optim. 50(3), 415–429.

and the optimization model was executed successively using the information of previous optimization stages. Fig. 6 compares the performance of static and changing optimal response actions in removing the contaminated water from a WDS.

In addition to performance improvements, the obtained results reveal a considerable difference in the optimal solutions of each optimization stage, which emphasizes the necessity of dynamic evaluation of optimal response actions.

Thousands of possible contamination scenarios can be identified in a WDS of medium- or large-scale. Two approaches of "contamination source identification" and "contamination scenarios categorization" have been proposed by the literature to determine accurate contamination response actions [10,80,81]. Identifying the contamination sources after their detection is associated with considerable uncertainties, and the provided WDS cleaning strategies may become ineffective, thus leading to worsening the situation [18,79]. On the other hand, by adopting the "contamination scenarios categorization" approach, various possible contamination scenarios should be modeled. The optimal contamination response actions are then determined through an iterative process considering the contaminant detection sensor activation and their activation order. By increasing the number of modeled decision variables and objectives, the considerable runtime of the GAs results in an impractical computation effort, which restricts the number of evaluated contamination scenarios through the optimization process [12,14,21]. Hence, some advancements are required to reduce the execution runtimes and improve the applicability of the obtained contamination response actions.

3.2 Optimization algorithm modifications

To address the shortcoming of common GA, researchers have applied various techniques to enhance its capabilities and achieve more reliable results. We discuss some of the modifications that were utilized to determine optimal contamination response actions in the following.

3.2.1 Noisy genetic algorithm

To address the impractical computational burden, Shafiee and Berglund [18] suggested the application of the noisy genetic algorithm (NGA), which enables the optimization algorithm to assess a wider range of contamination scenarios.

The NGA is a modified version of GA that can be utilized to determine optimal results in a noisy environment [82,83]. The NGA determines the most-fit candidate solutions by assessing a noisy fitness information in existence of inaccurate objective values, sampling errors, or input data uncertainty [84]. By developing a noisy fitness function, the NGA can utilize a sampling technique and calculate the fitness values of each candidate solution for several times and report the mean of the sampled values as the final value of the relevant fitness function. Among suggested methodologies, the Monte Carlo approach has been proposed as an applicable sampling technique [18,84,85].

Fig. 7 shows the capability of the developed NGA optimization model by Shafiee and Berglund [18] to increase the amount of removed contaminated water considering different sensor activation scenarios. Moreover, a comparison with the classical GA was made (as shown in Fig. 7), which showed the average superiority of the NGA up to 25%.

To develop contamination response actions of Fig. 7, the optimization model was executed with three trials. In every execution, the GA determines the optimal contamination response action by evaluating their performance for one selected contamination scenario, while the NGA assesses a set of three possible scenarios.

3.2.2 Modified NSGA-II

The original NSGA-II starts with a randomly generated initial population. In the case of extensive search spaces, random generation may reduce the algorithm speed to achieve the desired solution. Hence, several modifications have been proposed to generate proper initial population and reduce

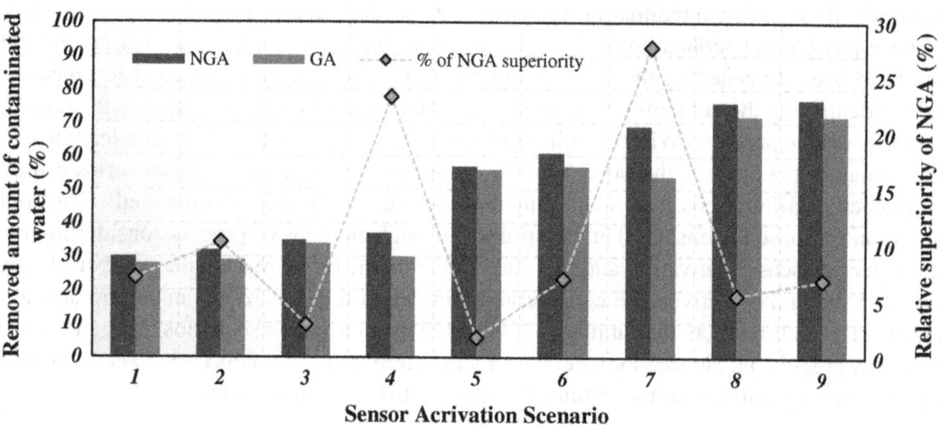

FIG. 7

The percentage of the removed amount of contaminated water for different sensor activation scenarios, and the relative superiority of NGA over GA.

Adopted from Shafiee, M.E., Berglund, E.Z., 2015. Real-time guidance for hydrant flushing using sensor-hydrant decision trees. J. Water Resour. Plan. Manag. 141(6), 04014079.

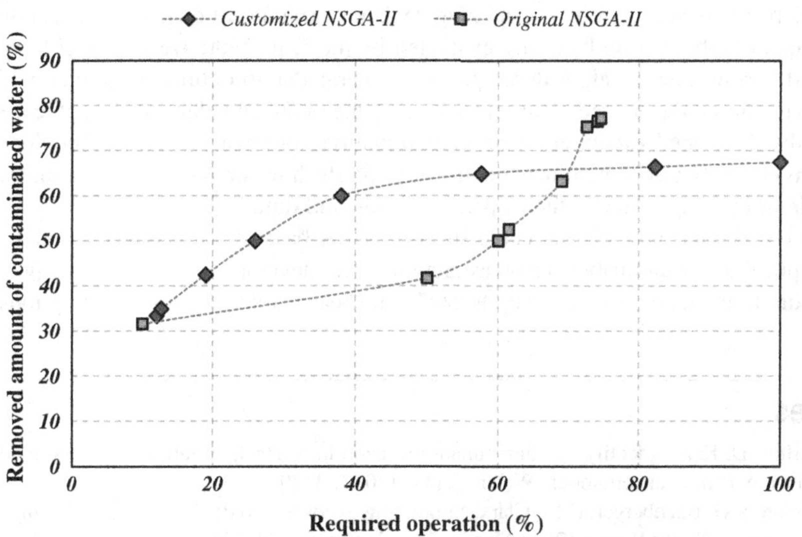

FIG. 8

Comparison of Pareto-front contamination response actions obtained by customized and original NSGA-II algorithms.

the required convergence duration [20]. For instance, Seok Jeong and Abraham [25] applied greedy heuristics [86] within the optimization model to produce feasible initial populations. The modified NSGA-II may not provide a globally optimal solution; however, acceptable response actions can be obtained by the optimization model with a significant reduction in computational time.

As another solution to reduce optimization time, various modifications have been made to enhance the performance of mutation and crossover operators (e.g., [78]). For instance, Hu et al. [11] proposed a customized NSGA-II algorithm by improving the mutation and crossover operators. Fig. 8 shows substantial improvements achieved by the customized optimization model.

4. Conclusion

Contamination of WDSs is an important issue that has attracted the attention of many researchers. EC is recommended as an efficient methodology to provide WDS cleaning strategies (contamination response actions).

EAs, as population-based heuristic search engines, are strong alternatives to customary mathematical problem-solvers in solving complex problems. They can be utilized within the optimization models to provide optimal solutions even in the presence of multiple uncertainties.

The EA-based optimization models can analyze a significant number of decision variables, which increases the chance of achieving real optimal solutions. By appropriate adjustment and modification of EA operators, EA-based optimization models can converge to the proper optimal solution with much less computation burdens compared to other applicable optimization engines.

The developed EAs are capable of analyzing decision variables of different types simultaneously. Multidimensional problems can be easily evaluated by the EAs. Moreover, some EA types, such as nondominated sorting genetic algorithms, provide several Pareto-optimal solutions, which enables the managers to choose the desired solution comparing the obtained values of each objective function. As a result, the developed EA-based WDS contamination optimization models provide reliable response actions consisting of contaminant controlling and flushing strategies by determining valve closure and hydrant opening patterns, their operation time, and durations.

The achieved results show the markable success of EA-based simulation-optimization models in providing applicable contamination response actions. The developed cleaning strategies can reduce public exposure to contaminated water up to 80% with reasonable modeling assumptions.

References

[1] M.S. Marlim, D. Kang, Identifying contaminant intrusion in water distribution networks under water flow and sensor report time uncertainties, Water 12 (11) (2020) 3179.

[2] K.A. Nilsson, S.G. Buchberger, R.M. Clark, Simulating exposures to deliberate intrusions into water distribution systems, J. Water Resour. Plan. Manag. 131 (3) (2005) 228–236.

[3] N. Pelekanos, D. Nikolopoulos, C. Makropoulos, Simulation and vulnerability assessment of water distribution networks under deliberate contamination attacks, Urban Water J. 18 (4) (2021) 209–222.

[4] D.T. Ramotsoela, G.P. Hancke, A.M. Abu-Mahfouz, Attack detection in water distribution systems using machine learning, HCIS 9 (1) (2019) 1–22.

[5] M.R. Bazargan-Lari, An evidential reasoning approach to optimal monitoring of drinking water distribution systems for detecting deliberate contamination events, J. Clean. Prod. 78 (2014) 1–14.

[6] C. Giudicianni, M. Herrera, A. Di Nardo, R. Greco, E. Creaco, A. Scala, Topological placement of quality sensors in water-distribution networks without the recourse to hydraulic modeling, J. Water Resour. Plan. Manag. 146 (6) (2020) 04020030.

[7] G. He, T. Zhang, F. Zheng, Q. Zhang, An efficient multi-objective optimization method for water quality sensor placement within water distribution systems considering contamination probability variations, Water Res. 143 (2018) 165–175.

[8] A. Preis, A. Ostfeld, Multiobjective contaminant response modeling for water distribution systems security, J. Hydroinf. 10 (4) (2008) 267–274.

[9] M. Zafari, M. Tabesh, S. Nazif, Minimizing the adverse effects of contaminant propagation in water distribution networks considering the pressure-driven analysis method, J. Water Resour. Plan. Manag. 143 (12) (2017) 04017072.

[10] O.S. Adedoja, Y. Hamam, B. Khalaf, R. Sadiku, Towards development of an optimization model to identify contamination source in a water distribution network, Water 10 (5) (2018) 579.

[11] C. Hu, X. Yan, W. Gong, X. Liu, L. Wang, L. Gao, Multi-objective based scheduling algorithm for sudden drinking water contamination incident, Swarm Evol. Comput. 55 (2020), 100674.

[12] M.A. Khaksar Fasaee, M.R. Nikoo, P. Hashempour Bakhtiari, S. Monghasemi, M. Sadegh, A novel dynamic hydrant flushing framework facilitated by categorizing contamination events, Urban Water J. 17 (3) (2020) 199–211.

[13] H.A. Mahmoud, Z. Kapelan, D. Savić, Real-time operational response methodology for reducing failure impacts in water distribution systems, J. Water Resour. Plan. Manag. 144 (7) (2018) 04018029.

[14] A. Rasekh, K. Brumbelow, Drinking water distribution systems contamination management to reduce public health impacts and system service interruptions, Environ Model Softw. 51 (2014) 12–25.

[15] D.E. Sheefa, B.D. Barkdoll, Feasibility of an environmentally friendly method of contaminant flushing in water distribution systems using containment ponds, Water Supply 22 (4) (2022) 3744–3755.

[16] M.E. Shafiee, E.Z. Berglund, M.K. Lindell, An agent-based modeling framework for assessing the public health protection of water advisories, Water Resour. Manag. 32 (6) (2018) 2033–2059.

[17] E.M. Zechman, Integrating complex adaptive system simulation and evolutionary computation to support water infrastructure threat management, in: Proceedings of the 12th Annual Conference Companion on Genetic and Evolutionary Computation, 2010, pp. 1809–1816.

[18] M.E. Shafiee, E.Z. Berglund, Real-time guidance for hydrant flushing using sensor-hydrant decision trees, J. Water Resour. Plan. Manag. 141 (6) (2015) 04014079.

[19] A. Afshar, E. Najafi, Consequence management of chemical intrusion in water distribution networks under inexact scenarios, J. Hydroinf. 16 (1) (2014) 178–188.

[20] L. Alfonso, A. Jonoski, D. Solomatine, Multiobjective optimization of operational responses for contaminant flushing in water distribution networks, J. Water Resour. Plan. Manag. 136 (1) (2010) 48–58.

[21] A. Preis, Y. Mayorchik, A. Ostfeld, Multiobjective contaminant detection response model, in: World Environmental and Water Resources Congress 2007: Restoring Our Natural Habitat, 2007, pp. 1–12.

[22] S.N. Bashi-Azghadi, A. Afshar, M.H. Afshar, Multi-period response management to contaminated water distribution networks dynamic programming versus genetic algorithms, Eng. Optim. 50 (3) (2018) 415–429.

[23] A. Poulin, A. Mailhot, P. Grondin, L. Delorme, N. Periche, J.P. Villeneuve, Heuristic approach for operational response to drinking water contamination, J. Water Resour. Plan. Manag. 134 (5) (2008) 457–465.

[24] A. Poulin, A. Mailhot, N. Periche, L. Delorme, J.P. Villeneuve, Planning unidirectional flushing operations as a response to drinking water distribution system contamination, J. Water Resour. Plan. Manag. 136 (6) (2010) 647–657.

[25] H. Seok Jeong, D.M. Abraham, Operational response model for physically attacked water networks using NSGA-II, J. Comput. Civ. Eng. 20 (5) (2006) 328–338.

[26] N. Sankary, A. Ostfeld, Incorporating operational uncertainty in early warning system design optimization for water distribution system security, Procedia Eng. 186 (2017) 160–167.

[27] Y. Wang, G. Zhu, Analysis of water distribution system under uncertainty based on genetic algorithm and trapezoid fuzzy membership, J. Pipeline Syst. Eng. Pract. 12 (4) (2021) 04021043.

[28] W. Bi, G.C. Dandy, H.R. Maier, Improved genetic algorithm optimization of water distribution system design by incorporating domain knowledge, Environ Model Softw. 69 (2015) 370–381.

[29] N. Elshaboury, T. Attia, M. Marzouk, Application of evolutionary optimization algorithms for rehabilitation of water distribution networks, J. Constr. Eng. Manag. 146 (7) (2020) 04020069.

[30] H. Monsef, M. Naghashzadegan, A. Jamali, R. Farmani, Comparison of evolutionary multi objective optimization algorithms in optimum design of water distribution network, Ain Shams Eng. J. 10 (1) (2019) 103–111.

[31] K. De Jong, Evolutionary computation: a unified approach, in: Proceedings of the 2016 on Genetic and Evolutionary Computation Conference Companion, 2016, pp. 185–199.

[32] A.E. Eiben, J.E. Smith, Introduction to Evolutionary Computing, vol. 53, Springer, Berlin, 2003, p. 18.

[33] A.P. Engelbrecht, Computational Intelligence: An Introduction, John Wiley and Sons, 2007.

[34] Z. Liang, T. Wu, X. Ma, Z. Zhu, S. Yang, A Dynamic Multiobjective Evolutionary Algorithm Based on Decision Variable Classification, IEEE Transactions on Cybernetics, 2020.

[35] K.A. DeJong, Evolutionary Computation: A Unified Approach, The MIT Press, Cambridge, Massachusetts, Landon, England, 2006.

[36] P.A. Vikhar, Evolutionary algorithms: a critical review and its future prospects, in: 2016 International conference on global trends in signal processing, information computing and communication (ICGTSPICC), IEEE, 2016, pp. 261–265.

[37] T. Hanne, R. Dornberger, Computational intelligence, in: Computational Intelligence in Logistics and Supply Chain Management, Springer, Cham, 2017, pp. 13–41.

[38] A. Konar, Computational Intelligence: Principles, Techniques and Applications, Springer Science & Business Media, 2006.

[39] N. Siddique, H. Adeli, Computational Intelligence: Synergies of Fuzzy Logic, Neural Networks and Evolutionary Computing, John Wiley and Sons, 2013.
[40] B. Xue, M. Zhang, W.N. Browne, X. Yao, A survey on evolutionary computation approaches to feature selection, IEEE Trans. Evol. Comput. 20 (4) (2015) 606–626.
[41] A.I. Diveev, S.V. Konstantinov, Study of the practical convergence of evolutionary algorithms for the optimal program control of a wheeled robot, J. Comput. Syst. Sci. Int. 57 (4) (2018) 561–580.
[42] W.E. Hart, N. Krasnogor, J.E. Smith, Memetic evolutionary algorithms, in: Recent Advances in Memetic Algorithms, Springer, Berlin, Heidelberg, 2005, pp. 3–27.
[43] L. Waltman, N.J. van Eck, R. Dekker, U. Kaymak, Economic modeling using evolutionary algorithms: the effect of a binary encoding of strategies, J. Evol. Econ. 21 (5) (2011) 737–756.
[44] R.B. Abdessalem, S. Nejati, L.C. Briand, T. Stifter, Testing vision-based control systems using learnable evolutionary algorithms, in: 2018 IEEE/ACM 40th International Conference on Software Engineering (ICSE), IEEE, 2018, pp. 1016–1026.
[45] S. Rahnamayan, H.R. Tizhoosh, M.M. Salama, A novel population initialization method for accelerating evolutionary algorithms, Comput. Math. Appl. 53 (10) (2007) 1605–1614.
[46] J. Brest, M. Sepesy Maučec, Population size reduction for the differential evolution algorithm, Appl. Intell. 29 (3) (2008) 228–247.
[47] G. Karafotias, M. Hoogendoorn, Á.E. Eiben, Parameter control in evolutionary algorithms: trends and challenges, IEEE Trans. Evol. Comput. 19 (2) (2014) 167–187.
[48] D.E. Goldberg, J.H. Holland, Genetic Algorithms and Machine Learning, Springer, 1988.
[49] J.H. Holland, Genetic algorithms, Sci. Am. 267 (1) (1992) 66–73.
[50] H.A. El-Ghandour, E. Elbeltagi, Comparison of five evolutionary algorithms for optimization of water distribution networks, J. Comput. Civ. Eng. 32 (1) (2018) 04017066.
[51] A. Haghighi, H. Samani, Z. Samani, GA-ILP method for optimization of water distribution networks, Water Resour. Manag. 25 (7) (2011) 1791–1808.
[52] P. Khatavkar, L.W. Mays, Optimization-simulation model for real-time pump and valve operation of water distribution systems under critical conditions, Urban Water J. 16 (1) (2019) 45–55.
[53] J. Carr, An introduction to genetic algorithms, Senior Project 1 (40) (2014) 7.
[54] A.K. Sarma, Introduction to genetic algorithm with a simple analogy, in: Nature-Inspired Methods for Metaheuristics Optimization, Springer, Cham, 2020, pp. 27–34.
[55] S. Mirjalili, Genetic algorithm, in: Evolutionary Algorithms and Neural Networks, Springer, Cham, 2019, pp. 43–55.
[56] O. Kramer, Genetic algorithms, in: Genetic Algorithm Essentials, Springer, Cham, 2017, pp. 11–19.
[57] A. Shukla, H.M. Pandey, D. Mehrotra, Comparative review of selection techniques in genetic algorithm, in: 2015 International Conference on Futuristic Trends on Computational Analysis and Knowledge Management (ABLAZE), IEEE, 2015, pp. 515–519.
[58] R.G. Baldovino, M.G.A.C. Bautista, A.U. Aquino, E.J. Calilung, E. Sybingco, E.P. Dadios, GA optimization of coconut sugar cooking process: a preliminary study using stochastic universal sampling (SUS) technique, in: Proceedings of the 9th International Conference on Computer and Automation Engineering, 2017, pp. 346–349.
[59] N.M. Kwok, G. Fang, W. Zhou, in: Evolutionary Particle Filter: Re-Sampling from the Genetic Algorithm Perspective, 2005 IEEE/RSJ International Conference on Intelligent Robots and Systems, IEEE, 2005, pp. 2935–2940.
[60] Y. Gao, L. Shi, P. Yao, Study on multi-objective genetic algorithm, in: Proceedings of the 3rd World Congress on Intelligent Control and Automation (Cat. No. 00EX393), vol. 1, IEEE, 2000, June, pp. 646–650.
[61] Y. Jin, M. Olhofer, B. Sendhoff, Dynamic weighted aggregation for evolutionary multi-objective optimization: why does it work and how? in: Proceedings of the genetic and evolutionary computation conference, 2001, pp. 1042–1049.

[62] S. Mirjalili, J.S. Dong, Multi-Objective Optimization Using Artificial Intelligence Techniques, Springer, 2020.

[63] K. Deb, S. Agrawal, A. Pratap, T. Meyarivan, A fast elitist non-dominated sorting genetic algorithm for multi-objective optimization: NSGA-II, in: International Conference on Parallel Problem Solving From Nature, Springer, Berlin, Heidelberg, 2000, pp. 849–858.

[64] H.C. Fu, P. Liu, A multi-objective optimization model based on non-dominated sorting genetic algorithm, Int. J. Simul. Model. 18 (3) (2019) 510–520.

[65] K. Deb, A. Pratap, S. Agarwal, T.A.M.T. Meyarivan, A fast and elitist multiobjective genetic algorithm: NSGA-II, IEEE Trans. Evol. Comput. 6 (2) (2002) 182–197.

[66] S. Chatterjee, S. Sarkar, N. Dey, A.S. Ashour, S. Sen, Hybrid non-dominated sorting genetic algorithm: II-neural network approach, in: Advancements in Applied Metaheuristic Computing, IGI Global, 2018, pp. 264–286.

[67] Q. Wang, L. Wang, W. Huang, Z. Wang, S. Liu, D.A. Savić, Parameterization of NSGA-II for the optimal design of water distribution systems, Water 11 (5) (2019) 971.

[68] USEPA, Example Exposure Scenarios, National Center for Environmental Assessment U.S. Environmental Protection Agency Washington, DC, 2004.

[69] USEPA, Risk Assessment Guidance for Superfund, Human Health Evaluation Manual, Pt E, Supplemental Guidance for Dermal Risk Assessment, vol. 1, 2004. US EPA, EPA 540-R-99-005, Washington, DC. Available through: Springfield, VA.

[70] S.G. Vrachimis, R. Lifshitz, D.G. Eliades, M.M. Polycarpou, A. Ostfeld, Active contamination detection in water-distribution systems, J. Water Resour. Plan. Manag. 146 (4) (2020) 04020014.

[71] B. Farley, S.R. Mounce, J.B. Boxall, Field testing of an optimal sensor placement methodology for event detection in an urban water distribution network, Urban Water J. 7 (6) (2010) 345–356.

[72] A. Moghaddam, M. Afsharnia, R. Peirovi Minaee, Preparing the optimal emergency response protocols by MOPSO for a real-world water distribution network, Environ. Sci. Pollut. Res. 27 (24) (2020) 30625–30637.

[73] E. Shafiee, E. Zechman, Integrating evolutionary computation and sociotechnical simulation for flushing contaminated water distribution systems, in: Proceedings of the 14th Annual Conference Companion on Genetic and Evolutionary Computation, 2012, pp. 315–322.

[74] E. Bradford, A.M. Schweidtmann, A. Lapkin, Efficient multiobjective optimization employing Gaussian processes, spectral sampling and a genetic algorithm, J. Glob. Optim. 71 (2) (2018) 407–438.

[75] C. Guerrero, I. Lera, C. Juiz, Genetic algorithm for multi-objective optimization of container allocation in cloud architecture, J. Grid Comput 16 (1) (2018) 113–135.

[76] N. Gunantara, A review of multi-objective optimization: methods and its applications, Cogent Eng. 5 (1) (2018) 1502242.

[77] S.S. Naserizade, M.R. Nikoo, H. Montaseri, A risk-based multi-objective model for optimal placement of sensors in water distribution system, J. Hydrol. 557 (2018) 147–159.

[78] M. Gavanelli, M. Nonato, A. Peano, S. Alvisi, M. Franchini, Genetic algorithms for scheduling devices operation in a water distribution system in response to contamination events, in: European Conference on Evolutionary Computation in Combinatorial Optimization, Springer, Berlin, Heidelberg, 2012, pp. 124–135.

[79] M.A.K. Fasaee, S. Monghasemi, M.R. Nikoo, M.E. Shafiee, E.Z. Berglund, P.H. Bakhtiari, A K-sensor correlation-based evolutionary optimization algorithm to cluster contamination events and place sensors in water distribution systems, J. Clean. Prod. 319 (2021), 128763.

[80] J. Deuerlein, L. Meyer-Harries, N. Guth, Efficient online source identification algorithm for integration within a contamination event management system, Drink. Water Eng. Sci. 10 (2) (2017) 53–59.

[81] L. Liu, S.R. Ranjithan, G. Mahinthakumar, Contamination source identification in water distribution systems using an adaptive dynamic optimization procedure, J. Water Resour. Plan. Manag. 137 (2) (2011) 183–192.

[82] L. Jun-hua, L. Ming, An analysis on convergence and convergence rate estimate of elitist genetic algorithms in noisy environments, Optik 124 (24) (2013) 6780–6785.

[83] S.N. Sivanandam, S.N. Deepa, Genetic algorithms, in: Introduction to Genetic Algorithms, Springer, Berlin, Heidelberg, 2008, pp. 15–37.
[84] H. Kita, Y. Sano, Genetic algorithms for optimization of noisy fitness functions and adaptation to changing environments, in: 2003 Joint Workshop of Hayashibara Foundation and 2003 Workshop on Statistical Mechanical Approach to Probabilistic Information Processing (SMAPIP), 2003.
[85] J.B. Smalley, B.S. Minsker, D.E. Goldberg, Risk-based in situ bioremediation design using a noisy genetic algorithm, Water Resour. Res. 36 (10) (2000) 3043–3052.
[86] A. Fügenschuh, Parametrized greedy heuristics in theory and practice, in: International Workshop on Hybrid Metaheuristics, Springer, Berlin, Heidelberg, 2005, pp. 21–31.

CHAPTER

5 Metaheuristic technique for solving fuzzy nonlinear equations

Hisham M. Khudhur

Department of Mathematics, College of Computers Sciences and Mathematics, University of Mosul, Mosul, Iraq

1. Introduction

Iterative approaches for solving nonlinear equations such as $F(x)=0$ have received a lot of interest in recent years.

The concept of fuzzy numbers, as well as the mathematical operations that can be performed on them, were first proposed and researched by Zadeh [1]. One of the most common applications of fuzzy number arithmetic is nonlinear equations whose parameters are completely or partially represented by fuzzy numbers [2–4]. Buckley and Qu's standard analytical procedures [5–8] are only suitable for linear equations and quadratic cases of nonlinear equations and can't be used to solve equations like

I. $Ay^3+By^2-Hy=\Theta$
II. $\Theta e^y-Ey=\Lambda$
III. $\Phi y\csc(y)+\Upsilon y=\Omega$
IV. $\Psi y^5-\Omega\cot(y)=\Phi$

where y, A, B, H, Θ, E, Λ, Υ, Ω, Ψ, and Φ are fuzzy numbers. In 2004, Abbasbandy and Asady used Newton's approach for solving a fuzzy nonlinear equation [9]. In 2010, Ramli, Abdullah, and Mamat employed a quasi-Newton technique for solving a fuzzy nonlinear equation [10]. This method is not particularly useful in practice because it necessitates the storage of the $n\times n$ matrix $[H_i]$, the computation of the elements of the matrix $[H_i]$ is extremely difficult and often impossible, It requires the inversion of the matrix $[H_i]$ at each step, and It requires the evaluation of the quantity $[H_i]^{-1}\nabla f_i$ at each step. Because of these flaws, the strategy is ineffective for problems with a complex objective function with many variables. Abbasbandy and Jafarian invented the steepest descent approach for solving fuzzy nonlinear equations [11] in 2004. However, because the sharpest descent direction is a local feature, this strategy is weak and inefficient in most applications. In 2021, Hisham and Khalil developed two conjugate gradient (CG) methods for solving fuzzy nonlinear equations [12] [13]. A disadvantage of conjugate gradient algorithms is that they are not always effective. Therefore, artificial intelligence numerical algorithms are needed to find the roots of these equations, in general as $F(x)=0$. In chapter, we use a gradient-based optimizer (GBO) metaheuristic algorithm. This algorithm is efficient and quick in finding the roots of equations as well as has global convergence. This algorithm was used to solve optimization functions by researchers in Ref. [14]. Here we use it to solve nonlinear fuzzy equations.

Comprehensive Metaheuristics. https://doi.org/10.1016/B978-0-323-91781-0.00023-5

This chapter is divided into five sections. Section 1 is the introduction, Section 2 presents the basics of arithmetic operations for fuzzy numbers, Section 3 is an explanation of the GBO technique, Section 4 presents examples and numerical results, and Section 5 concludes.

2. Preliminaries

Here we present some basic definitions and arithmetic operations for fuzzy numbers. For more information, we refer interested readers to Ref. [15].

Definition 1 A fuzzy number is defined as a set $j:\mathbb{R}\to I=[0,1]$ that meets the following conditions [16]:

- **a.** j denotes a semi-continuous upper boundary
- **b.** $j(x)=0$ outside some range $[r,t]$
- **c.** there exist $e, f\in\mathbb{R}$ such that $r\leq p\leq q\leq t$ and
  - **i.** $j(x)$ is increasing monotonically on $[r,p]$
  - **ii.** $j(x)$ is decreasing monotonically on $[q,t]$
  - **iii.** $j(x)=1, p\leq x\leq q$

Definition 2 j: $\mathbb{R}\to I=[0,1]$ in parametric form refer to the pair $\left(\underline{j},\bar{j}\right)$ of $\underline{j}(\mu), \bar{j}(\mu), 0\leq\mu\leq 1$. satisfying [16,17],

(1) $\underline{j}(\mu)$ is a monotonically bounded growing left continuous function
(2) $\bar{j}(\mu)$ is a monotonically bounded growing right continuous function
(3) $\underline{j}(\mu)\leq\bar{j}(\mu), 0\leq\mu\leq 1$

Definition 3 A classical fuzzy number h refers to the triangular number $j=(p,q,r)$, given as follows

$$j(x)=\begin{cases}\dfrac{(x-p)}{(r-p)}, & p\leq x\leq r\\ \dfrac{(x-q)}{(r-q)}, & r\leq x\leq q\end{cases}$$

with (x) representing the membership function and $r\neq p, r\neq q$ [9,18]. This function can be written in its parameterized form as follows

$$\bar{j}(\mu)=q+(r-q)\mu$$
$$\underline{j}(\mu)=p+(r-p)\mu$$

Assume $TF(\mathbb{R})$ denotes the set of all trapezoidal fuzzy number. The following extension principle can be used to extend the scalar multiplication and addition operations to fuzzy numbers [9].

Let $=\left(\underline{j}(\mu),\bar{j}(\mu)\right)$, $k=\left(\underline{k}(\mu),\bar{k}(\mu)\right)$ with $w>0$, the addition $(j+k)$ and multiplication by scalar w are defined as

$$\left(\overline{j+k}\right)(\mu)=\bar{j}(\mu)+\bar{k}(\mu)$$

$$\left(\underline{j+k}\right)(\mu) = \underline{j}(\mu) + \underline{k}(\mu)$$
$$\left(\overline{wj}\right)(\mu) = w\overline{j}(\mu)$$
$$\left(\underline{wj}\right)(\mu) = w\underline{j}(\mu)$$

3. A brief description of the GBO algorithm

A GBO solves complex fuzzy nonlinear equations by mixing the gradient technique and population-based approaches. The GBO technique employs Newton's method to regulate the search direction while exploring the search domain with a collection of vectors and two major operators: the gradient search rule (GSR) and the locale escaping operator (LEO) (for more see [14]).

3.1 Initialization phase

The GBO, like most metaheuristic techniques, begins the optimization operation by generating an initial population from a uniform random distribution. In a D-dimensional search space, each population agent is called a "vector," and the population has a total of N vector agents. The following steps are taken to complete the initialization process:

$$X_n = X_{\min} + \text{rand}(0, 1) \times (X_{\max} - X_{\min}) \tag{1}$$

where the boundaries of decision variables $X = \left(\overline{\chi}, \underline{\chi}\right)$ are $X_{\min} = \left(\overline{\chi}_{\min}, \underline{\chi}_{\min}\right)$ and $X_{\max} = \left(\overline{\chi}_{\max}, \underline{\chi}_{\max}\right)$, and rand (0; 1) is a random number defined in the ambit [0; 1] for more see [14].

3.2 Gradient-based optimizer phases

3.2.1 Gradient search rule (GSR) phase

The suggested GSR allows exploration and escapes from local optima by assisting the GBO in accounting for random conduct through the optimization phase. The GBO method's convergence speed is boosted by establishing an appropriate local search trend based on movement direction (DM). The following equation is utilized to modernization the present vector's location based on the GSR and DM $\left(x_n^m = \left(\overline{\chi}_n^m, \underline{\chi}_n^m\right)\right)$

$$X1_n^m = x_n^m - randn \times \rho_1 \times \frac{2\Delta x \times x_n^m}{(x_{worst} - x_{best} + \varepsilon)} + \text{rand} \times \rho_2 \times \left(x_{best} - x_n^m\right) \tag{2}$$

where $X1_n^m = \left(\overline{\text{X1}}_n^m, \underline{\text{X1}}_n^m\right)$, $x_{\text{worst}} = \left(\overline{x_{\text{worst}}}_n^m, \underline{x_{\text{worst}}}_n^m\right)$, and $x_{best} = \left(\overline{x_{\text{best}}}, \underline{x_{\text{best}}}\right)$

In which

$$\rho_1 = 2 \times \text{rand} \times \alpha - \alpha \tag{3}$$

$$\alpha = \left|\beta \times \sin\left(\frac{3\pi}{2} + \sin\left(\beta \times \frac{3\pi}{2}\right)\right)\right| \tag{3.1}$$

$$\beta = \beta_{\min} + (\beta_{\max} - \beta_{\min}) \times \left(1 - \left(\frac{m}{M}\right)^3\right)^2 \tag{3.2}$$

The number of iterations is m, and the total number of iterations is M, with 0.2 and 1.2 as the minimum and maximum values, respectively. ε is a small number in the range $[0, 0.1]$, and rand is a randomly generated number with a normal distribution. The value of ρ_2 is determined by

$$\rho_2 = 2 \times \text{rand} \times \alpha - \alpha$$

$$\Delta x = \text{rand}(1:N) \times |step| \tag{4}$$

$$step = \frac{(x_{best} - x_{r1}^m) + \delta}{2} \tag{4.1}$$

$$\delta = 2 \times \text{rand} \times \left(\left|\frac{x_{r1}^m + x_{r2}^m + x_{r3}^m + x_{r4}^m}{4} - x_n^m\right|\right) \tag{4.2}$$

where $rand(1:N)$ N-dimensional random number, $\tau1$, $\tau2$, $\tau3$, and $\tau4$ $(\tau1 \neq \tau2 \neq \tau3 \neq \tau4 \neq n)$ are different integers selected at random selection from $[1,N]$, $step$ is the size of a step dictated by x_{best} and x_{r1}^m.

By swapping out the optimal vector's position (x_{best}) with using the current vector (x_n^m) in Eq. (2), the new vector ($X2_n^m$) can be generated as follows:

$$X2_n^m = x_{best} - randn \times \rho_1 \times \frac{2\Delta x \times x_n^m}{(yp_n^m - yq_n^m + \varepsilon)} + \text{rand} \times \rho_2 \times (x_{r1}^m - x_{r2}^m) \tag{5}$$

where $X2_n^m = \left(\overline{X2}_n^m, \underline{X2}_n^m\right)$, $x_{r1}^m = \left(\overline{x}_{r1}^m, \underline{x}_{r1}^m\right)$, and $x_{r2}^m = \left(\overline{x}_{r2}^m, \underline{x}_{r2}^m\right)$

in which

$$yp_n = \text{rand} \times \left(\frac{[z_{n+1} + x_n]}{2} + \text{rand} \times \Delta x\right) \tag{5.1}$$

$$yq_n = \text{rand} \times \left(\frac{[z_{n+1} + x_n]}{2} - \text{rand} \times \Delta x\right) \tag{5.2}$$

Based on the positions $X1_n^m$, $X2_n^m$, and the present position (X_n^m), the following iteration's new solution (x_n^{m+1}) can be characterized as:

$$x_n^{m+1} = \tau_a \times (\tau_b \times X1_n^m + (1 - \tau_b) \times X2_n^m) + (1 - \tau_a) \times X3_n^m \tag{6}$$

$$X3_n^m = X_n^m - \rho_1 \times (X2_n^m - X1_n^m) \tag{6.1}$$

where $X3_n^m = \left(\overline{X3}_n^m, \underline{X3}_n^m\right)$

3.2.2 Local escaping operator (LEO) phase [14]

The LEO is used to help the suggested GBO technique solve complicated problems more efficiently. The LEO creates a superior solution (X_{LEO}^m) by combining numerous solutions, including the best position (x_{best}), the solutions $X1_n^m$ and $X2_n^m$, two random solutions x_{r1}^m and x_{r2}^m, and a new randomly generated solution (x_k^m). The following scheme generates the solution $X_{LEO}^m = \left(\overline{X}_{LEO}^m, \underline{X}_{LEO}^m\right)$.

$$
\begin{aligned}
&\textbf{\textit{if}}\ rand < p\tau \\
&\quad \textbf{\textit{if}}\ rand < 0.5 \\
&X_{LEO}^{m} = X_{n}^{m+1} + f_1 \times (u_1 \times x_{best} - u_2 \times x_{k}^{m}) + f_2 \times \rho_1 \times (u_3 \times (X2_{n}^{m} - X1_{n}^{m}) + u_2 \times (x_{r1}^{m} - x_{r2}^{m}))/2 \\
&\quad X_{n}^{m+1} = X_{LEO}^{m} \\
&\quad \textbf{else} \\
&X_{LEO}^{m} = x_{best} + f_1 \times (u_1 \times x_{best} - u_2 \times x_{k}^{m}) + f_2 \times \rho_1 \times (u_3 \times (X2_{n}^{m} - X1_{n}^{m}) + u_2 \times (x_{r1}^{m} - x_{r2}^{m}))/2 \\
&X_{n}^{m+1} = X_{LEO}^{m} \\
&\quad \textbf{End} \\
&\textbf{End}
\end{aligned} \tag{7}
$$

where f_1 is a random distribution that is uniform integer between -1 and 1, f_2 is a random number chosen from a normal distribution with a mean of 0 and a standard deviation of 1, pr is the probability, and u_1, u_2, and u_3 are three random integers defined as

$$u_1 = \begin{cases} 2 \times \text{rand} & \text{if } \mu_1 < 0.5 \\ 1 & \text{otherwise} \end{cases} \tag{7.1}$$

$$u_2 = \begin{cases} \text{rand} & \text{if } \mu_1 < 0.5 \\ 1 & \text{otherwise} \end{cases} \tag{7.2}$$

$$u_3 = \begin{cases} \text{rand} & \text{if } \mu_1 < 0.5 \\ 1 & \text{otherwise} \end{cases} \tag{7.3}$$

where *rand* is a number between 0 and 1 and μ_1 is a number between 0 and 1. The preceding equations can be simplified as

$$u_1 = L_1 \times 2 \times \text{rand} + (1 - L_1) \tag{7.4}$$

$$u_2 = L_1 \times \text{rand} + (1 - L_1) \tag{7.5}$$

$$u_3 = L_1 \times \text{rand} + (1 - L_1) \tag{7.6}$$

where L_1 is a binary parameter that can be either 0 or 1. L_1 has a value of 1 if parameter u_1 is less than 0.5, else it has a value of 0. To find the x_k^m answer in Eq. (7), the following plan is recommended:

$$x_k^m = \begin{cases} x_{rand} & \text{if } \mu_2 < 0.5 \\ x_p^m & \text{otherwise} \end{cases} \tag{7.7}$$

$$x_{rand} = X_{min} + \text{rand}(0,1) \times (X_{\max} - X_{\min}) \tag{7.8}$$

where x_{rand} is a novel solution, x_p^m is a population solution that was chosen at random ($p \in [1, 2, \ldots, N]$), and μ_2 is a number between [0,1] that is chosen at random. Eq. (7.7) can be summarized as:

$$x_k^m = L_2 \times x_p^m + (1 - L_2) \times x_{rand} \tag{7.9}$$

where L_2 is a binary parameter that takes one of two values: 0 or 1. L_2 has a value of 1 if μ_2 is less than 0.5, else it has a value of 0. Table 1 shows the GBO algorithm's pseudo code (for more see [14]).

Table 1 The GBO algorithm's pseudocode.

Stage [1]. Input
Values for the parameters pr, ε, and M should be assigned
Create a starting population $X_0 = (\overline{X}_0, \underline{X}_0) = [(\overline{X}_{0,1}, \underline{X}_{0,1}), (\overline{X}_{0,2}, \underline{X}_{0,2}) \ldots, (\overline{X}_{0,D}, \underline{X}_{0,D})]$
The objective function's value should be calculated. $f(X_0)$, n=1, …, N
Make a list of the best and worst solutions $x^m_{best} = \left(\overline{X}^m_{best}, \underline{X}^m_{best}\right)$ and $x^m_{worst} = \left(\overline{X}^m_{worst}, \underline{X}^m_{worst}\right)$
Stage [2]. The prime loop
If (m < M)
  for $n = 1 : N$
    for $i = 1 : D$
      Select randomly $\tau 1 \neq \tau 2 \neq \tau 3 \neq \tau 4 \neq n$ in the range of [1, N]
      Compute the position $x^{m+1}_{n,i}$ using Eq. 6
    end for
    Local fugitive operator
    if $rand < p\tau$
      Evaluate the position x^m_{LEO} using Eq. 7
      $X^{m+1}_n = x^m_{LEO}\left(\overline{X}^m_{LEO}, \underline{X}^m_{LEO}\right)$
    end
    Positions must be updated $x^m_{best} = \left(\overline{X}^m_{best}, \underline{X}^m_{best}\right)$ and $x^m_{worst} = \left(\overline{X}^m_{worst}, \underline{X}^m_{worst}\right)$
  end for
  $m = m + 1$
end
Stage [3]. go to back $x^m_{best} = \left(\overline{X}^m_{best}, \underline{X}^m_{best}\right)$

4. Numerical examples

In this section, we use numerical examples to demonstrate the efficiency and applicability of the GBO technique for solving fuzzy nonlinear equations. The following are some instances from [5]. All computations are performed in MATLAB 2009b on a Windows 8 HP system with an Intel Core i5 CPU, 4GB of RAM, and 500GB of hard disk space, with a maximum iteration count of 250 for solution stopping. Table 2 and 3 indicate that the GBO technique is convergent towards a comprehensive solution. The numerical solutions were also plotted in Figs. 1–3.

Iterations	Number of iterations
P-Best	Best variable
G-Best	Best function value

Table 2 Examples of fuzzy nonlinear equations solved by the CE algorithm.

Examples	Range	$f_{\min}$
$f_1=(3,4,5)\chi^2+(1,2,3)\chi=(1,2,3)$ [5] $\begin{cases}(3+\varsigma)\underline{\chi}^2(\varsigma)+(1+\varsigma)\underline{\chi}(\varsigma)-(1+\varsigma)=0,\\(5-\varsigma)\overline{\chi}^2(\varsigma)+(3-\varsigma)\overline{\chi}(\varsigma)-(3-\varsigma)=0.\end{cases}$ For $\varsigma=1$ $\begin{cases}4\underline{\chi}^2(1)+2\underline{\chi}(1)-2=0,\\4\overline{\chi}^2(1)+2\overline{\chi}(1)-2=0,\end{cases}$ For $\varsigma=0$ $\begin{cases}3\underline{\chi}^2(0)+\underline{\chi}(0)-1=0,\\5\overline{\chi}^2(0)+\overline{\chi}(0)-3=0\end{cases}$	[0,2]	0
$f_2=(3,4,5)\chi^2+(1,2,3)\chi=(1,2,3)$ [5] $\begin{cases}(4+2\varsigma)\underline{\chi}^2(\varsigma)+(2+\varsigma)\underline{\chi}(\varsigma)-(3+3\varsigma)=0,\\(8-2\varsigma)\overline{\chi}^2(\varsigma)+(4-\varsigma)\overline{\chi}(\varsigma)-(9-3\varsigma)=0.\end{cases}$ For $\varsigma=1$ $\begin{cases}6\underline{\chi}^2(1)+3\underline{\chi}(1)-6=0,\\6\overline{\chi}^2(1)+3\overline{\chi}(1)-6=0,\end{cases}$ *For* $\varsigma=0$ $\begin{cases}4\underline{\chi}^2(0)+2\underline{\chi}(0)-3=0,\\8\overline{\chi}^2(0)+4\overline{\chi}(0)-9=0,\end{cases}$	[0,2]	0
$f_3=(1,2,3)\chi^3+(2,3,4)\chi^2+(3,4,5)=(5,8,13)$ [5] $\begin{cases}(1+\varsigma)\underline{\chi}^3(\varsigma)+(2+\varsigma)\underline{\chi}^2(\varsigma)-(2+2\varsigma)=0,\\(3-\varsigma)\overline{\chi}^3(\varsigma)+(4-\varsigma)\overline{\chi}^2(\varsigma)-(8-4\varsigma)=0.\end{cases}$ For $\varsigma=1$ $\begin{cases}2\underline{\chi}^3(1)+3\underline{\chi}^2(1)-4=0,\\2\overline{\chi}^3(1)+3\overline{\chi}^2(1)-4=0,\end{cases}$ For $\varsigma=0$ $\begin{cases}\underline{\chi}^3(0)+2\underline{\chi}^2(0)-2=0,\\3\overline{\chi}^3(0)+4\overline{\chi}^2(0)-8=0.\end{cases}$	[0,2]	0

Table 3 Numerical results.

Example	Iterations	x_optimal	f_optimal
1	250	0.43426 0.5 0.5 0.53066	0
2	250	0.65139 0.78078 0.78078 0.83972	0
3	250	0.83929 0.91082 0.91082 1.0564	3.5499e−030

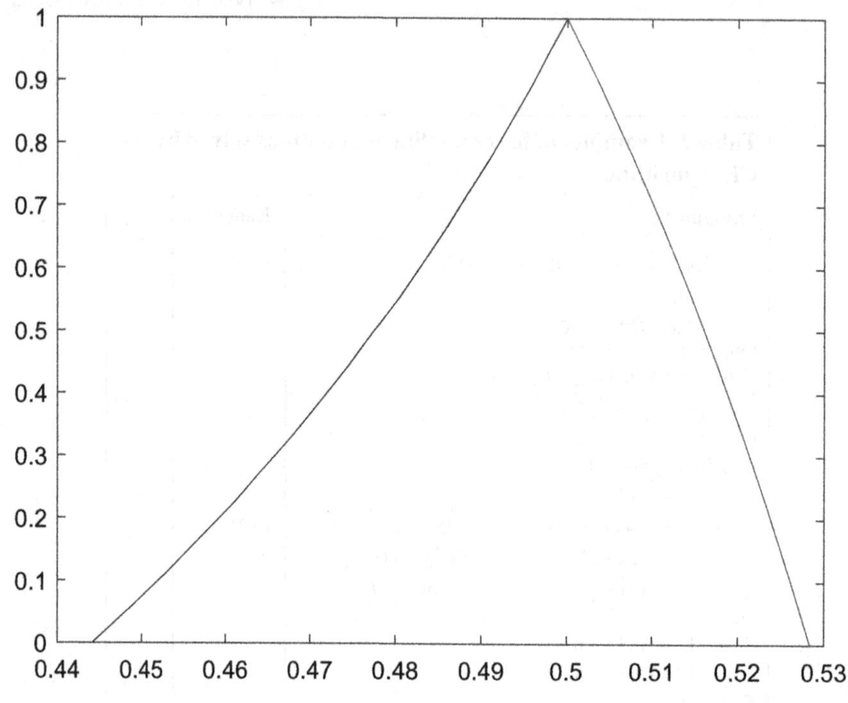

FIG. 1

Solution to Example 1 using the GBO technique.

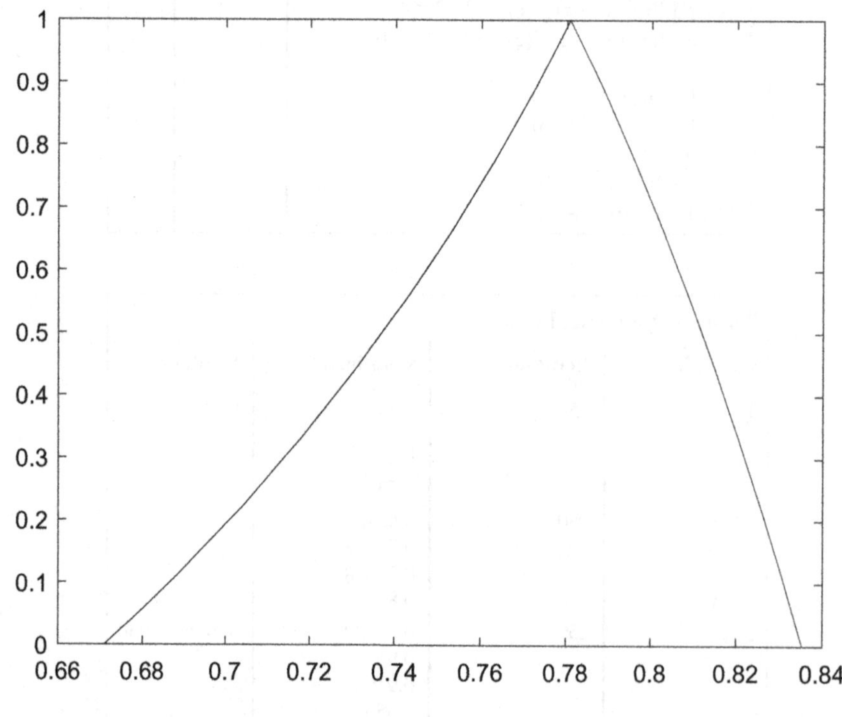

FIG. 2

Solution to Example 2 using the GBO technique.

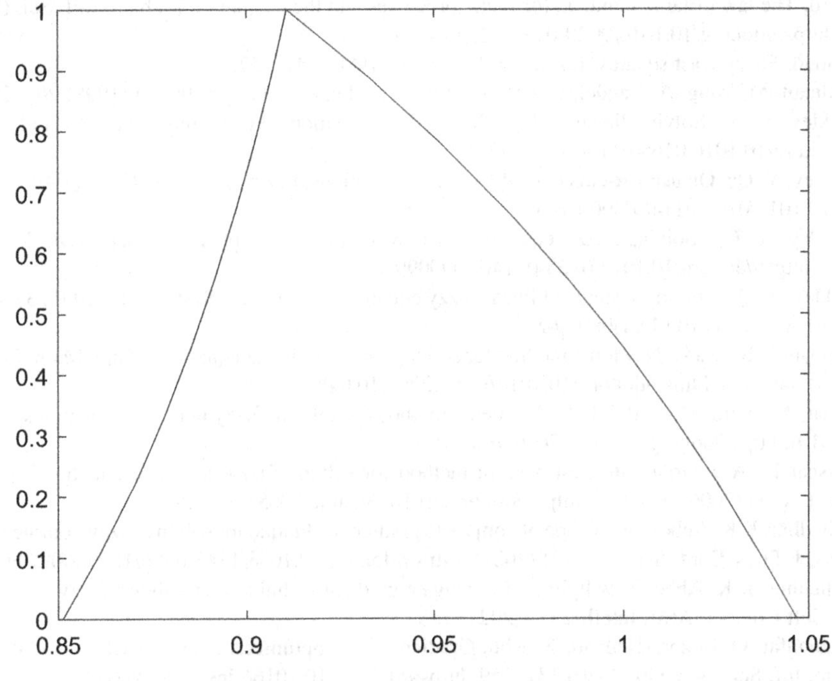

FIG. 3

Solution to Example 3 using the GBO technique.

5. Conclusion

In this chapter, we applied a metaheuristic technique that can be used to solve fuzzy nonlinear equations as an alternative to the usual analytical technique. We converted the fuzzy nonlinear problem into parametric form and then solved it using the GBO algorithm. The numerical findings show that GBO worked well for solving all test problems.

Acknowledgments

I would like to thank the College of Computer Science and Mathematics, University of Mosul, Iraq, for providing resources that helped me improve the quality of my work. Prof. Dr. Khalil K. Abbo, I truly thank you for your insightful remarks, recommendations, and encouragement.

References

[1] L.A. Zadeh, The concept of a linguistic variable and its application to approximate reasoning-III, Inf. Sci. (Ny) 9 (1) (1975) 43–80, https://doi.org/10.1016/0020-0255(75)90017-1.

[2] R. Badard, The law of large numbers for fuzzy processes and the estimation problem, Inf. Sci. (Ny) 28 (3) (1982), https://doi.org/10.1016/0020-0255(82)90046-9.
[3] P. Diamond, Fuzzy least squares, Inf. Sci. (Ny) 46 (3) (1988) 141–157.
[4] M. Friedman, M. Ming, A. Kandel, Fuzzy linear systems, Fuzzy Sets Syst. 96 (2) (1998) 201–209.
[5] J.J. Buckley, Y. Qu, Solving linear and quadratic fuzzy equations, Fuzzy Sets Syst. 38 (1) (1990) 43–59, https://doi.org/10.1016/0165-0114(90)90099-R.
[6] J.J. Buckley, Y. Qu, On using α-cuts to evaluate fuzzy equations, Fuzzy Sets Syst. 43 (1) (1991) 125, https://doi.org/10.1016/0165-0114(91)90026-M.
[7] J.J. Buckley, Y. Qu, Solving fuzzy equations: a new solution concept, Fuzzy Sets Syst. 39 (3) (1991) 291–301, https://doi.org/10.1016/0165-0114(91)90099-C.
[8] J.J. Buckley, Y. Qu, Solving systems of linear fuzzy equations, Fuzzy Sets Syst. 43 (1) (1991) 33–43, https://doi.org/10.1016/0165-0114(91)90019-M.
[9] S. Abbasbandy, B. Asady, Newton's method for solving fuzzy nonlinear equations, Appl. Math. Comput. 159 (2) (2004) 349–356, https://doi.org/10.1016/j.amc.2003.10.048.
[10] M. Mamat, A. Ramli, M.L. Abdullah, Broyden's method for solving fuzzy nonlinear equations, Adv. Fuzzy Syst. (2010), https://doi.org/10.1155/2010/763270.
[11] S. Abbasbandy, A. Jafarian, Steepest descent method for solving fuzzy nonlinear equations, Appl. Math. Comput. 174 (1) (2006) 669–675, https://doi.org/10.1016/j.amc.2005.04.092.
[12] H.M. Khudhur, K.K. Abbo, A new type of conjugate gradient technique for solving fuzzy nonlinear algebraic equations, J. Phys. Conf. Ser. 1879 (2) (2021), https://doi.org/10.1088/1742-6596/1879/2/022111.
[13] H.M. Khudhur, K.K. Abbo, New hybrid of conjugate gradient technique for solving fuzzy nonlinear equations, J. Soft Comput. Artif. Intell. 2 (1) (2021) 1–8.
[14] I. Ahmadianfar, O. Bozorg-Haddad, X. Chu, Gradient-based optimizer: a new metaheuristic optimization algorithm, Inf. Sci. (Ny) 540 (2020) 131–159, https://doi.org/10.1016/j.ins.2020.06.037.
[15] L.A. Zadeh, Fuzzy sets, Inf. Control. 8 (3) (1965) 338–353.
[16] D.J. Dubois, Fuzzy Sets and Systems: Theory and Applications, vol. 144, Academic Press, 1980.
[17] H. Zimmermann, Fuzzy set theory and its applications, Int. Ser. Manag. (1991).
[18] C.T. Kelley, Iterative Methods for Linear and Nonlinear Equations, SIAM, Philadelphia, 1995.

CHAPTER

Metaheuristic algorithms in network intrusion detection

Ibrahim Hayatu Hassan[a,b,#], Abdullahi Mohammed[b,#], and Mansur Aliyu Masama[c]

[a]*Institute for Agricultural Research, Ahmadu Bello University, Zaria, Nigeria,* [b]*Department of Computer Science, Ahmadu Bello University, Zaria, Nigeria,* [c]*E-Library, Kebbi State University of Science and Technology, Aliero, Nigeria*

1. Introduction

Many technologies and techniques, such as encryption, firewalls, and secure network protocols, have been implemented to prevent the unauthorized use of computer or network systems. However, even as security technologies advance, hackers create new methods for breaching the security of computer or network systems. In addition to the ever-changing attacker methods, novel network forms such as wireless sensor networks (WSNs), mobile ad hoc networks (MANETs), and software defined networking (SDN) have emerged, resulting in a more difficult state of network security. Most traditional security mechanisms are incapable of providing adequate security over these networks. As a result, there is a high demand for security mechanisms capable of detecting any unauthorized attempt to breach a computer or network system. This resulted in the development of the intrusion detection system (IDS), which is now regarded as an important component of security mechanisms. An IDS is an information security that can recognize an unauthorized attempt to compromise the safety of a computer system or network, thereby breaching security goals [1,2]. Integrity, availability, accountability, confidentiality, and assurance are all important security goals. The IDS supervises specific sources of activity in a network and computer system, like audit and network traffic data, and utilizes a variety of methods to discover fraudulent activity such as invasions of privacy. The primary goal of IDS is to detect all intrusions as quickly as possible. The use of IDS enables network administrators to detect security objective violations. External attackers attempting to gain unauthorized access to network security infrastructure or making resources unavailable to insiders abusing their access to system resources are examples of security goal violations. The IDS can be divided into two types: anomaly detection and misuse detection [3,4]. Misuse-based IDS, such as Snort, can effectively detect known attacks. This type of IDS has a low false alarm rate, but it does not detect new attacks that do not include any rules in the records. The anomaly-based intrusion detection system creates a normal behavior model and then identifies any significant deviations from this model as intrusions. This type of IDS can detect new or unknown attacks, but it has a high false alarm rate.

[#]Senior contributing author.

Comprehensive Metaheuristics. https://doi.org/10.1016/B978-0-323-91781-0.00006-5

However, designing an efficient and effective IDS is a real challenge because it must meet the requirements of a high true positive rate and a low false positive rate for the ever-changing trend of intrusions while using as few computing resources as possible. The ability to detect intrusions quickly may aid in reducing the loss of unauthorized access to a computer system.

Machine learning (ML) techniques have recently become the hot cake for developing IDSs. ML IDSs use a smart learning algorithm to classify network data as normal or malicious based on its attributes. However, the ever-increasing number of devices, as well as the resulting large and high-dimensional data, necessitates the search for more efficient searching and learning algorithms. This resulted in the development of metaheuristic algorithms in IDSs.

Metaheuristic algorithms are a type of stochastic search algorithm that encompasses both a random sampling and a local search aspect. Metaheuristic algorithms are the most popular procedures for resolving difficult optimization problems quickly. Various metaheuristic techniques, such as ant colony optimization (ACO) [5], particle swarm optimization (PSO) [6], and artificial bee colony (ABC) [7], are currently being used to maximize the efficiency of IDS. Also, various hybrid/ensemble classification and clustering techniques have been proposed such as simplified swarm optimization (SSO) [8], genetic algorithm-support vector machine (GA-SVM) [9], genetic fuzzy system [10], and support vector machine–k nearest neighbor–particle swarm optimization (SVM-KNN-PSO) [11].

This chapter provides a detailed survey of various research papers that employed metaheuristic algorithms for IDS. Moreover, the chapter highlights challenges and future directions.

The remainder of this chapter is organized as follows: Section 2 examines optimization metaheuristics algorithms. Section 3 details our methodology. Section 4 presents various research works on metaheuristic algorithms used in the design of IDSs and how they are used. Section 5 discusses challenges and future directions, and Section 6 concludes.

2. Metaheuristic algorithms

Metaheuristic algorithms are global optimization techniques that are planned based on simulations and nature-inspired approaches [12,13]. These nature-inspired procedures have been well known for a few decades. Understanding more about social swarm conduct in fish, birds, ants, and other animals has aided researchers in developing optimization search algorithms that have been successfully applied to a wide range of real-world optimization problems [14,15]. Surprisingly, these creatures can work in groups to find optimal solutions and complete missions competently. As a result, it stands to reason that we encourage their behavior in optimization problems.

The metaheuristic-based algorithms can be classified into three groups: swarm-based algorithms, evolutionary algorithms (EAs), and trajectory-based algorithms [16]. A broad outline of this groups and examples of each is shown in Fig. 1. EAs are regarded as the fastest emergent algorithms to handle optimization problems [17]. Algorithms in this group are motivated by natural selection practices and biological evolution. EAs usually start with a set of solutions, called the population. At each generation, the EAs recombine the better features of the existing population to form a new population that will be selected based on a set of evolution operators such as mutation and recombination operators. Optimization algorithms developed in this category includes genetic algorithm (GA) [18], genetic programming (GP) [19], evolutionary strategies (ES) [20], harmony search (HS) [21], differential

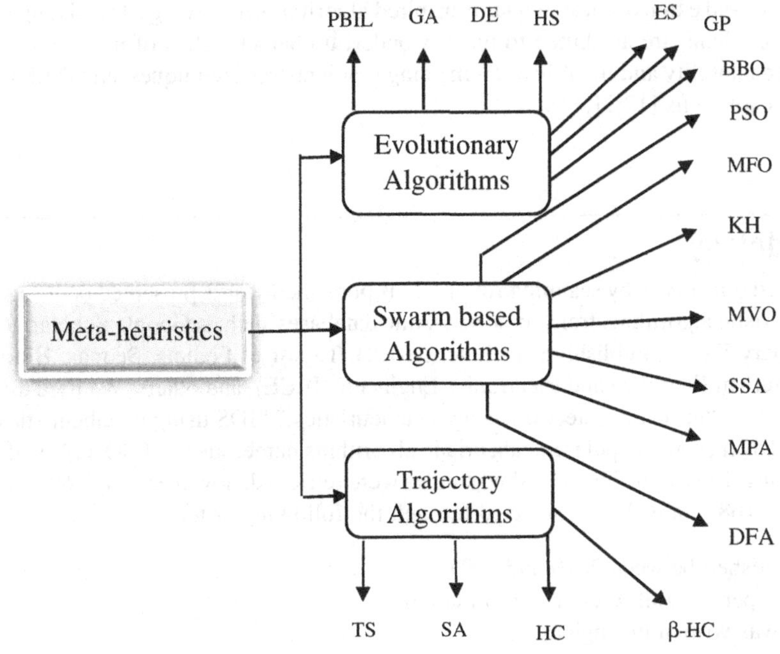

FIG. 1

Categorization of metaheuristic algorithms.

evaluation (DE) [22], population-based incremental learning (PBIL) [23], biography-based optimizer (BBO) [24], and many more.

Swarm intelligence-based algorithms are another powerful form of metaheuristics employed to handle optimization problems in different domains. Optimization algorithms belonging to this category mimic the natural swarms or systems like fish schools, bacterial growth, bird swarm's, insect's associations, and animal herds. Most of the swarm intelligence algorithms focus on the conduct of group's associates and their lifestyle in addition to the relations, and interactions between the group's associates to discover food sources. Some of the algorithms in this group include particle swarm optimization (PSO) [25], krill herd (KH) [26], moth-flame optimizer (MFO) [27], whale optimization algorithm (WOA) [28], multi-verse optimizer (MVO) [29], dragonfly algorithm (DFA) [30], salp swarm algorithm (SSA) [31], manta ray foraging optimizer (MFRO) [32], marine predators algorithm (MPA) [33], and many others. Moreover, trajectory-based algorithms begin with a particular temporary solution. At each iteration, that solution will be relocated to its close solution, which lives in the same search space area, based on a particular neighborhood structure. Algorithms in this category include tabu search (TS) [34], simulated annealing (SA) [35], hill climbing (HC) [36], and β-hill climbing (βHC) [37].

Metaheuristic algorithms have achieved promising performance in addressing problems in several areas such as biomedical signal processing [38], process control [39], image processing [40,41], text clustering [42], classification problems [43], and many other contemporaneous problems [44–46]. This broad popularity and extensive use of nature-inspired algorithms in solving a broad range of real-world optimization problems are attributed to many wonderful characteristics of metaheuristic algorithms such as their adaptability and flexibility, them being gradient-free techniques, and their ability to avoid local optimum solutions [13,31,47].

3. Methodology

We commenced our survey by scanning for previous published works presented on intrusion detection using metaheuristic algorithms from some academic databases such as Google Scholar, ResearchGate, Multidisciplinary Digital Publishing Institute (MDPI), Taylor & Francis, Scopus, Hindawi, Web of Science, Institute of Electrical and Electronics Engineers (IEEE), and others. We used different search parameters such as "intrusion detection using metaheuristics," "IDS using metaheuristics," "intrusion detection" and using some popular metaheuristic algorithms name, such as PSO, GA, and many others. After completing the search, 210 published studies were retrieved, however, only 168 were retained for this study. The 168 retained were selected through the following criteria.

1. studies published between 2010 and 2021
2. the study paper has full reference information
3. the paper was written in English

Furthermore, the 168 published studies were grouped into different categories. Metaheuristic algorithms with more than five published studies in IDS were grouped together with the algorithm name as subheading, studies that used hybrid metaheuristic approaches were combined as a single group, and other metaheuristic algorithms with a smaller number of published studies in IDS were grouped together and labeled as "others." Fig. 2 gives the number of published studies for the common metaheuristic algorithms in the intrusion detection domain.

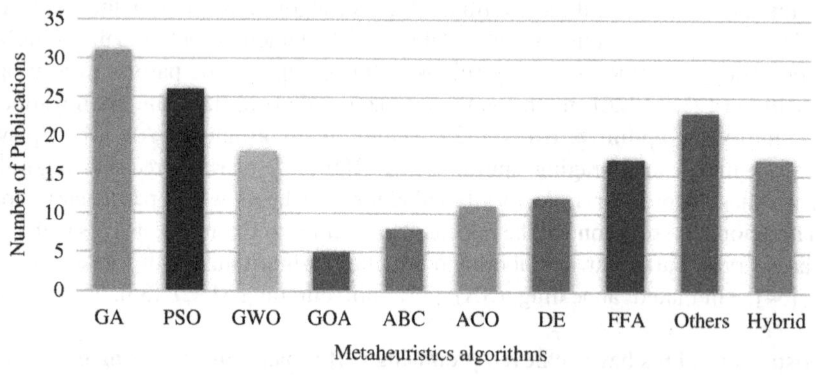

FIG. 2

Number of studies for the common metaheuristic algorithms in IDS.

4. Metaheuristic algorithms in IDS

In this section, we present and analyze various published research papers on different metaheuristic algorithms. In addition, we include the most common metaheuristic approaches applied to intrusion detection in the literature.

4.1 Genetic algorithm (GA)

GA is one of the dominant evolutionary metaheuristic algorithms employed for designing IDS in the literature. Publications using GA include the work of Hoque et al. [48], who implemented IDS using GA to efficiently detect various kinds of network interferences. This approach employs an evolution mechanism to information evolution to the traffic data. This IDS was implemented and evaluated using the KDD Cup 99 standard dataset and the results indicate a sensible detection rate. However, this IDS was compared with many other methods.

A similar work was proposed in Ref. [49] based on GA fuzzy-class association mining. The GA is used to generate many rules that are needed for an intrusion discovery model. The association rule mining technique is employed to find adequate number of vital rules for the user's purpose rather than to generate all the rules meeting the measures that are beneficial for misuse detection. The experimental research using the KDD Cup 99 intrusion detection dataset shows that the proposed method has a higher detection rate than crisp data mining approach.

Sindhu et al. [50] designed an efficient IDS that uses a GA wrapper-based attribute selection technique to improve the IDS's sensitivity and specificity by employing a neural ensemble decision tree (DT) iterative procedure as a learning method. Experimental evaluation of the developed method was compared with a family of six DT classifiers, namely, C4.5, decision stump, naïve Bayes (NB) tree, random forest (RF), representative tree, and random tree (RT).

A hybrid method for effective anomaly detection was proposed in Ref. [51], who combined GA and detectors generated using a multi-start metaheuristic method. The main idea of this method is based on using the k-means algorithm to choose a reduced training dataset to reduce time and processing intricacy. Similarly, the k-means algorithm is used to provide variation in the selection of opening start points used by the multi-start approach. Furthermore, the radius of the hyper-sphere detectors created by the multi-start approach is later enhanced using the GA. Finally, the reduction rule is used to remove unnecessary redundant detectors. The proposed approach was tested on the NSL-KDD dataset, and the results show that it is effective at generating suitable detectors with 96.1% accuracy when compared to other learning techniques.

Kuang et al. [52] developed a novel SVM-based IDS combining GA and kernel principal component analysis (KPCA). Here, a multilayer SVM is assumed to assess whether the activity is an attack or not and the KPCA is used as a preprocessing mechanism in SVM to diminish the dimension of attribute vectors and reduce training time. To decrease the noise triggered by attribute dissimilarities and enhance the effectiveness of SVM, an enhanced kernel function (N-RBF) is suggested by inserting the mean square difference and mean value of feature attributes in RBF kernel. Meanwhile, the GA is engaged to enhance the SVM parameters such as tube size, kernel, and penalty factor. In contrast with PCA-GA-SVM, SVM, RBFNN, and KPCA-GA-SVM, the proposed N-KPCA-GA-SVM attained greater predictive accuracy, earlier convergence, and better generalization.

A feature selection approach based on GA with NB algorithm for efficient IDS was presented in Desale and Ade [53]. The GA is employed as search technique though picking attributes from the NSL-KDD benchmark dataset together with the intersection mechanism of choosing only those that appear all over in the experiment. The experimental analysis revealed greater accuracy with minimum number of attributes in shorter time in contrast with other classifiers.

Narsingyani and Kale [54] designed an IDS model using GA to optimize the detection false positive rate. The proposed method was tested using the standard KDD Cup 99 dataset and results show its effective detection rate. However, the method was not compared with any other methods.

Hamamoto and Carvalho [55] proposed a digital signature of network segment using flows analysis (DSNSF) derived from GA for network anomaly detection. The DSNSF classifies network traffic on a given day based on the previous week's network behavior. The proposed GA-based approach was compared against ACO.

Patel and Sharma [56] implemented an IDS by applying GA to correctly spot various classes of invasive events in a computer or network system. The NSL-KDD benchmark dataset was used for experimental analysis and evaluation. The results indicate that the presented IDS obtained better accuracy compared to a heuristic algorithm.

El-Alfy and Alshammari [57] presented an accessible rough set centered feature subset selection for intrusion detection using parallel GA in MapReduce to estimate the lowest reduct that has the same visibility control as the initial feature set in the judgment table. The proposed approach was tested on four cyber security datasets: Spam-base, Kyoto, NSL-KDD, and CDMC2012. The experimental analysis showed that the presented model can be a great instrument to increase the effectiveness of finding features in the smallest deduct in large-scale judgment schemes.

Aslahi-shahri et al. [9] proposed a hybrid technique combining GA and SVM for designing IDS. The GA is used to lessen the number of attributes in the network traffic dataset and the SVM is used as the learning algorithm. The proposed method was able to reduce the number of features from 45 to 10. The GA is also used to categorize these reduced attributes into three priorities in deceasing importance. The distribution of the attribute is achieved in such a way that four attributes are put in the first priority, another four in the second, and two in the third priority. The performance of the proposed IDS method was measured using the KDD Cup 99 dataset in terms of precision, recall, f-measure, true positive rate, false positive, and ROC area. However, the authors did not compare this method with any other method.

Hamamoto et al. [10] presented an approach hybridizing the GA with fuzzy logic for network anomaly detection. The GA is used to generate a digital signature of a network segment based on flow analysis, in which information mined from network traffic data is used to classify the behavior of network traffic over a given time interval. Additionally, the fuzzy logic method is applied to decide whether a given traffic is normal or not. The presented approach was tested on six attributes generated from IP traffic data. The results show this approach achieved a false positive rate of about 0.56% and an accuracy of 96.5%. Furthermore, this approach achieved greater efficiency in comparison with other methods.

Tao et al. [58] proposed a hybrid IDS combining the capabilities of SVM and GA. This approach first enhances the crossover probability and mutation probability of GA giving to the population evolution algebra and fitness value; then, it subsequently uses an attribute selection technique using the GA with an improvement in the fitness function that decreases the SVM error rate and increases the true positive rate. Lastly, according to the selected attributes subset, the attributes' weights and parameters of SVM are concurrently improved. The results of the analysis show that the proposed method accelerates algorithm convergence, increases true positive rate, decreases error rate, and shortens prediction time. In comparison to other SVM-based IDS approaches, the presented method has a higher detection rate and a lower rate of false negatives.

An adaptive IDS based on GA and profiling was implemented by Resende and Drummond [59]. In this study, an adaptive methodology centered on GA is applied to select attributes for profiling and variables for an anomaly-based intrusion detection technique. Furthermore, the researchers presented two anomaly-based approaches to be combined with the suggested method. One is on the common statistics and the other is on an estimated clustering technique. Simulations were conducted using the CICIDS2017 dataset based on results, the proposed approach achieved a discovery level of 92.85% and a false positive rate of 0.69%.

A parallel GA was applied for feature selection in intrusion detection in Ref. [60]. In the proposed approach, the GA's traditional fitness function was improved to evaluate the efficiency and effectiveness of the algorithm. The features generated using the GA were then used to develop the IDS model using a self-organizing feature map (SOFM) artificial neural network (ANN). The KDD Cup 99 and 1998 DARPA datasets were used to evaluate the proposed IDS. The outcomes show how well the suggested technique performs in spotting both regular activity and other attacks.

GA was used to optimize an ANN for IDS in Ref. [61]. The GA is used for both feature selection and to optimize the self-organizing map ANN for intrusion detection using pretreatment of dataset and classifications. The proposed method was evaluated using the KDD Cup 99 and UNSW-nb15 standard datasets.

Reddy et al. [62] also used GA to implement fuzzy logic for detecting network anomalies in a wireless network. The proposed approach proved effective with a detection rate of 80%.

Ren et al. [63] built an efficient intrusion detection system using GA, RF, and the isolation forest (iForest). The iForest is used to eliminate outliers, GA is used to optimize the sampling ratio, and RF is used as the evaluation criteria to obtain the optimum data for training. Furthermore, the proposed method uses the GA and RF again to obtain the optimum feature subset. Lastly, the intrusion detection system was designed using RF based on the optimum training data acquired by data sampling and the attributes selected by feature selection. Simulation was carried out using the UNSW-NB15 benchmark dataset. The proposed IDS showed a clear advantage in identifying infrequent anomaly actions compared to KNN, SVM, MLP, NB, EM, DT, and RF.

Suhaimi et al. [64] applied GA for design of IDS to identify harmful types of connections in a network. Diverse features of connection histories like duration and kinds of connections in a network were investigated to produce a set of prediction rules. The KDD Cup 99 benchmark dataset was used to examine the efficiency of the proposed technique. The rules generated contain eight features that were simulated in the training procedure to discover any malevolent connection that can lead to a network intrusion. A good efficiency was reached in detecting corrupt connections using the proposed approach.

Chiba et al. [65] applied and improved GA, called (IGA), which is based on fitness value hashing and parallel processing, to optimize the back propagation neural network parameters such as the momentum term and learning rate for an effective network anomaly detection in cloud environment. The proposed approach was evaluated using the DARPA's KDD Cup 99 datasets on CloudSim simulator 4.0. The results clearly show the effectiveness of this method over compared methods for anomaly detection.

A different study by Zhang et al. [66] presented an IDS for IoT using enhanced GA and deep belief network (DBN). The GA is used to find the optimum number of hidden layers and the number of neurons in each layer, so that the IDS using DBN reached a better detection level. The experimental outcomes using the NSL-KDD dataset demonstrate that the proposed IDS successfully enhanced the discovery level of attacks.

In another work, Suhaimi et al. [67] applied immune GA for designing a network IDS. In the suggested IDS, the GA is hybridized with an immune algorithm process to improve the system's ability to identify likely intrusions in the computer system. Simulation results show that the proposed IDS can identify the occurrence of anomalies in a network.

An intrusion detection in IoT applications using GA was implemented in Ref. [68]. In the proposed work, the GA is used as the classifier for detecting different types of network attacks in IoT applications using the KDD Cup 99 dataset.

Nguyen and Kim [69] suggested a novel IDS algorithm based on an enhanced attribute subset selected directly by GA and fuzzy C-means clustering (FCM). The proposed algorithm recognized the bagging (BG) algorithm and the convolutional neural network (CNN) as an efficient extractor by applying the GA in combination with fivefold cross validation (CV) to select the CNN system structure. Experimental analysis shows that this approach considerably increases the final discovery performance.

In another study, Rai [70] applied GA to select the most important features for network intrusion detection system. The proposed approach employed various ensemble ML techniques such as gradient boosting, distributed random forest, and XGBoost for detecting intrusions in a network. The performance of this approach was tested using the NSL-KDD dataset and compared with a deep neural network (DNN). The result obtained show that the proposed algorithm outperformed the compared algorithm.

Fauzi and Hanuranto [71] employed GA for feature selection and KNN for classification of DoS attacks. The KDD Cup 99 dataset was used for performance analysis. The proposed approach selected 18 features out of 41 present in the dataset and achieved an accuracy of 97.5%.

Onah et al. [72] presented a GA wrapper-based feature selection and NB classifier for network anomaly detection called GANBADM in a fog computing setting. In a wrapper-based feature selection approach, the GA is used as a search strategy alongside the NB classifier as a learning algorithm to select the most relevant attributes from the NSL-KDD network traffic dataset. Furthermore, the NB algorithm is used to develop the IDS based on the attributes that have been chosen. The effectiveness of this approach was measured and the result indicated the superiority of this method in contrast with DT, Bayesian network, and SVM in terms of accuracy and execution time. However, this method was outperformed by random forest, SVM, and DT in terms of f-measure.

A similar work combining GA with NB and J48 algorithm for malicious network traffic detection was proposed by Al-Saqqa et al. [73]. To detect attacks, the GA is used to select the most relevant feature subsets. The NB and J48 are used to compare the system's performance before and after feature selection. The NSL-KDD dataset was used for the experimental evaluation. The results show that using the GA to select features improves the intrusion detection system in terms of detection accuracy and detection of unknown attacks. Furthermore, training time is shortened, which has a positive impact on overall system performance.

A parallel GA for attribute selection in intrusion detection was implemented in MapReduce by Mehanovic et al. [74]. For attribute selection, ML algorithms such as SVM, ANN, RT, NB, and logistic regression were embedded in the proposed approach. The NSL-KDD benchmark dataset was applied for feature selection and the proposed method was able to reduce the number of attributes from 40 to 10 with 90.45% accuracy. The results of this analysis show that the proposed implementation performed better than the compared works from the literature.

Tally and Amintoosi [75] presented a hybrid approach for IDS combining GA and SVM. The GA is applied to select the most appropriate attribute subset, while the SVM is used to categorize the network activities into normal and attacks using the selected attributes from GA. Simulations using NSL-KDD dataset show that the proposed approach outperformed the traditional SVM by attaining F1-score of 92.7%.

In Ref. [76], an enhanced GA-based dimensionality reduction method, termed as GA-based feature selection (GbFS), is implemented with a new objective function for solving attribute selection difficulties in network security and attack detection. The proposed work utilized three classifiers: XgBoost, SVM, and KNN to develop the IDS using the best attribute subset generated by the GbFS. The performance of this approach was tested and evaluated using CIRA-CIC-DOHBrw-2020, Bot-IoT, and UNSW-NB15 standard network traffic datasets. The result verify the superiority of this method over other techniques from the literature with an accuracy of 99.8%.

4.2 Particle swarm optimization (PSO)

A binary PSO with RF algorithm for detection of PROBE attacks in a network was proposed by Malik et al. [77]. The binary PSO is used for selecting the most relevant features and the RF is used for classification. The proposed approach efficiency was demonstrated using the KDD Cup 99 dataset. In comparison to RF, c4.5, bagging, NB tree, and Jrip, the results show that this approach performed better.

Anomaly intrusion detection technique using a hybrid of PSO and k-means clustering algorithm called PSO-KM algorithm was proposed by Li et al. [78]. The proposed approach combines the k-means and PSO to solve the problem of initial value selection and local convergence of the k-means algorithm. The experimental assessment conducted on the KDD Cup dataset demonstrates that the proposed PSO-KM is not influenced by initial centers and has better global search efficiency than the traditional k-means algorithm. Experimental outcomes also established that the proposed approach could realize an optimum cluster and well-disjointed clusters and increase the high detection rate and decrease the false positive rate at the same time in comparison with original k-means, PSO, and other works from the literature.

Karami and Guerrero-Zapata [79] proposed a fuzzy intrusion detection system using hybrid PSO-k-means algorithm as content-centric networks. The PSO is first hybridized with the k-means algorithm to achieve the best number of clusters using two concurrent objective functions, well-separated clusters, and local optimization. The fuzzy approach is then employed by combination of two distance-based approach as outlier and classification to detect anomalies in a novel monitoring dataset.

An improved chaotic PSO combined with KPCA was presented in Ref. [80] to optimize the SVM parameters including tube size, kernel parameters, and penalty. The suggested multi-layer SVM is then applied to estimate whether the network activity is an attack or not. Investigation using the KDD Cup 99 benchmark data proved that the enriched SVM system attained quicker computational time and greater accuracy, reduced the training time, and increased the efficiency of the SVM for intrusion detection in comparison with other approaches.

PSO was used by Manekar and Waghmare [81] for parameter optimization as well as feature selection. The authors first use the PSO algorithm to optimize the values of SVM parameters such as the cost (C) and the gamma (g). Then, PSO is used to select the most valuable from the the NSLKDD dataset. Finally, the features selected and optimized parameters value are used for training and testing the SVM model.

A combined negative selection algorithm-PSO-based email spam detection model was proposed by Idris et al. [82]. In this case, the PSO is used to improve the random generator in the negative selection algorithm (NSA). During the random detector generation phase of the NSA, the algorithm generates detectors. For detector generation, the combined NSA-PSO employs a local outlier factor (LOF) as the fitness function. When the expected spam coverage is reached, the detector generation process is terminated. After the detector is generated, a distance measure and a threshold value are used to improve the distinguishability of the nonspam and spam detectors. The researchers examined the models' implementation and evaluation and found that the accuracy of the proposed NSA-PSO model is better than the accuracy of the standard NSA model.

In Ref. [6], the authors employed a time-varying chaos PSO, called TVCPSO, to perform parameter optimization and feature selection simultaneously using SVM and multiple criteria linear programming (MCLP) to design an efficient network intrusion detection scheme. In the suggested technique, a weighted cost function is suggested that considers tradeoff among improving the detection rate and reducing the false alarm rate, as well as the number of features. The efficiency of the suggested technique was tested and evaluated using the NSL-KDD standard dataset. The simulation outcomes indicate that the proposed technique outperforms better in terms of having a high detection rate and a low false alarm rate when compared with the achieved results with all features.

A feature selection method based on PSO in intrusion detection was presented in Ref. [83]. In the proposed work, the authors applied principal component analysis (PCA) for feature transformation, then employed PSO to select the most important features from the transformed features. To design the IDS, a modular neural network (MNN) is applied to the selected feature from the PSO. The proposed method gives better performance on the KDD Cup dataset compared with a GA-based approach.

An intrusion detection system based on fast learning network (FLN) optimized with PSO was presented by Ali et al. [84]. The PSO was used to optimize the weights of the FLN. The proposed FLN-based PSO was tested using the KDD Cup 99 benchmark dataset and compared with several metaheuristic algorithm using training extreme learning machine and FLN algorithm. The comparison shows that the PSO-FLN model outperformed compared techniques.

A PSO algorithm with RF for IDS was also presented by Li et al. [85]. The authors propose a hybrid feature system using RF and PSO that makes use of both an independent measure and learning algorithm to evaluate feature measure. The proposed method employs an independent measure to determine the best subset of features for a given cardinality, and then employs a learning algorithm to select the final best subset from among the best subsets across different cardinalities. The KDD Cup 99 dataset was exploited to evaluate the performance of the proposed model with respect to true positive and false positive rates. In comparison to CFS and SVM, the experimental analysis demonstrates the capability of the presented system.

Bayu and Kyung-Hyune [86] presented an intrusion detection model combining an RF and PSO. The PSO is used as a selection method in the wrapper feature extraction technique to choose the most descriptive feature subset, and the RF is used as the learning algorithm. The presented model's efficiency was assessed using the NSL-KDD dataset in terms of precision, recall, false alarm rate, and accuracy. The statistical significance test revealed significant differences between the presented model and other classifiers of rotation forest (RoF) and DNN. The test shows that the presented model outperforms the compared algorithm.

A similar work was presented by Tama and Rhee [87], integrating PSO and RF for feature selection and design of IDS in an IoT network. The proposed IDS was evaluated using the NSL-KDD network traffic dataset and contrasted with RoF and DNN based on a statistically significant test. The statistical test shows that the proposed approach significantly outperforms the other algorithms it was compared with.

The authors in Ref. [88] proposed a k-means clustering-based intrusion detection system optimized with PSO. In the proposed work, a multi-criterion optimization technique built on the weighted PSO algorithm is suggested. In this technique minimizing the intra-cluster distance and maximization, the inter-cluster distance is selected as optimization criteria. In addition, a novel clustering algorithm based on the PSO technique is designed, with the k-means algorithm formulating the initial population. The superiority of the proposed model was demonstrated in comparison with a standard k-means clustering algorithm.

Ghosh et al. [89] proposed a hybrid IDS in a cloud environment by CS and PSO to form CS-PSO. The CS-PSO was used to extract the relevant features from the NSL-KDD dataset, and three classifiers including AdaBoost, RF, and linear regression were introduced to classify the different network traffics into either normal or attack.

For designing an IDS, a double PSO-based technique was used for feature selection as well as hyper-parameter fine tuning [90]. To investigate its performance, three deep neural networks including DNN, long short-term memory recurrent neural network (LSTM-RNN), and DBN were trained using the double PSO based on the NSL-KDD dataset. Experimental results demonstrate that using this strategy improves network intrusion detection by increasing detection rate by 4%–6% and decreasing false alarm rate by 1%–5% over the corresponding values of the same models without pretraining on the same dataset.

Shokoohsaljooghi and Mirvaziri [91] presented an improved IDS hybridizing neural networks (NNs) and PSO. The PSO is employed here to optimize the training parameters of the NNs, such as the weights and biases. In addition, the suggested technique used a Kolmogorov-Smirnov correlation-based fast redundancy removal filter (KS-CBF) to select appropriate features from the KDD Cup 99 and NSL-KDD intrusion detection benchmark datasets. In comparison with the ANN, the proposed approach offers a better accuracy and efficiency in identifying various types of attacks.

A variant of PSO called multi-PSO was applied for SVM parameter optimization by Kalita et al. [92] for developing an IDS. The multi-PSO is used to select the optimum values of SVM parameters including the penalty factor (C) and gamma. Then, the optimized SVM is used for the optimum design of the IDS. The proposed SVM-multi-PSO was tested using KDD Cup 99 dataset and compared with classical PSO, GA, gradient descent, and grid search. The results indicated the better performance of the proposed approach compared to the other methods.

Eesa [93] proposed a max and min boundary mining method using PSO to handle intrusion detection problems. The PSO is applied to find the best max and min boundary for each attribute in each class from the training dataset. The generated maximum and minimum boundaries are then used as discovery rules to distinguish attacks from normal traffic using a test dataset. The proposed was tested and compared with PART, SMO, J48, and NB using KDD Cup 99 and NSL-KDD datasets. Results obtained reveal the superiority of this approach over the contrasted techniques in terms of some standard evaluation metrics.

Dickson and Thomas [94] applied and improved PSO for enhancing the effectiveness of an IDS. The authors present a method to solve the tradeoff between the true positive and false positive conflicting objective functions using a multiobjective PSO scheme. The simulation results conducted on the NSL-KDD standard dataset demonstrate that the proposed method in conjunction with J48 offers the largest global best value of 10.77 with a false positive of 0.02 and a true positive of 0.995 in contrast with other algorithms including RF, SMO, NB, and logistic regression.

In another study by Kunhare et al. [95], the authors applied PSO algorithm on NSL-KDD dataset features selected by RF algorithm. A comparative analysis of this technique with KNN, SVM, LR, DT, and NB using accuracy, detection rate, precision, and false positive rate as performance indicators was conducted. The results show that the proposed method attained 99.32% competence, low computational complexity, and 99.26% detection level on the 10 features selected by RF out of the total number of features in the NSL-KDD dataset.

In Ref. [96], the authors present an IDS method using modified multi-objective PSO using Lévy flight mechanism called MOPSO-Lévy to handle feature selection as a multi-objective problem. The MOPSO-Lévy is then used with KNN algorithm to classify real IoT network data. The proposed method achieved better results in contrast with other multi-objective algorithms.

PSO was employed as an attribute selection technique in Ref. [97] for developing an intrusion detection model. The selected features are passed to a DT and KNN classifiers for classification procedure. The two intrusion detection models were coded as PSO+DT and PSO+KNN. Simulations were conducted using the KDD Cup 99 intrusion dataset to verify the success of the two models in terms of success indicators like F1-score, accuracy, specificity, precision, and consistency on security dataset for different categories of attacks. The outputs reveal the superiority of PSO+KNN over PSO+DT in recognizing network threats.

Bamhdi et al. [98] presented an ensemble-based intrusion detection model combining SVM, KNN, RF, multilayer perceptron (MLP), extra-tree classifier (ETC), and PSO. In the proposed approach, the output from SVM, KNN, RF, MLP, and ETC were combined based on majority voting procedure to enhance the detection rate. To identify significant feature, a DT classifier was employed with the aim to decrease the training time and complexity of the dataset. Finally, the PSO was used to optimize the performance of the ensemble method. The success of the proposed model was assessed using the NSL-KDD dataset. The experimental analysis showed that the efficiency of the model was 99%, 97.2%, 97.2%, and 93.2% for DoS, Probe, R2L, and U2R, respectively.

Ajdani and Ghaffary [99] proposed a hybrid IDS combining a PSO algorithm and RF classifier. The authors here focus on the applicability of the novel cosmology inspired PSO algorithm to train RF. The performance of this method was measured using UNSW-NB15and KDD Cup 99 benchmark datasets using AUC. The results indicate an AUC of 96.3% and 96.5% for UNSW-NB15 and KDD Cup 99 datasets, respectively. These values are greater than those obtained by PSO-XGboost, PSO-DT, and CS-PSO.

Another paper by Talita et al. [100] proposed an intrusion detection system based on PSO for feature selection and an NB classifier for feature subset evaluation and detection model development using the KDD Cup 99 dataset. The best experimental result was obtained with 38 features and an accuracy of 99.12%. The proposed method, however, was not compared to any other method.

4.3 Gray wolf optimization algorithm (GWO)

The authors in Ref. [101] presented a novel dynamic and effective method of industrial control network IDS, employing a cloud gray wolf optimization (CGWO) algorithm that balances exploration and exploitation capabilities while also performing parameter fine-tuning for semi-supervised learning and the one-class support vector machine (OCSVM). Results show that the proposed approach outperforms other techniques.

An unsupervised intelligent model using GWO and OCSVM botnet discovery in IoT was presented in Ref. [102]. The GWO is used here to adjust the hyper-parameters of the OCSVM and to discover the features that best define the IoT botnet problem. The efficiency of this approach is evaluated using the N-BaIoT dataset and compared with OCSVM, IF, and LOF. The simulation outcomes depict that the proposed approach overtakes all other algorithms in terms of false positive rate, true positive rate, and G-mean for all IoT device types. Equally, it achieves the shortest recognition time while significantly reducing the number of selected attributes.

In Ref. [103], the authors applied GWO to select the most important feature subset for optimum intrusion detection. The subset of features from the GWO is then used to design an IDS using generalized regression neural network (GRNN), SVM, and KNN. Experimental results using the KDD Cup 99 dataset show that GWO-KNN performed better than the other two algorithms in terms of accuracy and sensitivity. However, with respect to specificity, the GWO-GRNN shows better performance.

A similar study in Srivastava et al. (2019), presented an analysis of various hybrid approaches for IDS. The authors examine the relative performance of GWO+entropy-based graph (EBG), GWO+SVM, GWO+KNN, and GWO+generalized regression neural network (GRNN) using the KDD Cup 99 benchmark dataset. The analysis shows the performance of each method in detecting various types of attacks.

In Ref. [104], an enriched anomaly-based IDS using multi-objective GWO algorithm was presented. The GWO was hired as a feature selection tool to find the most appropriate variables from the dataset that help in obtaining high prediction accuracy. Moreover, an SVM classifier was employed to assess the ability of the selected attributes in predicting the intrusion correctly. To determine the ability of the proposed IDS, 20% of the NSL-KDD dataset was used. The proposed technique obtained a significant result in contrast with standard GWO+SVM.

In another study, Anitha and Kaarthick [105] proposed a new IDS model for extracting the most relevant features for intrusion detection based on the oppositional-based Laplacian gray wolf optimization algorithm (OLGWO). The features chosen are used to train the SVM classifier, which is then used to categorize the malicious activities. The KDD Cup 99 dataset was used to evaluate the proposed system's performance, and the results were compared to existing methods in terms of detection time, detection rate, false positive, and false negative.

In Ref. [106], the authors designed an IDS in SCADA system for identifying bias injection intrusions by combining modified GWO (MGWO), entropy extreme learning machine (EELM), and hybrid elliptical curve cryptography (HECC). The MGWO is first employed to extract the variables required for prediction and to identify the optimal weight of the EELM. The EELM is then used to extort the variables and identify the intruded data with its intruded time, file location, and date. Lastly, the data are encrypted using the HECC to avert later attacks. Simulations' outcomes depict good accuracy for both prevention and discovery of attacks.

In another study, Benisha and Ratna [107] presented a data reliability-based IDS combining GWO, deviation forest (d-forest) classifier, and black forest (BF) algorithm. In this procedure, the d-forest classifier is used to remove the barriers, the GWO is used for selection ratio optimization, and the BF is used to obtain the best training data. Furthermore, GWO and BF are repeatedly used to select the optimum features. Lastly, the IDS is constructed using BF based on the best features and training data obtained. Experimental outcomes were analyzed using the UNSW-NB15 dataset with a sampling ratio of 22%. The performance outcomes depict a greater detection level with small false-positive level.

Vatambeti et al. [108] proposed an approach for identifying and detecting gray hole and black hole intrusions in mobile ad hoc networks (MANETs) using GWO. To accomplish this, the authors present the GWO with trust arrangement data combination called gray wolf trust combination in wireless ad hoc network structure. Using the strategy of GWO, the proposed approach was able to identify the gray hole and black hole intrusions at a rate of 98.5% and packet delivery rate of 98.2% with 10m/s.

Mukaram et al. [109] proposed a GWOSVM-IDS intrusion detection model for WSNs, which combines modified binary GWO with SVM. The proposed method is intended to improve intrusion detection accuracy, detection rate, and processing time in the WSN environment. The KDD Cup 99 dataset was used to demonstrate the effectiveness of the proposed GWOSVM-IDS. According to the experimental results, the GWOSVM-IDS has the best performance in terms of feature generation, accuracy, false alarm rate, detection rate, and execution time.

Jawhar and Alelah [110] used a GWO algorithm to develop a model for discovering DoS intrusion. The GWO is used in identifying intrusions by picking the data from the packet of network. After examination, it is passed into the wolf algorithm to detect whether it is an attack. The authors declared that the results obtained through rigorous analysis were acceptable and good.

Safaldin et al. [111] presented an IDS in a WSN combining improved binary GWO and SVM, termed as GWOSVM-IDS. This method uses three, five, and seven wolves to find the optimum number of wolves. The NSL KDD dataset was used to prove the effectiveness of this approach and compare it with PSO. The outcomes of this study increased accuracy and recognition level rate and decreased execution time in the WSN setting by cutting the false alarms rate and the number of attributes from the IDSs in the WSN setting.

Swarna et al. [112] used a combination of GWO, PCA, one-hot encoding, and DNN for developing IDS in IoMT architecture. The categorical attributes are converted into numerical values using one-hot encoding. The GWO and the PCA are employed to reduce the number of features in the dataset. Finally, the DNN is used to design an efficient and effective IDS in the IoMT setting to categorize and predict unexpected cyberattacks. Through experimental analysis on the NSL-KDD dataset, it was established that the proposed DNN model achieves better performance than KNN, RF, SVM, and NB.

A study by Singh and Mahajan [113] proposed a new advanced GWO called AGWO to discover cyber intrusions and avert their spread to evade breakdown in the smart grid substation. In the proposed approach, each node of the wireless sensor is displayed based on graph theory. The proposed approach plays an important role in identifying the doubtful nodes in less time so that cyber breakdown and data damages will decrease. This method takes nodes as wolves and groups them using trust values during cyberattacks. The nodes with trust weight less than the set value will be seen as malevolent nodes and separated from the network. This safeguards the network against the influence of cyberattacks. The experimental analysis performed on smart grid substation with 100 sensors reveals that the AGWO has superior performance in comparison with GA, PSO, and firefly algorithm (FA) with respect to faster discovery and smaller data damage.

Amaran and Mohan [114] created a novel IDS for grouped WSN using enhanced GWO and fuzzy SVM (FSVM) classifier. The enhanced GWO proposed by including differential evolution (DE) for population initialization is used to selects the optimum attributes subset of features. Then, the FSVM classifies the data cases into either normal or attack. The effectiveness of the proposed approach was tested using the KDD Cup 99 intrusion dataset and compared with HIDS, SVM, ELM, and MK-ELM. The simulation results show that this approach has better results compared to other methods.

Wang et al. [115] presented a network intrusion detection model through combination of enhanced GWO (EGWO), DBN, and kernel extreme learning machine (KELM). The EGWO is designed with optimization strategy combining the inner and outer hunting mechanisms to optimize the parameters of the KELM. The DBN is used to develop the IDS trained with the optimized KELM to improve the IDS performance. Simulations using CICIDS2017, KDD Cup 99, UNSW-NB15, and NSL-KDD intrusion detection standard datasets demonstrate that the presented DBN-EGWO-KELM approach achieved greater leads with respect to accuracy, true positive level, precision, and false positive level in contrast with RBF, BP, SVM, LIBSVM, KELM, DBN-KELM, and CNN.

In another study by Shakya [116], a modified gray wolf optimizer (MGWO) with SVM was used for feature selection and classification for developing IDS using the NSL-KDD dataset in a WSN. Here, the best number of wolves is obtained by running analysis with several wolves in the system. The results indicate that the performance of the proposed model increases as the number of wolves increases.

Sharma and Tyagi [117] presented a hybrid model that combined GWO and ANN. The GWO is used to finetune the ANN's weight parameter. The proposed approach was validated in terms of various performance indicators such as recall, accuracy, F1-score, and precision using the MIT DARPA 1998 intrusion detection dataset.

4.4 Grasshopper optimization algorithm (GOA)

Dwivedi et al. [118] implemented an adaptive system for anomaly detection by hybridizing the adaptive grasshopper optimization algorithm (AGOA) and ensemble of feature selection (EFS) techniques, termed EFSAGOA. In the proposed system, the EFS is employed to rank the variable for choosing the high-rated subset of variables. The AGOA is then applied to find the most significant variables from the dataset reduced by the EFS, which can assist to identify the activities of the network traffic. Moreover, the AGOA makes use of an SVM classifier as a learning algorithm to pick the very capable variables to enhance the prediction efficiency. The AGA is also used to adjust the SVM parameters such as penalty factor, tube size, and kernel parameter. The proposed approach was tested and evaluated using the ISCX 2012 dataset in terms of false alarm rate, detection level, and accuracy.

In another study, Vardhan and Tripathi [119] used GOA with a GOIDS classifier for feature selection to design an IDS for DoS intrusion detection. These selected features are then used with other classifiers such as SVM, DT, MLP, and NB to discover various intrusion types. Experimental analysis using the KDD Cup 99 and CICIDS 2017 standard datasets reveals that DT achieved better accuracy and detection and lower false positive level.

Hosseiny et al. [120] suggested an enhanced IDS based on GOA and DT classifier. The GOA algorithm is used here to remove attributes that are not important in the dataset. Based on the remaining attributes, the DT classifier is employed to identify different network traffic as either normal or malicious. The results of experimental analysis using the NSL-KDD dataset show that this approach achieved better precision, accuracy, and speed of intrusion discovery in contrast with ACO, PSO, and GA.

A study by Moghanian et al. [121] presented an IDS using GOA and MLP. The aim of the GOA is to reduce the NN detection error by choosing the most important parameters like bias and weights of the network. The proposed approach was simulated using KDD Cup 99 and UNSW network traffic datasets, and the results indicate that this method performed better than XGBoost, RF, and embedded training of ANN with BWO, BOA, and HHO in anomaly discovery.

A new feature selection approach combining EFS and chaotic adaptive grasshopper optimization algorithm (CAGOA) techniques, termed ECAGOA for IDS, was proposed in Ref. [122]. The CAGOA is used to adjust parameters such as kernel, tube size, and punishment factor of the SVM algorithm. The proposed approach was validated on standard intrusion detection datasets such ISCX 2012, CIC-IDS2017, and NSL-KDD. The results demonstrate the capability of this approach over other methods with respect to discovery accuracy and false alarm rate.

4.5 Artificial bee colony (ABC)

Ghanem and Jantan [123] proposed a new method using multi-objective ABC for variable selection in intrusion detection. This method was achieved by generating the variable subsets of the Pareto front of nondominated solutions and then combining ABC, PSO, and feed-forward neural network (FFNN) as learning methods to evaluate the variable subsets. Moreover, the FFNN optimized with a hybrid of ABC and PSO algorithms is used to classify the reduced dataset to determine an intrusion packet and identify known and unknown attacks.

Rais and Mehmood [124] presented a dynamic ABC model with three-level update feature selection for intrusion detection. The presented method uses a different level of pheromones that help ants to find robust features. The approach also uses the information of the individual ant in the feature selection process and integrates the accuracy of the ML algorithm. Results obtained show that the proposed feature selection approach performed better than other methods.

Mazini et al. [125] combined ABC with AdaBoost algorithms for a reliable intrusion detection system. The ABC is used to select the most relevant attributes for detecting attacks in network traffic and AdaBoost is used as a learning algorithm to validate the attributes. The simulation outcomes based on ISCXIDS2012 and NSL-KDD datasets affirm the reliability of this approach in detecting different kinds of attacks.

Yang et al. [126] proposed a modified naïve Bayes (MNB) classifier based on ABC, called ABCMNB. The ABC is used in the MNB to optimize its feature weight. In this approach, MNB gives greater weight to individuals that have greater impact on the results and lesser weights to individuals with lesser impact. The prediction accuracy of MNB is used as the fitness value of ABC. The proposed method was evaluated on the NSL-KDD dataset and compared with GAMNB, GWOMNB, and WWOMNB. The results show that ABCMNB successfully increased the network intrusion detection rate compared to other methods.

Elsaid and Albatati [127] presented an optimized collaborative IDS (OCIDS). The proposed system uses an enhanced ABC to improve the hierarchical IDS applied to WSNs with respect to the detection accuracy and the consumption complexity. Moreover, the ABC is used to optimize the weighted SVM to increase the detection accuracy and diminish false alarm rates. The proposed method was analyzed and validated to verify its efficacy and robustness using the NSL-KDD dataset. The results indicate the superiority of this method in comparison with the classical ACO, back propagation (BP), and the original OCIDS.

Zhang and Song [128] developed an enhanced network IDS based on NN and ABC. The ABC is used to determine the optimal network weights and thresholds to improve the network's self-learning capability and convergence time. However, the proposed work has not presented any empirical result to buttress its performance.

ABC was used in Ref. [129] together with RF for attribute selection and classification, respectively, in intrusion detection. This approach was evaluated using NSL KDD and UNSW-NB15 datasets, and the results show that the proposed method could select good candidates attribute subset from both datasets using 80.83% and 88.17% accuracy respectively.

Al-Safi et al. [130] utilized ABC for feature selection as adjusting parameters of an SVM for anomaly intrusion detection. Efficiency of the proposed approach was measured using the NSLKDD dataset. Simulation depicts that this approach outperforms other methods with high accuracy.

4.6 Ant colony optimization (ACO)

Abadeh and Habibi [131] proposed a hybrid intrusion detection approach combining ACO and evolutionary fuzzy system to produce good-quality fuzzy rules for classification. This hybrid method was applied to DARPA and KDD Cup 99 network traffic datasets. In contrast with NB, C4.5, KNN, Ripper, SVM, MOGF-IDS, and PNrule, the proposed method obtained better detection accuracy.

Li et al. [132] developed a network intrusion detection system to identify network packets into normal or malicious combining ACO, clustering method, and SVM classifier. Experimental analysis with the KDD Cup 99 dataset reveals that the proposed approach achieved 98.6% accuracy and average Mathew's correlation coefficient (MCC) of 0.861 using 10-fold cross validation.

Mehmood and Rais [133] applied ACO for attribute selection and SVM for intrusion detection. The proposed work uses ant scheme, a variant of ACO, to screen out the redundant and irrelevant attributes for SVM classification. Experimental analysis conducted using the KDD Cup 99 dataset demonstrated the superiority of the SVM trained with the selected subset.

Ravi et al. [134] implemented a variable selection approach using ACO and relative fuzzy entropy for real-time network attack detection. The implemented method was applied to a real-time network traffic dataset and showed an encouraging outcome.

Feature selection using ACO for intrusion detection system was also proposed in Ref. [135]. The ACO is used in this case to extract the most important variable subset from the KDD Cup 99 and NSL-KDD datasets. Simulation results confirmed the effectiveness of the ACO compared to other methods, achieving good accuracy and lower false alarm rates with smaller number of features. In Ref. [136], the authors proposed ant colony-induced DTs for anomaly detection. This method was tested and evaluated using the NSL-KDD dataset.

Alsaadi et al. [137] applied an ACO algorithm and PSO to generate a subset of most important attributes for improving IDS. Based on the features generated, this study employed three classifiers: SVM, NB, and KNN. The NSL-KDD and KDD Cup datasets were used to evaluate the proposed system. The empirical output reveals that performance improved with the use of the ACO.

Penmatsa et al. [138] proposed a method for determining the minimum variable subset for malware discovery using a rough set dependent variable impact combined with ACO. A dataset of malware called claMP was used to evaluate the efficacy of the proposed method with both integrated and raw variables. The empirical outcomes verify that 97.15% and 92.8% dataset size optimization was obtained with a minimum loss of accuracy for clamp-integrated and raw datasets, respectively.

In Ref. [139], an assessment of an efficient DoS attack detection and prevention model was presented using ACO and an SVM classifier for WSN. The ACO is applied here to select the optimal path between the complete adjacent possible paths from the source to receiver node for data communication, and the SVM is used for designing the IDS. The proposed method improves the lifespan of the network and detects attack nodes in a WSN.

Miryahyaie et al. [140] proposed a method for IDS by combining ACO and fuzzy association rules. ACO is implemented on a graph in which every node offers seven factors. The starting fuzzy rules are established based on the values in the most populated town in the graph associated to ACO using the

Apriori algorithm. To improve the correctness of attack discovery in networks, the key rules are adjusted using a method known as conventional weighted aggregation. The enhanced rules' efficacy is assessed via DT algorithm using the NSL-KDD standard dataset. The assessments show that this approach achieves better discovery accuracy and further decreases threat of error in contrast with other techniques.

Zhang [141] proposed an IDS by combining ACO and NN. In the proposed IDS, the ACO is applied to examine the best solution of the fitness function and find the optimum neural network parameters. Then, the IDS is constructed using NN self-organizing. Through experimental analysis, the results obtained demonstrate that the system can handle the issue of NN parameter optimization in network intrusion detection.

4.7 Differential evolution (DE)

Victoire and Sakthivel [142] presented a refined DE (RDE) search technique to produce fuzzy rules proficient at discovering intrusive activities. In the proposed approach, the universal population is split into subpopulations, each allocated to a different CPU and comprising the similar class of fuzzy procedures. A series of investigational outcomes using the KDD Cup 99 dataset affirm that the recommended method is more efficient and powerful than other methods presented in the literature.

Salek et al. [143] used the DE algorithm to optimize and train an NN for the development of an IDS. For experimental analysis, the proposed study used the standard KDD Cup dataset. The results obtained using the DE algorithm was contrasted with other algorithms such as probabilistic network, RBF, and MLP. In addition, this study also applied PCA for dimensionality reduction. The results of this study demonstrate greater detection accuracy in contrast to compared methods.

Popoola and Adewumi [144] presented a discretized DE and C4.5 classifier-based dimensionality reduction scheme for developing IDS using the NSL-KDD dataset. The outcome achieved indicates a substantial enhancement in discovery accuracy and a decrease in training and testing time based on the reduced feature set.

Ku and Zheng [145] proposed an enhanced learning approach called self-adaptive differential evolution extreme learning machine combined with Gaussian kernel (SaDE-KELM) for predicting and identifying invasions in a computer network. A comparative analysis was conducted with techniques such as DE-ELM and ELM and the experimental outcome reveals that the presented approach achieves better attack discovery accuracy.

Guo et al. [146] presented an IDS in ESN based on a DE constraint multi-objective optimization problem centered on negative selection algorithm, coded as DE-CMOP-based NSA, to enhance the efficiency and distribution of the detector. In the proposed technique, employing the DE algorithm was able to reduce the black hole attack successfully.

Cui et al. [147] combined a DE algorithm with a BP neural network (BPNN) for designing an IDS. By this combination, the thresholds and the weights of the BPNN are optimized so that the learning ability and difference of the BPNN and the global search benefit of the DE algorithm are totally utilized.

A DE wrapper-based attribute selection for IDS was presented in Ref. [148]. The main notion is to choose the best important attributes from NSL-KDD standard datasets using the DE algorithm and ELM. The DE is employed as a search mechanism to search the attribute search space to select the best subset with the help of the ELM as the learning approach. The simulation results depict a good detection level with minimum false alarm rate in binary and multiclassification.

4.8 Firefly optimization algorithm (FFA)

Adaniya et al. [149] proposed a k-harmonic means (KHM) clustering algorithm, hybridized with an FFA for network capacity intrusion detection. The KHM computes a weighting function of each instance to estimate new centers. It avoids the problem of initial center initialization problem associated with most center-based clustering approaches and makes use of the search ability of FFA from local optima avoidance. The authors confirmed that the results obtained by this approach are acceptable with maximum true positive level and moderate false positive level.

Kaur et al. [150] presented a new PCA-FFA-based XGBoost network intrusion detection model using GPU. After data transformation, the researchers employed the hybrid PCA-FFA to reduce the dimensionality of the dataset. The classification of the attacks was then performed using the XGBoost classifier on the reduced dataset. To justify the capability of this method in detecting different types of attacks, an evaluation and comparison was conducted using KNN, RF, NB, and SVM. The results prove the superiority of this approach over others.

In another work, Kaur et al. [150] developed a novel approach for anomaly detection that is a hybridization of FFA and k-means clustering algorithm, coded as k-means+FFA. The proposed approach uses a clustering technique to build the training model and uses classification to evaluate the test data. The effectiveness of the new approach was assessed using the NSL-KDD standard dataset. Additionally, a contrast analysis was conducted between the new design model with k-means+CS, Canopy, farthest first, k-means++, and k-means+BA. The output confirms that k-mean+FFA and k-means +BA greatly outperform the other approaches.

Devi and Suganthe [151] proposed an intrusion detection system by combining the Kalman filtering approach for handling missing values present in the data, and Gaussian FFA (GFFA) to choose the most significant and optimum attributes. Then, they applied improved relevance vector machine (IRVM) to identify the attacks effectively by mining more appropriate vectors and consequently categorizing maximum likelihood values. Experimental analysis conducted using the NSL-KDD dataset concludes that the proposed approach achieves good performance with respect to accuracy, specificity, recall, and precision.

Najeeb and Dhannoon [152] presented a new feature method for intrusion detection using FFA. Furthermore, the suggested technique used the NB algorithm to distinguish between malicious and normal network activity in network packages. The differentiating attributes are chosen by the FFA from the NSL-KDD standard dataset. The authors conclude that the proposed approach improved the IDS's effectiveness in terms of attack detection when compared to binary PSO, NB, and binary BA.

Selvakumar and Muneeswaran [153] deployed both filter and wrapper feature selection approaches employing FFA in the wrapper approach as a search strategy for choosing the features. The resulting features using the KDD Cup 99 dataset were then subjected to Bayesian networks (BNs) and C4.5 classifiers for evaluation and subsequently for identifying various attacks. The experimental results demonstrate that 10 features selected by the FFA are enough to discover the attacks, displaying enhanced accuracy.

Krishna and Prakash [154] combined FFA with a multi-layered network for an effective IDS. The FFA is employed for both feature selection and training of the multilayered network. The most widely used NSL-KDD dataset was used to test and evaluate the proposed scheme. Results show that the proposed scheme was able to improve the detection capability of the IDS.

Al-Yaseen [155] developed an attribute selection scheme, combining FFA and an SVM algorithm. The FFA is modified to become more suitable for attribute selection in IDS. Furthermore, the SVM is applied as the learning classifier to validate the selected attribute and detect various kinds of network attacks. The experimental results based on the NSL-KDD dataset proved the efficiency of this method in improving network IDS.

Badran et al. [156] proposed a new hybrid IDS combining FFA and fast learning network (FLN), called FFA-FLN. The FFA is used to optimize the FLN for IDS. However, the authors do not provide any empirical evidence to confirm the effectiveness of the proposed model.

Gunay and Orman [157] proposed a modified FFA together with a KNN classifier for feature selection in an intrusion detection system. In a wrapper-based feature selection approach, the modified FFA is used as a search method and KNN is used as the learning classifier. Four different datasets were extracted from the KDD Cup 99 dataset to measure the performance of this method and the results obtained show that the FFA successfully reduced the dimensionality of the dataset.

Rajakumar [158] proposed an IDS model that exploits the FFA-motivated optimizer to choose the attributes. The model combines filter and wrapper-based methods to increase the optimizer method of the important feature subsets. Furthermore, BNs and SVM classifiers are used to validate the efficiency of the selected feature subset. The proposed scheme was validated against the KDD Cup 99 dataset. Simulation outcomes demonstrate that this scheme outperforms other approaches with respect to better detection accuracy, FPR, and F-score.

Alwan et al. [159] proposed an IDS using FFA for feature selection. Firstly, the FFA is improved using a mutation operator to avoid trapping into local optimum by enhancing the exploration ability of the original FFA. This improved FFA is then applied to select the most relevant attributes for identifying different kinds of attacks. The importance of these features extracted was assessed using the NB algorithm. The results obtained based on the NSL-KDD dataset proved the dominance of the improved FFA over the original FFA with respect to the number of features selected and accuracy of detection in diverse situations.

Vishnupriya et al. [160] implemented an FFA for dimensionality reduction in Hadoop environment for IDS. After the feature selection using the KDD Cup 99 dataset, the selected feature subset is then used to train and test the IDS model using ensemble of J48 and NB classifiers. The obtained results show that the ensemble classifier outperformed the single and parallel classifiers by obtaining the best attribute subset with an enhanced level of detection.

Ghosh et al. [161] presented an IDS using a modified FFA in a cloud computing setting. The authors designed a modified FFA to perform dimensionality reduction and to help the NSL-KDD dataset to consume a reduced amount of storage location by decreasing the dimensions as well as training time to achieve better detection accuracy.

Kaushik [162] author proposed an enhanced black hole detection and prevention scheme using an FFA and NN. The FFA is applied here to optimize the route of the nodes using fitness function to determine suspected nodes or normal nodes. The ANN is then employed to spot and discover black hole nodes in the network so that the transmitted data from source to receiver is protected.

Wang et al. [163] designed an effective DDoS intrusion identification model on an improved IFFA to adjust convolution neural networks (CNNs). The IFFA is utilized to adjust the weights and biases of CNN to provide detection for DDoS attacks in software-defined network IoT structure. The results from experimental analysis show that this approach achieved more than 99% DDoS activity and benign attack detection accuracy.

Tian et al. [164] proposed a two-stage scheme for designing IDS in software-defined IoT networks. In the proposed approach, an FFA is modified with a DE algorithm mutation operator strategy to improve its performance. Then, the modified FFA is used in a wrapper feature selection-based method to select the most valuable features for intrusion detection in a network traffic dataset. These selected attributes are passed to a new ensemble classifier comprised of instance-based learning, C4.5, and MLP. The proposed IDS discovery effectiveness was assessed in binary and multiclass classifications using the UNSW-NB15 and NSL-KDD public datasets. Simulation output confirms the superiority of the proposed scheme over some existing method.

4.9 Other metaheuristic algorithms

Apart from the aforementioned metaheuristic algorithms, other algorithms have also been applied for intrusion detection. Therefore, we present a survey and summary of these algorithms in Table 1. The metaheuristic algorithms identified include harmony search algorithm (HAS) [165], moth-flame optimization (MFO) [166], lion optimization algorithm (LOA) [167], whale optimization algorithm

Table 1 Other metaheuristic algorithms used for intrusion detection.

Reference	Algorithm	Methodology	Dataset	Problem solved
[165]	HSA			
[166]	MFO	Weighted KNN+MFO	KDD Cup 99	Optimized the weighted distance (K) of KNN
[167]	LOA	LOA+CNN	NSL-KDD	Attribute selection
[168]	WOA	WOA based clustering.	600 requests generated using DDOSIM simulator	DDoS detection using WOA based clustering
[169]	WOA	WOA+ANN	Mississippi State University and Oak Ridge National Laboratory databases of power system attacks	Initialized and optimized the ANN weight vector
[171]	ALO	Quantum ALO+K-means	KDD Cup 99	Improve k-means to converge to global optimum
[170]	IWOA	SVM+IWOA	KDD Cup 99	Optimized SVM parameters
[172]	CSO	RNN+CSO	KDD Cup 99	Feature selection
[175]	BA	CFS+BA + ensemble (C4.5, RF, Forest PA) algorithm	NSL-KDD, CIC-IDS2017, AWID	Feature selection
[173]	PIO	PIO+DT	KDD Cup 99, NSL-KDD, UNSW-NB15	Feature selection
[174]	SMO	SMO+DNN	NSL-KDD, KDD Cup 99	Dimensionality reduction

Continued

Table 1 Other metaheuristic algorithms used for intrusion detection—cont'd

Reference	Algorithm	Methodology	Dataset	Problem solved
[176]	CS	CS+ANN	UNSW-NB15	Parameter optimization
[177]	BOA	BOA + MLP	NSL-KDD	Feature selection
[178]	JAYA	ITLBO + SVM +IPJAYA	NSL-KDD, CICIDS2017	Update the penalty factor (C) and gamma parameter of SVM
[179]	WOA	RVM+WOA	NSL-KDD, CICIDS2017	Optimized the parameter of kernel
[180]	FOA	Fuzzy C-Means Rough Parameter (FCMRP) +Auto-encoder +BPN +FOA	NSL-KDD, UNSW-NB15	Optimized the population of the neurons in the hidden layers of deep auto-encoder
[181]	BOA	BOA + ANN	NSL-KDD	Optimal feature subset selection
[182]	TS	TS+RF	UNSW-NB15	Search method
[183]	CSA	CSA+OBL+RNN	KDD Cup 99	Dimensionality reduction
[184]	MOA			
[185]	FOA	FOA+CNN	NSL-KDD	Class imbalance handling and CNN parameter optimization
[186]	BAS	ANN+BAS	KDD Cup 99	Neural network training
[187]	CSSA	CSSA + ELM	NSL-KDD	Feature selection

(WOA) [168], [169], improved whale optimization algorithm (IWOA) [170], ant lion optimization (ALO) [171], crow search optimization (CSO) [172], pigeon inspired-optimizer (PIO) [173], and spider monkey optimization (SMO) [174].

Other research works presented for intrusion detection using metaheuristic algorithms include bat algorithm (BA) [175], CS [176], butterfly optimization algorithm (BOA) [177], JAYA algorithm [178], WOA [179], fruit fly optimization algorithm (FOA) [180], BOA [181], TS [182], CSA [183], FOA [185], chaotic salp swarm algorithm (CSSA) [187], and beetle antennae search (BAS) [186].

4.10 Hybrid metaheuristics algorithms

In this section, we present publications in which various metaheuristic algorithms were hybridized for intrusion detection. Fig. 2 presents the summary of this work and how the papers applied the hybrid algorithms.

Hybrid metaheuristic algorithms applied for developing IDS provided in the literature include GA with GWO [188], hybrid ABC and artificial fish swarm (AFS) [189], hybrid GWOCS optimization [190], GA and PSO [191], hybrid whale-genetic algorithm [192], hybrid optimization scheme (HOS) [193] combining adaptive ABC (AABC) with adaptive PSO (APSO), hybrid GA and GWO (GA-GWO) [194], and GWO with CSA [195].

Other hybrid metaheuristics include ABC and dragonfly algorithm (DA) [196], DE and GSA [197], ACO and GA [198], FFA and ALO [199], GB-EBGWO [200], PSO and GWO [201], hybrid modified grasshopper optimization algorithm and genetic algorithm (HMGOAGA) [202], hybrid PSO and GWO [203], PSO and GWO [181], GWO and PSO [204], and FFA and CS [184,205] (Table 2).

Table 2 Hybrid metaheuristic algorithms in intrusion detection.

Reference	Algorithms	Methodology	Dataset	Problem handle
[189]	ABC, AFS	ABC+AFS+FCM +CFS	NSL-KDD, UNSW-NB15	Feature selection
[190]	GWO, CS	HGWCSO+ETSV classifier	NSL-KDD	Dimensionality reduction
[191]	PSO, GA	GA+PSO+SVM	KDD Cup 99	Parameter optimization and feature selection
[192]	WOA, GA	WOA+GA	Not stated	As a classifier
[188]	GA, GWO	GA+GWO+RBF neural system	DAPRA 99	Adjust the weights and width of RBF neural system
[193]	AABC, APSO	AABC+APSO +SVM	NSL-KDD	Dimensionality reduction
[194]	GA, PSO	GA-GWO+SVM	AWID	Feature selection and parameter optimization
[195]	GA, CSA	GWO-CSA-DSAE	NSL-KDD, UNSW-NB15, CICIDS 2017	Attribute selection
[196]	ABC, DA	ABC+DA+ANN	KDD Cup 99, NSL-KDD, ISCX2012, NSW-NB15	Training NN to adjust its weights and biases
[199]	FFA, ALO	FFA-ALO +classifiers (DT, SVM, KNN, Naïve Bayes)	KDD Cup 99, NSLKDD, UNSW-NB15	Feature selection
[201]	PSO, GWO	PSOGWO + SVM	NSL-KDD	SVM parameter optimization
[200]	GA, GWO	GB-BGWO + ANF, AAP	NSL-KDD	Feature selection
[202]	MGOA, GA	HMGOAGA + RF	CICIDS 2017	Feature selection
[203]	PSO, GWO	PSO+GWO+RF	NSL-KDD	Dimensionality reduction
[181]	PSO, GWO	PSO+GWO+SVM	NSL-KDD, UNSW-NB15	SVM parameter optimization
[204]	GWO, PSO	GWO-PSO-RF	KDD Cup 99, NSL–KDD, CICIDS-2017	Feature extraction
[205]	PSO, MVO, GWO, MFO, WOA, FFA, BAT	Various combination of this algorithms + SV, RF, and C4.5	UNSW-NB15	Variable selection

Continued

Table 2 Hybrid metaheuristic algorithms in intrusion detection—cont'd

Reference	Algorithms	Methodology	Dataset	Problem handle
[184]	FFA, CS	FFA+CS	Implemented with different threshold values NS-2 simulator	To find the ideal route that protects the data from intruders and delivers it effectively
[197]	DE, GSA	DEGSA-HKELM	KDD Cup 99, UNSW-NB15	Parameter optimization

5. Challenges and future direction

Metaheuristic algorithms have been widely studied in recent years for developing reliable and effective IDS schemes. For the creation of IDS, the metaheuristic procedures were combined with several ML algorithms. Metaheuristic algorithms, for example, are combined with ML algorithms such as NN, XGBoost, SVM, RF, DT, NB, fuzzy logic, and k-means clustering to improve the effectiveness of the ML algorithms by adjusting the algorithm parameters. Dimensionality reduction is also accomplished using metaheuristic algorithms. These approaches can also distinguish between known and unknown actions to simply recognize intrusive patterns and predict the uncertainty and ambiguity of network activities, and thus enhance the capacity of an IDS. These methods are commonly used to maximize classifier parameters for attack classification, optimizing feature sets, and feature selection, according to the research presented in this chapter.

The applicability of metaheuristic algorithms has demonstrated that these approaches can be used in an ensemble or hybridized fashion to design an effective IDS with a high discovery rate and accuracy in real-world network settings. These approaches divide a massive problem into smaller ones by assigning entities to work in a collaborative, adaptive, and economical environment. Although metaheuristic algorithms have produced promising results for improving the effectiveness of IDS, there are challenges that require further research in this area.

1. Hybridization of ML algorithms with metaheuristic approaches is being abundantly used to accelerate the effectiveness of the algorithms. Yet, selecting proper parameters is a difficult action for accomplishing either parameter adjustment or dimensionality reduction.
2. Effectiveness of the ML algorithms rests on the magnitude and value of attributes used for detection and classification of attacks. Hence, creating the association between the network flow attributes and attack types is a difficult task. The time complexity increases with an upsurge in the amount of network attributes. However, if fewer number of attributes are selected, then there is a chance of losing important network information. Henceforth, it is essential to research the role of network attributes in the effectiveness of IDS.
3. Another difficult phase of intrusion detection is flexibility to continuously varying network situations. The network setting comprises intrusion events that continue to vary over time. In addition, the network scheme and user activities change over time. The IDS should have the ability to adjust to these changes to maintain the discovery accuracy of the model.

Based on these identified challenges, we recommend the following future work.

1. There is growth in end-to-end encryption in network communication, thus research handling encrypted network traffic should be conducted. The approaches should be able to mine network information from the encrypted packet for investigation.
2. An intrusion detection dataset is comprised of flow-based and packet-based network variables. Therefore, an assessment of the effectiveness of the approaches based on flow-based and packet-based network features should be carried out.
3. Furthermore, a dataset used for validating the effectiveness of an IDS is affected by the imbalance class problem. Therefore, metaheuristic-based approaches should be proposed to handle this problem.
4. Moreover, numerous novel metaheuristic optimization approaches have been suggested that can be exploited to handle intrusion detection. Novel hybrid and improved metaheuristic algorithms can also be proposed to handle intrusion detection problems.

6. Conclusion

The chapter conducted a literature survey on the application of metaheuristic algorithms in intrusion detection from 2010 to 2022. More than 200 research papers were extracted from various academic databases such as Google Scholar, MDPI, Taylor & Francis, Scopus, Web of Science, and IEEE. Based on the methodology employed, we identified 168 research papers relevant for our study, after eliminating those that lacked complete information and those written in languages other than English. The 168 papers selected were then analyzed and summarized. The results indicate that greater proportions of published papers on metaheuristic algorithms in intrusion detection discuss the following metaheuristic algorithms: GA, PSO, GWO, GOA, FFA, DE, and hybrid methods. Also, as per this study, these techniques have been widely used for optimizing the parameters of the ML classifiers for attack classification, adjusting feature set, and feature selection. Finally, we presented challenges and possible further future research directions on intrusion detection-based metaheuristic optimization algorithms.

References

[1] G. Stoneburner, Nist special publication 800-33: underlying technical models for information technology security, Gaithersburg, Estados Unidos de América: National Institute of Standards and Technology (NIST), 2001.

[2] K. Thakur, G. Kumar, Nature inspired techniques and applications in intrusion detection systems: recent progress and updated perspective, Arch. Comput. Meth. Eng. (2020) 1–23, https://doi.org/10.1007/s11831-020-09481-7.

[3] B. Dong, X. Wang, Comparison deep learning method to traditional methods for network intrusion detection, in: IEEE International Conference on Communication Software and Networks (ICCSN), 2016, pp. 581–585.

[4] X. Yao, The realisation of goal-driven airport enclosures intrusion alarm system, Int. J. Grid Util. Comput. (2017) 1–6.

[5] J.G. Femandes, L.E. Cavalho, J.J. Rodrigues, J.M. Proenca, Network anomaly detection using IP flows with principal component analysis and ant colony optimization, J. Netw. Comput. Appl. (2016) 1–11, https://doi.org/10.1016/J.JNCA.2015.11.

[6] H.S. Bamakan, H. Wang, T. Yingjie, Y. Shi, An effective intrusion dtection framework based on MCLP/SVM optimized by time varying chaos particle swarm optimization, Neurocomputing (2016) 90–102, https://doi.org/10.1016/J.neucom.2016.03.031.

[7] A. Rajasekhar, N. Lynnb, S. Das, P.N. Suganthan, Computing with collective intelligence of honey bees-a survey, Swarm Evol. Comput. (2017) 25–48, https://doi.org/10.1016/J.SWEVO.2016.06.001.

[8] Y.Y. Chung, N. Wahid, A hybrid network intrusion detection system using simplified swarm optimization (SSO), Appl. Soft Comput. (2012), https://doi.org/10.1016/j.asoc.2012.04.020.

[9] B.M. Aslahi-shahri, R. Rahmani, M. Chizari, A. Maralani, A. Eslami, M.J. Golkar, A. Ebrahami, A hybrid method consisting of GA and SVM for intrusion detection system, Neural Comput. Appl. (2016) 1669–1676, https://doi.org/10.1007/s00521-015-1964-2.

[10] A.H. Hamamoto, L.F. Carvalho, L. Sampaio, T. Abrao, M.L. Proenca, Network anomaly detection system using genetic algorithm and fuzzy logic, Expert Syst. Appl. (2017) 390–402, https://doi.org/10.1016/j-eswa.2017.09.013.

[11] A.A. Aburommman, M. Bin, I. Reaz, M.B. Reaz, A novel SVM-KNN-PSO ensemble method for intrusion detection system, Appl. Soft Comput. (2016) 360–372, https://doi.org/10.1016/j.asoc.2015.10.011.

[12] E. Nabil, A modified flower pollination algorithm for global optimization, Expert Syst. Appl. (2016) 192–203.

[13] G.G. Wang, Moth search algorithm: a bio-inspired metaheuristic algorithm for global optimization problems, Memetic Comput. (2018) 151–164.

[14] A. Colorni, M. Dorigo, V. Maniezzo, Distributed optimization by ant colonies, in: Proceedings of the first European Conference on Artificial Life, 1992, pp. 134–142. Cambridge.

[15] R. Eberhart, J. Kennedy, A new optimizer using particle swarm theory, in: Proceedings of the Sixth International Symposium on MHS'95, IEEE, 1995, pp. 39–43.

[16] C. Blum, A. Roli, Metaheuristics in combinatorial optimization: overviewand conceptual comparison, ACM Comput. Surv. (2003) 268–308.

[17] H. Faris, I. Aljarah, M.A. Al-Betar, S. Mirjalili, Grey wolf optimizer: a review of recent variants and application, Neural Comput. & Applic. (2017) 1–23, https://doi.org/10.1007/s00521-017-3272-5.

[18] Holland., Adaptation in Natural and Artificial Systems: An IntroductoryAnalysis with Applications to Biology, Control, and Artificial Intelligence, U Michigan Press, Michigan, 1975.

[19] J.R. Koza, Genetic Programming II: Automatic Discovery of Reusable Subprograms, MIT Press, Cambridge, 1994.

[20] N. Hansen, S. Kern, Evaluating the CMA evolution strategy on multimodal test functions, in: International Conference on Parallel Problem Solving from Nature, Springer, 2004, pp. 282–291.

[21] Z.W. Geem, J.H. Kim, G. Loganathan, A new heuristic optimizationalgorithm: harmony search, Simulation (2001) 60–68.

[22] R. Storn, K. Price, Differential evolution—a simple and efficient heuristic for global optimization over continuous spaces, J. Glob. Optim. (1997) 341–359.

[23] M. Hohfeld, G. Rudolph, Towards a theory of population based incremental learning, in: Proceedings of the 4th IEEE Conference on Evolutionary Computation, Citeseer, 1997.

[24] D. Simon, Biogeography-based optimization, IEEE Trans. Eval. Comput. (2008) 702–713.

[25] J. Kennedy, R. Eberhart, Particle swarm optimization, in: Proceeding of IEEE International Conference on Neural Network, 1995, pp. 1942–1948, https://doi.org/10.1109/ICNN.1995.48896 8.

[26] A.H. Gandomi, A.H. Alavi, Krill herd: a new bio-inspired Krill herd: a new bio-inspired, Commun. Nonlinear Sci. Numer. Simul. (2012) 4831–4845.

[27] S. Mirjalili, Moth-flame optimization algorithm: a novel nature-inspired heuristic paradigm, Knowl.-Based Syst. (2015) 228–249, https://doi.org/10.1016/j.knosy s.2015.07.006.

[28] S. Mirjalili, A. Lewis, The whale optimization algorithm, Adv. Eng. Softw. (2016) 51–67.
[29] S. Mirjalili, S.M. Mirjalili, A. Hatamlou, Multi-verse optimizer: a nature-inspired algorithm for global optimization, Neural Comput. Applic. (2016) 495–513.
[30] S. Mirjalili, Dragonfly algorithm: a new meta-heuristic optimization technique for solving single-objective, discrete, and multi-objective problems, Neural Comput. Applic. (2016) 51–67.
[31] S. Mirjalili, A.H. Gandomi, S.Z. Mirjalili, S. Saremi, H. Faris, S.M. Mirjalili, Salp swarm algorithm: a bio-inspired optimizer for engineering design problems, Adv. Eng. Softw. (2017) 163–191.
[32] W. Zhao, Z. Zhang, L. Wang, Manta ray foraging optimization: an effective bio-inspired optimizer for engineering applications, Eng. Appl. Artif. Intell. (2020) 103300.
[33] A. Faramarzi, M. Heidarinejad, S. Mirjalili, A.H. Gandomi, Marine predators algorithm: a nature-nspired metaheuristic, Expert Syst. Appl. (2020) 113377, https://doi.org/10.1016/j.eswa.2020.113377.
[34] F. Glover, Heuristics for integer programming using surrogate constraints, Decis. Sci. (1977) 156–166.
[35] S. Kirkpatrick, C. Gelatt, M.P. Vecchi, Optimization by simulated annealing, Science (1983) 671–680.
[36] S. Koziel, X.S. Yang, Computational Optimization, Methods and Algorithms, Springer, 2011.
[37] M.A. Al-Betar, B-hill climbing: an exploratory local search, Neural Comput. Applic. (2016) 1–16.
[38] P. Nguyen, J.M. Kim, Adaptive ecg denoising using genetic algorithm-based thresholding and ensemble empirical mode decomposition, Inf. Sci. (2016) 499–511.
[39] A. Noshadi, J. Shi, W.S. Lee, P. Shi, A. Kalam, Optimal pidtype fuzzy logic controller for a multi-input multi-output active magnetic bearing system, Neural Comput. Applic. (2016) 2031–2046.
[40] R. Arnay, F. Fumero, J. Sigut, Ant colony optimizationbased method for optic cup segmentation in retinal images, Appl. Soft Comput. (2017) 409–417.
[41] B. Karasulu, S. Korukoglu, A simulated annealing-based optimal threshold determining method in edge-based segmentation of grayscale images, Appl. Soft Comput. (2011) 2246–2259.
[42] H. Rashaideh, A. Sawaie, M.A. Al-Betar, L.M. Abualigah, M.M. Al-Laham, M. Ra'ed, M. Braik, A grey wolf optimizer for text document clustering, J. Intell. System (2018) 814–830.
[43] S.G. Devi, M. Sabrigiriraj, A hybrid multi-objective firefly and simulated annealing based algorithm for big data classificatio, Concurr. Comput. Pract. Exp. 13 (14) (2019) e4985, https://doi.org/10.1002/cpe.4985.
[44] M.R. Mosavi, M. Khishe, M.J. Naseri, G.R. Parvizi, A.A. Mehdi, Multi-layer perceptron neural network utilizing adaptive best-mass gravitational search algorithm to classify sonar dataset, Arch. Acoust. (2019) 137–151.
[45] Y. Ye, J. Li, K. Li, F. Hui, Cross-docking truck scheduling with product unloading/loading constraints based on an improved particle swarm optimisation algorithm, Int. J. Prod. Res. (2018) 5365–5385.
[46] A.A. Zaidan, A.M. Bayda, A. Bakar, B.B. Zaidan, A new hybrid algorithm of simulated annealing and simplex downhill for solving multiple-objective aggregate production planning on fuzzy environment, Neural Comput. Applic. (2019) 1823–1834.
[47] M. Yazdani, F. Jolai, Lion optimization algorithm (loa): a nature-inspired metaheuristic algorithm, J. Comput. Des. Eng. (2016) 24–36.
[48] M.S. Hoque, M.A. Mukit, M.A. Bikas, An implementation of intrusion detection, Int. J. Netw. Secur. Appl. (2012) 109–120, https://doi.org/10.5121/ijnsa.2012.4208.
[49] S. Dhopte, N.Z. Tarapore, Design of intrusion detection system using fuzzy class-association rule mining based on genetic algorithm, Int. J. Comput. Appl. (2012) 20–27.
[50] S.S. Sindhu, S. Geetha, A. Kannan, Decision tree based light weight intrusion detection using a wrapper approach, Expert Syst. Appl. (2012) 129–141, https://doi.org/10.1016/j.eswa.2011.06.013.
[51] T.F. Ghanem, W.S. Elkilani, H.M. Abdul-kader, A hybrid approach for efficient anomaly detection using metaheuristic methods, J. Adv. Res. (2014) 609–619, https://doi.org/10.1016/j.jare.2014.02.009.
[52] F. Kuang, W. Xu, S. Zhang, A novel hybrid KPCA and SVM with GA model for intrusion detection, Appl. Soft Comput. (2014) 178–184, https://doi.org/10.1016/j.asoc.2014.01.028.
[53] M.K. Desale, M.R. Ade, Genetic algorithm based feature selection approach for effective intrusion detection system, in: International Conference on Computer Communication and Informatics (ICCCI-2015), IEEE, Coimbatore, India, 2015, pp. 1–6. 978-1-4799-6805-3/15/$31.00.

[54] D. Narsingyani, O. Kale, Optimizing false positive in anomaly based intrusion detection using genetic algorithm, in: 2015 IEEE 3rd International Conference on MOOCs, Innovation and Technology in Education (MITE), IEEE, 2015, pp. 72–77. 978-1-4673-6747-9.
[55] A.H. Hamamoto, L.F. Carvalho, ACO and GA Metaheuristics for Anomaly Detection, IEEE, 2015, pp. 1–6.
[56] M.K. Patel, M.P. Sharma, An implementation of intrusion detection system based on genetic algorithm, Int. J. Adv. Res. Comput. Commun. Eng. (2016) 509–514, https://doi.org/10.17148/IJARCCE.2016.511109.
[57] E.-S.M. El-Alfy, M.A. Alshammari, Towards scalable rough set based attribute subset selection for intrusion detection using parallel genetic algorithm in MapReduce, Simul. Model. Pract. Theory (2016) 1–12.
[58] P. Tao, Z. Sun, Z. Sun, An improved intrusion detection algorithm based on GA and SVM, IEEE Access (2018) 1–8, https://doi.org/10.1109/ACCESS.2018.2810198.
[59] P.A. Resende, A.C. Drummond, Adaptive anomaly-based intrusion detection system using genetic algorithm and profiling, Secur. Priv. (2018) 1–13, https://doi.org/10.1002/spy2.36.
[60] J.T. Hounsou, T. Nsabimana, J. Degila, Detection system using soft computing algorithms (self organizing feature map and genetic algorithm), J. Inf. Secur. (2019) 1–24, https://doi.org/10.4236/jis.2019.101001.
[61] M. Moukhafi, K.E. Yassini, S. Bri, K. Oufaska, Artificial Neural Network Optimized by Genetic Algorithm for Intrusion Detection System, Springer Nature, 2019, pp. 393–404, https://doi.org/10.1007/978-3-030-11928-7_35.
[62] S.S. Reddy, P. Chatterjee, C. Mamatha, Intrusion Detection in Wireless Network Using Fuzzy Logic Implemented with Genetic Algorithm, Springer Nature, 2019, pp. 425–432, https://doi.org/10.1007/978-981-13-7150-9_45.
[63] J. Ren, J. Guo, W. Qian, H. Yuan, X. Hao, H. Jingjing, Building an effective intrusion detection system by using hybrid data optimization based on machine learning algorithms, Secur. Commun. Netw. (2019) 1–11, https://doi.org/10.1155/2019/7130868.
[64] H. Suhaimi, S.I. Suliman, I. Musirin, A.F. Harun, R. Mohamad, Network intrusion detection system by using genetic algorithm, Indones. J. Electr. Eng. Comput. Sci. (2019) 1593–1599, https://doi.org/10.11591/ijeecs.v16.i3.pp1593-1599.
[65] Z. Chiba, N. Abghour, K. Moussaid, A.E. Omri, M. Rida, New anomaly network intrusion detection system in cloud environment based on optimized back propagation neural network using improved genetic algorithm, Int. J. Commun. Netw. Inf. Secur. (2019) 61–84.
[66] Y. Zhang, P. Li, X. Wang, Intrusion detection for IoT based on improved genetic algorithm and deep belief network, IEEE Access (2019) 31711–33172, https://doi.org/10.1109/ACCESS.2019.2903723.
[67] H. Suhaimi, S.I. Suliman, I. Musirin, A. Harun, R. Mohamad, M. Kassim, S. Shahbudin, Network intrusion detection system using immune-genetic algorithm (IGA), Indones. J. Electr. Eng. Comput. Sci. (2020) 1060–1065, https://doi.org/10.11591/ijeecs.v17.i2.pp1060-1065.
[68] V. Jain, M. Agrawal, Applying genetic algorithm in intrusion detection system of iot applications, in: Proceedings of the Fourth International Conference on Trends in Electronics and Informatics (ICOEI 2020), IEEE Xplore, 2020, pp. 284–287.
[69] M.T. Nguyen, K. Kim, Genetic convolutional neural network for intrusion detection systems, Futur. Gener. Comput. Syst. (2020) 1–10, https://doi.org/10.1016/j.future.2020.07.042.
[70] A. Rai, Optimizing a new intrusion detection system using ensemble methods and deep neural network, in: Proceedings of the Fourth International Conference on Trends in Electronics and Informatics (ICOEI 2020), IEEE, 2020, pp. 527–532.
[71] M.A. Fauzi, A.T. Hanuranto, Intrusion detection system using genetic algorithm and K-NN algorithm on dos attack, in: 2020 2nd International Conference on Cybernetics and Intelligent System (ICORIS), IEEE, 2020, pp. 1–6, https://doi.org/10.1109/ICORIS50180.2020.9320822.
[72] J.O. Onah, S.M. Abdulhamid, A. Mohammed, I.H. Hassan, A. Al-Ghusham, Genetic algorithm based feature selection and Naïve Bayes for anomaly detection in fog computing environment, Mach. Learn. Appl. (2021) 100156, https://doi.org/10.1016/j.mlwa.2021.100156.

[73] S. Al-Saqqa, M. Al-Fayoumi, M. Qasaimeh, Intrusion detection system for malicious traffic using evolutionary search algorithm, Recent Adv. Comput. Sci. Commun. (2021) 1381–1389.

[74] D. Mehanovic, D. Kec˘o, J. Kevric, S. Jukic, A. Miljkovic, Z. Mas˘etic, Feature selection using cloud-based parallel genetic algorithm for intrusion detection data classification, Neural Comput. Applic. (2021) 1–13, https://doi.org/10.1007/s00521-021-05871-5.

[75] M.T. Tally, H. Amintoosi, A hybrid method of genetic algorithm and support vector machine for intrusion detection, Int. J. Electr. Comput. Eng. (2021) 900–908, https://doi.org/10.11591/ijece.v11i1.pp900-908.

[76] Z. Halim, M.N. Yousaf, M. Waqas, M. Sulaiman, A. Ghulam, M. Hussain, et al., An effective genetic algorithm-based feature selection method for intrusion detection systems, Comput. Secur. (2021) 102448, https://doi.org/10.1016/j.cose.2021.102448.

[77] A.J. Malik, W. Shahzad, F.A. Khan, Binary PSO and Random Forests Algorithm for PROBE Attacks Detection in a Network, IEEE, 2011, pp. 662–668.

[78] Z. Li, L. Yongzhong, L. Xu, Anomaly intrusion detection method based on K-means clustering algorithm, in: International Conference of Information Technology, Computer Engineering and Management Sciences, IEEE, 2011, pp. 157–161, https://doi.org/10.1109/ICM.2011.184.

[79] A. Karami, M. Guerrero-Zapata, A fuzzy anomaly detection system based on hybrid PSO-Kmeans algorithmincontent-centricnetworks, Neurocomputing (2014) 1–17, https://doi.org/10.1016/j.neucom.2014.08.070.

[80] F. Kuang, S. Zhang, Z. Jin, W. Xu, A novel SVM by combining kernel principal component analysis and improved chaotic particle swarm optimization for intrusion detection, Soft Comput. (2014) 1–13, https://doi.org/10.1007/s00500-014-1332-7.

[81] V. Manekar, K. Waghmare, Intrusion detection system using support vector machine (SVM) and particle swarm optimization (PSO), Int. J. Adv. Comput. Res. (2014) 808–812.

[82] I. Idris, A. Selamat, N.T. Nguyen, S. Omatu, O. Krejcar, K. Kuca, M. Penhaker, A combinednegative selectionalgorithm–particle swarmoptimization for an email spam detection system, Eng. Appl. Artif. Intell. (2015) 33–44, https://doi.org/10.1016/j.engappai.2014.11.001.

[83] A. Iftikhar, Feature selection using particle swarm optimization in intrusion detection, Int. J. Distrib. Sens. Netw. (2015) 1–8, https://doi.org/10.1155/2015/806954.

[84] M.H. Ali, B.A. Mohammed, A. Ismail, M.F. Zolkipli, A new intrusion detection system based on fast learning network and particle swarm optimization, IEEE Access (2018) 20256–20261, https://doi.org/10.1109/ACCESS.2018.2820092.

[85] H. Li, W. Guo, G. Wu, Y. Li, A RF-PSO based hybrid feature selection model in intrusion detection system, in: Third International Conference on Data Science in Cyberspace, IEEE, 2018, pp. 795–802, https://doi.org/10.1109/DSC.2018.00128.

[86] A.T. Bayu, R. Kyung-Hyune, An integration of PSO-based feature selection and random forest for anomaly detection in IoT network, EDP Sci. (2018) 01053.

[87] B.A. Tama, K.-H. Rhee, An integration of PSO-based feature selection and random Forest for anomaly detection in IoT network, in: MATEC Web of Conferences, 2018, pp. 1–6, https://doi.org/10.1051/matecconf/201815901053.

[88] R.M. Alguliyev, R.M. Aliguliyev, F.J. Abdullayeva, PSO+K-means algorithm for anomaly detection in big data, Stat. Optim. Inf. Comput. (2019) 348–359, https://doi.org/10.19139/soic.v7i2.623.

[89] P. Ghosh, A. Karmakar, J. Sharma, S. Phadikar, CS-PSO based intrusion detection system in cloud environment, in: Emerging Technologies in Data Mining and Information Security, Advances in Intelligent Systems and Computing, 2019, pp. 261–269, https://doi.org/10.1007/978-981-13-1951-8_24.

[90] W. Elmasry, A. Akbulut, A.H. Zaim, Evolving deep learning architectures for network intrusion detection using a double PSO metaheuristic, Comput. Netw. (2019), https://doi.org/10.1016/j.comnet.2019.107042.

[91] A. Shokoohsaljooghi, H. Mirvaziri, Performance improvement of intrusion detection system using neural networks and particle swarm optimization algorithms, Int. J. Inf. Technol. (2019) 1–12, https://doi.org/10.1007/s41870-019-00315-9.

[92] D.J. Kalita, V.P. Singh, V. Kumar, SVM Hyper-Parameters Optimization using Multi-PSO for Intrusion Detection, Springer, 2020, pp. 227–240, https://doi.org/10.1007/978-981-15-2071-6_19.
[93] A.S. Eesa, Rule mining using particle swarm optimization for intrusion detection systems, Acad. J. Nawroz Univ. (2020) 222–223.
[94] A. Dickson, C. Thomas, Improved PSO for optimizing the performance of intrusion detection, J. Intell. Fuzzy Syst. (2020) 1–11, https://doi.org/10.3233/JIFS-179734.
[95] N. Kunhare, R. Tiwari, J. Dhar, Particle swarm optimization and feature selection for IDS, Indian Acad. Sci. (2020) 1–14, https://doi.org/10.1007/s12046-020-1308-5.
[96] M. Habib, I. Aljarah, H. Faris, A modified multi-objective particle swarm optimizer-based Lévy, Arab. J. Sci. Eng. (2020) 1–28, https://doi.org/10.1007/s13369-020-04476-9.
[97] R.O. Ogundokun, J.B. Awotunde, P. Sadiku, E.A. Adeniyi, A. Moses, O.I. Dauda, An enhanced intrusion detection system using particle swarm optimization feature extraction technique, Procedure Comput. Sci. (2021) 504–512, https://doi.org/10.1016/j.procs.2021.10.052.
[98] A.M. Bamhdi, I. Abrar, F. Masoodi, An ensemble based approach for effective intrusion detection using majority votinf, Telkomnika (2021) 664–671, https://doi.org/10.12928/TELKOMNIKA.v19i2.18325.
[99] M. Ajdani, H. Ghaffary, Introduced a New Method For Enhancement of Intrusion Detection With Random Forest and PSO Algorithm, Wiley, 2021, pp. 1–10, https://doi.org/10.1002/spy2.147.
[100] A.S. Talita, O.S. Nataza, Z. Rustam, Naïve Bayes classifier and particle swarm optimization feature selection method for classifying intrusion detection system dataset, J. Phys. (2021) 012021.
[101] H. Yang, Z. Zhou, A novel intrusion detection scheme using Cloud Grey Wolf Optimizer, in: Proceedings of the 37th Chinese Control Conference, 2018, pp. 8297–8302. Wuhan, China.
[102] A.A. Shorman, H. Faris, I. Aljarah, Unsupervised intelligent system based on one class support vector machine and grey wolf optimization for IoT botnet detection, J. Ambient. Intell. Humaniz. Comput. (2019) 1–17, https://doi.org/10.1007/s12652-019-01387-y.
[103] D. Srivastava, R. Singh, V. Singh, An intelligent gray wolf optimizer: a nature inspired technique in intrusion detection system (IDS), J. Advancements Robot. 6 (1) (2019) 18–27.
[104] T.A. Alamiedy, M. Anbar, Z.N. Alqattan, Q.M. Alzubi, Anomaly-based intrusion detection system using multi-objective grey wolf optimization algorithm, J. Ambient. Intell. Humaniz. Comput. (2019) 1–22, https://doi.org/10.1007/s12652-019-01569-8.
[105] P. Anitha, B. Kaarthick, Oppositional based Laplacian grey wolf optimization algorithm, J. Ambient. Intell. Humaniz. Comput. (2019) 1–12, https://doi.org/10.1007/s12652-019-01606-6.
[106] R.B. Benisha, S.R. Ratna, Design of intrusion detection and prevention in SCADA system for the detection of Bias injection attacks, Secur. Commun. Netw. (2019) 1–12.
[107] R.B. Benisha, S.R. Ratna, Detection of data integrity attacks by constructing an effective intrusion detection system, J. Ambient. Intell. Humaniz. Comput. (2020) 1–12, https://doi.org/10.1007/s12652-020-01850-1.
[108] R. Vatambeti, K.S. Supriya, S. Sanshi, Identifying and detecting black hole and gray hole attack in MANET using gray wolf optimization, Int. J. Commun. Syst. (2020) 1–16, https://doi.org/10.1002/dac.4610.
[109] S. Mukaram, O. Mohammed, A. Laith, Improved binary gray wolf optimizer and SVM for intrusion detection system in wireless sensor networks, J. Ambient. Intell. Humaniz. Comput. (2020) 1–18.
[110] M.M. Jawhar, M.A. Alelah, The detects DoS attack from network traffic using gray wolf optimization algorithm, J. Eng. Sci. Technol. (2020) 3639–3648.
[111] M. Safaldin, M. Otair, L. Abualigah, Improved binary gray wolf optimizer and SVM for intrusion detection, J. Ambient. Intell. Humaniz. Comput. (2020) 1–18, https://doi.org/10.1007/s12652-020-02228-z.
[112] P.R. Swarna, K.R.M. Praveen, M. Parimala, K. Srinivas, R.G. Thippa, L.C. Chiranji, A. Mamoun, An effective feature engineering for DNN using hybrid PCA-GWO for intrusion detection in IoMT architecture, Comput. Commun. (2020) 139–149, https://doi.org/10.1016/j.comcom.2020.05.048.
[113] N.K. Singh, V. Mahajan, Detection of cyber cascade failure in smart grid substation using advance grey wolf optimization, J. Interdiscip. Math. (2020) 69–79, https://doi.org/10.1080/09720502.2020.1721664.

[114] S. Amaran, R.M. Mohan, An optimal Grey wolf optimization with fuzzy support vector machine based intrusion detection system in clustered wireless sensor networks, Int. J. Adv. Trends Comput. Sci. Eng. (2020) 2655–2661.

[115] Z. Wang, Y. Zeng, Y. Liu, D. Li, Deep belief network integrating improved kernel based extreme learning machine for network intrusion, IEEE Access 9 (2021) 16062–16091, https://doi.org/10.1109/ACCESS.2021.3051074.

[116] S. Shakya, Modified gray wolf feature selection and machine learning classification for wireless sensor network intrusion detection, J. Sustain. Wireless Syst. (2021) 118–127, https://doi.org/10.36548/jsws.2021.2.006.

[117] A. Sharma, U. Tyagi, A Hybrid approach of ANN-GWO Technique for Intrusion Detection, in: 2021 International Conference on Recent Trends on Electronics, Information, Communication & Technology (RTEICT), IEEE Xplore, 2021, pp. 467–472.

[118] S. Dwivedi, M. Vardhan, S. Tripathi, A.K. Shukla, Implementation of adaptive scheme in evolutionary technique for anomaly-based intrusion detection, Evol. Intel. (2019) 1–15, https://doi.org/10.1007/s12065-019-00293-8.

[119] S.M. Vardhan, S. Tripathi, Defense against distributed DoS attack detection by using intelligent evolutionary algorithm, Int. J. Comput. Appl. (2020) 1–12, https://doi.org/10.1080/1206212X.2020.1720951.

[120] S.M. Hosseiny, A.I. Rahman, M. Derakhshan, Improve intrusion detection using grasshopper optimization algorithm and decision trees, Int. J. Saf. Secur. Eng. (2020) 359–364, https://doi.org/10.18280/ijsse.100307.

[121] S. Moghanian, B.F. Saravi, G. Javidi, E.O. Sheybani, GOAMLP: network intrusion detection with multilayer perceptron and grasshopper optimization algorithm, IEEE Access (2020) 215202–215213, https://doi.org/10.1109/ACCESS.2020.3040740.

[122] S. Dwivedi, M. Vardhan, S. Tripathi, An effect of chaos grasshopper optimization algorithm for protection of network infrastructure, Comput. Netw. (2020) 107251, https://doi.org/10.1016/j.comnet.2020.107251.

[123] W.A. Ghanem, A. Jantan, New multi-objective artificial bee colony optimization for wrapper based feature selection in intrusion detection, Int. J. Adv. Soft Comput. Appl. (2016) 70–81.

[124] H.M. Rais, T. Mehmood, Dynamic ant Colony system with three level update feature selection for intrusion detection, Int. J. Netw. Secur. (2018) 184–192, https://doi.org/10.6633/IJNS.201801.20(1).20.

[125] M. Mazini, B. Shirazi, I. Mahdavi, Anomaly network-based intrusion detection system using a reliable hybrid artificial bee colony and AdaBoost algorithms, J. King Saud Univ.—Comput. Inf. Sci. (2018), https://doi.org/10.1016/j.jksuci.2018.03.011.

[126] J. Yang, Z. Ye, L. Yan, W. Gu, R. Wang, Modified naive bayes algorithm for intrussion detection on artificial bee colony algorithm, in: 4th IEEE International Symposium on Wireless Systems withing the International Conference on Intelligents Data Acquisition and Advanced Computing System, IEEE, Lviv, Ukraine, 2018, pp. 35–40.

[127] S.A. Elsaid, N.S. Albatati, An optimized collaborative intrusion detection system for wireless, Soft. Comput. (2020) 1–15, https://doi.org/10.1007/s00500-020-04695-0.

[128] L. Zhang, W. Song, Research on intrusion detection algorithm based on smart campus network security, in: ICASIT 2020: 2020 International Conference on Aviation Safety and Information Technology, ACM, Wuhan, China, 2020, pp. 446–449, https://doi.org/10.1145/3434581.3434627.

[129] M. Rani, Employing artificial bee colony algorithm for feature selection in intrusion detection system, in: 2021 8th International Conference on Computing for Sustainable Global Development (INDIACom), IEEE, 2021, pp. 496–500.

[130] A.H. Al-Safi, Z.I. Hani, M.M. Zahra, Using A hybrid algorithm and feature selection for network anomaly intrusion detection, J. Mech. Eng. Res. Dev. (2021) 253–262.

[131] M.S. Abadeh, J. Habibi, A hybridization of evolutionary fuzzy systems and ant colony optimization for intrusion detection, ISC Int. J. Inf. Secur. (2010) 33–46.

[132] Y. Li, J. Xia, S. Zhang, J. Jan, X. Ai, K. Dai, An efficient intrusion detection system based on support vector machines and gradually feature removal method, Expert Syst. Appl. (2012) 424–430, https://doi.org/10.1016/j.eswa.2011.07.032.

[133] T. Mehmood, H.B. Rais, SVM for network anomaly detection using ACO feature subset, in: 2015 International Symposium on Mathematical Sciences and Computing Research (iSMSC), IEEE, 2015, pp. 121–126.

[134] K.V. Ravi, V. Valli Kumari, K.S. Srinivas, Feature selection using relative fuzzy entropy and ant colony optimization applied to real-time intrusion detection system, Procedia Comput. Sci. (2016) 503–510.

[135] M.H. Aghdam, P. Kabiri, Feature selection for intrusion detection system using ant colony optimization, Int. J. Netw. Secur. (2016) 420–432.

[136] F.H. Botes, L. Leenen, R. De La Harpe, Ant colony induced decision trees for intrusion detectio, in: 16th European Conference on Cyber Warfare and Security, ACPI, 2017, pp. 53–62.

[137] H.I. Alsaadi, R.M. Almuttairi, O. Bayat, O.N. Uçan, Computational intelligence algorithms to handle dimensionality reduction for enhancing intrusion detection system, J. Inf. Sci. Eng. (2020) 293–308, https://doi.org/10.6688/JISE.202003_36(2).0009.

[138] R.K. Penmatsa, A. Kalidindi, S.K. Mallidi, Feature reduction and optimization of malware detection system using ant colony optimization and rough sets, Int. J. Inf. Secur. Priv. (2020) 1–20.

[139] K. Parabakaran, N. Kumaratharan, G. Suresh, P. Epsiba, An evaluation of effective intrusion DoS detection and prevention system based on SVM classifier for WSN, Mater. Sci. Eng. (2020) 1–11, https://doi.org/10.1088/1757-899X/925/1/012068.

[140] S. Miryahyaie, H. Ebrahimpour-komleh, A.M. Nickfarjam, ACO-based intrusion detection method in computer networks using fuzzy association rules, in: 2021 5th International Conference on Pattern Recognition and Image Analysis (IPRIA), IEEE, 2021, pp. 1–5.

[141] C. Zhang, Network security state identification based on neural network optimized by ant colony, Converter (2021) 473–479.

[142] T.A. Victoire, M. Sakthivel, A refined differential evolution algorithm based fuzzy classifier for intrusion detection, Eur. J. Sci. Res. (2011) 246–259.

[143] Z. Salek, F.M. Madani, R. Azmi, Intrusion detecetion using neural networks trained by differential evaluation algorithm, in: 10th Internationa ISC Conference on Information Security and Cryptology (ISCISC), IEEE, 2013, pp. 1–6.

[144] E. Popoola, A.O Adewumi, Efficient feature selection technique for network intrusion detection system using discrete differential evolution and decision, Int. J. Netw. Secur. (2017) 660–669, https://doi.org/10.6633/IJNS.201709.19(5).02.

[145] J. Ku, B. Zheng, Intrusion detection based on self-adaptive differential evolution extreme learning machine with Gaussian Kernel, in: International Symposium on Parallel Architecture, Algorithm and Programming, Springer, 2017, pp. 13–24.

[146] W. Guo, Y. Chen, Y. Cai, T. Wang, H. Tian, Intrusion detection in WSN with an improved NSA based on the DE-CMOP, KSII Trans. Internet Inf. Syst. (2017) 5574–5591, https://doi.org/10.3837/tiis.2017.11.022.

[147] K. Cui, L.-x. Fu, Y. Zhang, J.-L. Mao, Y.-J. Ren, Application of BP Neural Network Optimized by Differential Evolution Algorithm in Intrusion Detection, Software Guide, 2018.

[148] F.H. Almasoudy, W.L. Al-Yaseen, A.K. Idrees, Differential evolution wrapper feature selection for intrusion detection system, Procedia Comput. Sci. (2020) 1230–1239, https://doi.org/10.1016/j.procs.2020.03.438.

[149] M.H. Adaniya, M.F. Lima, L.D. Sampaio, T. Abr'ao, L.P. Mario, Anomaly detection using firefly harmonic clustering algorithm, in: Proceedings of the International Conference on Data Communication Networking and Optical Communication System (DCNET-2011), SCITEPRESS, 2011, pp. 63–68, https://doi.org/10.5220/0003525800630068.

[150] A. Kaur, S.K. Pal, A.P. Singh, Hybridization of K-means and firefly algorithm for intrusion detection system, Int. J. Syst. Assur. Eng. Manag. (2017) 1–10, https://doi.org/10.1007/s13198-017-0683-8.

[151] E.M. Roopa Devi, R.C. Suganthe, Improved relevance vector machine (IRVM) classifier for intrusion detection system, Soft Comput. (2018) 1–9, https://doi.org/10.1007/s00500-018-3621-z.

[152] R.F. Najeeb, B.N. Dhannoon, Improving detection rate of the network intrusion detection system based on wrapper feature selection approach, Iraqi J. Sci. (2018) 426–433, https://doi.org/10.24996/ijs.2018.59.1B.23.

[153] B. Selvakumar, K. Muneeswaran, Fireßy algorithm based feature selection for network intrusion detection, Comput. Secur. (2018), https://doi.org/10.1016/j.cose.2018.11.005.

[154] K.V. Krishna, B.B. Prakash, Intrusion detection system employing multi-level feed forward neural network along with firefly optimization (FMLF2N2), Int. Inf. Eng. Technol. Assoc. (2019) 139–145, https://doi.org/10.18280/isi.240202.

[155] W.L. Al-Yaseen, Improving intrusion detection system by developing feature selection model based on firefly algorithm and support vector machine, Int. J. Comput. Sci. (2019) 1–7.

[156] M.F. Badran, K. Moorthy, N.S. Zukifi, M.S. Mohamad, S. Deris, N.M. Sahar, Propose a new firefly-fast learning network model based intrusion-detection system, Int. J. Innov. Technol. Explor. Eng. (2019) 146–152, https://doi.org/10.35940/ijitee.L1027.10812S219.

[157] M. Gunay, Z. Orman, A modified firefly algorithm-based feature selection method and artificial immune system for intrusion detection, Uludağ Univ. J. Facul. Eng. (2020) 269–288, https://doi.org/10.17482/uumfd.649003.

[158] R. Rajakumar, RFA reinforced firefly algorithm to identify optimal feature subsets for network IDS, Int. J. Grid High Perform. Comput. (2020) 68–86, https://doi.org/10.4018/IJGHPC.2020070105.

[159] K.M. Alwan, A.H. Abu El-Atta, H.H. Zayed, Feature selection models based on hybrid firefly algorithm with mutation operator for network intrusion detection, Int. J. Intell. Eng. Syst. (2021) 192–202, https://doi.org/10.22266/ijies2021.0228.19.

[160] B. Vishnupriya, R.M. Tharsanee, R.S. Soundariya, M. Nivaashini, G. Pavithra, A meta-heuristic approach for intrusion detection system using cascaded classifier, in: Second International Conference on Image Processing and Capsule Network (ICIPCN 2021), Springer, 2021, pp. 719–731.

[161] P. Ghosh, N. Subhash, D. Sarkar, J. Sharma, S. Phadikar, An intrusion detection system using modified-firefly algorithm in cloud environment, Int. J. Digital Crime Forensics (2021) 77–93.

[162] P. Kaushik, An improved black hole detection and prevention mechanism in MANET using firefly and neural network, Int. J. Innov. Sci. Res. Technol. (2021) 749–754.

[163] J. Wang, Y. Liu, H. Feng, IFACNN: efficient DDoS attack detection based on improved firefly algorithm to optimize convolutional neural networks, Math. Biosci. Eng. 19 (2) (2021) 1280–1303, https://doi.org/10.3934/mbe.2022059.

[164] Q. Tian, D. Han, M.-Y. Hsieh, K.-C. Li, A. Castiglione, A two-stage intrusion detection approach for software-defined IoT networks, Soft. Comput. (2021) 1–18, https://doi.org/10.1007/s00500-021-05809-y.

[165] M. Daoud, A. Boukra, A new off-line intrusion detection system, J. Inf. Secur. Res. (2013) 53–62.

[166] H. Xu, C. Fang, Q. Cao, C. Fu, L. Yan, S. Wei, Application of a distance-weighted KNN algorithm improved by moth-flame optimization in network intrusion detection, in: 4th IEEE International Symposium on Wireless Systems within the International Conferences on Intelligent Data Acquisition and Advanced Computing Systems, IEEE, Lviv, Ukraine, 2018, p. 166.

[167] D. Arivudainamb, K.K. Varun, S.S. Chakkaravarthy, LION IDS: a meta-heuristics approach to detect DDoS attacks against, Neural Comput. & Applic. (2018) 1–11, https://doi.org/10.1007/s00521-018-3383-7.

[168] M. Shakil, A.F. Mohammed, R. Arul, A.K. Bashir, J.K. Choi, A novel dynamic framework to detect DDoS in SDN using metaheuristic clustering, Trans. Emerg. Telecommun. Technol. (2019) 1–18, https://doi.org/10.1002/ett.3622.

[169] L. Haghnegahdar, Y. Wang, A whale optimization algorithm-trained artificial neural network for smart grid cyber intrusion detection, Neural Comput. & Applic. (2019) 1–15, https://doi.org/10.1007/s00521-019-04453-w.

[170] D. Wang, G. Xu, Research on the detection of network intrusion prevention with SVM based optimization algorithm, Informatica (2020) 269–273, https://doi.org/10.31449/inf.v44i2.3195.

[171] J.Q. Chen, L. Chen, F. Chen, G. Cheng, Quantum-inspired ant lion optimized hybrid k-means for cluster analysis and intrusion detection, Knowl.-Based Syst. (2020) 106167, https://doi.org/10.1016/j.knosys.2020.106167.

[172] A.A. Lateef, S.T. Al-Janabi, B. Al-Khateeb, Hybrid intrusion detection system based on deep learning, in: 2020 International Conference on Data Analytics for Business and Industry: Way Towards a Sustainable Economy (ICDABI), IEEE, 2020, pp. 1–5.

[173] H. Alazzam, A. Sharieh, K. Sabri, A feature selection algorithm for intrusion detection system based on pigeon inspired optimizer, Expert Syst. Appl. (2020) 113249, https://doi.org/10.1016/j.eswa.2020.113249.

[174] N. Khare, P. Devan, C.L. Chowdhary, S. Bhattacharya, G. Singh, S. Singh, B. Yoon, SMO-DNN: spider monkey optimization and deep neural network hybrid classifier model for intrusion detection, Electronics (2020) 1–18.

[175] Y. Zhou, G. Cheng, S. Jiang, M. Dai, Building an efficient intrusion detection system based on feature selection and ensemble classifier, Comput. Netw. (2020) 107247, https://doi.org/10.1016/j.comnet.2020.107247.

[176] A. Gupta, M. Kalra, Intrusion detection and prevention system using cuckoo search algorithm with ANN in cloud computing, in: 2020 Sixth International Conference on Parallel, Distributed and Grid Computing (PDGC), IEEE, 2020, pp. 1–7, https://doi.org/10.1109/PDGC50313.2020.9315771.

[177] A.S. Mahboob, M.R. Moghaddam, An anomaly-based intrusion detection system using butterfly optimization, in: 2020 6th Iranian Conference on Signal Processing and Intelligent Systems (ICSPIS), IEEE, 2020, pp. 1–6, https://doi.org/10.1109/ICSPIS51611.2020.9349537.

[178] M. Aljanabi, M.A. Ismail, V. Mezhuyev, Improved TLBO-JAYA algorithm for subset feature selection and parameter optimisation in intrusion detection system, Complexity (2020) 1–18, https://doi.org/10.1155/2020/5287684.

[179] P. Gao, M. Yue, Z. Wu, A novel Intrusion detection method based on WOA optimized hybrid kernel RVM, in: 2021 IEEE 6th International Conference on Computer and Communication System, IEEE, 2021, pp. 1063–1069.

[180] R. Shekhar, K. Sasirekka, P. Raja, K. Thangavel, A novel GPU based intrusion detection system using autoencoder woth fruitfly optimization, SN Appl. Sci. (2021) 1–16.

[181] K. Li, Y. Zhang, S. Wang, An intrusion detection system based on PSO-GWO hybrid optimized supprot vector machine, in: 2021 International Joint Conference on Neural Networks, IEEE, 2021, pp. 1–7.

[182] A. Nazir, R.A. Khan, A novel combinatorial optimization based feature selection method for network intrusion detection, Comput. Secur. (2021), 102164, https://doi.org/10.1016/j.cose.2020.102164.

[183] R. SaiSindhuTheja, G.K. Shyam, An efficient metaheuristic algorithm based feature selection and recurrent neural network for DoS attack detection in cloud computing environment, Appl. Soft Comput. J. (2021) 106997, https://doi.org/10.1016/j.asoc.2020.106997.

[184] J. Kaur, R. Talwar, A.K. Goel, System based on meld optimization algorithm to mitigate amalgam attacks, Indian J. Sci. Technol. (2021) 1622–1634, https://doi.org/10.17485/IJST/v14i20.601.

[185] J. Hu, C. Liu, Y. Cui, An improved CNN approach for network intrusion detection system, Int. J. Netw. Secur. (2021) 569–575, https://doi.org/10.6633/IJNS.202107 23(4).03.

[186] Z. Pei, Z. Yinyan, A BAS algorith based neural network for intrusion detection, in: 11th Internal Conference on Intelligent Control and Information Processing, IEEE, 2021, pp. 22–27.

[187] R.N. Hadi, R.O. Mahmoud, A.S. Eldien, Feature selection method based on chaotic salp swarm algorithm and extreme learning machine for network intrusion detection systems, in: Special Issue on Computing Technology and Information Management, Webology, 2021, pp. 626–642, https://doi.org/10.14704/WEB/V18SI04/WEB18154.

[188] S. Velliangiri, R. Christin, R. Karthikeyan, Genetic gray wolf improvement for distributed denial of service attacks in the cloud, J. Comput. Theor. Nanosci. (2018) 1–6, https://doi.org/10.1166/jctn.2018.7463.
[189] V. Hajisalem, S. Babaie, A hybrid intrusion detection system based on ABC-AFS algorithm for misuse and anomaly detection, Comput. Netw. (2018) 37–50, https://doi.org/10.1016/j.comnet.2018.02.028.
[190] E.M. Roopa Devi, R.C. Suganthe, Enhanced transductive support vector machine classification with grey wolf optimizer cuckoo search optimization for intrusion detection system, Concurr. Comput. Pract. Exp. 32 (4) (2018) e4999, https://doi.org/10.1002/cpe.4999.
[191] M. Moukhafi, K.E. Yassini, S. Bri, A novel hybrid GA and SVM with PSO feature selection for intrusion detection system, Int. J. Adv. Sci. Res. Eng. (2018) 129–134, https://doi.org/10.31695/IJASRE.2018.32724.
[192] R. Bilaiya, R.M. Sharma, Intrusion detection system based on hybrid whale-genetic algorithm, in: Proceedings of the 2nd International Conference on Inventive Communication and Computational Technologies (ICICCT 2018), IEEE, 2018, pp. 822–825.
[193] S. Velliangiri, P. Karthikeyan, Hybrid optimization scheme for intrusion detection using considerable feature selection, Neural Comput. & Applic. (2019) 1–15, https://doi.org/10.1007/s00521-019-04477-2.
[194] A. Davahli, M. Shamsi, G. Abaei, Hybridizing genetic algorithm and grey wolf optimizer to advance an intelligent and lightweight intrusion detection system for IoT wireless networks, J. Ambient. Intell. Humaniz. Comput. (2020) 1–29, https://doi.org/10.1007/s12652-020-01919-x.
[195] P.K. Keserwani, M.C. Govila, E. Shubhaka, An optimal intrusion detection system using GWO-CSA-DSAE model, Cyber-Phys. Syst. (2020) 1–24, https://doi.org/10.1080/23335777.2020.1811383.
[196] W.A. Ghanem, A. Jantan, S.A. Ghaleb, A.B. Naseer, An efficient intrusion detection model based on hybridization of artificial bee colony and dragonfly algorithms for training multilayer perceptrons, IEEE Access (2020) 130452, https://doi.org/10.1109/ACCESS.2020.3009533.
[197] L. Lv, W. Wang, Z. Zhang, X. Liu, A novel intrusion detection system based on an optimal hybrid kernel extreme, Knowl.-Based Syst. (2020) 105648.
[198] G. Suresh, R. Balasubramanian, An ensemble feature selection model using fast convergence ant Colony optimization algorithm, Int. J. Emerg. Trends Eng. Res. (2020) 1417–1423.
[199] M.S. Bonab, A. Ghaffari, F.S. Gharehchopogh, P. Alemi, A wrapper-based feature selection for improving performance of intrusion detection systems, Int. J. Commun. Syst. (2020) 1–26, https://doi.org/10.1002/dac.4434.
[200] T. Yerriswamy, M. Gururaj, An efficient algorithm for anomaly intrusion detection in a network, Global Trans. Proc. (2021) 255–260, https://doi.org/10.1016/j.gltp.2021.08.066.
[201] C. Chen, L. Song, C. Bo, W. Shuo, A support vector machine with particle swarm optimization grey wolf optimizer for network intrusion detection, in: 2021 International Conference on Big Data Analysis and Computer Science (BDACS), IEEE, 2021, pp. 199–204.
[202] S. Mohammadi, M. Babagoli, A hybrid modified grasshopper optimization algorithm and genetic algorithm to detect and prevent DDoS attacks, Int. J. Eng. (2021) 811–824, https://doi.org/10.5829/ije.2021.34.04a.07.
[203] E.S. Krishna, T. Arunkumar, Hybrid particle swarm and gray wolf optimization algorithm for IoT intrusion detection system, Int. J. Intell. Eng. Syst. (2021) 66–76, https://doi.org/10.22266/ijies2021.0831.07.
[204] P.K. Keserwani, M.C. Govil, E.S. Pilli, P. Govil, A smart anomaly-based intrusion detection system for the Internet of things (IoT) network using GWO–PSO–RF model, J. Reliab. Intell. Environ. (2021) 1–19, https://doi.org/10.1007/s40860-020-00126-x.
[205] O. Almomani, Hybrid model using bio-inspired metaheuristic algorithms for network intrusion detection system, Mater. Contin. (2021) 409–429, https://doi.org/10.32604/cmc.2021.016113.

CHAPTER 7

Metaheuristic algorithms in text clustering

Ibrahim Hayatu Hassan[a,b,#], Abdullahi Mohammed[b], Yusuf Sahabi Ali[b], Isuwa Jeremiah[b], and Silifat Adaramaja Abdulraheem[b]

[a]*Institute for Agricultural Research, Ahmadu Bello University, Zaria, Nigeria,* [b]*Department of Computer Science, Ahmadu Bello University, Zaria, Nigeria*

1. Introduction

Clustering is a popular machine learning technique that has been widely investigated in the context of text, and it is used to organize large amounts of text documents into different groups or clusters. It has a wide range of applications, including text document organization, visualization, and classification [1]. Text document clustering is essential for retrieving information, indexing data, and handling and mining large amounts of text data on the Internet and in business information systems.

The central concept behind text clustering is dividing a set of text documents into a subset of distinct clusters based on similar subjects and contents. As a result, while each cluster contains relevant documents, other clusters contain irrelevant documents [2]. This method is used in many areas of research, including data mining, machine learning, pattern recognition, data compression, image segmentation, and wireless sensor networks [3].

Metaheuristic algorithms have emerged as an effective solution to the text document clustering problem over the years [2]. Depending on the number of solutions handled in each iteration, these algorithms are classified as either local search-based algorithms or population-based algorithms. For solving text clustering problems, several metaheuristic algorithms have been proposed, including the genetic algorithm (GA) [4], harmony search (HS) [5], particle swarm optimization (PSO) [6], cuckoo search (CS) [7], ant colony optimization (ACO) [8], and artificial bee colony (ABC) [9].

This chapter presents a comprehensive and up-to date literature review of metaheuristic algorithms in the context of text document clustering and examines how these algorithms are applied in this domain. In addition, the chapter discusses various challenges and possible future research directions in the use of metaheuristic algorithms in text document applications. Moreover, the chapter refers to all of the previous research presented on text clustering based on metaheuristic algorithms. Fig. 1 presents the distribution of these publications.

[#]Senior contributing author.

Comprehensive Metaheuristics. https://doi.org/10.1016/B978-0-323-91781-0.00007-7

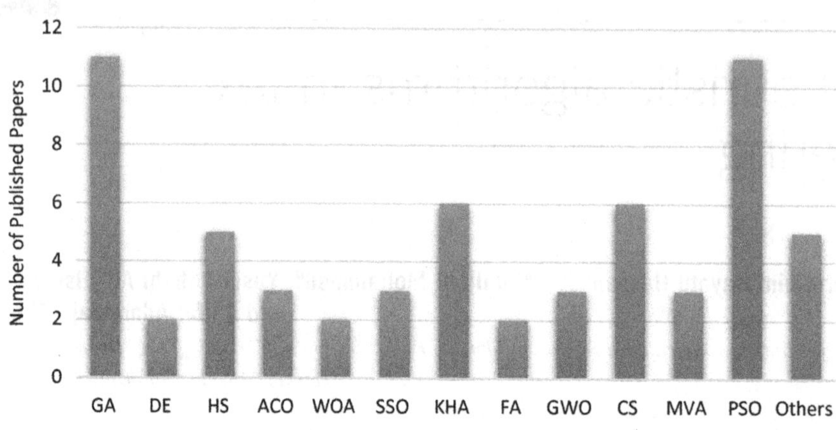

FIG. 1

Number of published papers in text clustering using metaheuristics algorithms.

Section 2 presents the major text clustering procedures. Section 3 highlights the different research that applied metaheuristics for solving text clustering problems. Section 4 discusses challenges and possible future research directions. Finally, Section 4 provides the conclusion and possible research focus.

2. Text clustering formulation procedure

The goal of text clustering is to create optimal clusters that include related text (entities). Clustering is based on dividing a collection of text into a known number of related classes, each of which contains several similar entities. This section discusses the various steps involved in text clustering. The subsections describe these steps, which include problem description, preprocessing, text representation, solution representation, and fitness function.

2.1 Design and problem descriptions

Here, we describe problems, descriptions, and designs of text clustering.

Given a collection of text *TD* containing m entities is divided into a known number of groups or clusters (*P*), the text *TD* can be represented as a vector of entities as,

$TD=(td_1, td_1, td_1, \ldots, td_i, \ldots, td_m)$, where *td* defines a text document entity, *i* represents the number of the entity, and *m* defines the total number of entities in the complete document.

Each cluster consist of center c_l that is represented as a vector of word term weights as $C_l=(c_{l1}, c_{l2}, c_{l3}, \ldots, c_{lj}, \ldots, c_{lp})$, where c_l defines the centroid of the l_{th} cluster, c_{l1} is the position two value in the *l* cluster center, and *p* is the number of all distinctive centroid terms in the given entity. The distance or similarity metrics are utilized to move each entity to the closest cluster based on centroid [9–11].

2.2 Preprocessing

The primary goal of clustering is to generate classes based on the inherent subjects of entities. To cluster a text, it must first be converted into a format that the clustering technique can process; this is known as preprocessing. Text clustering preprocessing procedures include tokenization, stop word removal, stemming, term weighting, and document representation [12,13]. We describe these steps in detail in the subsections that follow.

2.2.1 Tokenization

Tokenization is a technique for separating a string of letters in a text into tokens. Any characters compressed between two spaces, such as words, keywords, phrases, symbols, and other elements, are considered tokens.

2.2.2 Stop word removal

The stop words removal technique eliminates specific words that appear frequently in a text. The website http://www.unine.ch/Info/clef/ contains a list of 571 such words as "a," "the," "an," 'are," and "that."

2.2.3 Stemming

Stemming is a method for removing prefixes and suffixes from actual words to obtain the basic version of the words. Prefixes are letters that are added at the beginning of words, whereas suffixes are letters that are added at the end of words. Prefixes and suffixes, in particular, are well defined and easily identified in English. For example, "ed" is used to refer to a verb's past tense; additionally, "ing" and etc. are dropped. This process is typically carried out using the Porter stemmer, which removes some of the extremities such as prefixes and suffixes from each term such as "ed," "ly," and "ing." For example, the words or terms "connection," "connective," "connected," "connecting," and "connected" are all mutually exclusive.

2.3 Document representation

The data from the normal text arrangement must be transformed to numerical form during the text clustering process. The vector space model (VSM) is the most widely used numerical representation of text [14]. The VSM is a numerical model that allows text documents to be represented as vectors. The contents of a document are designed in a multidimensional location in this model.

The VSM determines term weighting to depict the text document in a standard format, as shown in Eq. (1). As shown in Eq. (2), this model displays each text document as a vector of words. Using the VSM, Eq. (1) displays n documents and p terms in a normal arrangement.

$$\begin{bmatrix} tw_{1,1} & \cdots & tw_{1,p} \\ \vdots & \ddots & \vdots \\ tw_{n,1} & \cdots & tw_{n,\ p} \end{bmatrix} \tag{1}$$

$$td_i = \left(tw_{i,1}, tw_{i,2}, tw_{i,3}, \ldots, j, \ldots, tw_{i,p}\right) \tag{2}$$

Table 1 Solution representation example.

Text document (*td*)	1	2	3	4	5	6	7	8
Clusters (*C*)	2	1	2	2	1	1	2	2

2.4 Solution representation

Text document clustering is classified as an optimization problem that is overcome using optimization algorithms. Optimization algorithms deal with a variety of solutions to a specific problem. Each solution describes a possible solution to the clustering problem (row). The solution is viewed as an n-dimensional vector that describes the content of each document in the provided dataset, with each location leading to a document. Table 1 provides an example of a solution design problem in text clustering. The solution's i_{th} location contributes to the decision about the i_{th} text. If the number of the clusters given is C, then the value in the range $(1,\ldots,C)$ is at each position of the solution. The set of C centroids fits each component. The number of clusters is normally given in advance.

In the example showcase in Table 1, eight text documents and two clusters are offered. Each solution projects where the text documents fit in. In this case, documents 2, 5, and 6 belong to cluster 1. Text documents 1, 3, 4, 7, and 8 belong to cluster 2.

2.5 Objective function

The objective value is computed to evaluate each solution based on its locations. Each collection of text documents belongs to a collection of C centroids $C=(c_1,c_2,\ldots,c_q)$, where c_q is the centroid of cluster q. The objective function value for each possible solution is computed as the average similarity of documents to the cluster centroid (ASDC) using Eq. (3). The similarity metric inside the objective function in Eq. (3) can be transformed into different distance or similarity metric

$$\left[\frac{\sum_{j=1}^{C}\frac{\sum_{i=1}^{n}\cos\left(d_i,c_j\right)}{m_i}}{C}\right], \tag{3}$$

where C is total number of clusters in the document, m_i is number of documents that correctly belong to cluster i, and $\cos(d_i,c_j)$ is the similarity value between document i and the cluster j centroid. Each solution is presented in binary matrix $a_{i,\,j}$ of size $n * C$ to compute the clusters centroid, as depicted in Eq. (4).

$$a_{i,j}=\begin{cases}1, & \text{if } d_i \text{ is allocated to the } j\text{th cluster}\\ 0, & \text{otherwise}\end{cases} \tag{4}$$

Eq. (5) is used to compute the k_{th} cluster centroids, which is given as,

$$c_k=\left(c_{k1},c_{k2},c_{k3},\ldots,c_{kj},\ldots,c_{kt}\right).$$

$$c_{kj}=\frac{\sum_{i=1}^{n}a_{ij}\left(d_{ij}\right)}{\sum_{i=1}^{n}a_{ij}}, \tag{5}$$

where a_{ij} is a matrix containing the clustered data, d_{ij} is the j_{th} attribute weight of the document i, and n is the total number documents in the dataset used.

3. Metaheuristic algorithms application in text clustering

In this section, we discuss the most popular metaheuristic algorithms used to handle text clustering problems.

3.1 Multiverse optimizer (MVO)

Abasi et al. [23] described a technique for text document clustering based on link-based multiverse optimizer (LBMVO). The original MVO was adjusted in the proposed technique by incorporating a neighborhood operator to improve its exploration capability via a neighborhood selection scheme. Using CSTR, 20 newsgroups, Tr12, Tr41, Wap, and Classic4 standard text document datasets, the performance of the LBMVO was tested and compared with agglomerative, DBSCAN, k-means++, k-means clustering methods, and other metaheuristi algorithms such as PSO, GA, HS, KHA, covariance matrix adaptation evolution strategy (CMAES), COA, and standard MVO. The findings show the superiority of the proposed method over the other algorithms.

Abasi et al. [24] proposed a hybrid scheme for clustering text documents that combines MVO with the k-means algorithm. The k-means algorithm is used in this study to generate the initial population of the MVO algorithm to improve the quality of the starting candidate solution as well as the best solution. The proposed study was tested using CSTR, 20 newsgroups, NIPS2015, AAAI, Tr411, Tr12, and Wap standard text document clustering datasets, and evaluated using error rate, accuracy, entropy, recall, precision, and F-measure. The experimental results validate the proposed method's effectiveness in comparison to other metaheuristic algorithms such as KHA, PSO, HS, GA, H-GA, H-PSO, original MVO, and clustering approaches such as k-means++, k-means, DBSCAN, and spectral.

Another study by Abasi et al. [67] employed MVO to solve the text document clustering problem. In this study, text document clustering is viewed as a discrete problem, and a fitness function based on Euclidean distance is used as a similarity metric. This study was evaluated and compared to other cutting-edge techniques. The results show that this method produces significant results when compared to the other algorithms.

3.2 Particle swarm optimization (PSO)

Hasanzadeh et al. [15] proposed PSO+LSI, a text clustering technique based on PSO and latent semantic indexing (LSI). The LSI was used to reduce the dimensionality of the text document, while the PSO was used to cluster the data. To balance exploration and exploitation in the search space, the PSO was first modified with an adaptive inertia weight in the proposed technique. The modified PSO was then combined with the LSI for effective web corpus clustering. The results show that the proposed technique outperforms PSO+k-means in all dimensions tested.

Karol and Mangat [16] presented two hybrid methods for text document clustering, called KPSO and FCPSO, which combine PSO with k-means and fuzzy c-means clustering algorithms. The k-means algorithm is used in the KPSO to initialize the PSO algorithm to effectively cluster the text documents.

Like KPSO, the FCM is used to generate the initial clusters in FCPSO and then the PSO is used to generate the global best clusters. The proposed methods were tested on the Reuters-21587 and 20 newsgroups datasets and compared to the standard k-means and FCM methods.

Cagnina et al. [17] proposed a better PSO for short text clustering. The proposed method was tested on texts involving science-based abstracts, news, and short legal documents retrieved from the Internet. The experimental results indicate that the proposed method is a good clustering technique for small- and medium-sized short-text corpora.

Sarkar et al. [18] presented a performance assessment of PSO, k-means, and a hybrid PSO+ k-means for text document clustering. The simulation was carried out using a text document written in Nepali. The texts in the simulation are denoted in terms of synsets matching a word and synonyms from WordNet to cluster semantically related terms. The results show that the PSO+k-means algorithm outperformed the single algorithms.

Aghdam and Heidari [19] used PSO to select the most useful attribute for representing text documents in text clustering. For greater efficiency, the selected attributes subset is then used as input to the k-means algorithm. The proposed method was tested on the Reuters-21578 dataset. The results show that using the PSO as a feature selection method before applying the k-means algorithm improves performance.

Song et al. [2] proposed a hybrid evolutionary computation approach to solving the text clustering problem by combining quantum behaved PSO (QPSO) and GA. The authors of the proposed study used GA's robust global search capability to improve the particle initialization approach in QPSO. A new position strategy was also proposed to normalize the particle search region. Simulations were run on four subsets of the 20 newsgroups and Reuter-21578 text document datasets to test the efficacy of the proposed method. The results show that the proposed approach outperformed the original GA, QPSO, and improved QPSO.

Bharti and Singh [12] proposed a hybrid feature selection approach for text clustering, combining binary PSO with chaotic map, opposition-based learning, mutation, and fitness-based dynamic inertia weight. To validate the efficacy of the proposed method, simulations were run on WebKB, Classic4, and Reuters-21578 standard datasets. The obtained results show that the proposed approach finds a more appropriate attribute subset than other methods.

Abualigah et al. [20] also used PSO to select the most useful attribute for representing text documents in text clustering. For greater efficiency, the selected attributes subset is then used as input to the k-means algorithm. The proposed method was tested on the Reuters-21578 and 20 newsgroups datasets. The results show that using the PSO as a feature selection method before applying the k-means algorithm improves performance.

Daoud et al. [21] combined PSO with classical k-means clustering to solve the problem of initial seed selection by k-means in Arabic text clustering. The proposed method was tested using data from three different sites: BBC, CNN, and OSAC. The experimental results show that the proposed approach outperformed the classical k-means algorithm in all three datasets in terms of precision, recall, f-measure, and accuracy.

Kushwaha and Pant [22] used a link-based PSO (LBPSO) for feature selection in big data text clustering. To select the most informative attributes, the LBPSO employs a novel neighbor selection mechanism. These informative attributes are then evaluated using k-means to improve the algorithm's efficiency and reduce its time complexity. The experimental results using TDT2, Reuter 21578, and Tr11 show that the LBPSO is superior in terms of selecting the most informative attributes from these datasets.

Janani and Vijayarani [6] proposed an improved text document clustering approach that combines PSO with spectral clustering, abbreviated as SCPSO. This method was tested on standard datasets from Reuters, 20 newsgroups, and TDT2 and compared to the traditional PSO, expected maximization method (EMM), and spherical k-means clustering techniques. In terms of clustering accuracy, the simulation results show that the SCPSO outperformed the other techniques.

3.2.1 Gray wolf optimizer (GWO)

Rashaideh et al. [26] presented a text clustering approach using a gray wolf optimizer (GWO) to solve the text clustering problem. The proposed method uses the average distance of documents to the cluster center as the fitness function to iteratively optimize the distance between the document clusters. The efficacy of this approach was demonstrated using a standard dataset and metrics such as accuracy, precision, and F-score. According to the results, the proposed method outperformed the other two hill-climbing techniques proposed in the literature.

Vidyadhari et al. [63] proposed a particle gray wolf optimizer (PGWO) combined with semantic word processing for automatic text clustering. The text documents are provided as input to the preprocessing stage in this approach, which provides the valuable keyword for feature selection and clustering. The resulting keyword is then sent to the Wordnet ontology to determine the synonyms and hyponyms of each keyword. Following that, the frequency of each keyword used to build the text feature library is determined. Because the text feature library has a larger dimension, the entropy is used to select the most important features. Finally, the PGWO is used to assign class labels to generate the various clusters of text documents. The experimental results obtained using 20 newsgroups and Reuter's datasets shows that the proposed method achieved better accuracy than ABC, PSO, and GWO.

Abasi et al. [23] proposed an improved text feature selection for clustering using binary gray wolf optimizer (BGWO). The BGWO is used to handle the text feature extraction problem in this case. This method introduces a new design of the GWO algorithm by selecting useful features from text documents. To reduce algorithm time complexity and improve clustering efficiency, the useful features are evaluated using the k-means clustering algorithm. Furthermore, the effectiveness of the proposed approach was investigated using the benchmark datasets Tr41, Tr12, Classic4, CSTR, 20 newsgroups, and Wap. The simulation results show that the proposed method outperforms GA and BPSO in terms of average purity and f-measure.

3.2.2 Cuckoo search (CS)

Zaw and Mon [28] suggested a web-based document clustering approach that combines the advantages of cuckoo search (CS) and PSO. The solutions of new cuckoos in this approach are based on the solution of PSO. Out of these solutions, the algorithm will replace some eggs with successful solutions in the absence of fitness until a best solution appears. The proposed approach was tested on a standard dataset of Web documents, and the results show that it performs well in web document clustering.

Boushaki et al. [29] introduced a text document clustering strategy based on an improved CS algorithm. The original CS algorithm was improved with a local search strategy in the proposed study to address the problem of the traditional CS algorithm in text clustering. The proposed method was tested using the datasets Classic300, Classic400, Tr23, Tr12, and Iris. The results demonstrate the robustness of the proposed method in terms of improving clustering quality in terms of purity, F-measure, and fitness function.

Boushaki et al. [30] created CS-LSI, a dynamic and incremental text lustering technique that combines cuckoo search optimization (CSO) and LSI. The authors used the LSI scheme to select the most relevant features from a text dataset. The CS-LSI was used to automatically determine the number of clusters by employing a new index based on the significance distance metric, as well as to discover outlier documents by maintaining more coherent clusters. Experimentation was carried out on Sports1045, Review1557, K1b, and La2 text datasets to assess the effectiveness of the proposed technique. In comparison to other clustering approaches from the literature, the results obtained confirm the superiority of the proposed method with good clustering quality.

Vaijayanthi [31] provided a comparison study of cluster quality results obtained by solving the issue of text document clustering with the ACO and CSO algorithms. Experimental analysis performed on the Classic4 benchmark dataset show that both algorithms are equally good at producing high-quality clusters. Cluster quality is measured using the f-measure and the DB index.

Boushaki et al. [7] created a hybrid text clustering algorithm that combines the CS and k-means algorithms. The CS algorithm is used to initialize the cluster centers in the k-means algorithm to solve the local optima solution problem caused by random initialization of cluster centroids in the k-means algorithm. The developed algorithm's effectiveness was validated using the Reuters 21+578 benchmark dataset and compared to three other hybrid methods. The results of the experiment show that the new hybrid algorithm can produce more compact clusters with higher F-measure and purity.

3.2.3 Genetic algorithm (GA)

Wang et al. [52] suggested a text clustering algorithm based on a hybrid approach that combines fuzzy concepts with a rough set and a genetic algorithm (GA). The GA is used to describe weight parameters through clustering, which makes parameters more reasonable and operational while avoiding some problems associated with describing weight parameters in other work, such as subjectivity and unreliability. The results demonstrated the viability of the proposed algorithm.

Song and Park [53] used an improved GA on a LS model to implement a text clustering approach rather than the most used VSM for text representation. The improved GA in the proposed approach is based on variable string length, and the LS aids in exploring the latent semantic implied by a query through representation in a reduced dimensional space. The simulation results obtained using the Reuters-21578 standard dataset confirm the effectiveness and superiority of clustering over traditional GA.

Premalatha and Natarajan [54] created a text clustering method based on GA that includes a ranked mutation rate and a simultaneous mutation operator. In simultaneous mutation, the GA used multiple mutation operators at the same time to create the next generation. Each operator's mutation ratio varies in relation to the value of the offspring it produces. The mutation rate on the chromosome in ranked strategy is adjusted based on the fitness rank of the previous population. The simulation results were compared to datasets such as the CISI, Cranfield, and ADI document corpora. The results show that the proposed method outperformed k-means and traditional GA in terms of statistical performance.

Song et al. [55] implemented GA based on ontology science, where the algorithm can work in a self-organizing manner to solve text clustering problems. To solve the text clustering problems, the authors used ontology concepts, thesaurus-based concepts, and corpus-based concepts. The authors presented two hybrid schemes that used various similarity metrics, including a thesaurus-based metric and a transformed LS index-based metric that outperformed the traditional similarity metric. Experiment results using the Reuters-21578 dataset show that the proposed text clustering method is more effective than k-means and the original GA.

Lee et al. [56] created a multiobjective GA for text document clustering approach to reduce the leaning of the local smallest value. The Reuters-21578 standard dataset was used to test this method. The experimental results show that the multiobjective GA outperformed the compared techniques by 20%.

Chun-hong et al. [57] introduced an efficient method for text document clustering-based GA and the most recent semantic analysis (LSA). The proposed method addresses the issue of the VSM and the k-means algorithm. The proposed method was tested and compared with the VSM using data from the multiple document automatic abstract corpus in the language technology platform (LTP). The results confirm the proposed method's effectiveness in terms of recall and precision metrics.

Shi and Li [58] suggested a text clustering approach using GA and k-means. The GA is used as the fitness function in the proposed approach, and k-means is used as the convergence principle. This method was tested with both Chinese and English text. The proposed approach for Chinese text clustering constructs an advanced selection technique of initial GA centers and recommends the role of characteristics of numerous parts of speech. The simulation results demonstrated the superiority of CGAM over standard GA and the k-means algorithm.

Casillas et al. [59] presented a GA for text document clustering with a sampling approach to reduce time complexity. The GA calculates an approximation of the best value of k and solves the optimum grouping of the text documents into the number of clusters in this approach. The performance of this approach was evaluated using sets of text documents returned by a search engine query. The proposed work demonstrates that GA with sampling achieves clustering in a time that allows interaction with an examine engine. Furthermore, the GA method with entity-based representation of text documents outperforms lemmas-only representation.

Divyashree and Rayar [60] conducted a comparison of the GA and k-means clustering algorithms in clustering text documents. The presented work was evaluated in terms of f-measure, entropy, and purity using a standard text document clustering dataset.

Karaa et al. [61] proposed a new method for clustering MEDLINE abstracts by combining GA with an agglomerative algorithm and a VSM. MEDLINE is the world's largest biomedical database. The experimental results show that the proposed approach is applicable to any textual dataset and any type of electronic document.

Garg and Gupta [4] used a text document clustering approach in which GA is used early on to determine the initial centroids. The k-means clustering technique is then used to cleanse and generate the appropriate clusters solutions. The clustering results on text clustering standard datasets such as Classic and 20 newsgroups show that the proposed technique outperformed the k-means technique in terms of accuracy.

Mustafi et al. [68] presented a novel text document clustering technique based on GA and employing the nearest neighbor heuristic. To improve cluster separation, the presented technique employs a weighted combination of several criteria as its objective function, with the influence of the nearest neighbor concept. The objective function is proposed to give more weight to traditional metrics on intra and inter-cluster distance. The efficiency of the proposed method was tested using BBC and 20 newsgroups datasets and compared with k-means clustering algorithm.

3.2.4 Firefly algorithm (FA)

Mohammed et al. [32] proposed a hierarchical text clustering approach that involved replacing k-means in a bisect k-means clustering with the firefly algorithm (FA), resulting in a bisect FA for hierarchical clustering. FA works to produce the best clusters at each level of the proposed bisect FA. The proposed bisect FA was validated using the 20 newsgroups dataset. The experimental results show that the proposed method produced more compact and dense clusters than standard k-means, bisect k-means, and c-FA.

Mohammed et al. [33] presented gravity firefly clustering (GF-CLUST), a new clustering algorithm that uses the FA for dynamic document clustering. The GF-CLUST can determine the appropriate number of clusters for a given text collection, which is a difficult problem in document clustering. It identifies documents with high force as centers and generates clusters based on cosine similarity. Following that, potential clusters are chosen, and small clusters are merged to them. Experiments on various document datasets, including 20Newgroups, Reuters-21578, and the TREC collection, were carried out to assess the performance of the proposed GF-CLUST. The purity, F-measure, and entropy results of GF-CLUST outperform those of existing clustering techniques such as k-means, PSO, and practical general stochastic clustering method (pGSCM). Furthermore, when compared to pGSCM, the number of obtained clusters in GF-CLUST is close to the actual number of clusters.

3.2.5 Krill herd algorithm (KHA)

Abualigah et al. [3] proposed a krill herd algorithm (KHA)-based web text clustering approach. Two approaches based on the KHA were presented in this work. The first approach employs the original KHA and all its operators, whereas the second ignores the genetic operators present in the original KHA. The effectiveness of the proposed approaches was evaluated using standard datasets from CSTR, Tr23, Tr4, and Reuters-21578 and compared to the k-means clustering algorithm. The experimental results revealed that the proposed KHA-based text clustering approach outperformed the k-means algorithm in terms of cluster quality, as measured by two commonly used clustering measures: purity and entropy.

Abualigah et al. [20] created a hybrid approach for document and text clustering that combines KHA with the HS algorithm. The goal of this hybridization is to improve KHA's global search capability by incorporating the HS algorithm's global search operator. This improved KHA's exploration search capability by incorporating the distance feature as a new probability feature. The developed method was tested on a variety of benchmark datasets. The experimental results show an increase in accuracy and a faster convergence speed.

Abualigah and Khader [35] proposed a hybrid text clustering method based on improved KHA. The authors modified two variants of the standard KHA to solve the text document clustering problem and obtain accurate clusters. Furthermore, by enhancing the basic KHA, the proposed work presented three variants of the KHA. Because genetic operators insignificantly improve the global search ability in the basic KH, the modification occurs in the ordering of genetic operators, where the crossover and mutation processes are invoked after updating the position of krills by the KH motion calculations. Finally, a hybrid function based on the global best concept was presented for improving the enhanced KHA. The proposed hybrid improved KH algorithm (MMKHA) produced the best results in comparison with k-means, PSO, HS, and GA.

Abualigah et al. [36] proposed a combination of objective functions and modified KHA, dubbed MKHA, for dealing with text document clustering issues. The proposed method inherits the starting population of the KHA from the k-means clustering algorithm, and the clustering decision is based on two combined fitness functions. The proposed method's accuracy, precision, recall, and f-measure were evaluated using LABIC datasets and compared to other clustering techniques such as HHS, HGS, HPSO, and k-means++.

Abualigah [37] presented a novel approach for dealing with text clustering using four different variants of KHA: original KHA, modified KHA, hybrid KHA, and multiobjective hybrid KHA. Each variant is an incremental improvement on the one before it. The proposed approaches' performance was evaluated using the benchmark datasets CSTR, SyskillWebert, tr32, tr12, tr11, tr41, and oh15. The simulation results show that the multiobjective hybrid variant outperforms other variants and outperforms other comparative approaches obtained from the literature.

3.2.6 Social spider optimization (SSO)

Chandran et al. [39] used social spider optimization (SSO) to solve the text document clustering problem. This method's performance was evaluated using datasets from Patent Corpus500 and compared to the k-means clustering algorithm. The results show that SSO-based clustering produces better clusters.

In another study, the same authors [41] presented SSO-based text document clustering using a single cluster approach. In this approach, each spider is considered as a single cluster. This cluster section of spider comprises of two sub modules: centroid and list of text documents closer to centroid. The proposed method returns spiders with best K objective values whose union is equivalent to all text documents existing in the dataset. The effectiveness of the proposed method was tested and validated using standard metrics.

The result shows that the SSO-k-means performed better than k-means-SSO and other methods such as k-means-PSO, k-means—GA, improved bee colony optimization (IBCO), and k-means ABC.

In yet another study, the same authors [41] presented an SSO-based text document clustering using a single cluster approach. Each spider is treated as a separate cluster in this method. This spider's cluster section is made up of two submodules: the centroid and a list of text documents that are closest to the centroid. The proposed method returns spiders with the best K objective values, whose union is equal to the union of all text documents in the dataset. The proposed method's efficacy was tested and validated using standard metrics.

3.2.7 Whale optimization algorithm (WOA)

Gopal and Brunda [42] proposed a text clustering method that incorporates whale optimization algorithm (WOA) and fuzzy c-means (FCM). The WOA was used at first to generate the cluster centroid, which was then passed to the FCM to aid in the creation of robust textual clusters. Using WebKB, re0, and 20 newsgroups standard text datasets, an experimental analysis was performed to assess the effectiveness of the proposed method in text clustering in terms of purity, F-measure, recall, and precision. The obtained results demonstrate the superiority of the proposed method over WOA and FCM.

Sawarn and Deepak [43] presented a method for clustering text documents relying on WOA and semantic similarity. First, TF-IDF was used in the presented approach to select the most informative features from the text dataset. WebPMI and KL divergence were used to assess the semantic similarity of queries. Finally, WOA was used to optimize the entire structure to improve the accuracy of the proposed method. The proposed method was validated and compared to other techniques in the literature using the English language version of the CHIC heritage dataset.

3.2.8 Ant colony optimization (ACO)

Cobo and Rocha [8] used ACO to solve the problem of fuzzy text document clustering. The proposed research utilizes thesaurus mining to generate a language-independent attribute vector for computing the similarities of documents written in various languages. The pheromone trails present in the ACO are used to select the participation values in the clustering procedure in the proposed study. The effectiveness of this method was evaluated using a dataset of research papers in various domains in English and Spanish. The results show that the presented method is more efficient than the compared techniques.

Azaryuon and Fakhar [45] proposed an ACO-based text clustering approach. The authors begin by changing the behavior of the ant movement to improve the original ant's clustering algorithm. In the original algorithm, the movement of the ants is completely random. Furthermore, the algorithm's effectiveness was improved by making ant movements more focused. By changing the rules of ant movement, a condition was provided so that the carrier ant moves to a location with intense similarity to the carried component, and the noncarrier ant moves to a location where a component is surrounded by dissimilar components. Text documents mined from the Reuters 21578 standard dataset were used to test the proposed method. The experimental results demonstrate that the proposed method gives a better average performance compared to the k-means and original ants clustering algorithm.

Nema and Sharma [44] used the ACO algorithm in conjunction with the relevance clustering classification technique for multilabel text grouping. The ant colony algorithm was used to optimize text data features. The proposed method has been tested on several datasets, including the WebKB of the CMU test learning group, the Yahoo dataset, and the Reuters Corpus Volume 1 dataset (RCV1). Other algorithms, such as MLFRC and rank SVM (RSVM), were used to evaluate the performance of the proposed method. The results of the experiments show that the proposed method outperforms both MLFRC and RSVM.

3.2.9 Harmony search (HS)

Cobos et al. [46] proposed a text clustering method based on the hybridization of the global optimum HS algorithm with k-means. The proposed approach algorithm was tested on the Reuters-21578 and DMOZ datasets. The obtained results demonstrated that the proposed algorithm produced promising precision results when compared to other similar methods.

By modeling clustering as an optimization problem, Forsati et al. [47] proposed an HS-based clustering approach. The authors first created a clustering algorithm based on the HS algorithm. The developed HS clustering algorithm was then combined with k-means in three different ways. The first method replaced the k-means algorithm's localized searching mechanism with a global searching mechanism. The k-means algorithm was improved in the second method by making it less reliant on the initial random selection of initial clusters. In the third method, the proposed method's population variance behavior was modeled as a Markov chain. The proposed method's performance was validated using datasets from Politics, TIPSTER, DMOZ collection, Usenet newsgroups, and 20 newsgroups, and compared to standard k-means and GA. The simulation outcome depict that the proposed HS-based approach find better clusters in terms of F-measure, entropy, average distance of document to the centroids, and purity.

Devi and Shanmugam [48] used k-means and the HS algorithm to effectively cluster text. The proposed hybrid method used TF-IDF to calculate the weight of the features, the coverage factor, which defined the percentage of documents containing at least one of the extracted features, and concept factorization. The Reuters dataset was used to test this approach.

Devi and Shanmugam [5] presented another text clustering method based on the use of the HS algorithm in conjunction with the theory factorization method. According to the results of the experiment, theory factorization can improve the effectiveness of the HS algorithm.

Al-Jadir et al. [49] developed a new text document clustering approach based on the differential evolution (DE) crossover with the differential HS algorithm (CDHS) for better exploitation in the search region was presented in. The proposed CDHS achieves highly competitive results. Furthermore, when compared to other related approaches, the proposed method achieves better results.

3.2.10 Differential evolution (DE)

Mustafa et al. [50] proposed a memetic differential evolution algorithm (MDETC) to solve the text clustering problem, with the goal of addressing the effect of combining the DE mutation strategy with the memetic algorithm (MA). This hybridization aims to improve the algorithm's exploitation and exploration capabilities as well as the quality of text clustering. Based on AUC metric, F-measure, and statistical analysis, simulation results using CSTR, tr41, tr23, tr12, tr11, and oh15 standard datasets obtained from the Laboratory of Computational Intelligence (LABIC) demonstrated that the MDETC algorithm outperformed other compared clustering algorithms. Furthermore, the MDETC is compared to state-of-the-art text clustering algorithms and achieves nearly the best results for the standard benchmark datasets.

Saini et al. [51] simulated the clustering of search results as a multiview problem and approached it from an optimization standpoint. Several assessments based on syntactic and semantic similarity metrics were used to perform the clustering. In comparison to the existing approach, the proposed approach incorporates three new assessments based on word mover distance, textual-entailment, and universal sentence encoder, which measure semantics while performing clustering. Simulations were run to evaluate the performance of the proposed approach on standard multiview datasets MORESQUE, AMBIENT, and ODP-2396, and the results were compared to other existing techniques in terms of F-measure.

3.3 Other metaheuristic algorithms

Mirhossein [62] proposed GSA-KHM, a text document clustering approach based on the gravitational search algorithm (GSA) and k-harmonic means (KHM). The proposed GSA-KHM algorithm first applies the KHM algorithm to the dataset, and the cluster centroids obtained are used as individuals in the initial population of the GSA. The dataset's minimum, maximum, and average values are used as the other components of the initial population. The remainder of the starting population is generated at random. The proposed method was tested using text datasets such as Reu01, Re0, and Re1 obtained from Reuters-21587, as well as datasets from the UCI machine learning repository such as Iris, Wine, Glass, CMC, and Cancer. The experimental analysis reveals that the GSA-KHM performs better than GSA-k-means and other methods in both text document and data clustering applications.

Vidyadhari et al. [63] presented an approach for automatic text clustering based on semantic word processing and the enhanced cat swarm optimization (CSO) algorithm. In this approach, the input documents are first passed through the preprocessing stage, which provides the appropriate keyword for feature extraction and clustering. Following that, the appropriate keywords are applied to the wordnet ontology to determine the hyponyms and synonyms of each keyword. The frequency is then calculated for each keyword used to model the text feature library. The entropy is then used to determine which attribute is most important. As a result, the suggested approach is used to assign class labels to generate different clusters of text documents. The performance of the proposed method was tested using the 20 newsgroups and Reuters datasets and compared with ABC, PSO, and GA.

Thirumoorthy and Muneeswaran [64] proposed a hybrid JAYA optimization (HJO) algorithm with crossover and mutation approach for text document dlustering (DC). The model is referred to as HJO-DC. The silhouette index was used by the authors to evaluate the quality of the solution in this case. The proposed method was tested and compared to k-means, k-means, GA, CS, PSO, FA, and GWO on some text clustering benchmark datasets. The experimental results show that the proposed approach outperformed the other approaches in terms of quality.

Abasi et al. [65] proposed a new hybrid optimization approach for effective text clustering, combining the salp swarm algorithm (SSA) and hill climbing algorithm (β-HC). The goal of this hybridization was to improve the quality of the initial solution and the SSA in terms of its local search capability and convergence rate when attempting to split the cluster optimally. The proposed hybrid approach was tested on six standard text clustering datasets and two datasets from scientific articles. The simulation outcome reveals that the proposed hybrid method boosted the solutions with a better performance than other approaches used for text clustering in the literature such as DBSCAN, k-means ++, agglomerative, k-means, and spectral method. Others include the original SSA, PSO, KHA, HS, GA, COA, CMAES, and MVO in terms of recall, f-measure, entropy, accuracy, purity, recall, and convergence rate.

Perumal and Mathivanan [66] developed a type 2 intuitionistic fuzzy clustering and seagull optimization algorithm (SOA) for text document clustering and topic detection, dubbed Type 2 IFCSOA. The Type 2 IFCSOA was used to group documents. The proposed clustering method's performance was evaluated using 20 newsgroups dataset and compared to k-means, FCM, FCM–GA, and FCM-PSO.

Table 2 summarizes the reviewed research works on metaheuristic algorithms in text clustering, and Fig. 1 shows the number of research papers for the most common metaheuristic algorithms used in text clustering.

Table 2 Summary of research works on metaheuristic algorithms in text clustering.

Algorithms	Reference	Methodology	Dataset	Application
PSO	Hasanzadeh et al. [15]	PSO+LSI	Extracted web copus	Clustering
	Karol and Mangat [16]	k-means + PSO, FCM +PSO	Reuters-21587, 20Newsgroup	Clustering
	Cagnina et al. [17]	PSO	Scientific abstracts, news, and short legal documents from the Web	Clustering
	Sarkar et al. [18]	k-means + PSO	Nepali language text documents	Initialized the initial centroids for k-means
	Aghdam and Heidari [19]	PSO+k-means	Reuters-21578	Feature selection
	Song et al. [2]	QPSO + GA	Reuters-21578, 20 newsgroups	Clustering
	Bharti and Singh [12]	PSO+chaotic map + OBL+inertia weight	WebKB, Classic4, Reuters-21578	Feature selection

Table 2 Summary of research works on metaheuristic algorithms in text clustering—cont'd

Algorithms	Reference	Methodology	Dataset	Application
	[20]	PSO+k-means	Reuters-21587, 20 newsgroups	Feature selection
	Daoud et al. [21]	PSO+k-means	Arabic news from BBC, CNN and OSAC	Initializing k-means centroids
	[22]	LBPSO + k-means	TDT2, Reuter 21578, Tr11	Feature selection
	[6]	PSO+spectral clustering	Reuters, 20 newsgroups, TDT2	Feature selection
MVA	[23]	MVO+neighborhood operator	CSTR, 20 newsgroups, Tr12, Tr41, Wap, Classic4	Clustering
	[24]	MVO+k-means	CSTR, 20 newsgroups, Tr12, Tr41, Wap, Classic4	Clustering
	Abasi et al. [25]	MVO+Euclidean distance	CSTR, 20 newsgroups, Tr12, Tr41, Wap, Classic4	Clustering
GWO	Rashaideh et al. [26]	GWO	Nill	Clustering
	Vidyadhari et al. [27]	PGWO + semantic word processing	20Newsgroup, Reuter's	Clustering
	[23]	BGWO + k-means	Tr41, Tr12, Classic4, CSTR, 20 newsgroups	Feature selection
CS	Zaw and Mon [28]	CS+PSO	Web documents	Clustering
	Boushaki et al. [29]	CS+local search method	Classic300, Classic400, Tr23, Tr12, and Iris	Clustering
	Boushaki et al. [30]	CS+LSI	Sports1045, Review1557, K1b, and La2	Outlier detection and clustering
	[31]	CS+ACO	Classic4	Clustering
	Vaijayanthi [7]	CS+k-means	Reuters 21578	k-means cluster centroids initialization
FA	Mohammed et al. [32]	Bisect-FA	20 newsgroups	Clustering
	Mohammed et al. [33]	GFA	20 newsgroups, Reuters-21578 and TREC	Determination number of clusters
KHA	Abualigah et al. [3]	KHA with all operators, KHA without genetic operator	CSTR, Tr23, Tr4, and Reuters-21578	Clustering
	Abualigah et al. [34]	KHA+HSA	20 newsgroups, Reuters-21578 CSTR	Clustering
	Abualigah and Khader [35]	Developed three variants of KHA	LABIC	Clustering
	Abualigah et al. [36]	k-means + MKHA	LABIC	Clustering

Continued

Table 2 Summary of research works on metaheuristic algorithms in text clustering—cont'd

Algorithms	Reference	Methodology	Dataset	Application
	Abualigah [37]	KHA, MKHA, HMKHA	Using CSTR, SyskillWebert, tr32, tr12, tr11, tr41, oh15	Clustering
	Abualigah et al. [38]	KHA+swap mutation scheme	–	Dimensionality reduction
SSO	Chandran et al. [39]	SSO	Patent Corpus500	Clustering
	Chandran et al. [40]	SSO-k-means, k-means-SSO	Patent Corpus500	Clustering
	Chandran et al. [41]	SSO with single cluster approach	Patent Corpus500	Clustering
WOA	Gopal and Brunda [42]	WOA+FCM	WebKB, re0 and 20 newsgroups	Determining cluster centroids
	Sawarn and Deepak [43]	WOA+semantic similarity	English language version of CHIC heritage	Clustering
ACO	Nema and Sharma [44]	ACO	WebKB, Yahoo web page dataset, (RCV1)	Clustering
	Cobo and Rocha [8]	ACO+thesaurus	English and Spanish documents dataset	Clustering
	Azaryuon and Fakhar [45]	ACO	Reuters 21578	Clustering
HS	Cobos et al. [46]	HS+k-means	Reuters-21578 and DMOZ	k-means initialization
	Forsati et al. [47]	HS algorithm + k-means	Politics, TIPSTER, DMOZ collection, Usenet newsgroups, and 20 newsgroups	k-means initialization
	Devi and Shanmugam [48]	k-means + HS algorithm	Reuters 21578	Clustering
	Devi and Shanmugam [5]	HS+Concept Factorization	–	Clustering
	Al-Jadir et al. [49]	DE crossover + Differential HS	–	Clustering
DE	Mustafa et al. [50]	DE+MA	CSTR, tr41, tr23, tr12, tr11, oh15	Clustering
	Saini et al. [51]	DE+word mover distance, textual-entailment, universal sentence encoder	MORESQUE, AMBIENT ODP-2396	Clustering

Table 2 Summary of research works on metaheuristic algorithms in text clustering—cont'd

Algorithms	Reference	Methodology	Dataset	Application
GA	Wang et al. [52]	Fuzzy concept (Rough set)+GA	–	Weight parameters description
	Song and Park [53]	GA+LSM	Reuters-21578	Clustering
	Premalatha and Natarajan [54]	GA with ranked mutation rate and simultaneous mutation operator	CISI, Cranfield and ADI document corpus	Clustering
	Song et al. [55]	GA+ontology, thesaurus-based and corpus-based concepts	Reuters-21578	Clustering
	Lee et al. [56]	Multiobjective GA	Reuters-21578	Clustering
	Chun-hong et al. [57]	GA+LSA	Extracted web copus	Clustering
	Shi and Li [58]	GA+k-means	Chinese and English text	AS a fitness function
	Casillas et al. [59]	GA+sampling technique	Query search data	Number of cluster determination and Clustering
	Divyashree and Rayar [60]	GA, k-means	–	Clustering
	Karaa et al. [61]	GA+Agglomerative algorithm + VSM	MEDLINE Abstract	Clustering
	Garg and Gupta [4]	GA+k-means	Classic, 20 newsgroups	Initialized the initial centroids for k-means
Others	Mirhossein [62]	GSA+KHM	Reu01, Re0, Re1, Iris, Wine, Glass, CMC, Cancer	Clustering
	Vidyadhari et al. [63]	semantic word processing + CSO	20 newsgroups, Reuters	Clustering
	Thirumoorthy and Muneeswaran [64]	JYO+Crossover and Mutation strategies	–	Clustering
	Abasi et al. [65]	SSA+β-HC	–	Clustering
	Perumal and Mathivanan [66]	SOA+Type 2 IFC	20 newsgroups	Clustering and topic detection

4. Conclusion and possible future research focus

Text document clustering is a vital data mining method that is used for unsupervised document categorization, information extraction, and automatic topic mining. In this chapter, we provided a brief overview of several previous research proposed in various databases such as Google scholar, Elsevier, MDPI, Taylor & Francis, Scopus, Web of science, IEEE, and others in the text document clustering domain using various metaheuristic algorithms until the end of 2021.

Metaheuristic optimization algorithms demonstrated effectiveness in handling different categories of text documents clustering difficulties. However, local optima can be trapped because of its focus on exploration instead of exploitation. This problem may be enhanced over time, as to how well the sets of rules governing various search algorithms work are better understood. There are two main issues in the text document clustering applications: the initial cluster centers and the number of clusters. Parameter tweaking will also play an important role in future studies since the parameters' values and settings govern the algorithm's overall performance. The analysis in this chapter shows metaheuristic optimization algorithms are strongly viable for continuing use in text clustering applications. This review will assist researchers in this field by describing how these algorithms have been applied in solving text clustering problems.

Lastly, we recommend some future research focus on text clustering-based metaheuristic optimization algorithms.

1. The most significant aspects of the most widely utilized metaheuristic algorithms in text document clustering may be combined for improved overall performance in dealing with text clustering problems.
2. To deal with text document clustering problems, novel hybrid and modified algorithms can be suggested.
3. Furthermore, in recent years, a slew of new metaheuristic optimization algorithms have been proposed, which can also be used to solve clustering problems.

References

[1] M. Allahyari, S. Pouriyeh, M. Assefi, S. Safaei, E.D. Trippe, J.B. Gutierrez, K. Krys, A brief survey of text mining: classification, clustering and extraction techniques, arXiv (2017) 1–13, https://doi.org/10.48550/arXiv.1707.02919.

[2] W. Song, Y. Qiao, S.C. Park, X. Qian, A hybrid evolutionary computation approach with its application, Expert Syst. Appl. (2015) 2517–2524, https://doi.org/10.1016/j.eswa.2014.11.003.

[3] L.M. Abualigah, A.T. Khader, M.A. Al-Betar, M.A. Awadallah, A krill herd algorithm for efficient text documents clustering, Proceedings of the 2016 IEEE Symposium on Computer Applications & Industrial Electronics (ISCAIE), IEEE, Batu Feringghi, Malaysia, 2016, pp. 67–72.

[4] N. Garg, R. Gupta, Performance evaluation of new text mining method based on GA and K-means clustering algorithm, in: C. R., et al. (Eds.), Advanced Computing and Communication Technologies, Advances in Intelligent Systems and Computing, Springer Nature, Singapore, 2018, pp. 23–30, https://doi.org/10.1007/978-981-10-4603-2_3.

[5] S. Devi, A. Shanmugam, Hybridized harmony search method for text clustering using concept factorization, Int. J. Adv. Comput. Technol (2016) 320–327.

[6] R. Janani, S. Vijayarani, Text document clustering using spectral clustering algorithm with particle swarm optimization, Expert Syst. Appl. (2019) 192–200, https://doi.org/10.1016/j.eswa.2019.05.030.

[7] S.I. Boushaki, O. Bendjeghaba, N. Brakta, Document clustering analysis based on hybrid cuckoo search and K-means algorithm, in: 2021 IEEE 12th Annual Information Technology, Electronics and Mobile Communication Conference (IEMCON), IEEE, Vancouver, BC, Canada, 2021, https://doi.org/10.1109/IEMCON53756.2021.9623204.

[8] A. Cobo, R. Rocha, Document management with ant Colony optimization metaheuristic: A fuzzy text clustering approach using pheromone trails, in: Soft Computing in Industrial Applications, Springer, Berlin/Heidelberg, Germany, 2011, pp. 261–270.

[9] K.K. Bharti, P.K. Singh, Chaotic gradient artificial bee colonyfor text clustering, Soft. Comput. (2015) 1113–1126.

[10] L. Abualigah, A.H. Gandomi, M. Abd Elaziz, H. Al Hamad, M. Omari, M. Alshinwan, A.M. Khasawneh, Advances in meta-heuristic optimization algorithms in big data text clustering, Electronics (2021) 1–29.

[11] K.A. Prabha, N.K. Visalakshi, Improved particle swarm optimization based k-means clustering, in: Proceedings of the IEEE 2014 International Conference Intelligent Computing Applications (ICICA), IEEE, Coimbatore, India, 2014, pp. 59–63.

[12] K.K. Bharti, P.K. Singh, Opposition chaotic fitness mutation based adaptive inertia weight BPSO for feature selection in text clusteringKusum, Appl. Soft Comput. (2016) 1–15, https://doi.org/10.1016/j.asoc.2016.01.019.

[13] N. Zhong, Y. Li, S. Wu, Effective pattern discovery for text mining, Knowl. Data Eng. IEEE Trans (2012) 30–44.

[14] G. Salton, A. Wong, C.S. Yang, A vector space model for automatic indexing, Commun. ACM (1975) 613–620, https://doi.org/10.1145/361219.361220.

[15] E. Hasanzadeh, M.P. Rad, H.A. Rokny, Text clustering on latent semantic indexing with particle swarm optimization (PSO) algorithm, Int. J. Phys. Sci. (2012) 116–120, https://doi.org/10.5897/IJPS11.692.

[16] S. Karol, V. Mangat, Evaluation of text document clustering approach based on particle swarm optimization, Cent. Eur. J. Comput. Sci. (2012) 69–90, https://doi.org/10.2478/s13537-013-0104-2.

[17] L. Cagnina, M. Errecalde, D. Ingaramo, P. Rosso, An efficient particle swarm optimization approach to cluster short texts, Inform. Sci. (2014) 36–49, https://doi.org/10.1016/j.ins.2013.12.010.

[18] S. Sarkar, A. Roy, B. Purkayastha, A comparative analysis of particle swarm optimization and K-means algorithm for text clustering using Nepali wordnet, Int. J. Nat. Lang. Comput. (2014), https://doi.org/10.5121/ijnlc.2014.3308.

[19] M.H. Aghdam, S. Heidari, Feature selection using particle swarm optimization in text categorization, JAISCR (2015) 231–238, https://doi.org/10.1515/jaiscr-2015-0031.

[20] L.M. Abualigah, A.T. Khader, M.A. Al-Betar, O.A. Alomari, Unsupervised text feature selection technique based on particle swarm optimization algorithm with genetic operators for the text clustering, J. Supercomput. (2017) 4773–4797.

[21] A.S. Daoud, A. Sallam, M.E. Wheed, Improving arabic document clustering using k-means algorithm and particle swarm optimization, in: Intelligent Systems Conference 2017, IEEE, London, UK, 2017, pp. 879–885.

[22] N. Kushwaha, M. Pant, Link based BPSO for feature selection in big data text clustering, Futur. Gener. Comput. Syst. (2017), https://doi.org/10.1016/j.future.2017.12.005.

[23] A.K. Abasi, A.T. Khader, M.A. Al-Betar, S. Naim, S.N. Makhadmeh, Z.A.A. Alyasseri, Link-based multi-verse optimizer for text documents clustering, Appl. Soft Comput. (2019) 1–39.

[24] A.K. Abasi, A.T. Khade, M.A. Al-Betar, S. Naim, M.A. Awadallah, O.A. Alomar, Text documents clustering using modified multi-verse optimizer, Int. J. Electr. Comput. Eng. (2020) 6361–6369, https://doi.org/10.11591/ijece.v10i6.pp6361-6369.

[25] A.K. Abasi, A.T. Khader, M.A. Al-Betar, S. Naim, S.N. Makhadmeh, A.Z.A. Alyasseri, A novel ensemble statistical topic extraction method for scientific publications based on optimization clustering, Multimed. Tools. Appl. (2020) 1–46, https://doi.org/10.1007/s11042-020-09504-2.

[26] H. Rashaideh, A. Sawaie, M.A. Al-Betar, L.M. Abualigah, M.M. Al-laham, R.M. Al-Khatib, M. Braik, A Grey wolf optimizer for text document clustering, J. Intell. Syst. (2018) 1–17, https://doi.org/10.1515/jisys-2018-0194.

[27] C. Vidyadhari, N. Sandhya, P. Premchand, Particle Grey wolf optimizer (PGWO) algorithm and semantic word processing for automatic text clustering, Int. J. Uncertain. Fuzz. Knowl.-Based Sys. (2019) 201–223, https://doi.org/10.1142/S0218488519500090.

[28] M.M. Zaw, E.E. Mon, Web document clustering by using PSO-based cuckoo search clustering algorithm, in: X.-S. Y. (Ed.), Recent Advances in Swarm Intelligence and Evolutionary Computation Studies in Computational Intelligence, Springer, Switzerland, 2015, pp. 263–281, https://doi.org/10.1007/978-3-319-13826-8_14.

[29] S. Boushaki, N. Kamel, O. Bendjeghaba, in: Improved Cuckoo Search Algorithm for Document Clustering, 5th International Conference on Computer Science and Its Applications(CIIA), HAL, Saida, Algeria, 2015, pp. 217–228, https://doi.org/10.1007/978-3-319-19578-0_18.

[30] S.I. Boushaki, N. Kamel, O. Bendjeghaba, High-dimensional text datasets clustering algorithm based on cuckoo search and latent semantic indexing, J. Inf. Knowl. Manag. (2018) 1850033, https://doi.org/10.1142/S0219649218500338.

[31] P. Vaijayanthi, Ant Colony and cuckoo search algorithms for document clustering, ARPN J. Eng. Appl. Sci. (2018) 2429–2437.

[32] A.J. Mohammed, Y. Yusof, H. Husni, GF-CLUST: A nature-inspired algorithm for automatic text clustering, J. Inf. Commun. Technol. (2016) 57–81.

[33] A.J. Mohammed, Y. Yusuf, H. Husni, Integrated bisect K-means and firefly algorithm for hierarchical text clustering, J. Eng. Appl. Sci. (2016) 522–527.

[34] L.M. Abualigah, A.T. Khader, M.A. Al-Betar, Multiobjectives-based text clustering technique using K-mean algorithm, in: *2016* 7th International Conference on Computer Science and Information Technology (CSIT), IEEE, Amman, Jordan, 2016, pp. 1–6.

[35] L.M. Abualigah, A.T. Khader, Hybrid clustering analysis using improved krill herd algorithm, Appl. Intell. 1-25 (2018), https://doi.org/10.1007/s10489-018-1190-6.

[36] L.M. Abualigah, A.T. Khader, E.S. Hanandeh, A combination of objective functions and hybrid krill herd algorithm for text document clustering analysis, Eng. Appl. Artif. Intel. (2018) 111–125, https://doi.org/10.1016/j.engappai.2018.05.003.

[37] L.M. Abualigah, Feature selection and enhanced krill herd algorithm for text document clustering, Springer, Berlin/Heidelberg, Germany, 2019, https://doi.org/10.1007/978-3-030-10674-4.

[38] L.M. Abualigah, B. Alsalibi, M. Shehab, M. Alshinwan, A. Khasawneh, H. Alabool, A Parallel Hybrid Krill Herd Algorithm for Feature Selection, Int. J. Mach. Learn, Cybern, 2020.

[39] T.R. Chandran, A.V. Reddy, B. Janet, A social spider optimization approach for clustering text documents, in: International conference on advances in electrical, electronics, information, communication and bio-informatics (AEEICB16), IEEE, 2016, pp. 2–5.

[40] T.R. Chandran, A.V. Redd, B. Janet, Text clustering quality improvement using a hybrid social spider optimization, Int. J. Appl. Eng. Res. (2017) 995–1008.

[41] T.R. Chandran, A.V. Reddy, B. Janet, An effective implementation of social spider optimization for text document clustering using single cluster approach, in: Proceedings of the 2nd international conference on inventive communication and computational technologies (ICICCT 2018), IEEE, 2018, pp. 1–4.

[42] J. Gopal, S. Brunda, Text clustering algorithm using fuzzy whale optimization algorithm, Int. J. Intell. Eng. Syst. (2019) 278–286, https://doi.org/10.22266/ijies2019.0430.27.

[43] S. Sawarn, G. Deepak, An approach for document clustering using semantic similarity and whale optimization: artificial intelligence systems and the internet of things in the digital era, in: European, Asian, Middle Eastern, North African Conference on Management & Information Systems (EAMMIS 2021), Springer, 2021, pp. 322–333.

[44] P. Nema, V. Sharma, Multi-label text categorization based on feature optimization using ant colony optimization and relevance clustering technique, in: 2015 International Conference on Computers, Communications, and Systems, IEEE, 2015, pp. 1–15.

[45] K. Azaryuon, B. Fakhar, A novel document clustering algorithm based on ant Colony optimization algorithm, J. Math. Comput. Sci. (2013) 171–180.

[46] C. Cobos, J. Andrade, W. Constain, M. Mendoza, E. León, Web document clustering based on global-best harmony search, K-means, frequent term sets and Bayesian information criterion, in: Proceedings of the IEEE Congress on Evolutionary Computation, IEEE, Barcelona, Spain, 2010, pp. 18–23.

[47] R. Forsati, M. Mahdavi, M. Shamsfard, M.R. Meybodi, Efficient stochastic algorithms for document clustering, Inform. Sci. (2013) 269–291, https://doi.org/10.1016/j.ins.2012.07.025.

[48] S.S. Devi, A. Shanmugam, A proficient method for text clustering using harmony search method E Dhivya Prabha, Int. J. Sci. Res. Sci., Eng. Technol. (2015) 145–150.

[49] I. Al-Jadir, K. Wong, C. Fung, H. Xie, Adaptive crossover memetic differential harmony search for optimizing document clustering, in: Proceedings of the International Conference on Neural Information Processing, Springer, Berlin/Heidelberg, Germany, 2018, pp. 509–518.

[50] H.M. Mustafa, M. Ayob, D. Albashish, S. Abu-Taleb, Solving text clustering problem using a memetic differential evolution algorithm, PLoS One (2020) 1–18, https://doi.org/10.1371/journal.pone.0232816.

[51] N. Saini, D. Bansal, S. Saha, P. Bhattacharyya, Multi-objective multi-view based search result clustering using differential evolution framework, Expert Syst. Appl. 114299 (2021), https://doi.org/10.1016/j.eswa.2020.114299.

[52] M.C. Wang, Z. Wang, O., Text Fuzzy Clustering Algorithm Based on Rough Set and Genetic Algorithm, J. Electron. Inf. Technol. (2005). Retrieved from http://en.cnki.com.cn/Article_en/CJFDTotal-DZYX200504011.htm.

[53] W. Song, S.C. Park, Genetic algorithm for text clustering based on latent semantic indexing, Comput. Math. Appl. (2009), https://doi.org/10.1016/j.camwa.2008.10.010.

[54] K. Premalatha, A.M. Natarajan, Genetic algorithm for document clustering with simultaneous and ranked mutation, Mod. Appl. Sci. (2009) 75–82.

[55] W. Song, C.H. Li, S.C. Park, Genetic algorithm for text clustering using ontology and evaluating the validity of various semantic similarity measures, Expert Syst. Appl. (2009) 9095–9104, https://doi.org/10.1016/j.eswa.2008.12.046.

[56] J.S. Lee, L.C. Choi, S.C. Park, Document clustering using multi-objective genetic algorithm with different feature selection methods, in: Proceedings of the International Workshop on Semantic Interoperability (IWSI-2011), SCITEPRESS, 2011, pp. 101–110.

[57] W. Chun-hong, N. Li-Li, R. Yao-Peng, Research on the text clustering algorithm based on latent semantic analysis and optimization, in: *Proceedings of the 2011* IEEE International Conference on Computer Science and Automation Engineering, China, Shanghai, 2011, pp. 1470–1473.

[58] K. Shi, L. Li, High performance genetic algorithm based text clustering using parts of speech and outlier elimination, Appl. Intell. (2013) 511–519.

[59] A. Casillas, M.T. Lena, R. Martinez, in: Sampling and Feature Selection in a Genetic Algorithm for Document Clustering, International Conference on Intelligent Text Processing and Computational Linguistics, Springer, 2004, pp. 601–612.

[60] G. Divyashree, G. Rayar, Comparison between K-means and genetic algorithm in text document clustering, Int. J. Eng. Res. Technol. (2015) 1–10.

[61] W.B. Karaa, A.S. Ashour, D.B. Sassi, P. Roy, N. Kausar, N. Dey, A.-E. H, et al., MEDLINE text mining: an enhancement genetic algorithm based approach for document clustering, in: Applications of Intelligent Optimization in Biology and Medicine, 2016, pp. 267–287. Switzerland https://doi.org/10.1007/978-3-319-21212-8_12.

[62] M. Mirhossein, A clustering approach using a combination of gravitational search algorithm and k-harmonic means and its application in text document clustering, Turk. J. Elec. Eng. & Comp. Sci. (2017) 1251–1262, https://doi.org/10.3906/elk-1508-31.

[63] C. Vidyadhari, N. Sandhya, P. Premchand, A semantic word processing using enhanced cat swarm optimization algorithm for automatic text clustering, Int. J. Multimed. Res. (2019) 23–32, https://doi.org/10.46253/j.mr.v2i4.a3.

[64] K. Thirumoorthy, K. Muneeswaran, A hybrid approach for text document clustering using Jaya optimization algorithm, Expert Syst. Appl. 115040 (2021), https://doi.org/10.1016/j.eswa.2021.115040.

[65] A.K. Abasi, A.T. Khader, M.A. Al-Betar, Z.A. Alyasseri, S.N. Makhadmeh, M. Al-laham, S. Naim, A hybrid Salp swarm algorithm with β-hill climbing algorithm for text documents clustering, in: I.I. Aljarah, H. Faris, S. Mirjalili (Eds.), Evolutionary Data Clustering: Algorithms and Applications. Algorithms for Intelligent Systems, Springer, Singapore, 2021, pp. 129–161, https://doi.org/10.1007/978-981-33-4191-3_6.

[66] P. Perumal, B. Mathivanan, Type2 IFC with SOA for topic detection and document clustering analysis, Res. Square (2021) 1–27, https://doi.org/10.21203/rs.3.rs-743089/v1.

[67] A.K. Abasi, A.T. Khader, M.A. Al-Betar, S. Naim, Z.A. Alyasseri, S.N. Makhadmeh, Clustering, A novel hybrid multi-verse optimizer with K-means for text documents clustering, Neural Comput. Applic. (2020) 1–27, https://doi.org/10.1007/s00521-020-04945-0.

[68] D. Mustafi, A. Mustafi, G. Sahoo, A novel approach to text clustering using genetic algorithm based on the nearest neighbour heuristic, Int. J. Comput. Appl. (2020) 1–12, https://doi.org/10.1080/1206212X.2020.1735035.

CHAPTER

Application of metaheuristic algorithms in optimal design of sewer collection systems

Mohammadali Geranmehr[a], Mohammad Reza Nikoo[b], Ghazi Al-Rawas[b], Khalifa Al-Jabri[b], and Amir H. Gandomi[c,d]

[a]*Department of Civil and Structural Engineering, University of Sheffield, Sheffield, United Kingdom,* [b]*Department of Civil and Architectural Engineering, College of Engineering, Sultan Qaboos University, Muscat, Oman,* [c]*Faculty of Engineering and Information Technology, University of Technology Sydney, Ultimo, NSW, Australia,* [d]*University Research and Innovation Center, Obuda University, Budapest, Hungary*

1. Introduction

Sewer collection systems are critical infrastructures of modern cities due to their vital role in public health. In general, sewage collection networks are responsible for collecting sewage from different sectors of the city and transferring it to the treatment plant. Depending on the type of wastewater, these networks are typically either separated (i.e., include only sanitary wastewater) or combined (i.e., sanitary wastewater and surface runoff). These systems also collect sewage by gravity and are therefore based on open hydraulic channel principles. Herein, we focus only on gravity-based systems since other types of sewer networks, such as pressurized sewer networks or vacuum sewer networks, are out of the scope of the chapter.

Designing gravity networks for collecting sanitary wastewater has always been a challenge for engineers due to their high construction costs and the need for an optimal design. The smaller the diameter and slope of the network's pipes, the more economical the network will be. However, the flow hydraulics, constraints, regulations, and the network's layout, especially in cities with an almost flat topography, are fundamental variables that can complicate matters.

In this field, researchers aim to achieve an optimal network design based on one or both hydraulic factors (i.e., diameter and slope of pipes) and network layout. Accordingly, research on optimizing sewer network designs can be divided into two categories. The first category includes the use of mathematical and classical methods, which is beyond the scope of this chapter. The second category is based on metaheuristic algorithms, which are discussed here.

Comprehensive Metaheuristics. https://doi.org/10.1016/B978-0-323-91781-0.00008-9

2. Sewer collection systems

2.1 Basics

The hydraulic flow in a sewer pipe is based on the flow of its incoming sewage. In this case, assuming a uniform flow, the Manning equation can be used:

$$Q = VA = \frac{1}{n} R^{\frac{2}{3}} S^{\frac{1}{2}} A \tag{1}$$

where Q is the flow discharge (m^3/s), V is the velocity (m/s), A is the cross-sectional area of the flow (m^2), S is the slope, and n is the Manning roughness coefficient that is related to the pipe material. R is the hydraulic radius (m), which is equal to the ratio of A to P, the wetted perimeter (m) as follows:

$$A = 0.25D^2\left(\cos^{-1}(1-2e) - 2(1-2e)\sqrt{e(1-e)}\right) = 0.125D^2(\theta - \sin\theta) \tag{2}$$

$$P = D\cos^{-1}(1-2e) = 0.5\theta D \tag{3}$$

$$R = \frac{A}{P} = \frac{D}{4}\left(1 - \frac{\sin\theta}{\theta}\right) \tag{4}$$

where D is the diameter of the pipe (m) and θ is the central angle of the flow section (Fig. 1). e is the relative flow depth (RFD), which is equal to the ratio of the flow depth (y) to the diameter of the pipe (D). y and θ can be determined as follows:

$$y = \frac{D}{2}\left(1 - \cos\frac{\theta}{2}\right) \tag{5}$$

$$\theta = 2\cos^{-1}(1-2e) \tag{6}$$

2.2 Layout design

Generally, gravity sewer collection networks are based on a branched layout that follows the city's topography. In many cases, especially in almost flat cities, choosing the optimal layout of pipes is essential to form a branch network. In this context, graph theory and innovative methods are typically used [1].

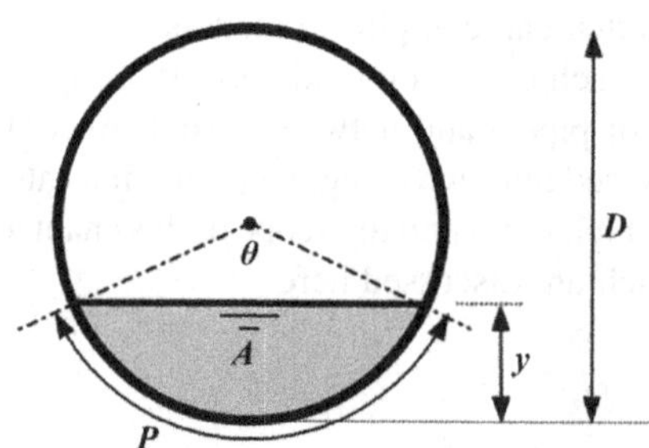

FIG. 1

Circular cross-section with semi-full flow.

2.3 Hydraulic design

The hydraulic design of the network, assuming a pre-determined layout, involves selecting the appropriate diameter and slope for every pipe. In this case, designing an economic network with suitable hydraulic performance is the primary goal. The network's hydraulic performance depends on hydraulic capacity, flow velocity, and relative flow depth in the pipes.

2.4 Objective function

Cost is the main and only objective function of almost all research on the optimal design of sewer networks. However, other objective functions, such as reliability and network performance [2], have also been reported. In general, the sewer network cost includes the costs of (1) pipes, (2) excavating and installing pipes, and (3) construction of manholes [3]. These three factors are usually considered in the objective function.

2.5 Constraints

Sewer network design constraints, which are typically specified by design codes, are divided into two categories as mechanical constraints and hydraulic constraints. Mechanical constraints require that:

(1) diameter of pipes are not reduced in the flow direction
(2) the crown of the outlet pipe from each manhole must not be higher than the crown of the inlet pipes
(3) minimum allowable cover depth on the pipe

Hydraulic constraints can also include the following:

(1) minimum and maximum velocity
(2) minimum and maximum relative flow depth
(3) minimum and maximum slope
(4) maximum Froude number

An optimal sewer network design can include one or more of the preceding constraints. However, applying all the constraints can pose a serious challenge to the optimization problem, especially when using metaheuristic algorithms.

3. Metaheuristic algorithms

Metaheuristic algorithms are mainly employed to find optimal solutions to problems that do not have a clear or straightforward mathematical solution. These methods are usually based on purposeful trial-and-error, which results in a specific number of repetitions or solutions. Subsequently, we describe some common algorithms used to optimize sewer collection networks in the sections that follow.

3.1 Ant colony optimization (ACO)

Ant colony optimization (ACO) is a metaheuristic algorithm that simulates the foraging behavior of an ant colony to discover the shortest path to the food. As ants search for food, they leave behind pheromones that attract other ants to follow their path. Specifically, ants are more likely to select a path with a higher concentration of pheromones. As more ants move in paths with higher pheromone intensities, the pheromone in those pathways increases, which increases the chance that other ants will choose the same path. In ACO, various solutions or paths are generated based on this behavior to find the optimal solution [4].

3.2 Genetic algorithm (GA)

The genetic algorithm (GA), which is based on the natural selection procedure, is one of the most effective optimization methods. In this technique, every decision variable is treated as a gene, and the aggregate of all the problem variables forms a chromosome. Each member of the population is evaluated by the objective function. The crossover operator is used on a population subset to produce new offspring, whereby the attributes of each child are a combination of those of their parents. The mutation operator also ensures that the algorithm does not become trapped in local optima. This operator is applied to a part of the mating population and randomly modifies some selected chromosome genes to generate new results. These operations are repeated based on a stop criterion for improving the objective function's value or the number of time steps [5].

3.3 Particle swarm optimization (PSO)

The particle swarm optimization (PSO) technique is based on swarm intelligence and is inspired by the swarming movement of birds and fish. Each particle in this method indicates a possible solution to the main optimization problem. First, an infinite number of particles are created and analyzed. Then, these particles move to the optimal place through an iterative process. Each particle's motion is determined by three factors: its prior motion, the location of the best position in the current time step, and the absolute best position of all particles recorded so far. A set number of iterations or a significant improvement in the objective function can be used to stop the algorithm [6].

3.4 Simulated annealing (SA)

The simulated annealing (SA) technique, which is motivated by the annealing process in metallurgy, is a simple and effective metaheuristic optimization algorithm for large search spaces that solves optimization problems. The SA method begins with an initial answer and is followed by an improvement process in an iterative loop to solve an optimization problem. The algorithm sets the neighbor's response as the current answer if it is better than the current solution; otherwise, it accepts the current answer as the solution with $\exp(-\Delta E/T)$ probability. In the latter equation, ΔE is the difference between the present answer and the neighbor's response, and T is a temperature parameter. The temperature is gradually decreased, and several repetitions are performed at each temperature. Initially, the temperature is set high in the first steps, so that it is more possible to accept poor responses. As the temperature drops, there is less possibility of accepting bad responses in subsequent steps, and thus the algorithm converges toward a good result [7].

3.5 Tabu search (TS)

The Tabu search (TS) algorithm starts with an initial response to discover the best solution to an optimization problem. The algorithm finds the best neighbor's answer among the current neighbors' answers. If this answer is not on the forbidden list, the algorithm continues to the next answer; if it is forbidden, the program checks a metric known as respiration measurement. According to the breathing criterion, the algorithm will move to the neighbor's solution if it is better than the best answer collected so far, even though it is on the forbidden list. The banned list is updated after the algorithm advances to the neighbor's answer. This means that the previous move to the neighbor's answer is placed in the forbidden list to prevent the algorithm from returning to it and establishing a cycle. In actuality, the banned list is a mechanism that prohibits the algorithm from being ideally situated locally. Following the placement of the previous move on the forbidden list, several previously banned moves will be removed from the list. Various termination conditions can be considered in the TS algorithm, such as limiting the number of moves based on a neighbor's answer [8].

4. Applications

Table 1 summarizes some recent studies on sewer network optimization using ACO, GA, PSO, SA, and TS algorithms. Details of each study include the metaheuristic algorithm(s) used, objective function(s), decision variables, constraints, and how constraints were satisfied. As shown, cost is the objective

Table 1 Overview of sewer network optimization using metaheuristic algorithms.

Research	Category	Objective function(s)	Decision variables	Constraints	Constraints handling	Details
Liang et al. [9]	GA, TS	Cost	Pipes' diameters and slopes	Min V, Max V, Min Cover, Max Cover, Min Slope	Systematically satisfied	Used both GA and TS
Izquierdo et al. [10]	PSO	Cost	pipes' diameters and slopes	Min V, Max V, Min RFD, Max RFD, Min Cover, Min Slope, Max Slope	Penalty functions	–
Pan and Kao [11]	GA	Cost	Pipes' diameters and slopes	Min V, Max V, Max RFD, Min Cover, Min Slope	Hybridizing	Developed a hybrid GA and quadratic programming (GA-QP)
Haghighi and Bakhshipour [12]	GA	Cost	Pipes' diameters and slopes	Min V, Max V, Max RFD, Min Cover, Min Slope	Systematically satisfied	Used an adaptive genetic algorithm (AGA)

Continued

Table 1 Overview of sewer network optimization using metaheuristic algorithms—cont'd

Research	Category	Objective function(s)	Decision variables	Constraints	Constraints handling	Details
Moeini and Afshar [13]	ACO	Cost	Layout, pipes' diameters and slopes	Min V, Max V, Min RFD, Max RFD, Min Cover, Max Cover, Min Slope	Systematically satisfied	Developed a constrained version of ACO coupled with a tree growing algorithm (TGA)
Yeh et al. [14]	SA, TS	Cost	Pipes' diameters and slopes	Min V, Max V, Max RFD, Min Slope	Systematically satisfied	Used both TS and SA
Karovic and Mays [15]	SA	Cost	Pipes' diameters and slopes	Min V, Max V, Min Cover, Min Slope	Penalty functions	Used SA optimization in Excel
Haghighi and Bakhshipour [16]	TS	Cost	Layout, pipes' diameters and slopes	Min V, Max V, Max RFD, Min Cover, Min Slope	Systematically satisfied	Used TS to find optimal layout, diameter, and slope simultaneously
Haghighi and Bakhshipour [17]	SA	Cost and reliability	Layout, pipes' diameters and slopes	Min V, Max V, Max RFD, Min Cover, Min Slope	Hybridizing	Used SA for layout optimization and discrete differential dynamic programming (DDDP) for hydraulic design
Navin and Mathur [18]	PSO	Cost	Layout, pipes' diameters and slopes	Miv V, Max V, Min Cover, Max Cover, Max RFD	Systematically satisfied	Used modified particle swarm optimization (MPSO) and spanning tree
Steele et al. [19]	SA	Cost	Layout, pipes' diameters and slopes	Min V, Max V, Min Cover, Min Slope	Hybridizing	Used SA optimization and general algebraic modeling system (GAMS)
Moeini and Afshar [20]	ACO	Cost	Layout, pipes' diameters and slopes	Min V, Max V, Min RFD, Max RFD, Min Cover, Max Cover, Min Slope	Systematically satisfied	Developed arc based ant colony optimization slgorithm (ABACOA)

Table 1 Overview of sewer network optimization using metaheuristic algorithms—cont'd

Research	Category	Objective function(s)	Decision variables	Constraints	Constraints handling	Details
de Villiers, et al. [21]	ACO	Cost	Layout, pipes' diameters and slopes	Min V, Max V, Min Cover, Max Cover, Min Slope, Min Capacity	Penalty functions	Used ACO for layout optimization and a heuristic algorithm for the optimal hydraulic design
Ahmadi et al. [22]	PSO	Cost	Pipes' diameters and slopes	Min V, Max V, Min RFD, Max RFD, Min Diameter, Max Diameter, Min Slope, Max Slope, Min Cover	Hybridizing	Used fly-back and harmony memory particle swarm optimization (HPSO) combined with dynamic programming (DP)
Moeini and Afshar [23]	ACO	Cost	Pipes' diameters and slopes	Min V, Max V, Min RFD, Max RFD, Min Cover, Max Cover, Min Slope	Penalty method and nonlinear programming (NLP)	Developed ACOA-NLP, an ACOA, hybridized with nonlinear programming (NLP)
Heydarzadeh et al. [2]	PSO	Cost and Performance indices	Pipes' diameters and slopes	–	–	Used a single-objective optimization model by combining six objective functions based on the weighting coefficients
Atiyah and Hassan [24]	GA	Cost	Pipes' diameters and slopes	Min V, Max V, Min Slope, Max Slope	Penalty functions	Proposed an AGA

function in all cases. Both the diameter and slope of pipes are applied as decision variables in all studies, while the layout of the network was also considered in some research.

As mentioned, the problem of optimizing sewer networks is highly constrained and satisfying these constraints with evolutionary algorithms has always been a significant challenge. In this context, three methods are used to achieve optimal results: (1) defining penalty functions added to the objective function, (2) customizing the optimization algorithm to satisfy constraints systematically, and (3) hybridization of algorithms.

5. Summary and conclusion

This chapter investigates the application of metaheuristic algorithms in the optimal design of sanitary sewer collection networks. Sewer collection networks are an essential urban infrastructure directly related to community health. However, the high cost of building these networks forces us to use optimization tools to design economical networks. Due to the nonlinearity of the governing hydraulic equations in such optimization problems, metaheuristic algorithms are commonly applied in this field. Nevertheless, the existence of many constraints on the optimal design of sewer networks presents a major challenge. In this regard, to achieve better performance, researchers use various methods including penalty functions, personalization of algorithms to systematically satisfy constraints, or hybridization with other algorithms or methods.

References

[1] A. Haghighi, Loop-by-loop cutting algorithm to generate layouts for urban drainage systems, J. Water Resour. Plan. Manag. 139 (6) (2013) 693–703, https://doi.org/10.1061/(ASCE)WR.1943-5452.0000294.

[2] R. Heydarzadeh, M. Tabesh, M. Scholz, Multiobjective optimization in sewer network design to improve wastewater quality, J. Pipeline Syst. Eng. Pract. 10 (4) (2019) 04019037, https://doi.org/10.1061/(ASCE)PS.1949-1204.0000416.

[3] H. Safavi, M.A. Geranmehr, Optimization of sewer networks using the mixed-integer linear programming, Urban Water J. 14 (5) (2017) 452–459, https://doi.org/10.1080/1573062X.2016.1176222.

[4] M. Dorigo, G. Di Caro, The ant colony optimization meta-heuristic, in: New Ideas in Optimization, McGraw-Hill, 1999.

[5] D.E. Goldberg, Genetic Algorithms in Search, Optimization and Machine Learning, Addison-Wesley Longman Publishing Co., Inc., 1989.

[6] J. Kennedy, R. Eberhart, Particle swarm optimization, in: Paper Presented at the International Conference on Neural Networks, Perth, Western Australia, 1995, https://doi.org/10.1109/ICNN.1995.488968.

[7] S. Kirkpatrick, C.D. Gelatt, M.P. Vecchi, Optimization by simulated annealing, Science 220 (4598) (1983) 671–680, https://doi.org/10.1126/science.220.4598.671.

[8] A. Hertz, D. de Werra, The tabu search metaheuristic: how we used it, Ann. Math. Artif. Intell. 1 (1) (1990) 111–121, https://doi.org/10.1007/BF01531073.

[9] L.Y. Liang, R.G. Thompson, D.M. Young, Optimising the design of sewer networks using genetic algorithms and tabu search, Eng. Constr. Archit. Manag. 11 (2) (2004) 101–112, https://doi.org/10.1108/09699980410527849.

[10] J. Izquierdo, I. Montalvo, R. Pérez, V.S. Fuertes, Design optimization of wastewater collection networks by PSO, Comput. Math. Appl. 56 (3) (2008) 777–784, https://doi.org/10.1016/j.camwa.2008.02.007.

[11] T. Pan, J. Kao, GA-QP model to optimize sewer system design, J. Environ. Eng. 135 (1) (2009) 17–24, https://doi.org/10.1061/(ASCE)0733-9372(2009)135:1(17).

[12] A. Haghighi, A. Bakhshipour, Optimization of sewer networks using an adaptive genetic algorithm, Water Resour. Manag. 26 (12) (2012) 3441–3456, https://doi.org/10.1007/s11269-012-0084-3.

[13] R. Moeini, M.H. Afshar, Sewer network design optimization problem using ant colony optimization algorithm and tree growing algorithm, in: Paper Presented at the EVOLVE—A Bridge Between Probability, Set Oriented Numerics, and Evolutionary Computation IV, Heidelberg, 2013.

[14] S. Yeh, Y. Chang, M. Lin, Optimal design of sewer network by tabu search and simulated annealing, in: Paper Presented at the 2013 IEEE International Conference on Industrial Engineering and Engineering Management, 2013, https://doi.org/10.1109/IEEM.2013.6962687.

[15] O. Karovic, L. Mays, Sewer system design using simulated annealing in excel, Water Resour. Manag. 28 (13) (2014) 4551–4565, https://doi.org/10.1007/s11269-014-0750-8.

[16] A. Haghighi, A. Bakhshipour, Deterministic integrated optimization model for sewage collection networks using tabu search, J. Water Resour. Plan. Manag. 141 (1) (2015) 04014045, https://doi.org/10.1061/(ASCE)WR.1943-5452.0000435.

[17] A. Haghighi, A.E. Bakhshipour, Reliability-based layout design of sewage collection systems in flat areas, Urban Water J. 1-13 (2015), https://doi.org/10.1080/1573062X.2015.1036085.

[18] P.K. Navin, Y.P. Mathur, Layout and component size optimization of sewer network using spanning tree and modified PSO algorithm, Water Resour. Manag. 30 (10) (2016) 3627–3643, https://doi.org/10.1007/s11269-016-1378-7.

[19] J.C. Steele, K. Mahoney, O. Karovic, L.W. Mays, Heuristic optimization model for the optimal layout and pipe design of sewer systems, Water Resour. Manag. 30 (5) (2016) 1605–1620, https://doi.org/10.1007/s11269-015-1191-8.

[20] R. Moeini, M.H. Afshar, Arc based ant Colony optimization algorithm for optimal design of gravitational sewer networks, Ain Shams Eng. J. 8 (2) (2017) 207–223, https://doi.org/10.1016/j.asej.2016.03.003.

[21] N. de Villiers, G.C. van Rooyen, M. Middendorf, Sewer network design layout optimisation using ant colony algorithms, J. South Afr. Inst. Civil Eng. 60 (2018) 2–15, https://doi.org/10.17159/2309-8775/2018/v60n3a1.

[22] A. Ahmadi, A. Zolfagharipoor Mohammad, M. Nafisi, Development of a hybrid algorithm for the optimal design of sewer networks, J. Water Resour. Plan. Manag. 144 (8) (2018) 04018045, https://doi.org/10.1061/(ASCE)WR.1943-5452.0000942.

[23] R. Moeini, M. Afshar, Hybridizing ant colony optimization algorithm with nonlinear programming method for effective optimal design of sewer networks, Water Environ. Res. 91 (2019), https://doi.org/10.1002/wer.1027.

[24] R.H. Atiyah, W.H. Hassan, Optimum design of sewer networks with pump station using genetic algorithms, J. Phys. Conf. Ser. 1973 (1) (2021), 012187, https://doi.org/10.1088/1742-6596/1973/1/012187.

CHAPTER

Space truss structures' optimization using metaheuristic optimization algorithms

9

Nima Khodadadi[a], Farhad Soleimanian Gharehchopogh[b], Benyamin Abdollahzadeh[b], and Seyedali Mirjalili[c,d]

[a]*Department of Civil and Environmental Engineering, Florida International University, Miami, FL, United States,* [b]*Department of Computer Engineering, Urmia Branch, Islamic Azad University, Urmia, Iran,* [c]*Centre for Artificial Intelligence Research and Optimisation, Torrens University Australia, Fortitude Valley, Brisbane, QLD, Australia,* [d]*University Research and Innovation Center, Obuda University, Budapest, Hungary*

1. Introduction

Optimization problems can be found in any field. There is typically a set of unknowns in such problems, which are often called variables or parameters. The goal is to find the best value for these variables to achieve a goal. This goal is usually to maximize or minimize an objective function. So, the output is a function of the inputs. In the literature, the objective function can also be defined as fitness, cost, or merit function. With just these two components, an unconstrained optimization problem can be formulated as a minimization problem as follows without the loss of generality:

$$\underset{x \in R^n}{\text{Minimize:}} f(x)$$

where $f: R^n \rightarrow R$, x is a real vector, and $n \geq 1$.

In this formulation R^n is the set of all possible solutions for the optimization problem and x is a member of this set. For this problem, there is one x in R^n that provides the least output, called the global minimum.

If a problem also has some constraints, it is no longer an unconstrained problem. A solution x is acceptable if it does not violate the constraint. The preceding formulation can be changed to represent such constrained problems as a minimization problem as follows:

$$\underset{x \in R^n}{\text{Minimize:}} f(x)$$

$$\text{Subject to: } g_i(x) \leq 0$$
$$h_j(x) = 0$$

where $f: R^n \rightarrow R$, x is a real vector, $n \geq 1$, g_i shows ith inequality constraint, and h_j indicates jth equality constraint.

Comprehensive Metaheuristics. https://doi.org/10.1016/B978-0-323-91781-0.00009-0

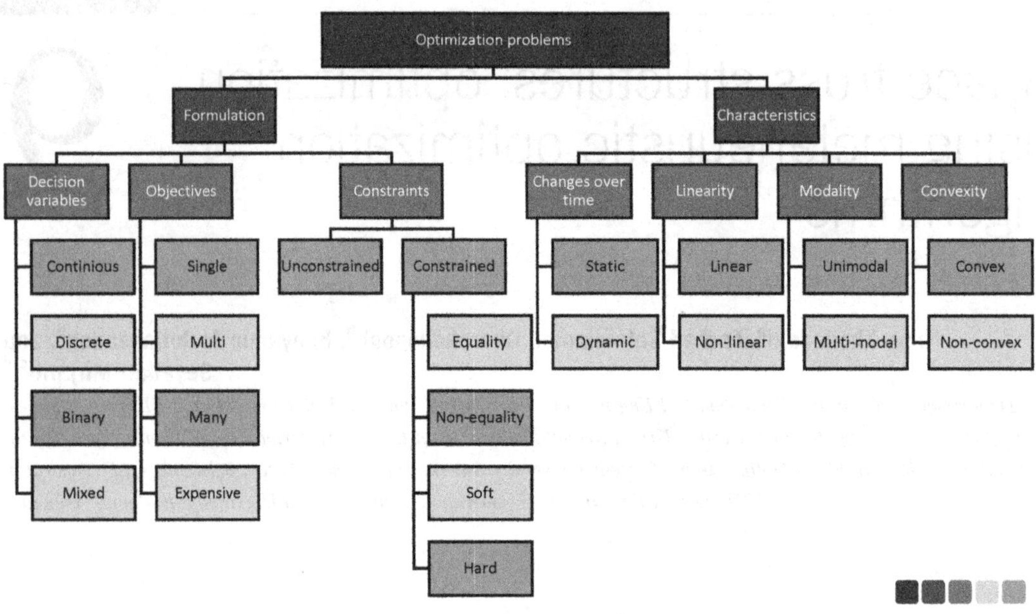

FIG. 1

A classification of optimization problems based on their formulations and characteristics [1].

In constrained optimization problems, we try to find the best solution (x) that gives the minimum f while not violating any of g_i and h_j. In the problem formulation, it is evident that both types of constraints are also functioning of x, so it is safe to say that the constraints are only applied to the inputs of an optimization system.

The preceding two examples classified optimization problems into two classes: unconstrained and constrained. However, they are other types as well. Fig. 1 provides a classification of optimization problems based on how we formulate them and their characteristics by nature [1]. It can be seen in this figure that when categorizing optimization problems based on the way we formula them, they can be divided into several classes based on how the formulation is done considering the three components mentioned previously: variables, objectives, and constraints.

From the variable's perspective, a problem can have continuous, discrete, binary, or mixed variables. Based on the objectives, a problem may be one objective (single), between two and five (multi), or more than six (many). Note that these numbers are arbitrary, but many-objective problems are increasingly emerging and there is no clear definition from them as to when multiple objectives are considered too many. The last notable type of objective function under this category is computationally expensive ones. Such problems require special mechanism to solve in a reasonable time and with reasonable computational resources. Fig. 1 also shows that constraints can be divided into inequality, equality, soft, and hard constraints.

On the right subtree of Fig. 1, optimization problems are categorized based on their characteristics, such as changing over time, linearity, modality, and convexity. Based on this classification, an optimization problem can be static, dynamic, linear, nonlinear, unimodal, multimodal, convex, or nonconvex. Understanding the type of an optimization problem is a steppingstone to choosing the right optimization algorithm for solving it.

Regardless of the type of optimization problem, the literature shows that metaheuristics is the choice of many scientists and industry experts to solve them. This is predominantly due to their gradient-free mechanism and a high chance of avoiding locally optimal solutions. A mathematical optimization algorithm usually requires calculating the definitive of an optimization problem to move an initial, random solution towards an optimum. If the derivative is unknown, the algorithm is no longer effective.

Heuristic algorithms use a heuristic function to converge towards an optimum in conjunction with the objective function. In other words, the heuristic function allows a heuristic algorithm to make educated decisions for updating a solution. However, the heuristic function is specific to the problem and cannot be generalized. Metaheuristics only rely on the objective function and consider an optimization problem as a black box. So, there is no need to calculate the derivative of the objective function.

Metaheuristics can be divided into three classes based on inspiration, search mode, and a number of solutions. (See Fig. 2.) From the inspiration perspective, an algorithm might have been inspired by evolution, swarm, physical rules, or other sources of inspiration (e.g., society, politics, etc.). A metaheuristic can be designed only to perform a local search, a global search, or both in terms of the search model. In challenging optimization problems (nonlinear, nonconvex, etc.). Algorithms that use one solution or a set of solutions create the final class of metaheuristics. It is considered a single-solution algorithm if an algorithm starts, improves, and finishes with a single solution. If more than one solution is involved in this process, the algorithm is called population based.

Some of the most popular single-solutions metaheuristics are Hill Climbing (HC) [2], Simulated Annealing (SA) [3], and Tabu Search (TS) [4]. Some of the most popular and recent population-based

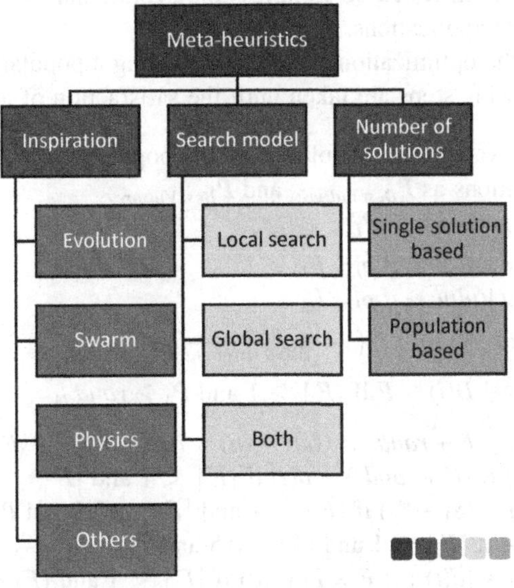

FIG. 2

A classification of metaheuristics based on inspiration, search model, and solution set.

algorithms are Genetic Algorithm (GA) [5], Differential Evolution (DE) [6], Evolution Strategy (ES) [7], Biogeography-Based Optimization (BBO) [8], Particle Swarm Optimization (PSO) [9], Artificial Bee Colony (ABC) [10], Stochastic Paint Optimizer (SPO) [11], and Farmland Fertility (FF) [12].

As discussed, the literature shows that metaheuristics are popular tools to estimate the global optimum for real-world optimization problems [13–15]. In this chapter, we use three population-based metaheuristics for weight optimization of space truss structures. These are African Vulture Optimization Algorithm (AVOA) [16], Artificial Gorilla Troops Optimizer (AGTO) [17], and Artificial Hummingbird Algorithm (AHA) [18].

The rest of the chapter is organized as follows: Section 2 discusses the details of the three metaheuristics used in this work. Section 3 presents the structure design problem and case studies. It also provides the results of the three metaheuristics on these case studies. Section 4 concludes.

2. African vulture optimization algorithm, artificial gorilla troops optimizer, and artificial hummingbird algorithms

This section provides a short overview of the AVOA, AGTO, AHA.

2.1 African Vulture's Optimization Algorithm (AVOA)

AVOA is a population-based metaheuristic algorithm inspired by the search and predation behaviors of African vultures [16]. It has been tested on Congress on Evolutionary Computation (CEC) standard benchmark functions and real applications.

This algorithm initiates the optimization process by creating a population of solutions (vultures): $\boldsymbol{P}_i(\boldsymbol{i} = \mathbf{1,2,\ldots,}\boldsymbol{N})$. The following steps are taken until the satisfaction of an end condition:

- Calculating the objective value of each solution in the population ($\boldsymbol{P}_i$)
- Choose the best two solutions as $\boldsymbol{P}_{Best\,Vulture_1}$ and $\boldsymbol{P}_{Best\,Vulture_2}$
- For each solution in the population ($\boldsymbol{P}_i$)
  - o Calculate $R(i) = \begin{cases} Best\,Vulture_1 \text{ if } p_i = L_1 \\ Best\,Vulture_2 \text{ if } p_i = L_2 \end{cases}$
  - o Calculate $F = (2 \times rand_1 + 1) \times z \times \left(1 - \frac{iteration_i}{\max\, iterations}\right) + t$
  - o Update $P(i+1) = R(i) - D(i) \times F$ if $|F| \geq 1$ and $P_1 \geq rand_{P1}$
  - o Update $P(i+1) = R(i) - F + rand_2 \times ((ub - lb) \times rand_3 + lb)$ if $|F| < 1$ and $P_1 < rand_{P1}$
  - o Update $P(i+1) = D(i) \times (F + rand_4) - d(t)$ if $|F| < 1$ and $|F| \geq 0.5$ and $P_2 \geq rand_{P2}$
  - o Update $P(i+1) = R(i) - (S_1 + S_2)$ if $|F| < 1$ and $|F| \geq 0.5$ and $P_2 < rand_{P2}$
  - o Update $P(i+1) = \frac{A_1 + A_2}{2}$ if $|F| < 1$ and $|F| < 0.5$ and $P_3 \geq rand_{P2}$
  - o Update $P(i+1) = R(i) - |d(t)| \times F \times Levy(d)$ if $|F| < 1$ and $|F| < 0.5$ and $P_3 < rand_{P2}$
- Return $\boldsymbol{P}_{Best\,Vulture_1}$ as the best estimation of the global optimum for the given optimization problem

2.2 Artificial gorilla troops optimizer (AGTO)

Metaheuristic algorithms play an important role in solving optimization problems, and most of them are inspired by the collective intelligence of natural beings. The AGTO algorithm [8] is a new meta-heuristic algorithm based on the social intelligence of gorilla forces in nature. In this algorithm, the collective life of gorillas is mathematically formulated, and new mechanisms are designed for exploration and exploitation. AGTO evaluation is performed by 52 standard functions and seven engineering problems. It is demonstrated that AGTO is very efficient in solving optimization problems and has merits among the existing meta-heuristics in the literature.

This algorithm is started with a random population of gorillas that is assumed to form a troop: $X_i(i = 1, 2, \ldots, N)$. It then iteratively performs the following steps until the satisfaction of an end condition:

- Calculate the fitness value of the troop.
- Find the best solution and name it $X_{silverback}$
- Calculate $C = F \times \left(1 - \frac{It}{MaxIt}\right)$ and $L = C \times l$
- For each gorilla in the troop:
  - Update $GX(t+1) = (UB - LB) \times r_1 + LB$ if $rand < p$
  - Update $GX(t+1) = (r_2 - C) \times X_r(t) + L \times H$ if $rand \geq 0.5$
  - Update $GX(t+1) = X(i) - L \times (L \times (X(t) - GX_r(t)) + r_3 \times (X(t) - GX_r(t)))$ if $rand < 0.5$
  - Update $X(t+1) = L \times M \times (X(t) - X_{silverback}) + X(t)$ if $|C| \geq 1$
  - **Update** $GX(i) = X_{silverback} - (X_{silverback} \times Q - X(t) \times Q) \times A$ if $|C| < 1$
  - if GX is better than X, replace them
- Return $X_{silverback}$

2.3 Artificial hummingbird algorithm (AHA)

The AHA is a new optimization algorithm proposed to solve complex optimization problems. It simulates the specific flight skills and intelligent search strategies of hummingbirds in the wild [18]. Three types of flight skills are used in food search strategies, including axial, oblique, and all-round flights.

Algorithm 1. The pseudocode of the AHA.

```
Input: n, d, f, Max_Iteration, Low, Up
Output: Global minimum, Global minimizer
Initialization:
For ith hummingbird from 1 to n,
    Do x_i=Low+r(Up-Low),
    For jth food source from 1 to n, Do
        If i≠i
        Then Visitable=1,
        Else Visit table_ij=null,
        End If
    End For
End For
While t≤ Max_Iteration Do
```

Continued

Algorithm 1. The pseudocode of the AHA—cont'd

```
For ith hummingbird from 1 to n, Do
    If rand ≤ 0.5 Then
        If r < 1/3 Then perform D^(i) = { 1  if i = randi([i, d]); i = 1,2,…,d
                                        { 0  else
        Else If r > 2/3
            Then perform D^(i) = { 1  if i = P(j), j∈[1, k], P = randperm(k), k∈[2, ⌈r_1.(d − 2)⌉ + 1]
                                 { 0  else
            Else perform D^(i) = 1; i = 1, …, d
            End If
        End If
Perform v_i(t + 1) = x_i,tar(t) + a. D. (x_i(t) − x_i,tar(t))
If f(v_i/(t+1)) <f(x_i(t))
Then xi/(t+1) =vi/(t+1),
    For jth food source from 1 to n(j≠tar, i), Do
    Visit table(i,j) =Visit table(i,j) +1,
    End For
    Visit table(i,tar)=0,
    For jth food source from 1 to n, Do
    Visit table(j,i)=  max          (Visit_table(j, l)) + 1,
                     l∈n and l≠j
    End For
  Else
    For jth food source from 1 to n (j≠tar,i), Do
    Visit table(i,j)=Visit table(i,1)+1,
    End For
        Visit table(i,tar)=0,
    End
Else
  Perform Eq. (9), v_i(t + 1) = x_i(t) + b. D. x_i(t)
  If f(vi/(t+1)) <f(xi(t)
  Then xi/(t+1) = vi/(t+1),
      For jth food source from 1 to n (j≠i), Do
      Visit table(I,j)=Visit table(i,j)+1,
     End For
      For jth food source from 1 to n, Do
      Visit table(j,i)=  max          (Visit_table(j, l)) + 1,
                       l∈n and l≠j
     End For
  Else
     For jth food source from 1 to n (j≠i), Do
     Visit table(i,j)=Visit table(i,j)+1,
     End For
  End If
  End If
End For
If mod(t,2n) ==0,
Then perform X_wor(t + 1) = Low + r. (Up − Low)
    For jth food source from 1 to n (j≠wor), Do
    Visit table(wor,j)=Visit table(wor,j)+1,
    End For
    For jth food source from 1 to n, Do
    Visit table(j,wor) =  max          (Visit_table(j, l)) + 1
                        l∈n and l≠j
    End For
End If
End While
```

The performance of the AHA shows that there is a strong balance between exploration and exploitation, and the AHA has excellent ability to find optimal solutions. The pseudocode is shown in Algorithm 1.

In the next section, we present the structure design problems and case studies as well as the results of the three metaheuristics on these case studies.

3. Structural design optimization

To demonstrate the effectiveness of AVOA, AHA, and AGTO, various common structural optimum design problems are investigated in this section. The mathematical formulas of the size optimization used in this chapter are presented in [14]. This minimum design also satisfies the constraints restricting design variable sizes and structural responses. When working with truss problems in MATLAB, the code was prepared and all runs were carried out on a MacBook Pro with a CPU running at 2.3 GHz (8-Core on an Intel Core i9 computer platform), RAM at 16 GB on a computer running Macintosh (macOS Big Sur). The total number of iterations in the proposed algorithm is set to 200, and a population of 50 particles is employed in each example. When it comes to having fair competition, all methods use the same number of iterations and agents in each case. In addition, each example runs 20 times.

3.1 The 25-bar space truss design problem

As a common optimization test problem, the 25-bar space truss has been solved using a variety of optimization approaches. Fig. 3 depicts the proposed 25-bar space truss. It is divided into eight groups, each of which uses the same materials and has the same cross-sectional properties. Table 1 assigns members based on their total number. The truss is subjected to two different load scenarios in Table 2. Table 3 shows the axial load constraints for each group for each node. It should be noted that in each direction, there is a maximum displacement limit of ± 0.35 in. The cross-sectional area ranges from 0.01 to 3.4 in^2. The mass density is 0.1 lb/in^3 and the modulus of elasticity is 10000 ksi for this material.

Table 4 shows the results of the algorithms for a 25-bar truss. As shown, the AGTO algorithm has a better performance compared to AHA and AVOA. The value of best weight by AGTO is 545.0501. The values of best weight by AHA and AVOA are 545.2713 and 545.5990, respectively.

Fig. 4 shows the results of the best convergence of algorithms for a 25-bar truss with 200 iterations. As shown, the AGTO algorithm has better convergence in the initial iterations compared to AHA and AVOA. Also, the convergence of the AHA algorithm is better than that of AVOA. Fig. 4 shows that the initial convergence of the AGTO algorithm is less than 700 iterations, whereas the initial convergence of AHA and AVOA is greater than 700 iterations and greater than 750 iterations, respectively. In iteration 112, it is observed that the convergence of the algorithms tends to an optimal value.

Fig. 5 shows the average convergence results of the algorithms for the 25-bar truss with 200 iterations. As shown, the AHA has better convergence than AVOA and AGTO. Also, the AVOA's convergence is better than that of the AGTO algorithm. Fig. 5 shows that the initial convergence of the AHA is greater than 700 iterations, whereas the initial convergence of AVOA and AGTO is greater than 720 iterations and less than 700 iterations, respectively. The convergence of the AHA in 50 iterations is better than that of AVOA and AGTO.

Fig. 6 shows the displacement ratio of AGTO for this example. As shown, the maximum displacement ratio is one which is equal to the maximum allowable ratio. The stress of the element is depicted in Fig. 7. In more detail, the stress of elements is very close to their lower bound lines.

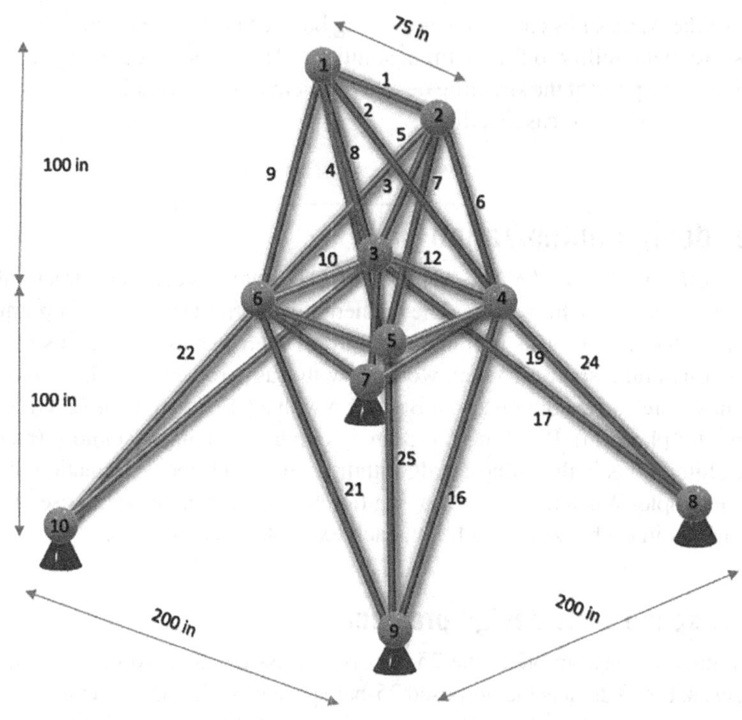

FIG. 3

The 25-bar space truss structure.

3.2 The 72-bar space truss design problem

The following example is of a 72-bar space truss, as shown in Fig. 8. The density of the material is assumed to be 0.1 lb/in^3 and the modulus of elasticity is 10,000 ksi. The 72-bar elements of this truss are organized into 16 groups. Tension and compression stress are equal to 25 ksi for all elements. The displacement of the top node in both the x and y directions is less than 0.25 in. There is a 0.10 in^2 minimum and a 4.00 in^2 maximum allowable cross-sectional area for each member in the system. The values and directions of the two space truss load scenarios are shown in Table 5.

Table 6 shows the results of the algorithms for a 72-bar truss. As shown, the AGTO algorithm has a better performance compared to AHA and AVOA. The value of best weight by AGTO is 380.3100. The values of best weight by AHA and AVOA are 384.4603 and 380.7010, respectively.

Table 1 Group numbers of elements on the 25-bar space truss problem.

Group number of the element							
1	2	3	4	5	6	7	8
1:(1,2)	2:(1,4) 3:(2,3) 4:(1,5) 5:(2,6)	6:(2,4) 7:(2,5) 8:(1,3) 9:(1,6)	10:(6,3) 11:(5,4)	12:(3,4) 13:(6,5)	14:(3,10) 15:(6,7) 16:(4,9) 17:(5,8)	18:(4,7) 19:(3,8) 20:(5,10) 21:(6,9)	22:(10,6) 23:(3,7) 24:(4,8) 25:(5,9)

Table 2 Loading condition for 25-bar space truss.

Node number	Load (kips)					
	Case 1			Case 2		
	P_x	P_y	P_z	P_x	P_y	P_z
1	0	20	−5	1	10	−5
2	0	−20	−5	0	10	−5
3	0	0	0	0.5	0	0
6	0	0	0	0.5	0	0

Table 3 Member stress limitation for the 25-bar spatial truss [19].

Element group		Compressive stress, limitations, ksi (MPa)	Tensile stress, limitations, ksi (MPa)
1	A1	35.092 (241.96)	40.0 (275.80)
2	$A_2 \sim A_5$	11.590 (79.0913)	40.0 (275.80)
3	$A_6 \sim A_9$	17.305 (119.31)	40.0 (275.80)
4	$A_{10} \sim A_{11}$	35.092 (241.96)	40.0 (275.80)
5	$A_{12} \sim A_{13}$	35.092 (241.96)	40.0 (275.80)
6	$A_{14} \sim A_{17}$	6.759 (46.603)	40.0 (275.80)
7	$A_{18} \sim A_{21}$	6.959 (47.982)	40.0 (275.80)
8	$A_{22} \sim A_{25}$	11.082 (76.410)	40.0 (275.80)

Table 4 Comparison of optimized 25-bar spatial truss.

Member group	AVOA	AGTO	AHA
1 (A_1)	0.0149	0.0104	0.0116
2 ($A_2 \sim A_5$)	2.1194	2.0199	2.0887
3 ($A_6 \sim A_9$)	2.9157	3.0115	2.9326
4 ($A_{10} \sim A_{11}$)	0.0131	0.0100	0.0165
5 ($A_{12} \sim A_{13}$)	0.0207	0.0100	0.0134
6 ($A_{14} \sim A_{17}$)	0.6418	0.6844	0.6870
7 ($A_{18} \sim A_{21}$)	1.6041	1.6384	1.6184
8 ($A_{22} \sim A_{25}$)	2.7545	2.6651	2.6857
Best weight (lb.)	545.5990	545.0501	545.2713
Average weight (lb.)	549.0180	554.0576	546.3708
Standard deviation	3.2023	18.5163	1.1052
Number of analyses	10,000	10,000	10,000

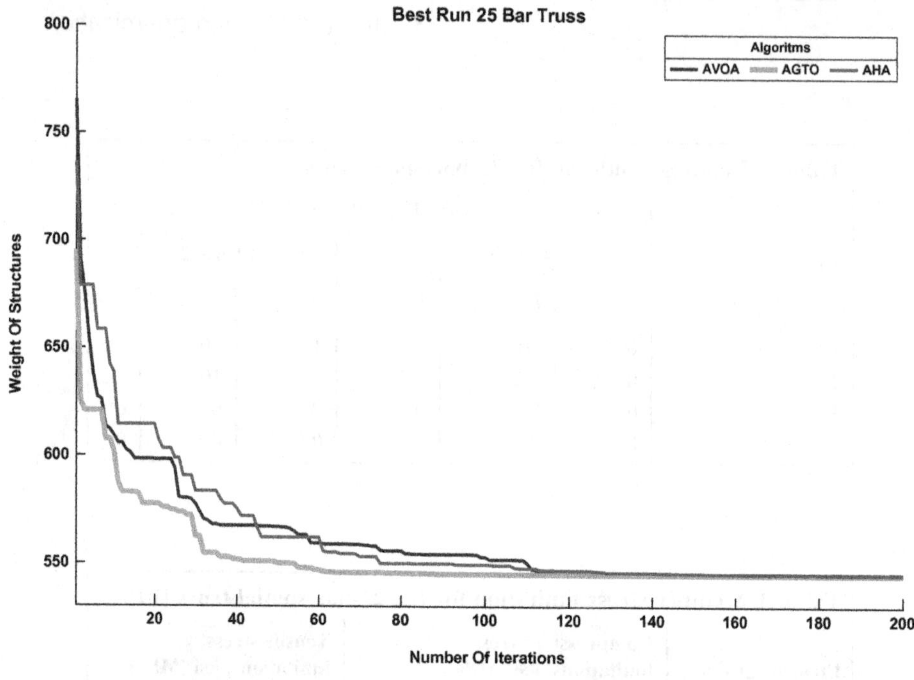

FIG. 4

Convergence curve of best run for 25 bar-truss of AVOA, AGTO, and AHA.

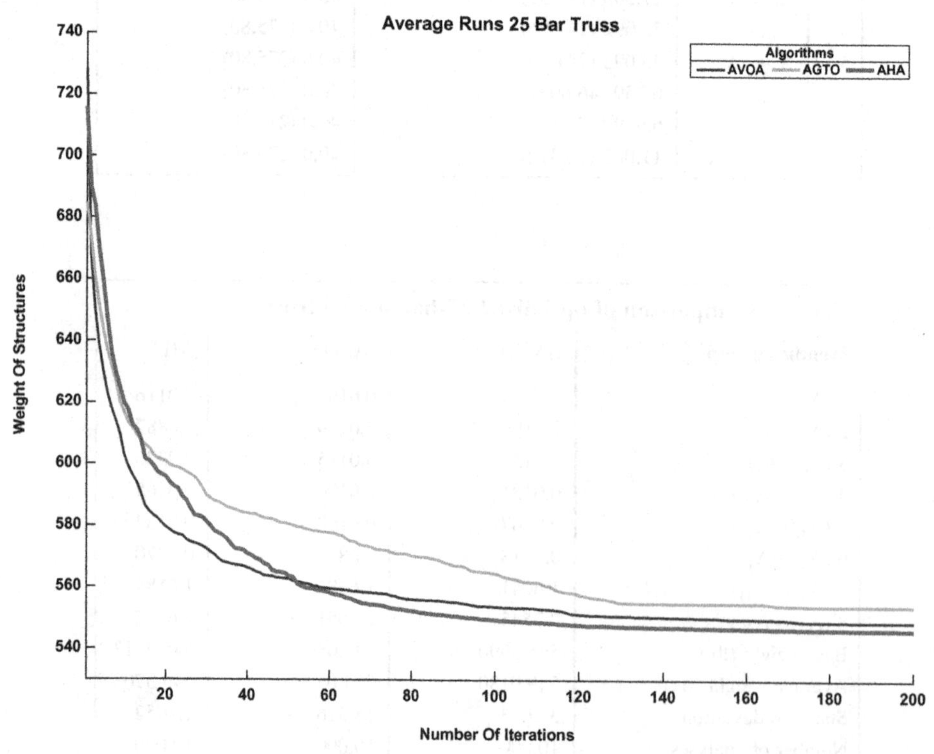

FIG. 5

Convergence curve of average runs for 25 bar-truss of AVOA, AGTO, and AHA.

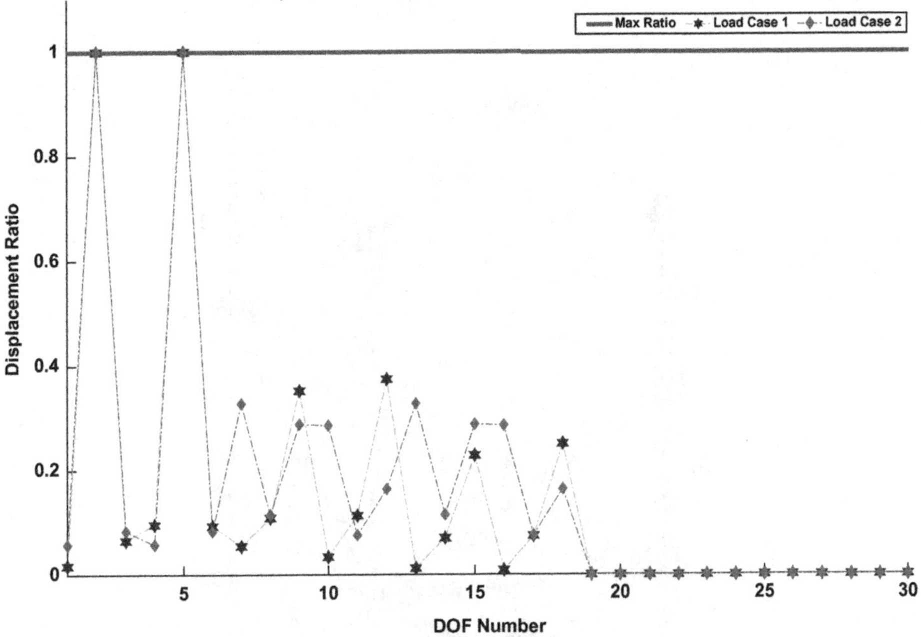

FIG. 6

Displacement ratio of best run for 25 bar-truss of AGTO algorithm.

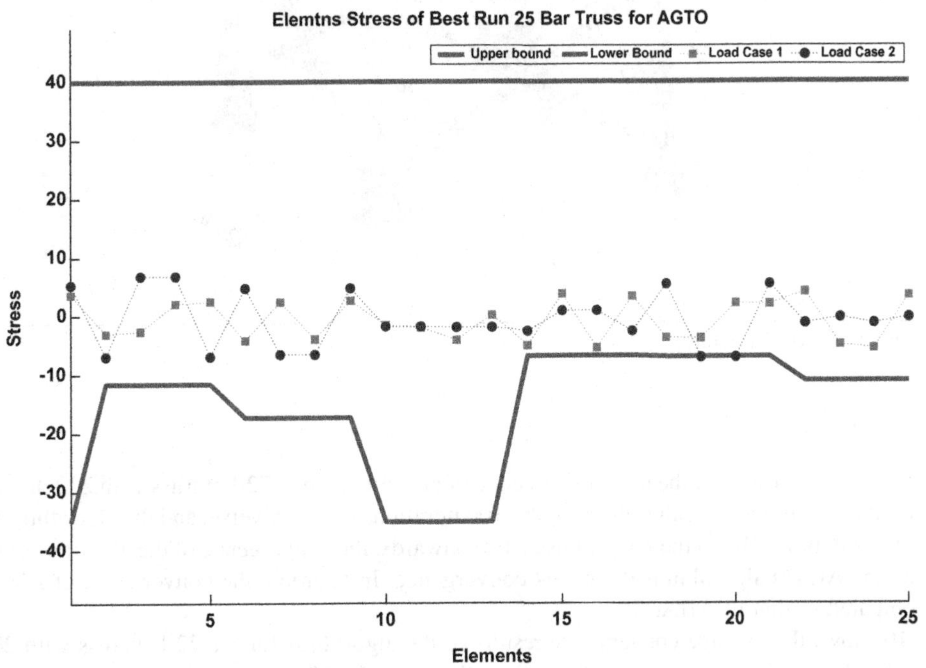

FIG. 7

Elements stress of best run for 25 bar-truss of AGTO algorithm.

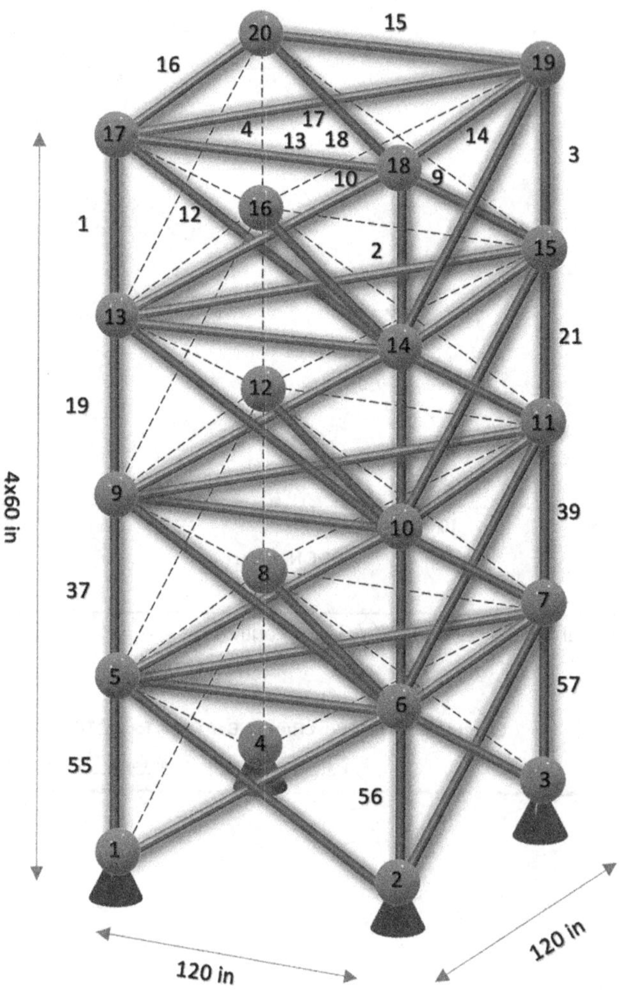

FIG. 8

The 72-bar space truss structure.

Fig. 9 shows the results of the best convergence of algorithms for a 72-bar truss with 200 iterations. As shown, the convergence of algorithms in 20–100 iterations is very diverse, and the algorithms try to find optimal solutions. But in the repetitions of 130 onwards, the convergence of the algorithms is very close, and the AGTO algorithm had the best convergence. In addition, the convergence of AVOA is better compared to that of AHA.

Fig. 10 shows the average convergence results of the algorithms for the 72-bar truss with 200 iterations. As shown, the convergence of the algorithms in 20–120 iterations is very diverse, and the

Table 5 Load condition for the 72-bar space truss.

	Load (kips)					
	Case 1			Case 2		
Node number	P_x	P_y	P_z	P_x	P_y	P_z
1	5	5	−5	0	0	−5
2	0	0	0	0	0	−5
3	0	0	0	0	0	−5
6	0	0	0	0	0	−5

Table 6 Comparison of optimized 72-bar space truss.

Member group	AVOA	AGTO	AHA
1 (A_1–A_4)	1.8859	1.9950	1.9483
2 (A_5–A_{12})	0.5103	0.4915	0.5142
3 (A_{13}–A_{16})	0.1003	0.1002	0.1123
4 (A_{17}–A_{18})	0.1000	0.1001	0.1053
5 (A_{19}–A_{22})	1.2738	1.2646	1.3575
6 (A_{23}–A_{30})	0.5049	0.5226	0.5519
7 (A_{31}–A_{34})	0.1054	0.1000	0.1025
8 (A_{35}–A_{36})	0.1001	0.1000	0.1063
9 (A_{37}–A_{40})	0.4754	0.5264	0.4414
10 (A_{41}–A_{48})	0.5442	0.4874	0.4759
11 (A_{49}–A_{52})	0.1000	0.1000	0.1198
12 (A_{53}–A_{54})	0.1000	0.1167	0.1111
13 (A_{55}–A_{58})	0.1613	0.1552	0.1688
14 (A_{59}–A_{66})	0.5545	0.5506	0.5487
15 (A_{67}–A_{70})	0.3911	0.4172	0.3925
16 (A_{71}–A_{72})	0.5618	0.5972	0.5902
Best weight (lb.)	380.7010	380.3100	384.4603
Average weight (lb.)	383.0660	386.0337	389.7981
Standard deviation	2.3955	13.1536	3.4577
Number of analyses	10,000	10,000	10,000

competition between AGTO and AHA is in the optimization discovery mode. At 140 iterations, the convergence of the algorithms is very close to each other, and AVOA has the best convergence. In addition, the convergence of AHA is better compared to that of AGTO.

Fig. 11 shows the displacement ratio of AGTO for a 72-bar truss. According to Fig. 6, there is no violation of the displacement limitation ratio. Fig. 12 shows the stress of elements for this example. In more detail, the maximum stress of elements for load case number 2 is equal to the lower bound limitation line.

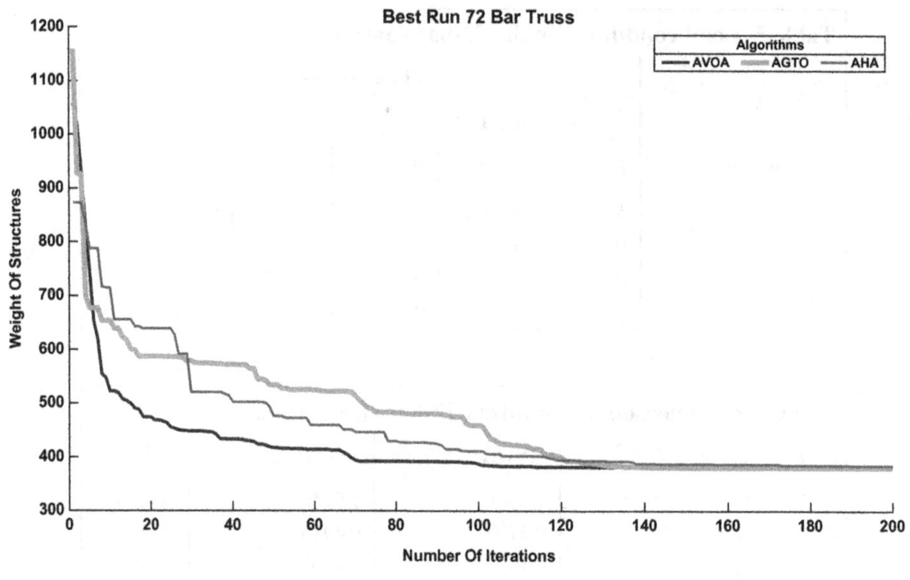

FIG. 9

Displacement ratio of best run for 25 bar-truss of AGTO algorithm.

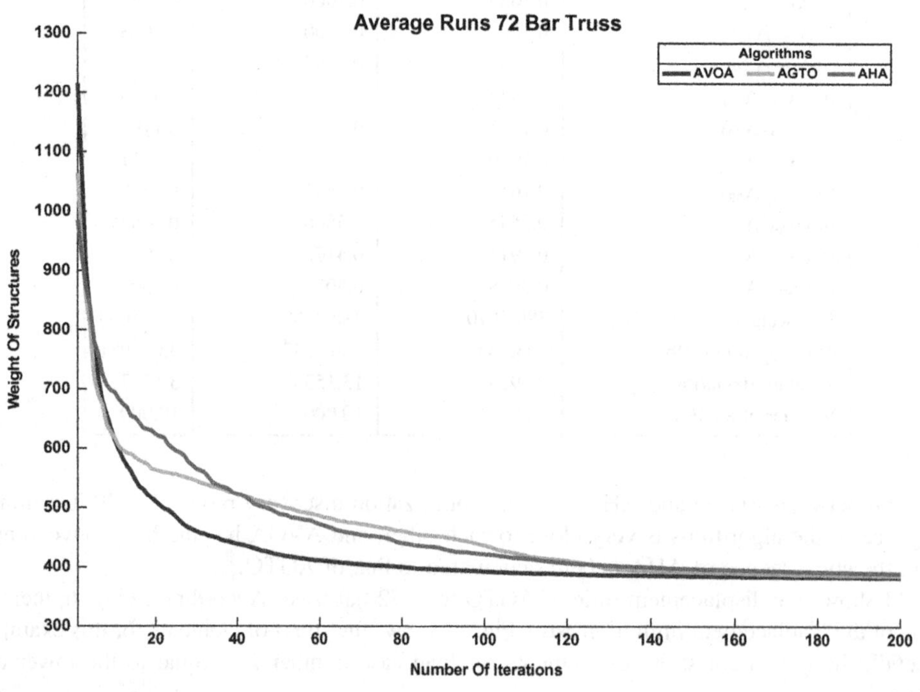

FIG. 10

Elements stress of best run for 72 bar-truss of AGTO algorithm.

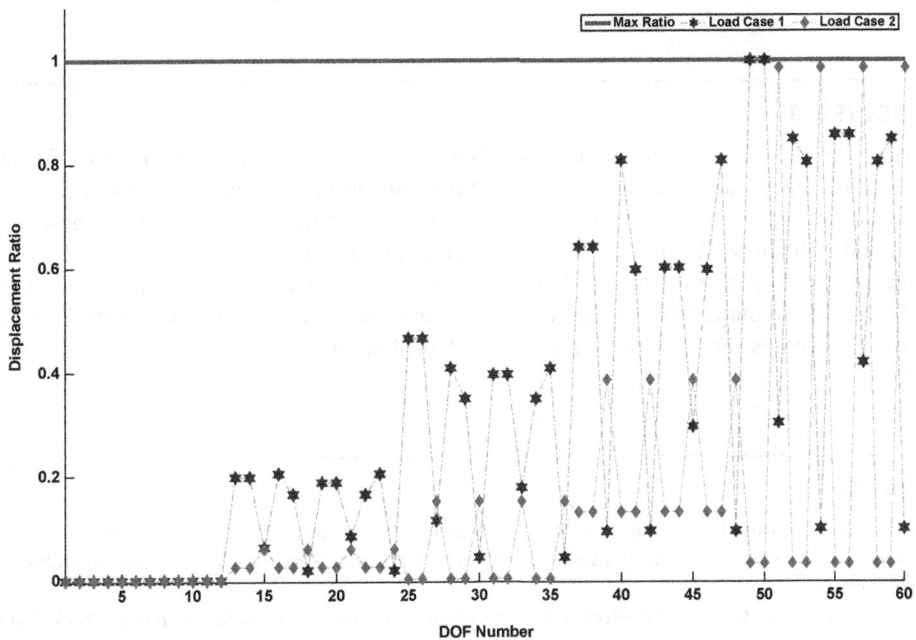

FIG. 11

Displacement ratio of best run for 72 bar-truss of AGTO algorithm.

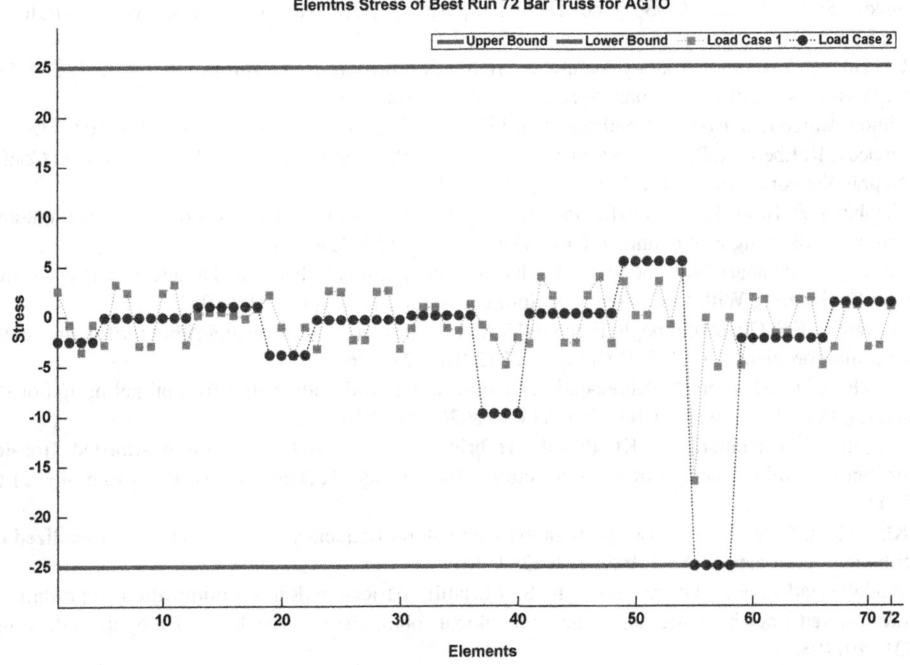

FIG. 12

Elements stress of best run for 72 bar-truss of AGTO algorithm.

4. Conclusion

This chapter evaluated the performance of three population-based metaheuristics for weight optimization of space truss structures: AVOA, AHA, and AGTO. These metaheuristic algorithms were used to optimize the weight of two benchmark space truss structures to demonstrate their capability and efficiency. When it comes to the weight optimization of space truss structures, these algorithms appear to perform well. According to an in-depth analysis of the optimization results, the AGTO algorithm outperformed the other algorithms by finding solutions with the least weight in all the case studies. For future works, these algorithms will be used to optimize steel frame structures.

References

[1] J.R. Martins, A. Ning, Engineering Design Optimization, Cambridge University Press, 2021.

[2] B. Selman, C.P. Gomes, Hill-climbing search, in: Encyclopedia of Cognitive Science, vol. 81, John Wiley & Sons, 2006, p. 82.

[3] S. Kirkpatrick, C.D. Gelatt, M.P. Vecchi, Optimization by simulated annealing, Science 220 (4598) (1983) 671–680.

[4] F. Glover, M. Laguna, Tabu search, in: Handbook of Combinatorial Optimization, Springer, 1998, pp. 2093–2229.

[5] J.H. Holland, Genetic algorithms, Sci. Am. 267 (1) (1992) 66–73.

[6] K. Fleetwood, An introduction to differential evolution, in: Proceedings of Mathematics and Statistics of Complex Systems (MASCOS) One Day Symposium, 26th November, Brisbane, Australia, 2004, pp. 785–791.

[7] I. Rechenberg, Evolution strategy: nature's way of optimization, in: Optimization: Methods and Applications, Possibilities and Limitations, Springer, 1989, pp. 106–126.

[8] D. Simon, Biogeography-based optimization, IEEE Trans. Evol. Comput. 12 (6) (2008) 702–713.

[9] J. Kennedy, R. Eberhart, Particle swarm optimization, in: Proceedings of ICNN'95-International Conference on Neural Networks, vol. 4, IEEE, 1995, pp. 1942–1948.

[10] D. Karaboga, B. Basturk, A powerful and efficient algorithm for numerical function optimization: artificial bee colony (ABC) algorithmnnnn, J. Glob. Optim. 39 (3) (2007) 459–471.

[11] A. Kaveh, S. Talatahari, N. Khodadadi, Stochastic paint optimizer: theory and application in civil engineering, in: Engineering With Computers, 38, Springer Nature, 2022, pp. 1921–1952.

[12] H. Shayanfar, F.S. Gharehchopogh, Farmland fertility: a new metaheuristic algorithm for solving continuous optimization problems, Appl. Soft Comput. 71 (2018) 728–746.

[13] A. Kaveh, A.D. Eslamlou, N. Khodadadi, Dynamic water strider algorithm for optimal design of skeletal structures, Period. Polytechn. Civil Eng. 64 (3) (2020) 904–916.

[14] A. Kaveh, S. Talatahari, N. Khodadadi, Hybrid invasive weed optimization-shuffled frog-leaping algorithm for optimal design of truss structures, Iran. J. Sci. Technol. Trans. Civil Eng. 44 (2) (2020) 405–420.

[15] N. Khodadadi, S. Mirjalili, Truss optimization with natural frequency constraints using generalized normal distribution optimization, Appl. Intell. (2022) 1–14.

[16] B. Abdollahzadeh, F.S. Gharehchopogh, S. Mirjalili, African vultures optimization algorithm: a new nature-inspired metaheuristic algorithm for global optimization problems, Comput. Ind. Eng. 158 (2021) 107408.

[17] B. Abdollahzadeh, F. Soleimanian Gharehchopogh, S. Mirjalili, Artificial gorilla troops optimizer: a new nature-inspired metaheuristic algorithm for global optimization problems, Int. J. Intell. Syst. 36 (10) (2021) 5887–5958.
[18] W. Zhao, L. Wang, S. Mirjalili, Artificial hummingbird algorithm: a new bio-inspired optimizer with its engineering applications, Comput. Methods Appl. Mech. Eng. 388 (2022) 114194.
[19] N. Khodadadi, S.M. Mirjalili, S. Mirjalili, Optimal design of truss structures with continuous variable using moth-flame optimization, Handbook of Moth-Flame Optimization Algorithm, CRC Press, 2022, pp. 265–280.

CHAPTER

Metaheuristics for solving the wind turbine placement problem 10

Ahmet Cevahir Cinar

Department of Computer Engineering, Faculty of Technology, Selçuk University, Konya, Turkey

1. Introduction

Wind energy is one of the most important renewable energy sources. Wind turbines are used for producing electricity and they should be located at wind farms with an optimized placement approach. This optimization process is time-consuming. The main aim is to maximize power and minimize installation cost [1]. Solving these types problems with canonical gradient-based optimization methods is difficult because of their discrete variables [2]. Metaheuristic algorithms are generally inspired by natural phenomena and used for solving these types of problems.

In literature, many metaheuristic algorithms have been used for solving wind turbine placement problems (WTTPs). Kunakote et al. [3] compared twelve metaheuristic algorithms that solve WTPP. These are grasshopper optimization algorithm, artificial bee colony (ABC), salp swarm optimizer, real-code ant colony optimization, differential evolution (DE), crow search optimization algorithm, particle swarm optimization (PSO), teaching-learning based optimization (TLBO), whale optimization algorithm, evolution strategy with covariance matrix adaptation, moth-flame optimization algorithm, and sine cosine algorithm. Azlan et al. [4] prepared a comprehensive survey about optimization of WTPP with metaheuristic algorithms. Genetic algorithm [5–10], evolutive algorithm (EA) [11], DE [12–14], lazy greedy algorithm [15], ant colony optimization [16], viral-based optimization algorithm [17], random search [18], greedy randomized adaptive search procedure-variable neighborhood search [19], simulated annealing [20,21], particle swarm optimization [13,22–27], TLBO [28,29], lightning search algorithm [30], artificial algae algorithm [31], JAYA algorithm [32,33], invasive weed optimization [34] elephant herding optimization [35], passing vehicle search [36], cuckoo search [37], binary negatively correlated search [13], Harris hawks optimization [38], sooty tern optimization algorithm [39], are other metaheuristic algorithms used in the literature.

The continuous optimization algorithms have continuous variables in a predetermined search space with upper and lower bounds and cannot solve binary optimization algorithms directly; they must be remodified for binary search space. Binary optimization problems have only two decision variables, 0 and 1, in binary search space. Transfer functions [40] are commonly used for transferring the continuous variables to the binary search space. In the literature to date, seventeen transfer functions [41] are used for mapping continuous variables to binary variables. The artificial gorilla troops optimizer (AGTO) [42] is a recently developed continuous optimization algorithm that models the life behavior of gorilla troops.

Comprehensive Metaheuristics. https://doi.org/10.1016/B978-0-323-91781-0.00010-7

In this chapter, we combine AGTO with the seventeen transfer functions mentioned in the literature. Consequently, we propose seventeen novel binary AGTOs (BAGTO1-BAGTO17). To determine the quality of these binary variants, we use the WTPP [33] as a benchmark problem. A 10×10 grid-type WTPP with 100 dimensions is solved by these seventeen binary variants of AGTO.

The remainder of the chapter is organized as follows. Sections 2 and 3 explain AGTO and its binary variants , respectively. Section 4 presents the experimental setup and Section 5 reports the experimental results. Section 6 concludes.

2. Artificial gorilla troops optimizers

AGTO [42] is a metaheuristic optimization algorithm recently developed by Abdollahzadeh et al. [42]. In AGTO, gorilla behaviors are modeled as mathematical functions for solving optimizations problems. AGTO, as a canonical metaheuristic optimization algorithm, has five different mathematic equations for conducting the exploration and exploitation phases. The exploration functions are given in Eq. (1).

$$GX_i(t+1) = \begin{cases} (UB-LB)\times r1 + LB, & \text{rand} < p \\ (r2-C)\times X_r(t) + L \times H, & \text{rand} \geq 0.5 \\ X_i(t) - (L \times (X_i(t) - GX_r(t)) + r3 \times (X_i(t) - GX_r(t))), & \text{rand} < 0.5 \end{cases} \tag{1}$$

where $GX_i(t+1)$ is the value of the candidate gorilla in the next iteration, UB is the upper bound of the optimization problem, LB is the lower bound of the optimization problem, $r1$, $r2$, and $r3$ are random numbers between 0 and 1, $X_r(t)$ is a random gorilla in the current iteration, $X_i(t)$ is the ith gorilla in the current iteration, and $GX_r(t)$ is the value of the candidate gorilla in the current iteration. C is shown and calculated as in Eq. (2), L is shown and calculated as in Eq. (3), and H is shown and calculated as in Eq. (4).

$$C = F \times \left(1 - \frac{t}{maxiter}\right) \tag{2}$$

$$F = \cos(2 \times r4) + 1 \tag{3}$$

$$L = C \times rand(-1, 1) \tag{4}$$

where t is the current iteration number, *maxiter* is the number of maximum iterations, $r4$ is a random number between 0 and 1, $rand(-1,1)$ is a random number between -1 and 1, and cos() is the cosine function. C starts with approximately 0.5 and stops with 0, thus, in the first half-iteration C provides the scaling of the exploration phase.

After the exploration phase, the fitness values of candidate gorillas and gorillas are compared. If a candidate gorilla is better than the current gorilla, the current gorilla is removed from the population and the candidate gorilla becomes a gorilla in the population.

In the exploitation phase, two mathematical functions are used. The first one is given in Eq. (5) and the second one is given in Eq. (6).

$$GX_i(t+1) = L \times \left(\left| \frac{1}{N} \sum_{i=1}^{N} GX_i(t) \right|^{2^L} \right)^{\frac{1}{2^L}} \times (X_i(t) - X_{SB}(t)) + X_i(t) \tag{5}$$

Step 1 - Initialization of the gorilla population
Step 2 - Calculation of the fitness values
Step 3 - Exploration Phase
Step 4 - Exploitation Phase
Step 5 - Calculation of the fitness values of candidate gorillas
Step 6 - Selection for next generation
Step 7 - Increment the iteration counter
Step 8 - Until the termination criterion is met, go to Step 3

FIG. 1

The general pseudocode of AGTO.

where $GX_i(t+1)$ is the value of the candidate gorilla in the next iteration, L is a value calculated with Eq. (4), $X_i(t)$ is the ith gorilla in the current iteration, $X_{SB}(t)$ is the best gorilla in the current iteration, $GX_i(t)$ is the value of the candidate gorilla in the current iteration, and N is the number of population.

$$GX_i(t+1) = X_{SB}(t) - (X_{SB}(t) \times (2 \times r5 - 1) - X_i(t) \times (2 \times r5 - 1)) \times \beta \times \begin{cases} N1, \text{rand} \geq 0.5 \\ N2, \text{rand} < 0.5 \end{cases} \tag{6}$$

where $GX_i(t+1)$ is the value of the candidate gorilla in the next iteration, $X_{SB}(t)$ is the best gorilla in the current iteration, $r5$ is a random number between 0 and 1, $X_i(t)$ is the ith gorilla in the current iteration, β is a constant value, $N1$ is a random normal distribution values vector, and $N2$ is the random normal distribution value.

After the exploitation phase, the fitness values of candidate gorillas and gorillas are compared. If a candidate gorilla is better than the current gorilla, the current gorilla is removed from the population and the candidate gorilla becomes a gorilla in the population. Fig. 1 gives the general pseudocode of AGTO.

3. Binary variants of artificial gorilla troops optimizers

3.1 Transfer functions

Table 1 lists the mathematical formulas of transfer functions. TF1, TF2, TF3, TF4, TF5, TF6, TF8, TF13, TF14, TF15, TF16, and TF17 are stationary transfer functions. TF7, TF9, TF10, TF11, and TF12 are nonstationary transfer functions. In Table 1, c is the continuous value, tv is the transferred value, e is the exponential function, tanh is the hyperbolic tangent function, atan is the inverse tangent in radians, erf is error function, *failure* is the number of failures, *Qmax* and *Qmin* are predetermined values for nonstationary transfer functions, *Max_iter* is the maximum number of iteration, *iter* is the current iteration number, *sin* is the sinus function, *cos* is the cosinus function, *mod* is the modulo

Table 1 The mathematical formulas of transfer functions.

Name	Formula
TF1	$tv = \frac{1}{(1+e^{-c})}$
TF2	$tv = \frac{1}{(1+e^{-2c})}$
TF3	$tv = \frac{1}{\left(1+e^{-\frac{c}{2}}\right)}$
TF4	$tv = abs(\tanh(2c))$
TF5	$tv = abs\left(\frac{2}{\pi} \times \text{atan}\left(\frac{\pi}{2} \times c\right)\right)$
TF6	$tv = \left(\text{erf}\left(\frac{\sqrt{\pi}}{2} \times c\right)\right)$
TF7	$tv = \text{erf}\left(\frac{failure}{Max\ iter}\right) + 1 - \text{erf}\left(\frac{failure}{Max\ iter}\right) \times abs(\tanh(c))$
TF8	$tv = abs\left(\frac{c}{\sqrt{1+c^2}}\right)$
TF9	$tv = \frac{1}{\left(1+e^{\frac{-2c}{Qmax - iter \times ((Qmax - Qmin)/Max\ iter)}}\right)}$
TF10	$tv = \frac{1}{\left(1+e^{\frac{-c}{Qmax - iter \times ((Qmax - Qmin)/Max\ iter)}}\right)}$
TF11	$tv = \begin{cases} 1 - \frac{2}{1+e^{\frac{-2c}{Qmax - iter \times ((Qmax - Qmin)/Max\ iter)}}}, & x \leq 0 \\ \frac{2}{1+e^{\frac{-2c}{Qmax - iter \times ((Qmax - Qmin)/Max\ iter)}}} - 1, & x > 0 \end{cases}$
TF12	$tv = \begin{cases} 1 - \frac{2}{1+e^{\frac{-c}{2 \times (Qmax - iter \times ((Qmax - Qmin)/Max\ iter))}}}, & x \leq 0 \\ \frac{2}{1+e^{\frac{-c}{2 \times (Qmax - iter \times ((Qmax - Qmin)/Max\ iter))}}} - 1, & x > 0 \end{cases}$
TF13	$tv = (c - (-10))/(20)$
TF14	$tv = abs(\tanh(C))$
TF15	$tv = abs(sin(2 \times \pi \times c \times cos(2 \times \pi \times c)))$
TF16	$tv = mod(round(abs(mod(c, 2))), 2)$
TF17	$tv = fix(abs(mod(x, 2)))$

function, *round* is the round function, *abs* is the absolute value function, and *fix* is the round towards zero function. The continuous variable is given as an input to transfer function and the output may be 0 or 1.

3.2 Binary variants of artificial gorilla troops optimizers

Table 1 lists the transfer functions used by the binary variants of the AGTO. Fig. 2 depicts the proposed binary variants of AGTO.

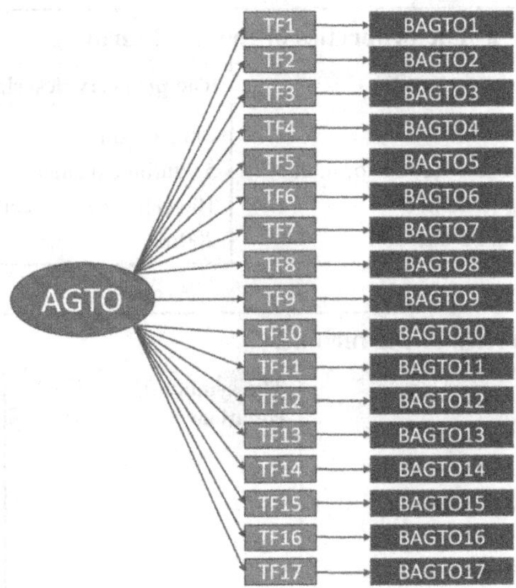

FIG. 2

The proposed binary variants of AGTO.

4. Experimental setup

In our experiments, thirty different runs were conducted to cope with the random nature of the metaheuristic algorithms. The upper bound of the continuous search space was set as 8 and the lower bound of the search space was set as −8. The continuous values were converted to binary values via transfer functions.

In this chapter, a 10×10 grid-type WTPP that has 100 dimensions is solved by seventeen binary variants of AGTO. These binary algorithms optimize 2^{100} possible solutions. The wind direction is fixed as 0 degrees and the wind speed is fixed as 12 m/s. The details of the WTPP can be found in [33].

The artificial gorilla troops population size is set as 50 and the maximum iteration number is set as 12,000. The maximum function evaluation number is fixed as 600,000, as in [33,34,39].

4.1 Wind turbine placement problem (WTPP)

Mosetti et al. [5] proposed the first mathematical model for WTPP that used the simple model for cluster efficiency [43]. The goal of WTPP is to minimize total cost and maximize total power. WTPP is a proven NP-hard problem due to its exponentially increasing nature [33,37,44]. The abbreviation NP means nonpolynomial, and NP-hard problems cannot be solved in polynomial time with a deterministic Turing machine.

Table 2 lists the properties of the wind farm in this mathematical model.

Table 3 lists the properties of the turbine used in this mathematical model.

Fig. 3 gives the wake model of Jensen [44], which calculates the wake effects of the wind on a wind turbine.

Table 2 The properties of the wind farm.

The property name	The property description
Area of wind farm	2 km × 2 km
The feasible turbine location	5 × turbine diameter
Total grid numbers	100 cells (10 × 10 cells)
The area of each cell	200 m^2

Table 3 The properties of the turbine.

The property name	The symbol in equations	The property value
Hub height	z	60 m
Diameter / Rotor Radius / Rotor Diameter	r_r	40 m
Thrust coefficient curve	C_t	0.88
Axial induction factor	a	0.33
Entertainment constant	α	0.094 kW
Downstream rotor radius	r_1	55.75

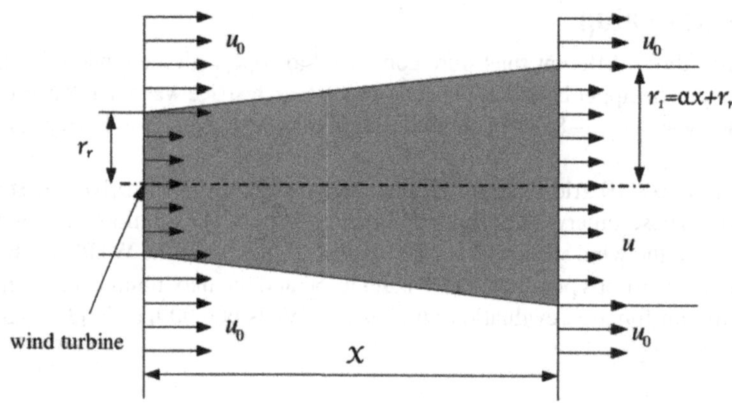

FIG. 3

Jensen waken model [33,37,44].

The downstream wind speed (**u**) is calculated by Eq. (7).

$$u = u_0 = \left[1 - 2a\left(1 + \frac{\alpha x}{r_r}\right)^{-2}\right] \tag{7}$$

where u_0 is the speed of free wind (m/s), α is the entertainment constant (kW), a is the factor of axial induction (a double value), r_r is the radius of turbine (m), and x is the distance between downstream and upstream (m).

The entertainment constant (α) is calculated by Eq. (8).

$$\alpha = \frac{0.5}{\ln \frac{z}{z_0}} \tag{8}$$

where z_0 is the site roughness value, set as 0.3 m, and z is the hub height of the turbine (m). The downstream rotor radius is calculated by Eq. (9).

$$r_1 = r_r \sqrt{\frac{1-a}{1-2a}} \tag{9}$$

where r_r is the radius of turbine (m) and a is the factor of axial induction (a double value).

The thrust coefficient curve is calculated by Eq. (10).

$$C_T = 4a(1-a) \tag{10}$$

where a is the factor of axial induction (a double value).

The total power (P_{Total}) is calculated by Eq. (11).

$$P_{Total} = \sum_{i=1}^{N} 0.3 \times u^3 \tag{11}$$

where N is the number of wind turbines.

The cost is calculated by Eq. (12).

$$Cost = N\left[\frac{2}{3} + \frac{1}{3}e^{-0.00174N^2}\right] \tag{12}$$

where N is the number of wind turbines.

The main objective function is given in Eq. (13).

$$Objective = \frac{Cost}{P_{Total}} \tag{13}$$

The main aim is to minimize the objective function value.

The efficiency is calculated by Eq. (14).

$$Efficiency = \frac{P_{Total}}{N \times 0.3 \times u_0^3} \tag{14}$$

WTPP is categorized into three cases in terms of wind direction. In C1, the wind direction is assumed to be 0 with a fixed speed of 12 m/s. In C2, the wind direction is assumed to be variable from 0 degrees to 350 degrees with a fixed speed of 12 m/s. In C3, the wind direction is assumed to be variable from 0 degrees to 350 degrees with a varying mean speed of 8, 12, and 17 m/s. Problems used in the literature generally include 10 × 10 Grid Case 1 (10 × 10C1), 10 × 10 Grid Case 2 (10 × 10C2), 10 × 10 Grid Case 3 (10 × 10C3), 20 × 20 Grid Case 1 (20 × 20C1), 20 × 20 Grid Case 2 (20 × 20C2), and 20 × 20 Grid Case 3 (20 × 20C3). A 10 × 10 grid includes 100 potential wind turbine positions and a 20 × 20 grid includes 400 potential wind turbine positions. This means there are 2^{100} and 2^{400} possible situations as binary optimization problems. The metaheuristics can solve these problems in an optimal or near optimal manner in a reasonable time. The binary decision variables (0 and 1) are used in the algorithms, where 0 means the wind turbine is not in this grid and 1 means the wind turbine is in this grid.

5. Results and discussion

The fitness value, the total produced power, the number of tribunes (NoT), the average power, and the efficiency values are given in Tables 4–8 for seventeen BAGTOs. BAGTO12 could not produce eligible solutions and produces completely zero fill candidate solutions, therefore the experimental results of BAGTO12 were removed. TF12 did not produce eligible solutions in a previous study [39], just like in this work. Table 4 lists the mean values of all the variants. As shown, the top three variants in terms of mean values are BAGTO10, BAGTO9, and BAGTO17. In terms of efficiency values, the top three variants are BAGTO16, BAGTO17, and BAGTO9.

Table 5 lists the maximum values of all the variants. As shown, the top three variants in terms of maximum values are BAGTO9, BAGTO16, and BAGTO10. In terms of efficiency values, the top three variants are BAGTO1, BAGTO16, and BAGTO9.

Table 6 lists the minimum values of all the variants. As shown, the top three variants in terms of minimum values are BAGTO9, BAGTO10, and BAGTO16. In terms of efficiency values, the top three variants are BAGTO16, BAGTO9, and BAGTO17.

Table 7 lists the median values of all the variants. As shown, the top three variants in terms of median values are BAGTO10, BAGTO16, and BAGTO9. In terms of efficiency values, the top three variants are BAGTO17, BAGTO16, and BAGTO9.

Table 8 lists the standard deviation values of all the variants. As shown, the top three variants in terms of standard deviation values are BAGTO16, BAGTO17, and BAGTO10.

Table 9 lists the full rank values of all the variants. As shown, the top three variants in terms of mean rank are BAGTO10, BAGTO9, and BAGTO16. In terms of the Friedman rank values, the top three variants are BAGTO10, BAGTO17, and BAGTO9. Fig. 4 shows a graph of all the rank values.

Table 4 The mean values of all BAGTO variants.

Variant	Fitness	Total power	NoT	Average power	Efficiency	Rank
BAGTO1	0.00158430	15,544.7042	40	398.3166	0.7684	5
BAGTO2	0.00158434	15,971.9371	41	403.7781	0.7789	6
BAGTO3	0.00161081	16,591.5494	46	364.2570	0.7027	7
BAGTO4	0.00162114	24,239.1367	88	285.4458	0.5506	10
BAGTO5	0.00164524	23,466.1975	81	289.5419	0.5585	14
BAGTO6	0.00161778	23,093.8097	81	300.2774	0.5792	9
BAGTO7	0.00168158	26,284.1378	100	262.8414	0.5070	16
BAGTO8	0.00163434	23,933.9334	85	285.7815	0.5513	12
BAGTO9	0.00157844	15,924.1402	39	414.8434	0.8002	2
BAGTO10	0.00157362	16,627.9518	45	381.7268	0.7364	1
BAGTO11	0.00162715	25,904.4828	97	266.8796	0.5148	11
BAGTO13	0.00166784	17,134.5365	49	346.9360	0.6692	15
BAGTO14	0.00161574	22,380.7930	77	308.3941	0.5949	8
BAGTO15	0.00163749	18,984.7245	58	330.3679	0.6373	13
BAGTO16	0.00158036	14,801.3510	33	450.7576	0.8695	4
BAGTO17	0.00158036	15,176.5765	36	435.6309	0.8403	3

Table 5 The maximum values of all BAGTO variants.

Variant	Fitness	Total power	NoT	Average power	Efficiency	Rank
BAGTO1	0.00162493	18,676.0954	56	473.3344	0.9131	5
BAGTO2	0.00164958	19,172.0331	60	456.4898	0.8806	6
BAGTO3	0.00167117	19,416.6376	60	434.1710	0.8375	11
BAGTO4	0.00166953	26,170.9849	99	420.8586	0.8118	10
BAGTO5	0.00168315	24,848.4236	90	302.1065	0.5828	14
BAGTO6	0.00166113	25,944.6791	97	435.6141	0.8403	8
BAGTO7	0.00174539	26,284.1378	100	262.8414	0.5070	16
BAGTO8	0.00166648	25,483.9273	94	426.6566	0.8230	9
BAGTO9	0.00160703	18,828.6551	56	469.1689	0.9050	1
BAGTO10	0.00161671	20,956.6428	65	451.5181	0.8710	3
BAGTO11	0.00165516	26,284.1378	100	295.7065	0.5704	7
BAGTO13	0.00168577	19,163.6434	58	375.7652	0.7249	15
BAGTO14	0.00167855	25,656.7742	95	442.0011	0.8526	12
BAGTO15	0.00168281	20,743.4869	66	359.0830	0.6927	13
BAGTO16	0.00161333	18,708.6968	54	469.4904	0.9057	2
BAGTO17	0.00161748	19,847.4870	60	468.6023	0.9039	4

Table 6 The minimum values of all BAGTO variants.

Variant	Fitness	Total power	NoT	Average power	Efficiency	Rank
BAGTO1	0.00155783	12,780.0289	27	328.7375	0.6341	4
BAGTO2	0.00155800	13,789.6638	31	317.1329	0.6118	5
BAGTO3	0.00158130	13,113.0141	31	317.6559	0.6128	9
BAGTO4	0.00159416	12,646.1271	31	263.1282	0.5076	11
BAGTO5	0.00161291	22,565.2040	76	275.8908	0.5322	14
BAGTO6	0.00157186	13,102.4879	32	266.2205	0.5135	7
BAGTO7	0.00163195	26,284.1378	100	262.8414	0.5070	16
BAGTO8	0.00159884	12,373.0415	29	268.8503	0.5186	12
BAGTO9	0.00155095	13,970.9594	30	336.2260	0.6486	1
BAGTO10	0.00155528	14,030.6694	32	322.4099	0.6219	2
BAGTO11	0.00158946	23,065.1105	78	262.8414	0.5070	10
BAGTO13	0.00161826	15,308.1607	43	325.8661	0.6286	15
BAGTO14	0.00157286	11,492.0287	26	268.1662	0.5173	8
BAGTO15	0.00160116	16,122.3305	47	300.2797	0.5792	13
BAGTO16	0.00155781	13,124.6290	28	346.4573	0.6683	3
BAGTO17	0.00156138	13,497.3712	29	330.1095	0.6368	6

Table 7 The median values of all BAGTO variants.

Variant	Fitness	Total power	NoT	Average power	Efficiency	Rank
BAGTO1	0.00158066	15,155.1385	36	413.6989	0.7980	6
BAGTO2	0.00157841	15,615.6708	36	429.0680	0.8277	4
BAGTO3	0.00160435	16,814.5802	47	353.7096	0.6823	7
BAGTO4	0.00161962	25,710.2331	96	267.6429	0.5163	10
BAGTO5	0.00164881	23,340.8051	80	289.4292	0.5583	14
BAGTO6	0.00161755	25,132.2583	92	275.4189	0.5313	9
BAGTO7	0.00168504	26,284.1378	100	262.8414	0.5070	16
BAGTO8	0.00163114	24,784.6595	90	276.6408	0.5336	12
BAGTO9	0.00157800	15,443.0267	35	438.9681	0.8468	3
BAGTO10	0.00156946	16,592.5167	47	361.7782	0.6979	1
BAGTO11	0.00162802	26,114.4084	99	263.7408	0.5088	11
BAGTO13	0.00167246	17,074.2316	51	346.2336	0.6679	15
BAGTO14	0.00161740	24,727.1305	90	275.0434	0.5306	8
BAGTO15	0.00163746	19,074.3078	58	327.8362	0.6324	13
BAGTO16	0.00157669	14,621.7910	32	454.9820	0.8777	2
BAGTO17	0.00157972	14,635.1105	32	457.4094	0.8823	5

Table 8 The standard deviation values of all BAGTO variants.

Variant	Fitness	Total power	NoT	Average power	Efficiency	Rank
BAGTO1	0.00001581	1693.0628	9	45.1385	0.0871	5
BAGTO2	0.00001787	1659.9082	10	48.3610	0.0933	7
BAGTO3	0.00002469	1542.3731	7	30.4689	0.0588	14
BAGTO4	0.00001884	3895.7953	21	45.3936	0.0876	10
BAGTO5	0.00001793	633.4725	4	7.2224	0.0139	8
BAGTO6	0.00002102	4174.9150	22	53.5309	0.1033	12
BAGTO7	0.00002734	0.0000	0	0.0000	0.0000	16
BAGTO8	0.00002135	3044.1497	15	33.2325	0.0641	13
BAGTO9	0.00001511	1406.2385	8	47.1968	0.0910	4
BAGTO10	0.00001369	1606.8180	9	46.2844	0.0893	3
BAGTO11	0.00001856	686.6096	5	7.6039	0.0147	9
BAGTO13	0.00001603	976.7689	4	11.9871	0.0231	6
BAGTO14	0.00002492	4637.8836	24	61.0574	0.1178	15
BAGTO15	0.00001975	1044.8581	5	11.9771	0.0231	11
BAGTO16	0.00001325	1000.1924	4	21.8005	0.0421	1
BAGTO17	0.00001350	1626.3951	9	48.4431	0.0934	2

Table 9 The full rank values of all BAGTO variants.

Variant	Mean	Max	Min	Median	Std. dev.	Mean rank	Friedman rank
BAGTO1	5	5	4	6	5	5.0	4.4
BAGTO2	6	6	5	4	7	5.6	4.5
BAGTO3	7	11	9	7	14	9.6	8.3
BAGTO4	10	10	11	10	10	10.2	9.9
BAGTO5	14	14	14	14	8	12.8	12.8
BAGTO6	9	8	7	9	12	9.0	9.4
BAGTO7	16	16	16	16	16	16.0	15.3
BAGTO8	12	9	12	12	13	11.6	11.2
BAGTO9	2	1	1	3	4	2.2	3.9
BAGTO10	1	3	2	1	3	2.0	3.0
BAGTO11	11	7	10	11	9	9.6	10.6
BAGTO13	15	15	15	15	6	13.2	14.6
BAGTO14	8	12	8	8	15	10.2	8.8
BAGTO15	13	13	13	13	11	12.6	11.7
BAGTO16	4	2	3	2	1	2.4	4.0
BAGTO17	3	4	6	5	2	4.0	3.7

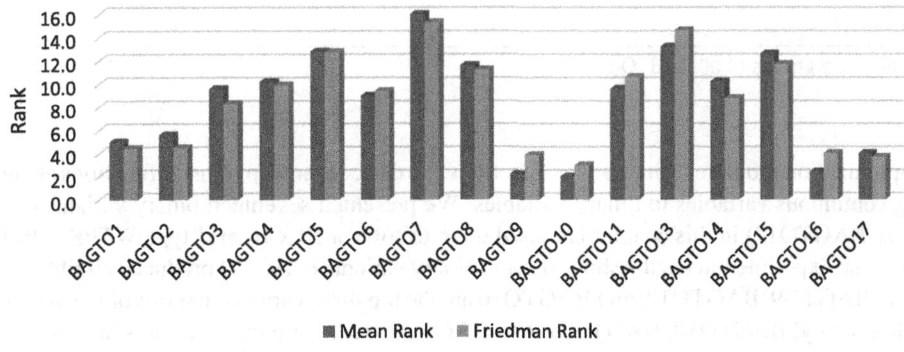

FIG. 4

The rank values of the variants.

Fig. 5 presents the convergence graphs of all BAGTO variants. As shown, the top three variants in terms of fast convergence are BAGTO9, BAGTO10, and BAGTO2.

6. Conclusion

The WTPP is a challenging binary optimization problem. The binary optimization problems have two binary decision variables: 0 and 1. The AGTO is a continuous optimization algorithm. AGTO cannot solve WTPP with its pure form. In the literature, continuous optimization algorithms are converted to

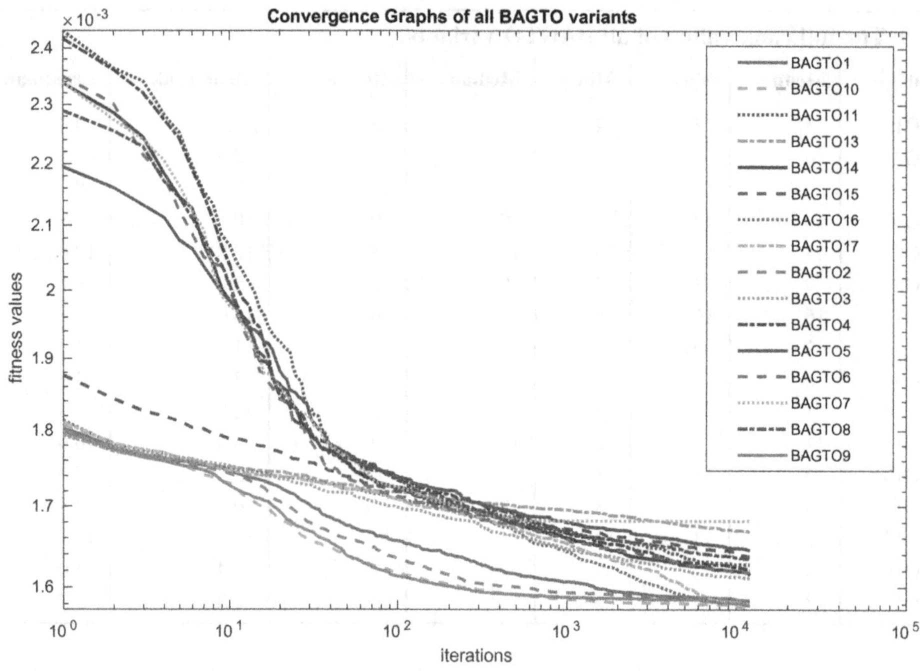

FIG. 5

The convergence graphs of all BAGTO variants.

binary optimization problems with transfer functions. To date, seventeen transfer functions are used for mapping continuous variables to binary variables. We presented seventeen binary variants of AGTO (BAGTO1-BAGTO17) in this chapter and used them to solve a 10×10 grid-type WTPP with 100 dimensions. The experimental results showed some binary variants of AGTO produced eligible solutions for WTPP. BAGTO9, BAGTO10, and BAGTO16 are the top three variants in terms of mean final rankings. Additionally, BAGTO10, BAGTO17, and BAGTO9 are the top three variants in terms of Friedman's test values. The convergence graphs support the experimental results.

Future research will study the peculiar parameter analysis of AGTO, the effects of the different iteration numbers, and the solution quality of the new binary variants on 400-dimensional (20×20 grid-type) WTPPs.

References

[1] A.A. Veisi, M.H.S. Mayam, Effects of blade rotation direction in the wake region of two in-line turbines using Large Eddy Simulation, Appl. Energy 197 (2017) 375–392.

[2] X. Gao, H. Yang, L. Lu, Optimization of wind turbine layout position in a wind farm using a newly-developed two-dimensional wake model, Appl. Energy 174 (2016) 192–200.

[3] T. Kunakote, N. Sabangban, S. Kumar, G.G. Tejani, N. Panagant, N. Pholdee, S. Bureerat, A.R. Yildiz, Comparative performance of twelve metaheuristics for wind farm layout optimisation, Arch. Comput. Methods Eng. (2021) 1–14.

[4] F. Azlan, J. Kurnia, B. Tan, M.-Z. Ismadi, Review on optimisation methods of wind farm array under three classical wind condition problems, Renew. Sust. Energ. Rev. 135 (2021), 110047.

[5] G. Mosetti, C. Poloni, B. Diviacco, Optimization of wind turbine positioning in large windfarms by means of a genetic algorithm, J. Wind Eng. Ind. Aerodyn. 51 (1) (1994) 105–116.

[6] S. Grady, M. Hussaini, M.M. Abdullah, Placement of wind turbines using genetic algorithms, Renew. Energy 30 (2) (2005) 259–270.

[7] C. Wan, J. Wang, G. Yang, X. Zhang, Optimal siting of wind turbines using real-coded genetic algorithms, in: Proceedings of European Wind Energy Association Conference and Exhibition, 2009, pp. 1–6.

[8] K. Chen, M. Song, X. Zhang, Binary-real coding genetic algorithm for wind turbine positioning in wind farm, J. Renew. Sustain. Energy 6 (5) (2014) 053115.

[9] L. Parada, C. Herrera, P. Flores, V. Parada, Wind farm layout optimization using a Gaussian-based wake model, Renew. Energy 107 (2017) 531–541.

[10] Q. Yang, J. Hu, S.-s. Law, Optimization of wind farm layout with modified genetic algorithm based on Boolean code, J. Wind Eng. Ind. Aerodyn. 181 (2018) 61–68.

[11] J.S. González, A.G.G. Rodriguez, J.C. Mora, J.R. Santos, M.B. Payan, Optimization of wind farm turbines layout using an evolutive algorithm, Renew. Energy 35 (8) (2010) 1671–1681.

[12] Y. Wang, H. Liu, H. Long, Z. Zhang, S. Yang, Differential evolution with a new encoding mechanism for optimizing wind farm layout, IEEE Trans. Ind. Inform. 14 (3) (2017) 1040–1054.

[13] Q. Niu, K. Jiang, B.A. Liu, Novel binary negatively correlated search for wind farm layout optimization, in: 2019 IEEE Congress on Evolutionary Computation (CEC), IEEE, 2019, pp. 191–196.

[14] H. Hakli, A new approach for wind turbine placement problem using modified differential evolution algorithm, Turkish J. Electr. Eng. Comput. Sci. 27 (6) (2019) 4659–4672.

[15] Z. Changshui, H. Guangdong, W. Jun, A fast algorithm based on the submodular property for optimization of wind turbine positioning, Renew. Energy 36 (11) (2011) 2951–2958.

[16] Y. Eroğlu, S.U. Seçkiner, Design of wind farm layout using ant colony algorithm, Renew. Energy 44 (2012) 53–62.

[17] C.M. Ituarte-Villarreal, J.F. Espiritu, Optimization of wind turbine placement using a viral based optimization algorithm, Proc. Comput. Sci. 6 (2011) 469–474.

[18] J. Feng, W.Z. Shen, Solving the wind farm layout optimization problem using random search algorithm, Renew. Energy 78 (2015) 182–192.

[19] P.-Y. Yin, T.-Y. Wang, A GRASP-VNS algorithm for optimal wind-turbine placement in wind farms, Renew. Energy 48 (2012) 489–498.

[20] M. Bilbao, E. Alba, Simulated annealing for optimization of wind farm annual profit, in: 2009 2nd International Symposium on Logistics and Industrial Informatics, IEEE, 2009, pp. 1–5.

[21] K. Yang, K. Cho, Simulated annealing algorithm for wind farm layout optimization: a benchmark study, Energies 12 (23) (2019) 4403.

[22] S. Chowdhury, J. Zhang, A. Messac, L. Castillo, Optimizing the arrangement and the selection of turbines for wind farms subject to varying wind conditions, Renew. Energy 52 (2013) 273–282.

[23] P. Asaah, L. Hao, J. Ji, Optimal placement of wind turbines in wind farm layout using particle swarm optimization, J. Modern Power Syst. Clean Energy 9 (2) (2021) 367–375.

[24] S. Pookpunt, W. Ongsakul, Optimal placement of wind turbines within wind farm using binary particle swarm optimization with time-varying acceleration coefficients, Renew. Energy 55 (2013) 266–276.

[25] S. Rehman, S.A. Khan, L.M. Alhems, The effect of acceleration coefficients in Particle Swarm Optimization algorithm with application to wind farm layout design, FME Trans. 48 (4) (2020) 922–930.

[26] C. Wan, J. Wang, G. Yang, X. Zhang, Optimal micro-siting of wind farms by particle swarm optimization, in: International Conference in Swarm Intelligence, Springer, 2010, pp. 198–205.

[27] C. Wan, J. Wang, G. Yang, H. Gu, X. Zhang, Wind farm micro-siting by Gaussian particle swarm optimization with local search strategy, Renew. Energy 48 (2012) 276–286.

[28] J. Patel, V. Savsani, V. Patel, R. Patel, Layout optimization of a wind farm to maximize the power output using enhanced teaching learning based optimization technique, J. Clean. Prod. 158 (2017) 81–94.

[29] J. Patel, V. Savsani, R. Patel, Maximizing energy output of a wind farm using teaching–learning-based optimization, in: Energy Sustainability, American Society of Mechanical Engineers, 2015. p. V002T019A004.

[30] S.R. Moreno, J. Pierezan, C.L. dos Santos, V.C. Mariani, Multi-objective lightning search algorithm applied to wind farm layout optimization, Energy 216 (2021) 119214.

[31] M. Beşkirli, İ. Koç, H. Haklı, H. Kodaz, A new optimization algorithm for solving wind turbine placement problem: binary artificial algae algorithm, Renew. Energy 121 (2018) 301–308.

[32] R.V. Rao, H.S. Keesari, Multi-team perturbation guiding Jaya algorithm for optimization of wind farm layout, Appl. Soft Comput. 71 (2018) 800–815.

[33] M. Aslan, M. Gunduz, M.S. Kiran, A Jaya-based approach to wind turbine placement problem, Energy Sources, Part A: Recov. Utiliz. Environ. Effects (2020) 1–20.

[34] M. Beşkirli, I. Koc, H. Kodaz, Optimal placement of wind turbines using novel binary invasive weed optimization, Tehnički vjesnik 26 (1) (2019) 56–63.

[35] H. Hakli, BinEHO: a new binary variant based on elephant herding optimization algorithm, Neural Comput. Appl. (2020) 1–21.

[36] J. Patel, V. Savsani, V. Patel, R. Patel, Exploring the effect of passing vehicle search (PVS) for the wind farm layout optimization problem, in: Reliability and Risk Assessment in Engineering, Springer, 2020, pp. 411–418.

[37] S. Rehman, S. Ali, S.A. Khan, Wind farm layout design using cuckoo search algorithms, Appl. Artif. Intell. 30 (10) (2016) 899–922.

[38] A. Beşkirli, İ. Dağ, A new binary variant with transfer functions of Harris Hawks Optimization for binary wind turbine micrositing, Energy Rep. 6 (2020) 668–673.

[39] A.C. Cinar, Binary sooty tern optimization algorithms for solving wind turbine placement problem, in: Paper Presented at the International Conference on Interdisciplinary Applications of Artificial Intelligence 2021 (ICIDAAI'21), Yalova, Turkey, 2021.

[40] M.A. Sahman, A.C. Cinar, Binary tree-seed algorithms with S-shaped and V-shaped transfer functions, Int. J. Intell. Syst. Appl. Eng. 7 (2) (2019) 111–117.

[41] Z. Beheshti, UTF: upgrade transfer function for binary meta-heuristic algorithms, Appl. Soft Comput. 106 (2021), 107346.

[42] B. Abdollahzadeh, F. Soleimanian Gharehchopogh, S. Mirjalili, Artificial gorilla troops optimizer: a new nature-inspired metaheuristic algorithm for global optimization problems, Int. J. Intell. Syst. 36 (10) (2021) 5887–5958.

[43] I. Katic, J. Højstrup, N.O. Jensen, A simple model for cluster efficiency, in: European Wind Energy Association Conference and Exhibition, 1986, pp. 407–410.

[44] N.O. Jensen, A Note on Wind Generator Interaction, vol. 2411, Risø National Laboratory, Roskilde, Denmark, 1983.

CHAPTER

Use of metaheuristics in industrial development and their future perspectives 11

Muhammad Najeeb Khan, Amit Kumar Sinha, and Ankush Anand
School of Mechanical Engineering, Shri Mata Vaishno Devi University, Katra, Jammu & Kashmir, India

1. Introduction

Many industrial and scientific optimization problems are intractable in terms of computing optimal solutions. In reality, "acceptable" results provided using heuristic or metaheuristic algorithms are frequently sufficient. Heuristic and metaheuristic algorithms are two types of approximate algorithms. The nature of the problem determines which heuristics are used. They are frequently created and utilized to address specific problems. Metaheuristics are more prevalent and are utilized to solve a variety of problems. They may be utilized to solve practically any problem. By seeking a broad space of solutions, the metaheuristic solves issues that appear to be difficult. These algorithms accomplish this by efficiently exploring space and shrinking the solution space.

Metaheuristics are algorithms that solve problems faster, address larger issues, and are more powerful [1]. They're also extremely adaptable and simple to design and implement. Metaheuristics are a group of approximate optimization approaches that have become increasingly prominent over the last two decades. Metaheuristic algorithms are a type of computational intelligence paradigm used to solve complex optimization issues. A metaheuristic is a problem-independent algorithmic framework that gives a set of recommendations or techniques for developing heuristic optimization algorithms at a high level. Metaheuristic optimization is concerned with the use of metaheuristic techniques to solve optimization issues. Metaheuristics, unlike exact optimization methods, do not ensure that the solutions obtained are optimal. Metaheuristics, unlike approximation techniques, do not specify how near the produced solutions are to the optimal ones.

The term heuristic comes from the ancient Greek word heuriskein, which means "the art of discovering new strategies (rules) for solving problems." The suffix meta, which is derived from the Greek word, signifies "higher level approach." The term metaheuristic was introduced by Fred Glover. Metaheuristic search methods are higher-level general procedures (templates) that can be used to construct underlying heuristics to tackle specific optimization issues.

Comprehensive Metaheuristics. https://doi.org/10.1016/B978-0-323-91781-0.00011-9

2. Classification of metaheuristics

We must distinguish between metaheuristic frameworks and algorithms when classifying metaheuristics. The classification of frameworks is critical for detecting innovation and connecting framework-specific aspects to optimization success and the problems that the metaheuristic is best suited to tackle. The algorithm categorization can be used to identify implementation specifics that are useful in resolving certain issues. These details are easily transferable to other algorithms and can be utilized to improve them. Overall, classification can be used to narrow down the number of metaheuristics to compare based on the characteristics specified in the classification criteria.

Fig. 1 illustrates the overall structure of a classification system. This system divides metaheuristics into seven categories as summarized by Moshtaghi et al. [1]: structure, behavior, search, algorithm, specific features, evaluation, and metaheuristic. Each level has its own set of related criteria that was previously applied in classification methods. However, the current criteria are insufficient for a full description of some levels or there is disagreement over their expressiveness. These may be useful for getting a general idea of metaheuristics, but they don't provide any information about their quality, functionality, or associated features.

In general, there are four types of methods: evolutionary, trajectory-based, nature-inspired, and ancient-inspired methods (see Fig. 2). The latter is a new addition to the metaheuristic classes and is based on ancient past ideology.

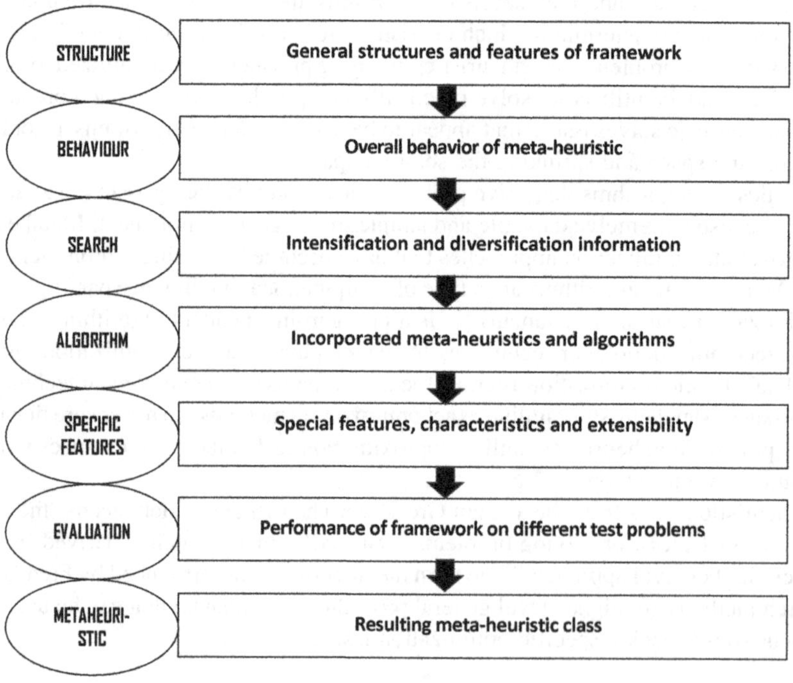

FIG. 1

Classification structure for metaheuristics.

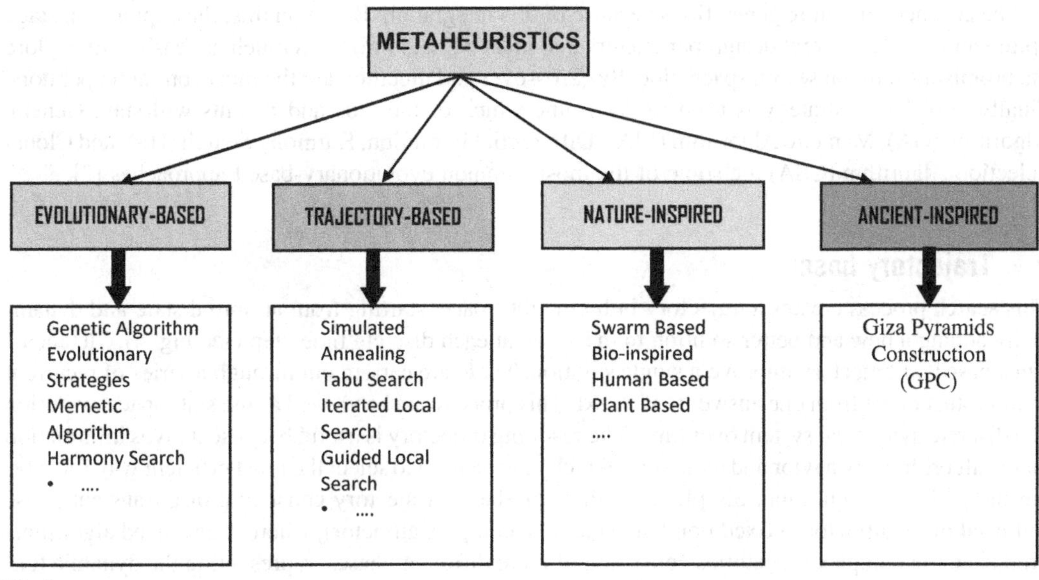

FIG. 2

Metaheuristics.

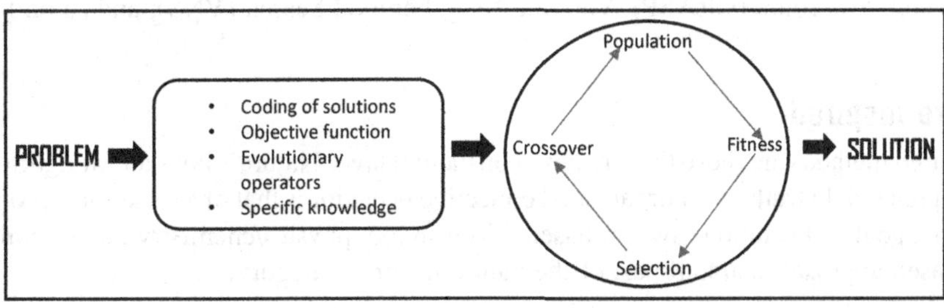

FIG. 3

Evolutionary algorithm structure.

2.1 Evolutionary algorithm

Natural selection or genetic diversity is widely applied to populations that are considered candidates for a solution in evolutionary models. The principle of competition is at the heart of evolutionary theory. Fig. 3 shows the structure of an Evolutionary Algorithm (EA). Metaheuristics of this type imitate species evolution. They are based on the population's evolution. This algorithm's search procedure is divided into two stages: exploration and exploitation. The exploration phase comes before the exploitation phase, which is the process of thoroughly investigating the search space. During the exploration stage, the search process is started with a randomly generated population that is subsequently evolved over several generations. The most applicable element of these heuristics is that the next generation of individuals is shaped by gathering and integrating the greatest individuals. The population will be enhanced in future generations because of this integration. Based on this, the exploration stage optimizer includes several design parameters that must be randomized as much as feasible to explore the promising solution search space globally. Crossover and mutation are the most common operators. Finally, a different strategy is used to determine which of the kids and parents will stay. Genetic Algorithm (GA), Memetic Algorithm (MA), Differential Evolution, Harmony Search (HS), and Clonal Selection Algorithm (CSA) are some of the most common evolutionary-based approaches [2].

2.2 Trajectory based

This search process creates a trajectory in the search space, starting from an initial state and dynamically adding a new and better solution to the curve at each discrete time-step (see Fig. 4). Metaheuristics based on trajectory improve a single solution. These are carried out through a series of recurrent routines that move from one answer to the next. This process can be viewed as the state space evolution of a discrete dynamical system over time. The resulting trajectory is useful because it gives information on the algorithm's behavior and dynamics, which may be used to select the most efficient way for solving the problem in question. Simple algorithms produce a trajectory consisting of a transient phase followed by an attractor (a fixed point, a cycle, or a complex attractor), whereas advanced algorithms produce more complex trajectories consisting of more different phases, representing the dynamic tuning between diversification and intensification during the search process. It has proven to be effective in a range of situations. Some of the popular techniques in this category are Simulated Annealing (SA), Tabu Search (TS), Iterated Local Search (ILS), Guided Local Search (GLS), Greedy Randomized Adaptive Search Procedure (GRASP), Variable Neighborhood Search (VNS), and so on [3].

2.3 Nature inspired

Nature-inspired methods are those that are based on natural laws. Nature's laws are straightforward and easy to comprehend. In truth, most organisms' collective behavior is that of a seeker in a problem space that leads to a goal and solution. Swarm-based, bio-inspired, physics/chemistry-based, human-based, and plant-based approaches are all part of the nature-inspired category.

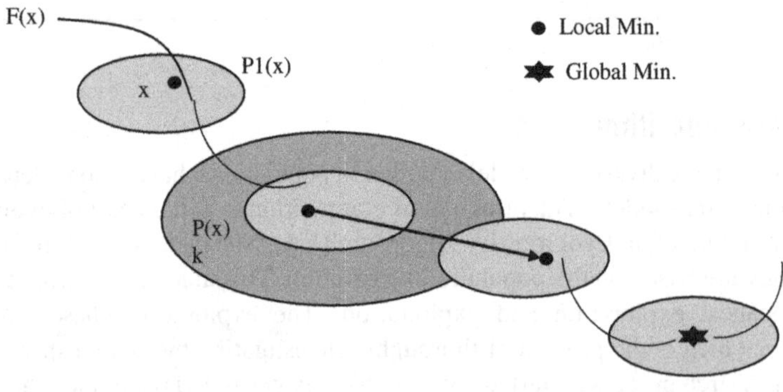

FIG. 4

Trajectory-based algorithm flow.

Swarm intelligence refers to the collective activity and intelligence of a collection of beings. A swarm is an organized group of agents or entities that collaborate. Particle Swarm Optimization (PSO), Firefly Algorithm (FA), Artificial Bee Colony (ABC), Ant Colony Optimization (ACO), and Emperor Penguins Colony (EPC) are presented as swarm-based algorithms. Swarm intelligence algorithms can be considered a subset of bio-inspired algorithms, which is correct. However, many bio-inspired algorithms do not exploit swarm behavior directly. As a result, it is preferable to categorize them as bio-inspired. Bio-inspired algorithms include the Krill Herd Algorithm (KHA), Crow Search Algorithm (CSA), Gray Wolf Optimizer (GWO), Owl Search Algorithm (OSA), Ant Lion Optimizer (ALO), and more. As previously stated, algorithms can fall into various categories. Physics/ chemical-based are generally made by copying physics or chemistry laws. Electrical charge, gravity, physical and chemical changes in materials, and so on are examples of these characteristics and rules. Chemical Reaction Optimization (CRO), Black Hole (BH), Multi-Verse Optimizer (MVO), Thermal Exchange Optimization (TEO), and so on are examples of this subset. Human-based approaches simulate all a person's individual and social behaviors. Imperialist Competitive Algorithm (ICA), Cultural Algorithm (CA), and other algorithms are among the most popular in this subset. Plant-based approaches are inspired by plant development, dissemination, root and plant expansion, and other natural processes. This subset includes any algorithm that models a plant in some way. Invasive Weed Optimization (IWO) and Artificial Root Foraging Algorithm (ARFA) are two examples of algorithms in this subset [4].

2.4 Ancient inspired

The study of ancient past was recently introduced as a new ideology and source of inspiration. There were countless limits in the ancient past, yet diverse humanmade structures reveal that these limitations, as well as a lack of hardware and software facilities, led to some form of optimization. Harifi et al. [5] The Giza Pyramids Construction (GPC) optimization algorithm is also described as the first metaheuristic algorithm influenced by ancient civilizations. All the wonderful features of nature and evolution are brought together in ancient inspiration.

The Giza pyramid complex, often known as the Giza necropolis, consists of three massive pyramids built during Egypt's Fourth Dynasty. The Khufu Pyramid is the largest of the Seven Wonders of the Ancient World. Khafre and Menkaure are the names of the other two pyramids. The process of construction differs from one pyramid to the next, according to archeologists, because pyramid construction has evolved over time. The most essential challenge in the construction of the pyramids was how to manage the workers. Their construction has been optimized due to a lack of hardware facilities, a relatively quick building time, and the enormous number of stone blocks utilized in the pyramids. This novel algorithm was made possible by the construction process, inspiration, and optimization methods of the Giza pyramids construction.

3. Optimization in industry

Industry optimization is the optimization of industrial processes that takes place on a regular basis in industries all around the world. Industrial processes, including design, product development, forecasting, scheduling, and so on, must be optimized to maximize profit and efficiency while minimizing time and effort. These processes will be frantic, troublesome, and chaotic if they are not created and implemented according to a specified algorithm, which in turn, will have an impact on the industry. These industrial processes can be made more sustainable by optimizing them. In today's world, the most pressing demand is for sustainability. As a result, optimization is no longer a choice, it is a requirement.

Engineering optimization entails creating mathematical models of engineering decisions and calculating the intended maximum and lowest values of objective functions quantitatively. Industrial engineers must transform their engineering change ideas into mathematical models to determine the best or optimal solution. However, in industrial engineering research, the initial step is to look for engineering adjustments that can be made. Then, both the existing configuration and the new feasible configuration are subjected to an engineering optimization technique to determine the best outcome, and a decision is made whether to keep the current solution as optimized or to try out the new option. Industrial planning, design, and control are main optimization applications in industrial areas [6]. Machine sequencing, stock control and scheduling, plant renewal, distribution, financial issues, and chemical process control and design are among the main areas [7]. The distance between those who devise the methods and those who need to apply them is bridged by integrating optimization approaches with their applications.

3.1 Metaheuristics in industrial optimization

Because of the increasing complexity of models and the need for speedy decision making in business, metaheuristics for decision making have attracted much interest and investigation. Transportation, storage (warehousing), production planning, supply chain, and scheduling are all examples of industrial operations where metaheuristics have been used [8,9]. The majority of problems were defined as single-objective mathematical models based on experimental or gathered data. Multi-objective optimization has recently gained popularity due to its ability to evaluate problems from multiple angles and produce more realistic conclusions.

Every day, businesses make a slew of optimization decisions in a competitive and changing environment. They use a variety of methods, ranging from trial and error to highly complicated mathematical models and algorithms. Optimization decisions are common in companies, and they are becoming more difficult as technology improves. The increasing number of variables under consideration, as well as the complexity of the objective functions that lead the optimization, are the causes of this increased complexity. Decision makers can use metaheuristic procedures to obtain near-optimal results in a shorter amount of time. These methodologies contribute to the academic literature on the creation of unique or enhanced industrial operations solution procedures [10].

As discussed, there are a few key sectors in industry where metaheuristic optimization has shown to be a key differentiator.

4. Future perspective of metaheuristics in industrial development

4.1 Industry 4.0

Industry 4.0, the fourth industrial revolution, has been considered the beginning of revolutionary concepts in production systems that enable smart factories to communicate vertically and horizontally to improve their performance. Many virtual systems can foresee bad weather, save energy, research specific circumstances, and so on. Industry 4.0 makes use of technologies that connect all stages of product development [11]. Interconnectivity, automation, machine learning, and real-time data are all part of Industry 4.0 [12].

Industry 4.0, also known as the Industrial Internet of Things (IIoT) or smart manufacturing, combines physical manufacturing and operations with smart digital technologies, machine learning, and big data to create a more holistic and linked environment for manufacturing and supply chain management firms. While each company and organization is unique, they all confront the same challenge: the need for connectivity and real-time information across processes, partners, products, and people. In addition, Industry 4.0 combines the physical and digital worlds to create what is called a smart factory. This factory will require components for monitoring, control, and, most likely, cloud connectivity. It would result in a factory that could be dispersed and optimized [13]. Fig. 5 illustrates the framework of Industry 4.0.

Industry 4.0's main goal is to create a smart network based on digitalization and automation in which machines and products interact with one another without the need for human intervention. Smart factory systems, which include smart machines, smart gadgets, smart manufacturing processes, smart engineering, smart logistics, smart suppliers, smart goods, and so on, are the result of Industry 4.0. Cyber-physical systems (CPS), the Internet of Things (IoT), the Internet of Services (IoS), robotics, big data, and cloud manufacturing are all promoted as part of Industry 4.0 [14]. As a result, gadgets, equipment, production modules, and products, among other things, are used in a variety of disciplines, including supply chain management, manufacturing, and management, particularly in real-time scenarios. Industry 4.0 has the potential to improve sales and operations planning as well as the logistics process. Real-time data can be shared across this digitalized process after Industry 4.0 is implemented, allowing for more informed decisions [15].

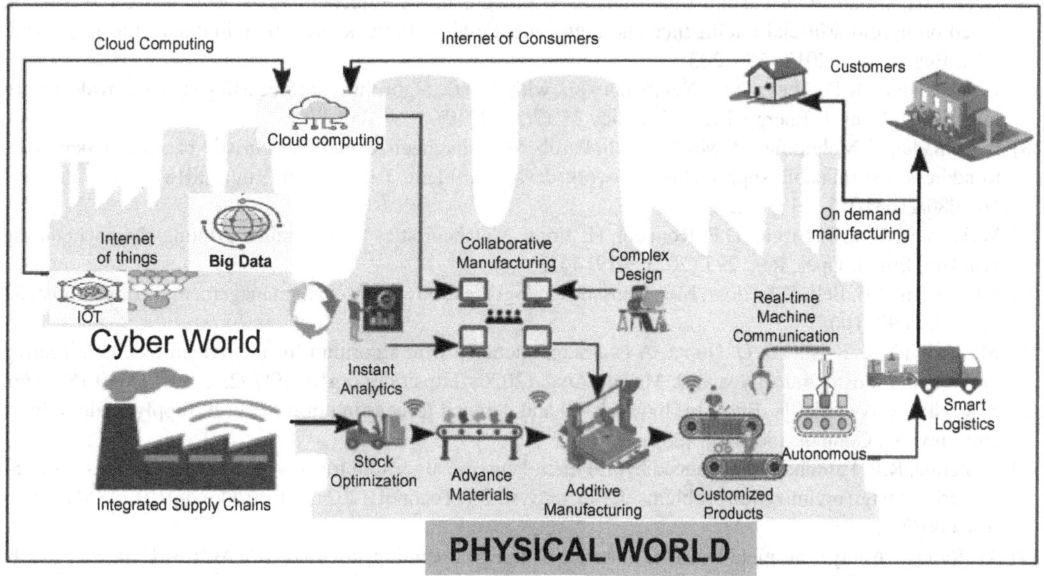

FIG. 5

Framework of Industry 4.0 in the physical world.

5. Conclusion

Industry 4.0 is a digital and automation-based smart network that ultimately improves productivity in the manufacturing sectors. A smart factory can handle complex situations with the help of metaheuristic algorithms. Metaheuristics provides optimal solutions without human intervention during any real-life complex situations in manufacturing. From raw materials to the end customer, total supply chain management of any product is easily controlled in the digital manufacturing platform and this digitization is possible only with the help of metaheuristic algorithms because they provide optimal solutions for each and every workstation.

References

[1] H.R. Moshtaghi, A.T. Eshlaghy, M.R. Motadel, A comprehensive review on meta-heuristic algorithms and their classification with novel approach, J. Appl. Res. Ind. Eng. 6 (3) (2021) 63–89.

[2] D. Greiner, J. Periaux, D. Quagliarella, J. Magalhaes-Mendes, B. Galván, Evolutionary algorithms and metaheuristics: applications in engineering design and optimization, Hindawi Math. Probl. Eng. (2018), https://doi.org/10.1155/2018/2793762. Article ID 2793762, 4 p.

[3] M. Samà, A. D'Ariano, F. Corman, D. Pacciarelli, A variable neighbourhood search for fast train scheduling and routing during disturbed railway traffic situations, Comput. Oper. Res. (2016).

[4] M. Zhao, M. Ghasvari, Product design-time optimization using a hybrid meta-heuristic algorithm, Comput. Ind. Eng. 155 (2021) (2021), 107177.

[5] S. Harifi, J. Mohammadzadeh, M. Khalilian, S. Ebrahimnejad, Giza pyramids construction: an ancient-inspired metaheuristic algorithm for optimization, Evol. Intel. (2021), https://doi.org/10.1007/s12065-020-00451-3.

[6] A. Goli, H.K. Zare, R. Tavakkoli-Moghaddam, A. Sadeghieh, A comprehensive model of demand prediction based on hybrid artificial intelligence and metaheuristic algorithms: a case study in dairy industry, J. Ind. Syst. Eng. 11 (4) (2018) 190–203.

[7] C. Jeenanunta, K.D. Abeyrathna, Neural network with genetic algorithm for forecasting short-term electricity load demand, Int. J. Energy Technol. Policy 15 (2/3) (2019).

[8] A. Samadia, N. Mehranfara, A.M. Fathollahi Fardb, M. Hajiaghaei-Keshtelib, Heuristic-based metaheuristics to address a sustainable supply chain network design problem, J. Ind. Prod. Eng. (2018), https://doi.org/10.1080/21681015.2017.1422039.

[9] W.T. Lunardi, E.G. Birgin, D.P. Ronconi, H. Voos, Metaheuristics for the online printing shop scheduling problem, Eur. J. Oper. Res. 293 (2021) 419–441.

[10] S.E. Griffis, J.E. Bell, D.J. Closs, Metaheuristics in logistics and supply chain management, J. Bus. Logist. 33 (2) (2012) 90–106.

[11] M. Abdirad, K. Krishnan, D. Gupta, A two-stage metaheuristic algorithm for the dynamic vehicle routing problem in Industry 4.0 approach, J. Manag. Anal. (2020), https://doi.org/10.1080/23270012.2020.1811166.

[12] A.R. Nia, A. Awasthi, N. Bhuiyan, Industry 4.0 and demand forecasting of the energy supply chain: a literature review, Comput. Ind. Eng. (2021).

[13] P. Verma, R.P. Parouha, An advanced hybrid meta-heuristic algorithm for solving small- and large-scale engineering design optimization problems, J. Electr. Syst. Inf. Technol. (2021), https://doi.org/10.1186/s43067-021-00032-z.

[14] M. Ramya, Analysison meta-heuristic Internet of things operation in cloud, ICTACT J. Data Sci. Mach. Learn. 02 (01) (2020).

[15] D. Balderas, A. Ortiz, E. Mendez, P. Ponce, A. Molina, Empowering Digital Twin for Industry 4.0 using metaheuristic optimization algorithms: case study PCB drilling optimization, Int. J. Adv. Manuf. Technol. 113 (2021) 1295–1306, https://doi.org/10.1007/s00170-021-06649-8.

CHAPTER

12 Lévy flight and Chaos theory based metaheuristics for grayscale image thresholding

Sajad Ahmad Rather[a], Aybike Özyüksel Çiftçioğlu[b], and P. Shanthi Bala[a]

[a]*Department of Computer Science, School of Engineering and Technology, Pondicherry University, Puducherry, India,*
[b]*Department of Civil Engineering, Faculty of Engineering, Manisa Celal Bayar University, Manisa, Turkey*

1. Introduction

Image segmentation is the process of separating an input image into multiple areas or objects, each with different properties, so that the object may be characterized by its intensity, color, size, shape, and texture. Image segmentation can be implemented using a variety of techniques including edge detection, thresholding, and so on. The technique of recognizing edges or line segments in an image is known as edge detection. It helps in extracting features from images and using them for further processing.

The thresholding technique is widely used for image segmentation due to its ease of implementation and robustness. The technique involves assigning a single pixel value above the threshold as foreground; all values below it are classified as background. Bi-level thresholding and multilevel thresholding are the two types of thresholding procedures. Bi-level thresholding sets all pixels in the image to either black or white based on some predefined threshold value, whereas multilevel thresholding sets different pixel values to different colors according to an adaptive thresholding algorithm.

One of the fundamental difficulties in computer vision is bi-level image thresholding. It requires determining an appropriate threshold value from a grayscale digital image that has both foreground and background regions in which the two regions are strongly separated. The threshold value is a key step for many image processing and pattern recognition tasks such as object segmentation, image enhancement, medical image analysis, and so on. Otsu's approach and Kapur's scheme are two well-known bi-level thresholding procedures. They are image segmentation algorithms that split a picture into sections of similar intensity. The two methods differ in their treatment of edge pixels: Otsu's method maximizes class variance, whereas Kapur's scheme maximizes histogram entropy. However, when the number of thresholds increases, so does the computing cost, and these strategies become ineffective. They also show poor performance in bi-level images with complex structures. On the contrary, due to their simplicity in design and rapid convergence speed, heuristic algorithms allow us to achieve solutions quickly [1–3]. They have been successfully used in finding the shortest paths in a graph or map, scheduling, and designing software algorithms.

In this study, we have developed a unique hybrid image segmentation method based on the gravitational search algorithm (GSA). Our proposed hybrid approach, called Lévy flight and chaos

Comprehensive Metaheuristics. https://doi.org/10.1016/B978-0-323-91781-0.00012-0

theory-based gravitational search algorithm (LCGSA), is designed to overcome the drawbacks of traditional thresholding methods and accomplish widely expected segmentations at a faster speed and lower computational cost. To enhance segmentation outcomes, the parameters of the algorithm are improved using Lévy flight and chaos theory. Lévy flight is used in the algorithm to provide exploration capabilities and create a suitable balance between exploration and exploitation stages. Furthermore, chaos theory prevents the algorithm from getting stuck in local optima. The efficiency of our approach is evaluated and compared with several cutting-edge methods. The presented method is fast, precise, stable, and can be used in a complex background environment.

2. Literature survey

Optimization is the process of finding the best solution by providing certain constraints. Many engineering problems in different fields can be defined as optimization problems. For example, computer engineers use image processing with optimum threshold level, civil engineers design structures that provide the lowest cost and highest durability, mechanical engineers design a machine that provides the lowest production cost or highest component life, chemical engineers design the process plant that ensures the optimum production rate, industrial engineers design systems that will minimize the total turnaround time, and electrical engineers are concerned with designing communication networks that provide the minimum time required for communication from one node to another. Therefore, the optimum design of any problem remains a research topic that has not lost its popularity for decades. In addition, the use of algorithms developed in this field is seen as an effective solution tool in many engineering problems and the search for optimization of more complex problems has led to the need to develop new algorithms.

Metaheuristic algorithms used in solving optimization problems are often inspired by simple concepts such as physical phenomena, behavior of animals, and evolutionary trends. Because of their simplicity, nonderivative mechanism, flexibility, and avoidance of local optima compared to traditional optimization techniques, they can find the optimum in a wide variety of problem types. However, there is no metaheuristic method that gives the optimum result for every problem type [4]. Some metaheuristics may work best for a sequence of problems but may be insufficient for a different sequence of problems. It is a fact that the most successful algorithm is different for each problem type, allowing for the creation of new metaheuristic algorithms or improvement of existing ones. In the literature, there are several studies on improved versions [5–7] of optimization algorithms, hybrid algorithms created by combining algorithms [8–10] and newly created algorithms [11,12].

Metaheuristic methods can generally be examined under two basic headings: single-solution based and population based. The search process for single-solution-based metaheuristics starts with initialization of the candidate solution. This solution is developed through iterations [3]. However, in population-based metaheuristics, the search process is initiated with a set of solutions (population) and this population is developed through iterations [13]. Population-based metaheuristics have a broader knowledge of the search space by having more than one solution. In addition, they have the advantage of avoiding local optima more easily and having a wider problem area compared to single-solution-based methods. Population-based metaheuristics can be categorized as swarm intelligence [2,12,14–16], physics-based [17–19], mathematics-based [1,11], and evolution-based algorithms, among others [1,20–22].

As there have been more studies on optimization, more optimization algorithms are being used for the development of image segmentation methods. Segmentation involves dividing an image into clusters of meaningful and homogeneous parts. Thresholding is a simple implementation of segmentation. First, a threshold value is defined, then each pixel in the image is compared with the defined threshold value. If the compared pixel is greater than the threshold value, then it is considered an object; otherwise, it is set as the background. In entropy-based thresholding, the entropy of the object and background parts are used to select the most appropriate threshold values. There are various entropies used as objective functions in the literature such as Kapur, Renyi, Otsu, Tsallis, and others. In the Kapur thresholding technique, the total entropy of both the object and the background is maximized. The desired threshold value for the total entropy is one. In the Renyi thresholding technique, the total entropy of the object and background parts are calculated for various intensity values and the appropriate level that provides the best thresholding results is selected. Otsu technique is a thresholding process that can be implemented for the gray-level images. When using this approach, the within-class variance value of the two color classes (foreground and background) is calculated for all threshold values. The threshold value, which obtains the smallest in-class variance value, is considered to be the most appropriate. In the Tsallis technique, each object on the image is selected as a subpart of the image and the entropies of these subparts are calculated. Kandhway and Bhandari utilized the water cycle algorithm to find and use the optimum threshold value for image segmentation. Tsallis and Masi entropies were used as the objective functions in the work. They argued that the use of water cycle algorithm and Masi entropy gave successful results when compared to Tsallis entropy for multilevel thresholding [23].

Tan et al. presented an enhanced cuckoo search method to perform color image segmentation. Using modified fuzzy entropy as an objective function, six different methods were compared with improved cuckoo search strategy on a set of color comparison test images. The superiority of the presented technique was defended in terms of statistical tests and finding the best objective function value [24]. Anitha et al. introduced an improved whale optimization method with Otsu and Kapur's objective functions to optimize threshold selection in color image segmentation. With the experiments, it was argued that the improved technique outperforms other techniques, considering image quality, convergence rate, feature preservation, and less CPU computation time [25].

He and Huang conducted experimental studies comparing seven different algorithms over 10 color benchmark images. They proposed the krill herd algorithm to specify the optimal thresholding value in color image segmentation. Otsu, Kapur, and Tsallis entropies were determined as the objective function. By experiments, it was claimed that the presented approach is superior to other methods in values such as optimal threshold and objective function. In addition, Kapur's entropy was found to be superior in terms of color image thresholding segmentation [26]. Bhandari aimed to calculate the optimal threshold value in color image segmentation, with the Kapur and Tsallis entropy by using the beta differential evolution algorithm. The presented approach was evaluated by comparing it with many different algorithms over five real-color and four satellite images. The performance of the presented approach is claimed to be more accurate and robust than other algorithms considering the best objective function value and minimum CPU computation time [27].

Xing aimed to optimize a multithreshold image segmentation technique that has Kapur objective function by using the improved emperor penguin optimization algorithm. For this purpose, experiments were implemented on Berkeley images, plant canopy images, and satellite images. The proposed approach was claimed to be an efficient method considering that it provides color segmentation accuracy with minimal CPU time [28]. Borjigin and Sahoo developed a multilevel thresholding model for the red-green-blue color image using Tsallis entropy. Optimal threshold values for each component of the image were found by particle swarm algorithm. The better performance of the algorithm developed was indicated by calculating the boundary displacement error, probability rand index, and global consistency error performance indices [29]. Houssein et al. aimed to achieve the best threshold value by using black widow optimization with Otsu and Kapur objective functions. In addition, the black widow optimization-based method was compared with six metaheuristic algorithms over a variety of benchmark images. It was claimed that the method showed the most successful performance in terms of segmentation accuracy [30].

Ding et al. aimed to perform multithreshold image processing over 24 benchmark functions with the fruit fly optimization algorithm. In addition, the hybrid adaptive-cooperative learning strategy and the modified fruit fly algorithm were combined. The experimental outcomes showed that the developed method has great promise in the field of image processing [31]. Sowjanya and Injeti investigated the butterfly and gases Brownian motion methods to locate the optimal threshold values using the Otsu objective function in image segmentation. The investigated methods were compared with seven different methods using various benchmark images. According to the experimental results, it was stated that the proposed methods were superior to other methods, and the butterfly algorithm was more successful than the gases Brownian motion algorithm [32].

Wu et al. developed a robust image segmentation approach using advanced teaching-learning-based method and applied it to X-ray image segmentation. They demonstrated the supremacy of the developed method by first comparing it with other evolutionary algorithms in a series of benchmark functions and then applied it in multithreshold image segmentation problems where Otsu and Kapur acted as objective functions [33]. Rahaman and Sing introduced a new image segmentation method with the adaptive cuckoo search algorithm using Otsu and Tsallis entropies as the objective functions. They compared their method with the cuckoo search algorithm that includes McCulloch's Lévy flight generation method using feature similarity (FSIM), peak signal-to-noise ratio (PSNR), structural similarity (SSIM), mean square error (MSE), and CPU measurement techniques [34].

Abd Elaziz et al. improved a multi-leader whale optimization method to specify the most appropriate threshold values in the process of segmenting a series of images. By using Otsu, Fuzzy, and Kapur entropies as fitness functions, they conducted a variety of experiments on diverse benchmarks to assess the performance of the improved technique. The developed method showed optimal values for the PSNR in the experiments for Kapur's and Otsu's methods. Besides, it was reported that fuzzy and Otsu entropies were very close in terms of SSIM criteria, but fuzzy entropy gives better results [35]. Bhandari and Rahul introduced a new method for image segmentation with moth swarm optimizer via the energy-based Masi entropy. Comparing the presented method with other algorithms, they argued that the technique shows better performance with regard to low computational cost and threshold quality [36]. Yue and Zhang developed a hybrid bat method with Otsu and Kapur entropies to select optimal thresholds. Comparing the proposed algorithm with different algorithms, the authors claimed that their algorithm gave better results [37].

Ahmedi et al. introduced a hybrid optimization technique for image segmentation by combining differential evolution and bird mating algorithms, with Kapur and Otsu entropies. The developed approach is evaluated on benchmark images and compared with various optimization techniques. It was reported that the segmentation results were better with the use of the developed hybrid method [38]. Mousavirad and Ebrahimpour proposed a new thresholding approach for image segmentation by the human mental search algorithm with Otsu and Kapur entropies. They evaluated the efficiency of the approach with images and compared them with different algorithms and argued that the method gave better results over various performance indexes [39]. Khairuzzaman and Chaudhury improved a new multilevel thresholding approach using the gray wolf optimizer (GWO) with Kapur and Otsu entropies. The efficiency of the presented method has been compared over a set of standard test images using the MSSIM index with two different algorithms. According to the results, the researchers argued that the proposed approach was superior to other methods [40]. Bhandari et al. developed an Otsu and Kapur objective-functional approach using the cuckoo search algorithm for the color image segmentation purpose. The performance of the introduced method was evaluated with three different methods over the benchmark images. Results of evaluating the efficiency of the approach with different performance indices suggest its superiority over other algorithms [41].

Raj et al. presented an approach based on a differential evolution algorithm with Tsallis-fuzzy entropy to calculate optimum threshold values in multilevel image thresholding. Experiments were conducted on different image sets and the outcomes were compared with two different algorithms. According to the Friedman and Wilcoxon test results, the superiority of the developed approach was defended [42]. Upadhyay and Chhabra investigated the crow search method with Kapur's entropy to achieve optimal values of multilevel thresholds. After experimenting on comparative images for different threshold values, they also compared the introduced approach with different optimization methods. The experimental outcomes were evaluated using PSNR, SSIM, and FSIM performance indices. It was reported that the performance of the developed approach was superior to the other methods [43]. For multilevel thresholding of different medical images, Kalyani et al. introduced an exchange market algorithm with Kapur and Otsu entropies. After evaluating the effectiveness of the presented approach on three different medical images at different threshold levels and comparing the method with different algorithms, the researchers claimed the superiority of the approach via validation indices such as PSNR, SSIM, and Wilcoxon test [44].

Zhao et al. developed a multithreshold image segmentation approach using an improved slime mold algorithm with Renyi entropy. The efficiency of the presented method was demonstrated by comparing different algorithms and using the segmentation dataset. They reported that the experimental outcomes demonstrate the superiority of the developed approach [45]. Table 1 presents a summary of the literature related to image segmentation.

3. Gravitational search algorithm

Every optimization algorithm is inspired by something, whether it be a natural process, physical phenomenon, anthropology, chemical activity, and so on. Most of the heuristic techniques are nature inspired, such as particle swarm optimization (PSO), ant colony optimization (ACO), biogeography-based optimization (BBO), and others. Likewise, GSA is one of the heuristic optimization approaches inspired by physics. It is based on the principle of Newton's law of universal gravitation and motion. According to the gravitation law, "the attractive gravitational force between two masses is directly proportional to the product of masses and inversely proportional to the square of the distance between them". The first step in the GSA optimization process is the initialization of masses because searcher agents are in the form of masses. The gravitational force at time "t" between masses "x" and "y" is calculated using Eq. (1).

Table 1 Literature survey related to heuristic methods for image thresholding.

Reference	Algorithm used	Procedure	Performance evaluation	Performance metrics
Kandhway and Bhandari [23]	Water cycle algorithm	Masi/Tsallis entropies	Convergence speed	PSNR, MSE, FSIM, and SSIM
Tan et al. [24]	Improved cuckoo search	Fuzzy Kapur entropy	Convergence speed	PSNR, FSM, and Wilcoxon test
Anitha et al. [25]	Modified whale optimization algorithm	Otsu/Kapur entropies	Improved computational efficiency	CPU time, PSNR, MSE, SSIM, FSIM, and Wilcoxon test
He and Huang [26]	Krill herd algorithm	Otsu/Kapur/ Tsallis entropies	Reduced complexity	PSNR, SSIM
Bhandari [27]	Beta differential evolution	Kapur/Tsallis entropies	Improved performance indices	CPU time, MSE, PSNR, SSIM, and FSIM
Xing [28]	Improved emperor penguin optimization algorithm	Kapur entropy	Increased accuracy and convergence speed	PSNR, FSIM, and CPU time
Borjigin and Sahoo [29]	Particle swarm optimization	Tsallis-Havrda-Charvát entropy	Best solution for optimal threshold	Wilcoxon test and run time
Houssein et al. [30]	Black widow optimization	Otsu/Kapur entropies	Improved computational efficiency	PSNR, SSIM, and FSIM
Ding et al. [31]	Fruit fly 7orithm	Otsu/Kapur entropies	Enhanced population diversity and Increase convergence speed	Interclass variance and standard deviation
Sowjanya and Injeti [32]	Bat algorithm and gases Brownian motion algorithm	Otsu/Kapur entropies	Increased accuracy and convergence speed	PSNR, RMSE, SSIM, and Wilcoxon test
Wu et al. [33]	Teaching learning-based method	Otsu/Kapur entropies	Enhanced population diversity and increased accuracy and convergence speed	Optimal threshold, class variance, and standard deviation
Abd Elaziz et al. [35]	Multi-leader whale optimization algorthm	Otsu/Fuzzy/Kapur entropies	Reduced complexity	PSNR and SSIM
Rahaman and Sing [34]	Adaptive cuckoo search	Otsu/Tsallis entropies	Convergence speed	PSNR, SSIM, MSE, FSIM, UIQI, and CPU time

Table 1 Literature survey related to heuristic methods for image thresholding—cont'd

Reference	Algorithm used	Procedure	Performance evaluation	Performance metrics
Bhandari and Rahul [36]	Moth swarm algorithm	Energy-based Masi/histogram-based Masi entropies	Improved computational efficiency and performance of threshold quality	PSNR, SSIM, and FSIM
Yue and Zhang [37]	Bat algorithm	Otsu/Kapur entropies	Best solution for optimal threshold	Standard deviation, PSNR, and MSSIM
Ahmadi et al. [38]	Bird mating optimizer and differential evolution	Otsu/Kapur entropies	Increased accuracy and speed	Standard deviation and CPU time
Mousavirad and Ebrahimpour-Komleh [39]	Human mental search	Otsu/Kapur entropies	Best solution for optimal threshold	PSNR, SSIM, FSIM, and the stability test
Khairuzzaman and Chaudhury [40]	Gray wolf optimizer	Otsu/Kapur entropies	Reduced complexity	SSIM
Bhandari et al. [41]	Cuckoo search	Otsu/Kapur entropies	Improved performance indices	PSNR, MSE, SSIM, and FSIM
Raj et al. [42]	Differential evolution	Tsallis Fuzzy/ Fuzzy entropies	Reduced complexity	CPU time, standard deviation, SSIM, PSNR, and SNR
Upadhyay and Chhabra [43]	Crow search algorithm	Kapur entropy	Improved computational efficiency and increased accuracy	PSNR, SSIM, FSIM, Wilcoxon test, and computational time
Kalyani et al. [44]	Exchange market algorithm	Otsu/Kapur/ Minimum Cross entropies	Increased accuracy and reduced complexity	PSNR, SSIM, and Wilcoxon test
Zhao et al. [45]	Slime mold algorithm	Renyi entropy	Improved PSNR, SSIM, and FSIM	PSNR, SSIM, FSIM

$$f_{xy} = G(t)\,\frac{m_{px}(t)\,m_{ay}(t)}{R_{xy}(t)+\in}\left(x_x^d(t)+x_y^d(t)\right) \tag{1}$$

In Eq. (1), $m_{ay}(t)$ and $m_{px}(t)$ are the active and passive attractive masses, respectively. Basically, $m_{ay}(t)$ measures the attractive force applied on a point mass. Alternatively, $m_{px}(t)$ measures the attractive force applied by a point mass in a gravitational field. Meanwhile, $R_{xy}(t)$ is the Euclidian distance and $\in$ is a small constant.

It is important to have a proper balance between diversification and intensification phases. In GSA, it is achieved through the gravitational constant "G". In addition, G helps in locating the feasible regions of the solution space. It is formulated in Eq. (2).

$$G(t) = G(t_0)\, e^{\left(-\alpha \frac{CI}{MI}\right)} \tag{2}$$

Such that $G(t_0)$ and $G(t)$ are the initial and final values of G, respectively. Besides, α is a small coefficient, CI is the current iteration, and MI represents the maximum number of iterations. Apart from active and passive mass, inertia mass is another category of gravitational mass that measures the particle(s) opposition towards external force. It is important to calculate the value of the masses present in the search space. It is evident that the heavier masses will have a greater value of attraction, meaning that they will maintain their position in the solution space. So, when the value of all the three masses namely active (m_{ax}), passive (m_{px}), and inertial (m_{ix}) is equal; the value of gravitational mass (M_x) is calculated as shown in Eq. (5).

$$m_{ax} = m_{px} = m_{ix} = M_x \tag{3}$$

$$m_x(t) = \frac{fit_x(t) - \text{worst}(t)}{\text{best}(t) - \text{worst}(t)} \tag{4}$$

$$M_x(t) = \frac{m_x(t)}{\sum_{y=1}^{m} m_x(t)} \tag{5}$$

In Eq. (4), $m_x(t)$ depicts the fitness function, while best(t) and worst(t) are variables whose value depends on the nature of the problem like minimization or maximization. Similarly, in Eq. (5), "m" indicates the number of masses in d-dimensional space. All the masses are interacting with each other through a gravitational force. Then, according to Newtonian mechanics, the total gravitational force is calculated using Eq. (6).

$$f_x^d(t) = \sum_{y=1, y \neq x}^{m} \gamma_y f_{xy}(t) \tag{6}$$

where γ_y is a random variable. It is also obvious that the masses that are heavier will have high gravitational field, according to Eq. (1). Moreover, in the whole search space, there will be the presence of feasible neighborhoods. Therefore, it is necessary to maintain the quality of the solutions. Due to this, *k*best (cardinality constraint) strategy is employed, as shown in Eq. (7).

$$f_x^d(t) = \sum_{y=k\text{best}, y \neq x}^{m} \gamma_y f_{xy}^d(t) \tag{7}$$

In a physical system, the force goes hand in hand with acceleration. Masses in the search space are always exerting forces on each other. As a result, acceleration gets generated in the candidate solutions and guides them towards feasible regions, as given by Eq. (8).

$$a_x^d(t) = \frac{f_x^d(t)}{m_{ix}(t)} \tag{8}$$

such that m_{ix}(t) is the inertial mass. In GSA, every point mass will have its position and velocity. However, there will be only one mass having a strong gravitational field at the end of iterations. In simpler terms, one heavy mass will attract all other masses. Thus, it is crucial to calculate the velocity ($v_x^d(t)$) and position ($x_x^d(t)$) of the heavy mass to find the optimal solution, as shown in Eqs. (9), (10).

$$v_x^d(t+1) = \gamma_y v_x^d(t) + a_x^d(t) \tag{9}$$

$$x_x^d(t+1) = x_x^d(t) + v_x^d(t+1) \tag{10}$$

4. Lévy flight and chaos theory-based gravitational search algorithm

In this section, we introduce the preliminary concepts of Lévy flight and chaos theory to provide a firm foundation for the proposed LCGSA algorithm. In fact, Lévy flight is employed to provide exploration capability and proper balance between exploration and exploitation in LCGSA. Furthermore, ten chaotic maps will help in the local search and speed up the convergence of the candidate solutions towards the global optimum.

4.1 Lévy flight

Lévy flight [46] is one of the random walking methods based on the probabilistic distribution of the position changes that occur during the movements of living things. In Lévy flight, the size of the step (change of position) of the moving element varies. This variability is determined by the Lévy distribution expressed based on the Fourier transform described in Eq. (11).

$$F(k) = \exp\left[-\alpha|k|^{\beta}\right], 0 < \beta \leq 2 \tag{11}$$

where α is a scale coefficient.

$$\text{If } \beta = 2, \text{then } F(k) = \exp\left[-\alpha k^2\right] \tag{12}$$

whose inverse Fourier transform is consistent with a Gaussian distribution.

$$\text{Or if } \beta = 1, \text{then } F(k) = \exp[-\alpha|k|] \tag{13}$$

which is consistent with a Cauchy distribution [47].

Lévy flights are more effective than Brownian random walks [48]. Flights or movements similar to Lévy flights have been noticed among some species of flies, monkeys, and so on. Moreover, there are many physical occurrences, including spreading of fluorescent molecules, cooling process, and more that exhibit Lévy-flight like properties. It has been observed that Lévy flights have the capacity to maximize the resource exploration efficiency in uncertain environments [49,50].

4.2 Chaos theory

Long-period random number sequences have an important place in most metaheuristic algorithms. By collecting randomly generated numbers in a specific area or generating the same values, the possibility of the algorithm getting stuck in local optima may increase. To overcome these drawbacks, the results produced should not be the same and should have a spread spectrum [51].

The basis of chaos-based approaches are functions called chaotic maps. Chaotic maps are discrete-time systems that exhibit stochastic behavior. In addition, it has been theoretically proven that the numbers produced by chaotic maps have unpredictable, spread-spectrum characteristics and are not periodic. Metaheuristic algorithms in which chaotic maps are embedded use values obtained from functions called chaotic maps instead of random variables, which are often preferred in standard approaches [52,53] (Table 2).

The sum of the chaotic variables used during a given iteration represents a chaotic sequence. The purpose of using chaotic sequences is to give the algorithm the ability to escape from local minima during the process of searching for the global minimum and hence indicating its flexibility. In this way, it is possible to escape from optimum points or reduce the risk of getting stuck in local optima by using chaotic maps in metaheuristic algorithms [63]. Consequently, it is predicted that determining the initial random number string of metaheuristic algorithms with chaotic maps will be faster and more effective in producing suitable solutions for optimization problems [64,65]. Among the many chaotic maps defined in the literature, we explain those used in this study in the sections that follow.

Table 2 Chaotic maps and their mathematical equations.

Chaotic function	Chaotic map	Limits
Chebyshev [54]	$x_{i+1} = \cos(i\cos^{-1}(x_i))$	(1,−1)
Circle [55]	$x_{i+1} = \mathrm{mod}\left(x_i + b - \left(\frac{a}{2\pi}\right)\sin(2\pi x_i), 1\right), a = 0.5, b = 0.2$	(0,1)
Gauss [56]	$x_{i+1} = \begin{cases} 1 & x_i = 0, \\ \frac{1}{\mathrm{mod}(x_i, 1)} & \text{otherwise} \end{cases}$	(0,1)
Iterative [57]	$x_{i+1} = \sin\left(\frac{a\pi}{x_i}\right) a = 0.7$	(−1,1)
Logistic [57]	$x_{i+1} = ax_i(1 - x_i),\ a = 4$	(0,1)
Piecewise [58]	$x_{i+1} = \begin{cases} \frac{x_i}{P} & 0 \le x_i \le P \\ \frac{x_i - P}{0.5 - P} & P \le x_i \le 0.5 \\ \frac{1 - P - x_i}{0.5 - P} & 0.5 \le x_i < 1 - P \\ \frac{1 - x_i}{P} & 1 - P \le x_i < 1 \end{cases}, P = 0.4$	(0,1)
Sine [59]	$x_{i+1} = \frac{a}{4}\sin(\pi x_i)\ a = 4$	(0,1)
Singer [60]	$x_{i+1} = \mu(7.86x_i - 23.31x_i^2 + 28.75 - 13.30x_i^4),\ \mu = 2.3$	(0,1)
Sinusoidal [61]	$x_{i+1} = ax_i^2 \sin(\pi x_i),\ a = 2.3$	(0,1)
Tent [62]	$x_{i+1} = \begin{cases} \frac{x_i}{0.7} & x_i < 0.7 \\ \frac{10}{3}(1 - x_i) & x_i \ge 0.7 \end{cases}$	(0,1)

4.3 Mathematical model of the LCGSA algorithm

In this section, we propose a modified version of standard GSA based on two mathematical tools, that is, Lévy flight and chaos theory. As stated earlier, the GSA has the drawbacks of slow convergence speed and entrapment in local minima. To resolve the aforementioned issues, two strategies have been employed.

In the first strategy, Lévy flight distribution has been utilized to solve the diversity issue of standard GSA. Infinite variance and variable step size of Lévy distribution help in the resolution of the local optima problem. In other words, it increases the diversity and domain of the search process. The Lévy flight is mathematically calculated as,

$$\text{Levy}\,(d, N) = \mathbf{c}' \frac{u}{|v|^{\frac{1}{\beta}}} \sigma_u \tag{14}$$

where d is the dimension of the search space and N is the number of candidate solutions. Moreover, $\mathbf{c}'$ is a multiplicative constant having a value of 0.01, u and v are normal distributions, and β is a Lévy index with a value of 1.5. Moreover, σ_u is the variance.

In the second strategy, ten different chaotic maps have been employed to overcome slow convergence and the local searching issue of standard GSA. Chaotic maps create huge changes in the output when the initial conditions of the chaotic maps are modified. This helps searcher agents to move out of the local minima traps. In addition, chaotic normalization [12] helps in the proper balance between exploration and exploitation. It is mathematically calculated as shown in Eq. (14).

$$C_i^{\text{norm}}\,(t) = \frac{(C_i(t) - a)*(d - c)}{(b - a)} + c \tag{15}$$

In Eq. (15), (a, b) is the range of the chaotic map, i represents the chaotic index, and (c, d) are chaotic normalized intervals in which c is having a value of zero while d is calculated using Eq. (16).

$$d = MI - \frac{CI}{MI}(\text{Max} - \text{Min}) \tag{16}$$

Here, MI and CI represent the maximum number of iterations and the current iteration. In addition, adaptive intervals are indicated by Max and Min with a value of 20 and 1e-10, respectively.

In standard GSA, the gravitational constant (G) is the main parameter that specifies the intensity of the gravitational field, as shown in Eq. (2). It is pivotal for providing a proper balance between exploration and exploitation phases. During the initial iteration phase, the value of G decreases exponentially, which results in the exploration of the search space. Moreover, at the last iterations, the value of G changes slowly, hence promoting the exploitation of the candidate solutions towards the global optimum. That is why we have chosen G, as it is a central controlling parameter of standard GSA that streamlines the exploration and exploitation phases.

In the proposed LCGSA, the Lévy flight and chaotic sequences have been combined with the gravitational constant of GSA. Hence, Lévy-chaotic gravitational constant ($G^{LC}(t)$) is the addition of Eqs. (2), (14), and (15).

$$G^{LC}\,(t) = \text{Levy}\,(d, N) + C_i^{\text{norm}}\,(t) + G\,(t_0)\, e^{\left(-\alpha \frac{CI}{MI}\right)} \tag{17}$$

It is evident in Eq. (17) that $G^{LC}(t)$ has all the essential characteristics required for solving intensification and diversification problems of GSA. Fig. 1 provides the flowchart of the Lévy-chaotic GSA.

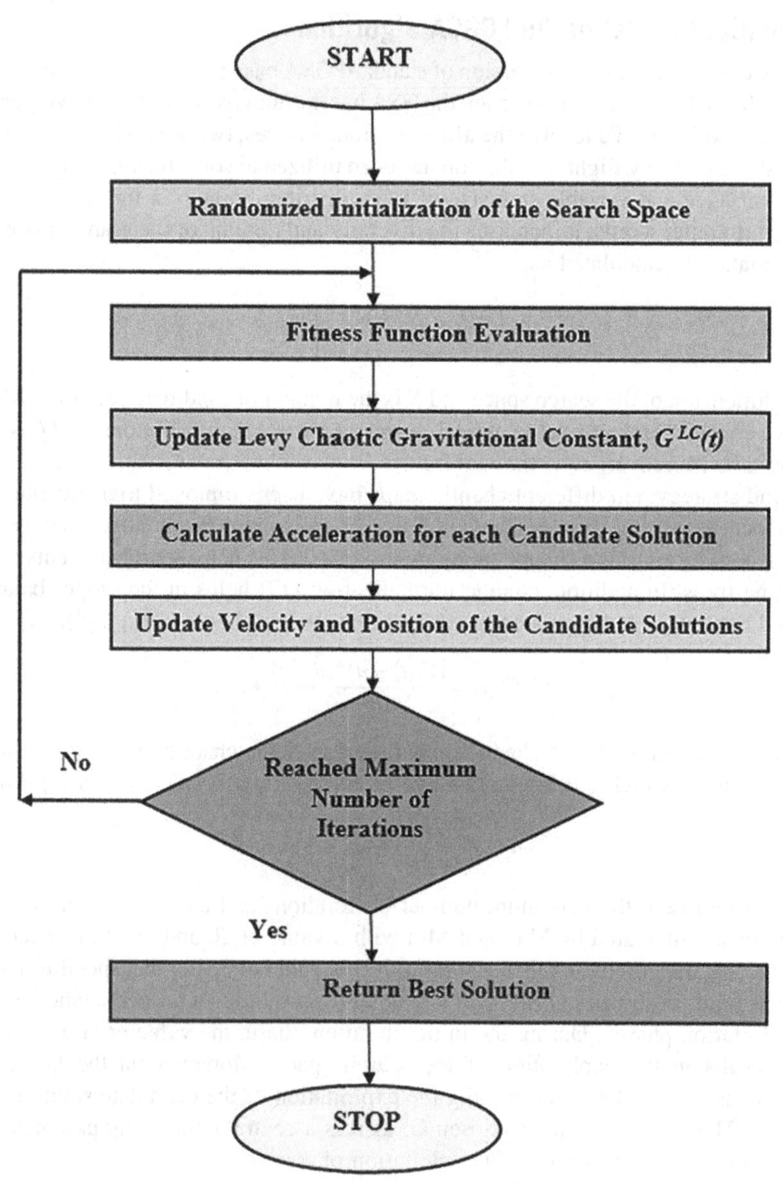

FIG. 1

Flow chart of LCGSA algorithm.

5. Image segmentation using LCGSA technique

In this study, we propose a hybrid LCGSA enriched with Lévy flight and chaos theory for image segmentation. Kapur's entropy is utilized to decide the best threshold values. Among the advantages of LCGSA are its success in not getting stuck in the local optima by increasing the diversity of the solution area, its appropriate balance between exploration and exploitation, and rapid convergence. The pseudocode of the LCGSA is introduced in Algorithm 1. Fig. 2 shows the LCGSA-based multilevel image thresholding of the test image(s).

6. Experimental results and discussion

The simulation analysis of image segmentation based on ten chaotic versions of LCGSA, that is, LCGSA1 to LCGSA10, has been benchmarked using two standard images: Cameraman and Lena from the USC-SIPI database. The images have symmetrical pixel distribution. Figs. 3 and 4 show the graphical representation of the standard test images. Both traditional and robust heuristic algorithms have been employed for the empirical analysis, including particle swarm optimization (PSO), gravitational search algorithm (GSA), gray wolf optimizer (GWO), moth-flame optimization (MFO), particle swarm optimization and gravitational search algorithm (PSOGSA), sine-cosine algorithm (SCA), salp swarm

Algorithm 1: LCGSA for multilevel thresholding

Obtain the details of the input benchmark image
Perform pixel intensity analysis of the image
Initialize the problem space
Calculate the value of fitness function
Set the values of coefficient α, chaotic index β, levy multiplicative constant c, total iterations (MI), $G(t_0)$, and adaptive intervals: Max and Min
Start the optimization process, $t=0$
while $t<MI$ **do**
    for each particle **do**
        Calculate $Levy\ (d,\ N)$ using Eq. (14)
        Compute $C_i^{norm}(t)$ using Eq. (15)
        Update $G^{LC}(t)$ using Eq. (17)
        Calculate the total force, $f_x^d(t)$ with the help of Eq. (7)
        Evaluate the acceleration, $a_x^d(t)$ for each point mass using Eq. (8)
        Compute the particle velocity, $v_x^d(t+1)$ using Eq. (9)
        Update the particle position, $x_x^d(t+1)$ using Eq. (10)
    end for
$t=t+1$
end while
Return the value of the best solution

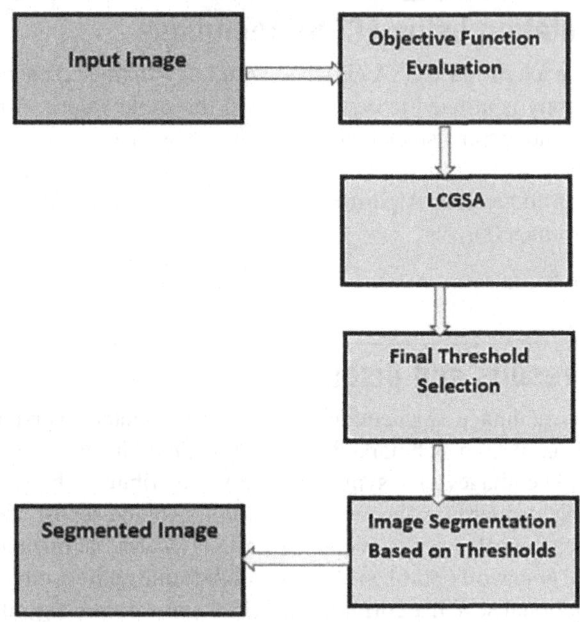

FIG. 2

LCGSA-based image segmentation.

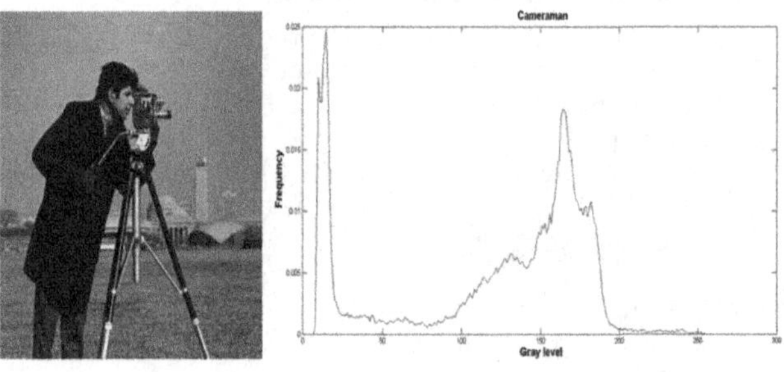

FIG. 3

Grayscale background and histogram of Cameraman image.

algorithm (SSA), and constriction coefficient-based particle swarm optimization and gravitational search algorithm (CPSOGSA).

The simulation outcomes have been recorded using uniform population size (15) and the same total number of maximum iterations (300). The experimental results were recorded after ten trials of the algorithms. The initial values of the competitive algorithms have been considered from the base papers

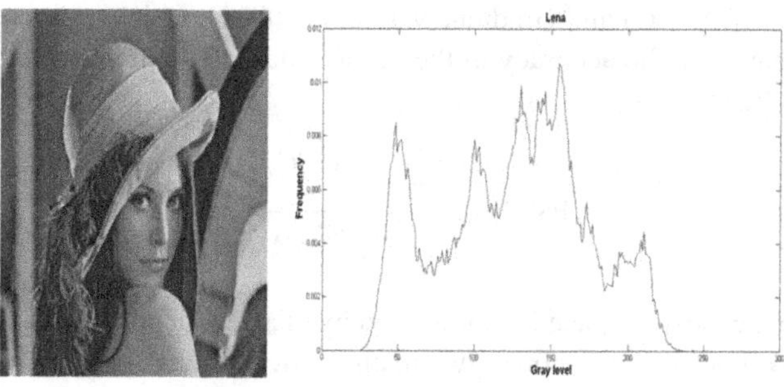

FIG. 4

Grayscale background and histogram of Lena image.

of those algorithms. In addition, the optimization process of the participating heuristic algorithms stops when they provide identical outcomes for 10% of the total iterations. The source codes will be available online on the author's GitHub website (https://github.com/SajadAHMAD1). We used the 2.2 GHz i5 Intel core processor and R2013a MATLAB version for the simulation analysis.

Different performance metrics such as PSNR, SSIM, and FSIM have been utilized to benchmark the quality of the output segmented image. PSNR deals with inspecting the reliability of the segmented image by using the values of the thresholds at successive iterations. The mathematical formulation of PSNR is shown in Eq. (18).

$$\mathrm{PSNR} = 10\log_{10}\frac{255^2}{MSE} \tag{18}$$

$$\mathrm{MSE} = \frac{1}{RC}\sum_{i=1}^{R}\sum_{j=1}^{C}(I(i,j) - O(i,j))^2 \tag{19}$$

In Eq. (19), the rows and columns of a matrix are depicted by R and C, respectively. In addition, I and O are the standard input image and segmented output image. Likewise, SSIM is an image quality metric that deals with accessing the uniformity and resemblance between the segmented image and the input standard image. The value of SSIM should be high, as it shows the presence of high-quality pixels in the segmented image. The mathematical formulation of SSIM is shown in Eq. (20).

$$\mathrm{SSIM}(x,y) = \frac{(2\mu_x\mu_y + c_1)(2\sigma_{xy} + c_2)}{(\mu_x^2 + \mu_y^2 + c_1)(\sigma_x^2 + \sigma_y^2 + c_2)} \tag{20}$$

where μ_x and μ_y are the mean intensities of input and segmented images, while σ_x^2 and σ_y^2 represent standard deviation, σ_{xy} is the covariance, and $<c_1, c_2>$ are the small constants.

FSIM is another image processing metric that deals with examining the quality of the segmented image by assessing the pixels in the local neighborhood of the image. In addition, FSIM inspects the values of Phase Congruency (PC) and Gradient Magnitude (GM), which are essential for locating the best pixel values to determine the accuracy of the output image. The mathematical formulation of the FSIM is shown in Eq. (21).

$$\text{FSIM}(x) = \frac{\sum_{x \in \Omega} S_L(x)\, PC_m(x)}{\sum_{x \in \Omega} PC_m(x)} \tag{21}$$

where domain of the pixel search space is represented by Ω and $S_L(x)$ is the similitude of the image. If PC_1 and PC_2 are the symmetrical phase(s) of two mathematical functions, then PC is given by Eq. (22).

$$PC_m(x) = \max\left(PC_1(x), PC_2(x)\right) \tag{22}$$

Likewise, similitude content of the segmented image, $S_L(x)$ is formulated in Eq. (23).

$$S_L(x) = [S_{PC}(x)]^{\alpha}\, [S_G(x)]^{\beta} \tag{23}$$

Such that

$$S_{PC}(x) = \frac{2PC_1(x)\, PC_2(x) + T_1}{PC_1^2(x)\, PC_2^2(x) + T_1} \tag{24}$$

$$S_G(x) = \frac{2G_1(x)\, G_2(x) + T_2}{G_1^2(x)\, G_2^2(x) + T_2} \tag{25}$$

where S_G and S_{PC} are the similitude of Gradient and Phase Congruency while constants are denoted by α, β, T_1, and T_2. As image segmentation task is a maximization problem, the higher value of FSIM shows capability of locating the best pixels in the solution space and having efficient segmentation capability.

It is also important to test the statistical significance of the experimental results. Therefore, the signed Wilcoxon rank-sum test has been used to validate the effectiveness and the superiority of the heuristic algorithms. The null and alternate hypotheses are given as:

Null hypothesis (H_0): LCGSA does not provide the best pixel values to the image.
Alternate hypothesis (H_1): LCGSA provides efficient pixel values to the image.

The *P*-values of all the competitive algorithms are computed through MATLAB simulation analysis. If a heuristic algorithm provides *P*-value less than .05, then, H_1 is accepted. Moreover, if the *P*-value is equal to one, it simply means the particular algorithm has been compared with itself or its performance is consistent with the best-performing algorithm.

6.1 Simulation results of cameraman image

Tables 3–6 present the simulation outcomes of competitive algorithms for the Cameraman image. It can be clearly seen that all 10 versions of LCGSA have better values for the image thresholds at $k = 2$, 4, 6, 8, and 10. The minimum values of SD and MSE for the LCGSA show stability in the optimization process and high quality of the pixels. Moreover, LCGSA versions, particularly LCGSA3 and

Table 3 Simulation results for Cameraman benchmark using classical heuristic algorithms.

Image	Algorithm	k	Optimal thresholds	SD	MSE	PSNR	SSIM	FSIM	BV	Run time (s)	P value
Cameraman	PSO	2	292,198	1.69	1.73E+04	5.73	0.02	0.58	3.14E+03	9.57	.0625
		4	6,7,5,4	1.89	1.77E+04	5.63	0.03	0.12	9.23	13.22	.1250
		6	7,8,34,66,57,140	1.05	5.23E+03	10.93	0.58	0.69	20.35	60.33	0
		8	6,12,22,31,45,45,45,89	1.06	3.46E+03	12.73	0.67	0.75	28.3	69.78	.0020
		10	16,21,25,28,30,84,102,83, 115,112	1.12	1.63E+03	16	0.67	0.75	0.81	34.93	.0020
	GSA	2	1,1	1.23	1.63E+03	5.63	0.03	0.23	0.32	7.89	.125
		4	1,1,1,1	0.98	1.77E+04	5.63	0.03	0.12	0.21	13.23	.2500
		6	194,230,158,195,128,180	0.13	2.23E+03	14.63	0.5	0.68	26.01	65.61	.0020
		8	98,112,95,109,160,121,15,125	0.22	5.47E+02	20.74	0.66	0.8	28.42	54.12	.0020
		10	175,148,130,171,108, 130,188,104,115,131	0.12	6.68E+02	19.88	0.63	0.79	34.65	102.34	.0020
	PSOGSA	2	84,88	27.32	3.75E+03	12.38	0.49	0.71	3.31E+03	8.66	.0020
		4	1,1,1,1	0.49	1.77E+04	5.63	0.03	0.11	10.21	13.12	.0020
		6	13,60,65,68,81,119	0.36	1.40E+03	16.66	0.74	0.81	23.55	57.26	.0020
		8	1,9,36,35,44,58,86,104	0.96	2.22E+03	14.65	0.73	0.78	21.53	60.38	.0020
		10	3,5,35,75,86,115,121,124, 143,168	0.53	2.23E+02	24.64	0.85	0.89	37.22	88.05	.0020
	CPSOGSA	2	1,1	1.57	1.77E+04	5.63	0.03	0.19	7.10E+00	22.38	.0020
		4	10,40,9,21	0.54	7.31E+03	9.48	0.47	0.66	15.22	38.7	.0059
		6	13,19,26,56,76,80	0.36	4.35E+03	11.74	0.55	0.72	24.39	57.22	.0020
		8	5,54,54,68,72,79, 118,143	0.51	5.86E+02	20.45	0.81	0.85	30.63	39.21	.0020
		10	4,8,10,17,9,32,42,77,98,1,11	0.5	1.82E+03	15.52	0.79	0.81	32.92	86.94	.0020
	SCA	2	68,120	246.83	1.54E+03	16.24	0.63	0.79	3.56E+03	9.42	.0020
		4	92,245,82,140	816.35	9.17E+02	18.5	0.65	0.8	4.31E+03	15.61	.0020
		6	65,31,109,131,230,77	2.41	8.17E+02	19	0.72	0.88	26.56	71.11	.0137
		8	158,183,44,72,23,136,151,205	2.66	5.64E+02	20.61	0.67	0.84	31.27	69.03	.0039
		10	54,121,1,30,75,91,139,128,5,64	3.2	5.98E+02	20.35	0.79	0.88	34.13	104.52	.0020

Table 4 Simulation results for Cameraman benchmark using recent heuristic algorithms and the first two LCGSA versions.

Image	Algorithm	k	Optimal thresholds	SD	MSE	PSNR	SSIM	FSIM	BV	Run time (s)	P value
Cameraman	SSA	2	204,191	1.39E +03	1.71E+04	5.78	0.03	0.59	3.64E+03	9.62	.0039
		4	238,255,225,32	1.53E +03	1.09E+03	7.71	0.22	0.65	3.81E+03	15.17	.0195
		6	200,210,19,238,239,135	4.56	2.50E+03	14.14	0.56	0.71	28.02	70.12	.0020
		8	226,67,102,57,1226,226,183	5.72	1.97E+03	15.16	0.62	0.76	32.96	84.17	.0020
		10	228,227,228,226,226,227,227,227,227,227	12.95	1.77E+04	5.64	0.01	0.57	40.91	108	.0020
	GWO	2	8,26	181.13	1.22E+04	7.24	0.3	0.59	3.28E+03	8.95	.0625
		4	8,9,5,10	150.23	1.79E+04	5.58	0.005	0.21	9.02E+00	13.37	.5000
		6	7,21,17,29,25,10	2.47	1.18E+04	7.37	0.3	0.59	17.27	60.06	.0020
		8	32,84,84,7,78,54,17,84	2.87	3.92E+03	12.19	0.63	0.73	32.03	79.08	.0020
		10	5,7,6,4,3,5,9,3,5,6	7.37	1.65E+04	5.93	0.13	0.12	24.1	89.09	.0020
	MFO	2	255,1	531.64	1.77E+04	5.63	0.03	0.11	9.83E+02	10.04	.0020
		4	255,255,124,1	1.22E +03	2.78E+03	13.68	0.55	0.67	2.60E+03	15.64	.0020
		6	255,1255,1222,1	1.88	1.74E+04	5.71	0.04	0.57	24.73	78.26	.0020
		8	35,255,67,256,173,255,1256	2.66	3.78E+03	12.35	0.48	0.71	33.28	85.56	.0020
		10	18,114,78,64,37,71,243,107,255,179	2.5	1.08E+03	17.76	0.65	0.83	34.84	110.78	.0020
	LCGSA1	2	249,7	0.38	1.63E+04	5.99	0.15	0.54	1.23E+01	0.0001	.0098
		4	69,131,197,253	0.54	1.17E+03	17.43	0.64	0.79	2.14E+01	9.75E−05	.0195
		6	123,194,31,225,19,104	1.81	1.20E+03	17.31	0.68	0.83	24.63	0.0001	.3750
		8	77,142,240,21,96, 222,24,222	1.95	6.29E+02	20.13	0.7	0.85	28.23	0.0002	.3750
		10	46,98,142,242,33, 234,25,114,243,56	2.73	5.32E+02	20.87	0.74	0.89	36.92	0.0001	.4316
	LCGSA2	2	247,6	0.23	1.63E+04	6	0.15	0.51	1.33E+01	5.59E−06	.0488
		4	236,22,90,146	1.15	7.87E+02	19.16	0.65	0.81	2.04E+01	1.86E−06	.0645
		6	189,22,96,218,19,102	1.87	2.35E+03	14.41	0.59	0.77	23.17	6.53E−06	.6250
		8	98,142,197,253,38,	1.79	6.09E+02	20.28	0.7	0.87	28.67	6.53E−06	.6250
		10	136,221,51,207,21, 101,246,142,26,222	1.63	5.54E+02	20.69	0.73	0.88	31.61	5.13E−06	.0645

Table 5 Simulation results for Cameraman benchmark using LCGSA3 – LCGSA7 versions.

Image	Algorithm	*k*	Optimal thresholds	SD	MSE	PSNR	SSIM	FSIM	BV	Run time (s)	*P* value
Cameraman	LCGSA3	2	75,140	0.42	1.16E+03	17.47	0.62	0.78	15.93	5.59E−06	.0020
		4	181,100,203,110	0.69	1.74E+03	15.7	0.55	0.72	1.91E+01	1.86E−06	1
		6	88,120,148,172,204,153	0.92	4.12E+02	21.97	0.67	0.83	2.43E+01	3.26E−06	.0039
		8	100,144,205,134,56, 162,73,129	0.81	2.54E+02	24.06	0.72	0.88	31.85	5.13E−06	.0039
		10	113,178,64,117,156, 193,139,82,137,193	1.88	2.54E+02	24.06	0.71	0.88	33.14	3.26E−06	.0020
	LCGSA4	2	89,145	0.13	9.78E+02	18.22	0.62	0.77	15.94	3.26E−06	.4922
		4	119,223,19,105	1.32	1.56E+03	16.18	0.64	0.79	1.94E+01	1.86E−06	.0273
		6	38,90,139,199,254,20	2.46	7.24E+02	19.53	0.7	0.86	2.27E+01	4.19E−06	.3223
		8	132,211,20,89,139, 214,24,103	3.16	5.99E+02	20.35	0.72	0.87	29.35	3.26E−06	.3223
		10	191,81,219,23,88, 146,230,72,222,76	2.99	7.14E+02	19.59	0.66	0.82	32.01	3.26E−06	.0840
	LCGSA5	2	123,67	0.45	9.54E+02	18.33	0.62	0.76	16.14	3.26E−06	1
		4	35,87,96,100	2.11	7.78E+02	19.21	0.67	0.79	1.96E+01	1.86E−06	.0195
		6	54,77,123,122,89,54	2.33	5.94E+02	20.39	0.69	0.84	2.38E+01	3.73E−06	.9219
		8	112,123,233,49,77,123,45,87	3.45	3.50E+02	22.68	0.72	0.87	28.42	4.66E−06	.0137
		10	123,200,178,189,190,54,77,64,156,189	3.00	5.35E+02	20.84	0.73	0.88	35.72	3.26E−06	.9219
	LCGSA6	2	90,144	0.51	9.77E+02	18.23	0.63	0.77	13.94	2.79E−06	1
		4	132,244,9248	0.65	2.81E+03	13.62	0.65	0.66	1.97E+01	1.86E−06	.375
		6	222,32,210,20,97,149	2.1	6.47E+02	20.01	0.67	0.83	2.23E+01	5.13E−06	.9219
		8	55,104,217,21,61, 107,159,231	2.56	5.77E+02	20.51	0.65	0.83	26.49	3.26E−06	.0098
		10	86,145,234,74,219, 19,64,102,149,199	3.27	4.50E+02	21.59	0.71	0.86	38.37	3.26E−06	.1055
	LCGSA7	2	88,143	0.22	9.80E+02	18.21	0.63	0.78	14.97	2.33E−06	.8457
		4	122,231,20,102	0.82	2.14E+03	14.81	0.64	0.78	2.00E+01	1.39E−06	.0273
		6	36,94,144,198,253,54	2.25	7.58E+02	19.33	0.67	0.82	2.72E+01	3.26E−06	.7695
		8	129,233,42,199,57, 226,22,102	2	4.06E+02	22.04	0.87	0.9	32.6	2.79E−06	.0488
		10	44,89,130,200,251, 122,21,104,222,62	4.7	2.47E+02	24.19	0.74	0.89	33.72	4.19E−06	1

Table 6 Simulation results for Cameraman benchmark using LCGSA8–LCGSA10 versions.

Image	Algorithm	*k*	Optimal thresholds	SD	MSE	PSNR	SSIM	FSIM	BV	Run time (s)	*P* value
Cameraman	LCGSA8	2	91,143	0.22	9.80E+02	18.21	0.63	0.78	14.97	2.33E−06	.8457
		4	48,107,243,16	0.82	2.14E+03	14.81	0.64	0.78	2.00E+01	1.39E−06	.0273
		6	80,142,249,20,90,144	2.25	7.58E+02	19.33	0.67	0.82	2.72E+01	3.26E−06	.7695
		8	52,116,229,8201, 20,96,144	2	4.06E+02	22.04	0.87	0.9	32.6	2.79E−06	.0480
		10	207,19,93,147,196, 253,21,80,123,160	4.7	2.47E+02	24.19	0.74	0.89	33.72	4.19E−06	1
	LCGSA9	2	247,16	0.41	1.44E+04	6.53	0.12	0.58	12.4	2.33E−07	.0371
		4	224,16,249,16	0.83	1.40E+04	6.66	0.13	0.61	1.85E+01	9.33E−07	.0020
		6	46,123,227,20,89,146	1.81	4.34E+02	21.75	0.74	0.89	2.70E+01	2.79E−06	1
		8	129,231,18,217,17,202,20,107	2.1	1.14E+03	17.55	0.67	0.81	30.99	2.79E−06	.1055
		10	202,22,92,146,229, 20,88,145,200,252	3.44	6.76E+02	19.82	0.67	0.82	33.65	4.19E−06	.1055
	LCGSA10	2	243,17	0.33	1.40E+04	6.66	0.12	0.58	12.46	3.30E+01	.0488
		4	74,143,247,7	0.74	1.08E+03	17.77	0.74	0.77	2.01E+01	4.82E+01	.0039
		6	127,222,13,218,18,101	1.6	1.16E+03	17.47	0.74	0.81	2.47E+01	7.11E+01	.6953
		8	213,20,100,208,44, 225,19,104	1.79	2.25E+03	14.6	0.63	0.8	30.68	7.95E+01	.1055
		10	57,123,205,66,137, 226,19,90,135,196	5	6.03E+02	20.32	0.74	0.9	29.22	1.01E+02	.1309

LCGSA7, have large values for PSNR, FSIM, and SSIM, which indicates high output image quality and segmentation capability. It can be seen that SCA, PSOGSA, CPSOGSA, and PSO also provide appreciable values to image quality metrics. Furthermore, LCGSA takes less run time in seconds (s) to find the best pixels in the search space. The signed Wilcoxon rank-sum test validated the best performance of LCGSA as its P values are mostly less than .05.

Fig. 5 shows the segmented and histogram curves of LCGSA. As the number of thresholds increases, the clarity and contrast of the output segmented image also increases. Figs. 6 and 7 depict the convergence curves and box plot graphs, respectively. The convergence curves show that LCGSA takes less computational time to find the best pixels in the problem space. Moreover, SCA also depicts appreciable exploitation capability. However, GWO, MFO, and CPSOGSA have small values for Kapur's fitness function, indicating issues in the optimization capability. The box plots also validated the optimal performance of LCGSA as it has the highest fitness value, whereas other peer algorithms have suboptimal values for the fitness function.

6.2 Simulation results of Lena image

Tables 7–10 present the experimental results of the Lena image. The LCGSA versions depicted the best threshold values at $k = 2, 4, 6, 8$, and 10. In addition, they have fewer error and image contrast issues because their MSE and SD values are small in magnitude. The PSNR, SSIM, FSIM, and fitness values of LCGSA are large than other peer algorithms indicating high segmentation capability. LCGSA takes less run time than its counterparts, while the performance of SCA is also appreciable in finding the feasible regions of the pixel search space.

Segmented images as shown in Fig. 8 also convey the optimal performance of LCGSA in handling the complex search space of the Lena benchmark image. The convergence curves as shown in Fig. 9 indicate that LCGSA has high fitness values during the optimization process, whereas other peer algorithms have small values for the fitness function. It can also be seen that SCA, GSA, CPSOGSA, and PSO show appreciable performance in finding the optimal pixels in the solution space. However, GWO, SSA, PSOGSA, and MFO displayed suboptimal performance indicating issues in handling local minima regions. Box plots as shown in Fig. 10 also validated the efficient performance of LCGSA.

The simulation results of both Cameraman and Lena images indicated that LCGSA has efficient optimization capability to handle complex and unknown search spaces. The high values of LCGSA for PSNR, SSIM, and FSIM convey its efficient segmentation power and image processing capability. Moreover, LCGSA provided superior values for optimal thresholds, which were better than standard GSA depicting its superiority over GSA as far as image segmentation is concerned. Furthermore, LCGSA took less run time while locating the quality pixels in both Cameraman and Lena images indicating fewer computational overhead issues.

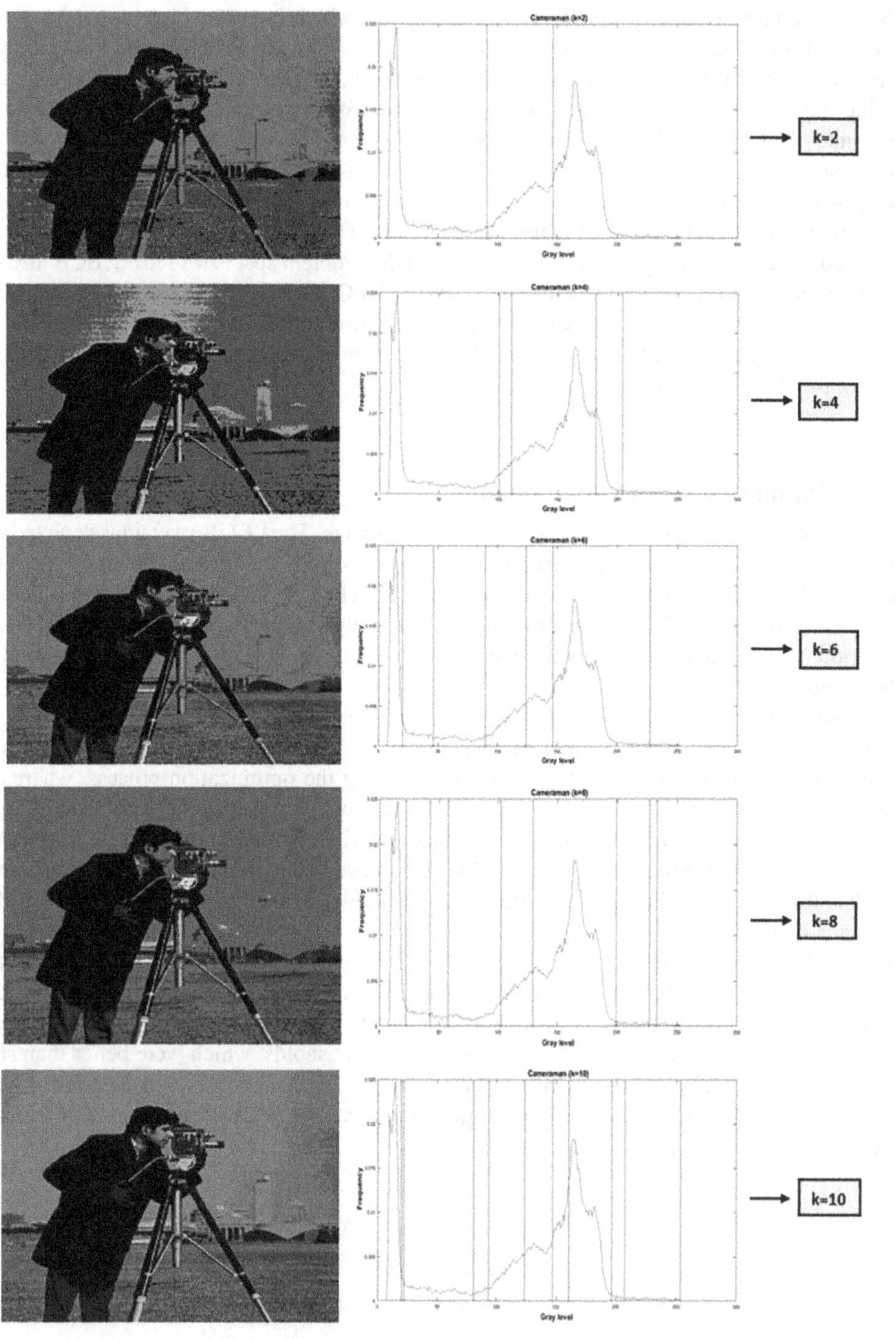

FIG. 5

LCGSA segmented images and histogram curves for Cameraman benchmark image at $k = 2, 4, 6, 8, 10$.

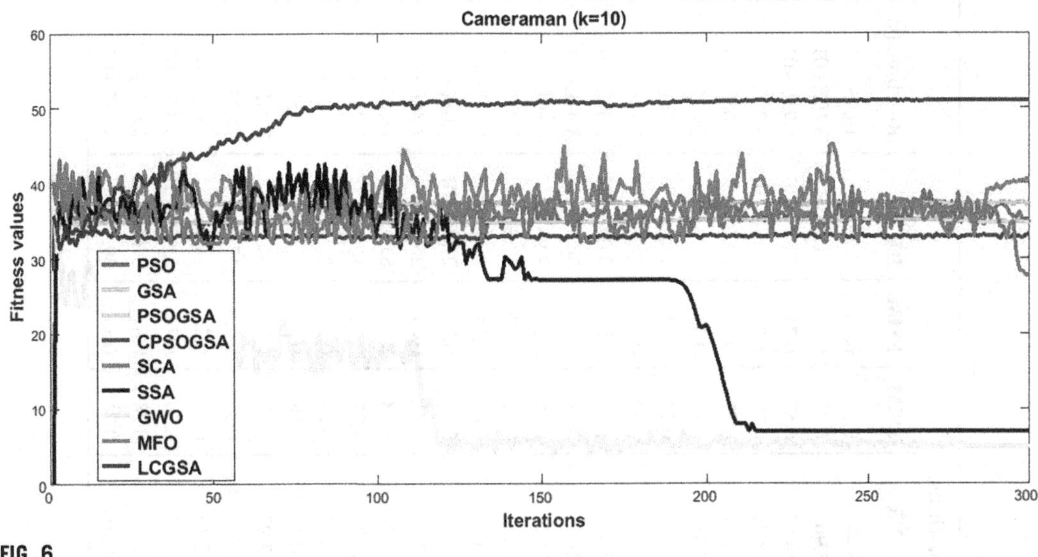

FIG. 6

Convergence curves of heuristic algorithms for Cameraman image.

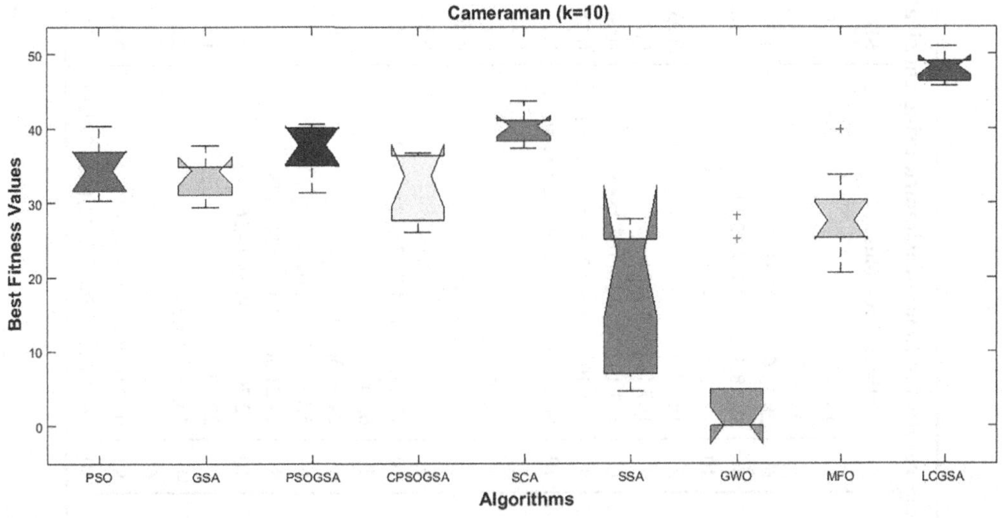

FIG. 7

Box plots of heuristic algorithms for Cameraman image.

Table 7 Simulation results for Lena benchmark using classical heuristic algorithms.

Image	Algorithm	*k*	Optimal thresholds	SD	MSE	PSNR	SSIM	FSIM	BV	Run time (s)	*P* value
Lena	PSO	2	13,10	1.79	1.46E+04	6.48	0.09	0.18	9.72	19.95	.0020
		4	5,5,4,4	1.89	1.64E+04	5.96	0.03	0.21	11.16	3.19E+01	.0020
		6	28,53,67,81,102,71	1.36	2.24E+03	14.62	0.67	0.71	2.84E+01	5.93E+01	.0020
		8	12,18,39,56,60,112,82,141	1.16	6.84E+02	19.77	0.79	0.82	3.17E+01	6.75E+01	.0020
		10	12,22,29,42,34,45,56,81,72,81	0.97	3.87E+03	12.24	0.57	0.66	32.38	7.65E+01	.0020
	GSA	2	164,228	0.03	1.07E+04	7.81	0.13	0.53	12.68	23.45	.0371
		4	167,187,210,215	0.2	1.10E+04	7.71	0.12	0.54	18.94	3.30E+01	.0039
		6	112,197,135,187, 167,188	0.06	2.49E+03	14.15	0.52	0.71	2.69E+01	6.17E+01	.0020
		8	155,196,152,168,233,156,175,162	0.31	8.03E+03	9.08	0.21	0.59	3.02E+01	7.49E+01	.0020
		10	50,151,82,158,52,119,39,55,80,56	0.24	4.87E+02	21.25	0.79	0.83	36.77	8.44E+01	.0020
	PSOGSA	2	10,21	0.28	1.28E+04	7.02	0.15	0.25	9.95	19.91	.0020
		4	11,17,16,34	0.87	1.03E+04	7.96	0.23	0.52	13.34	2.97E+01	.0020
		6	1,14,26,36,84,92	0.39	3.04E+03	13.29	0.6	0.64	2.31E+01	5.34E+01	.0020
		8	13,16,45,57,60,66,122,122	0.52	1.50E+03	16.36	0.68	0.72	2.74E+01	6.61E+01	.0020
		10	17,44,49,74,74,76,76,83,84,141	1.18	1.05E+03	17.89	0.71	0.74	26.48	7.84E+01	.0020
	CPSOGSA	2	6,10	0.51	3.40E+03	1.52E+04	6.28	0.07	8.4	18.93	.0020
		4	24,36,36,39	1.08	9.34E+03	8.42	0.27	0.52	10.32	2.95E+01	.0020
		6	5,10,38,41,96,102	0.28	2.42E+03	14.27	0.61	0.65	2.24E+01	5.48E+01	.0020
		8	18,21,36,82,94,104,119,148	0.4	6.03E+02	20.32	0.78	0.8	3.26E+01	6.45E+01	.0020
		10	12,20,33,35,36,38,44,56,88,148	0.4	1.04E+03	17.93	0.72	0.76	34.73	7.29E+01	.0020
	SCA	2	22,90	1.39	3.57E+03	1.26E+01	0.55	0.62	17.74	24.29	.9291
		4	73,136,181,23	2.01	9.98E+02	18.13	0.7	0.76	19.96	3.60E+01	.0059
		6	3,21,45,166,23,115	2.83	1.14E+03	17.52	0.69	0.74	2.54E+01	6.37E+01	.0039
		8	94,216,254,61,136,33,87,191	2.81	4.85E+02	21.27	0.78	0.81	3.19E+01	7.81E+01	.0020
		10	70,161,194,130,180,242,44,10,221,12	3.35	6.68E+02	19.88	0.74	0.8	39.11	8.73E+01	.0020

Table 8 Simulation results for Lena benchmark using recent heuristic algorithms and first two LCGSA versions.

Image	Algorithm	*k*	Optimal thresholds	SD	MSE	PSNR	SSIM	FSIM	BV	Run time (s)	*P* value
Lena	SSA	2	75,86	2.9	3.99E+03	1.21E+01	0.48	0.62	12.17	21.99	.0020
		4	109,85,186,56	3.74	1.25E+03	17.16	0.67	0.74	17.41	3.50E+01	.0020
		6	28,245,163,84,220,200	5.33	1.68E+03	15.87	0.64	0.7	2.78E+01	6.68E+01	.0020
		8	94,96,97,102,97,103,93,96	8.19	3.21E+03	13.06	0.49	0.62	3.09E+01	7.19E+01	.0020
		10	33,38,35,40,28,39,42,40,38,37	12.14	8.85E+03	8.66	0.28	0.52	39.82	8.68E+01	.0020
	GWO	2	28,10	1.43	1.15E+04	7.52E+00	0.19	0.52	14.48	22.67	.0020
		4	71,52,41,75	2.51	4.61E+03	11.48	0.52	0.62	20.93	3.34E+01	.0020
		6	5,3,7,2,4,3	4.27	1.74E+04	5.72	0.008	0.13	6.39E+00	5.77E+01	.0020
		8	7,11,3,23,3,4,10,9	2.66	1.24E+04	7.16	0.16	0.17	1.80E+01	6.94E+01	.0020
		10	3,2,1,4,3,6,7,4,1,4	6.32	1.74E+04	5.72	0.008	0.16	20.49	7.98E+01	.0020
	MFO	2	102,1	0.62	3.89E+04	1.22E+01	0.45	0.61	14.88	26.75	.0059
		4	1255,255,255	2.67	1.74E+04	5.72	0.008	0.15	22.34	3.60E+01	.0020
		6	255,134,1,1,1109	3.79	2.90E+03	13.49	0.49	0.66	2.77E+01	7.03E+01	.0020
		8	255,1255,1,1,2255,1	4.53	1.74E+04	5.72	0.008	0.16	3.30E+01	8.29E+01	.0020
		10	255,255,255,255,3,252,165,17,1124	5.51	2.62E+03	13.94	0.51	0.65	36.72	9.72E+01	.0020
	LCGSA1	2	97,166	0.54	2.37E+03	1.44E+01	0.52	0.66	16.45	9.89E−05	.9219
		4	227,26,98,161	1.85	1.48E+03	16.4	0.63	0.69	20.43	1.00E−04	.9219
		6	104,169,239,14,124, 234	1.57	1.57E+03	16.16	0.61	0.7	2.55E+01	1.00E−04	.6250
		8	225,12,118,229,16,97,168,246	2.35	1.24E+03	17.19	0.65	0.71	2.79E+01	1.00E−04	1
		10	100,175,247,31,131,243,17,98,167,235	2.93	7.53E+02	19.36	0.7	0.75	37.6	2.00E−04	.8457
	LCGSA2	2	237,28	0.43	1.13E+04	7.59E+00	0.2	0.52	14.81	1.86E−06	.4932
		4	119,252,17,120	1.56	3.03E+03	13.31	0.51	0.63	19.16	5.13E−06	1
		6	219,15,226,23,240,21	1.96	1.20E+04	7.3	0.18	0.51	2.23E+01	1.39E−06	.0020
		8	130,221,19,127,228,39,240,34	2.65	2.20E+03	14.69	0.56	0.64	3.05E+01	5.13E−06	.4316
		10	204,21,199,35,224,10,104,170,241,34	2.99	1.27E+03	17.06	0.65	0.7	35.31	6.04E−06	.8457

Table 9 Simulation results for Lena benchmark using LCGSA3–LCGSA7 versions.

Image	Algorithm	*k*	Optimal thresholds	SD	MSE	PSNR	SSIM	FSIM	BV	Run time (s)	*P* value
Lena	LCGSA3	2	120,232	0.24	4.18E+03	11.91	0.4	0.61	12.86	1.86E−06	.1309
		4	113,194,79,207	0.58	1.57E+03	16.15	0.6	0.72	20.51	3.26E−06	.0840
		6	180,84,148,216,91,161	1.71	1.34E+03	16.85	0.6	0.75	25.5	1.39E−06	.0098
		8	111,159,200,88,142,198,89,136	2.66	1.05E+03	17.89	0.65	0.79	30.83	2.33E−06	.0059
		10	133,81,130,181,104,174,116,192,105,160	2.62	839.55	18.89	0.7	0.82	34.55	4.66E−06	.0020
	LCGSA4	2	232,29	0.52	1.11E+04	7.66	0.21	0.52	12.94	1.39E−05	.4922
		4	224,13,99,160	1.78	1.84E+03	15.47	0.58	0.68	19.07	5.13E−06	.4922
		6	214,21,123,238,15,131	2.1	2.68E+03	13.83	0.51	0.63	27.16	1.39E−06	1
		8	97,170,238,33,239,18,236,36	3.39	1.39E+03	16.67	0.65	0.68	28.83	4.19E−06	.1934
		10	96,161,223,32,209,34,224,41,222,25	3.34	1.14E+03	17.56	0.66	0.71	33.47	3.73E−06	.8457
	LCGSA5	2	122,239	0.11	4.30E+03	11.78	0.38	0.61	12.21	1.39E−06	1
		4	131,235,25,241	0.83	3.27E+03	12.98	0.49	0.61	17.13	2.33E−06	.6953
		6	125,230,11,97,165,233	2.59	1.32E+03	16.9	0.62	0.72	22.57	2.33E−06	.6953
		8	116,234,113,26,101, 166,236,28	2.99	1.14E+03	17.53	0.66	0.7	25.09	3.26E−06	.3750
		10	120,227,22,83,128,179,247,42,143,232	3.5	3.41E+02	22.79	0.8	0.83	33.96	5.59E−06	1
	LCGSA6	2	97,165	0.35	2.32E+03	14.47	0.52	0.66	16.47	1.86E−06	1
		4	79,122,161,192	1.57	1.05E+03	17.91	0.64	0.78	18.2	4.19E−06	.6953
		6	116,233,16,235,23,118	2.58	2.67E+03	13.85	0.55	0.64	24.14	1.86E−06	.3750
		8	131,231,21,119,237,31,229,36	3	1.87E+03	15.4	0.6	0.66	29.94	2.79E−06	.7695
		10	122,240,32,229,40,249,23,106,165,243	3.01	8.82E+02	18.67	0.7	0.74	31.2	4.19E−06	.1602
	LCGSA7	2	97,166	0.49	2.38E+03	14.36	0.52	0.66	16	1.86E−06	.7695
		4	114,235,175,29	0.95	1.70E+03	15.8	0.61	0.67	18.55	2.79E−06	1
		6	228,25,229,18,122,246	2.12	2.83E+03	13.6	0.52	0.63	17.04	1.86E−06	.9219
		8	127,225,27,138,231,32,237,28	2.17	2.32E+03	14.47	0.54	0.64	33.87	2.79E−06	1
		10	77,124,178,238,42,120,176,240,53,165	3.27	4.81E+02	21.31	0.78	0.81	41.23	3.73E−06	.0840

Table 10 Simulation results for Lena benchmark using LCGSA8–LCGSA10 versions.

Image	Algorithm	*k*	Optimal thresholds	SD	MSE	PSNR	SSIM	FSIM	BV	Run time (s)	*P* value
Lena	LCGSA8	2	100,166	0.29	2.48E+03	14.18	0.51	0.66	16.9	1.39E−06	.4316
		4	127,232,16,237	0.9	3.56E+03	12.61	0.46	0.61	16.91	2.79E−06	.4316
		6	95,162,243,12,245,22	1.66	1.61E+03	16.06	0.62	0.69	25.29	1.39E−06	.7695
		8	114,228,30,231,17,83,126,171	2.48	5.90E+02	20.42	0.76	0.78	33.32	2.79E−06	1
		10	130,241,29,232,11,138,238,117,10,128	3.23	1.78E+03	14.61	0.6	0.67	32.4	4.19E−06	.3750
	LCGSA9	2	125,237	0.25	4.54E+03	11.55	0.37	0.6	12.72	1.39E−06	.8457
		4	245,25,115,173	2.54	1.79E+03	15.58	0.6	0.67	18.17	2.79E−06	1
		6	99,169,241,25,230,21	1.46	1.66E+03	15.92	0.62	0.68	27.51	1.39E−06	.3223
		8	118,229,15,234,32,126,241,13	2.82	2.07E+03	14.95	0.58	0.65	30.48	3.26E−06	.7695
		10	123,228,17,235,15,126,241,12,230,26	2.31	2.61E+03	13.96	0.53	0.63	39.55	5.13E−06	.4316
	LCGSA10	2	97,169	0.57	2.41E+03	14.3	0.51	0.66	15.77	2.29E+01	.7695
		4	241,24,121,235	0.72	2.86E+03	13.56	0.52	0.63	17.28	4.53E+01	.3223
		6	131,243,13,122,240,25	2.38	2.47E+03	14.19	0.54	0.64	22.14	6.78E+01	.0645
		8	78,122,160,194,240,31,144,246	2.64	4.12E+02	21.98	0.77	0.83	24.95	7.45E+01	.6953
		10	81,132,170,244,45,228,22,141,2 35,60,100,123,56	3.29	5.26E+02	20.92	0.79	0.82	39.09	8.57E+01	.6250

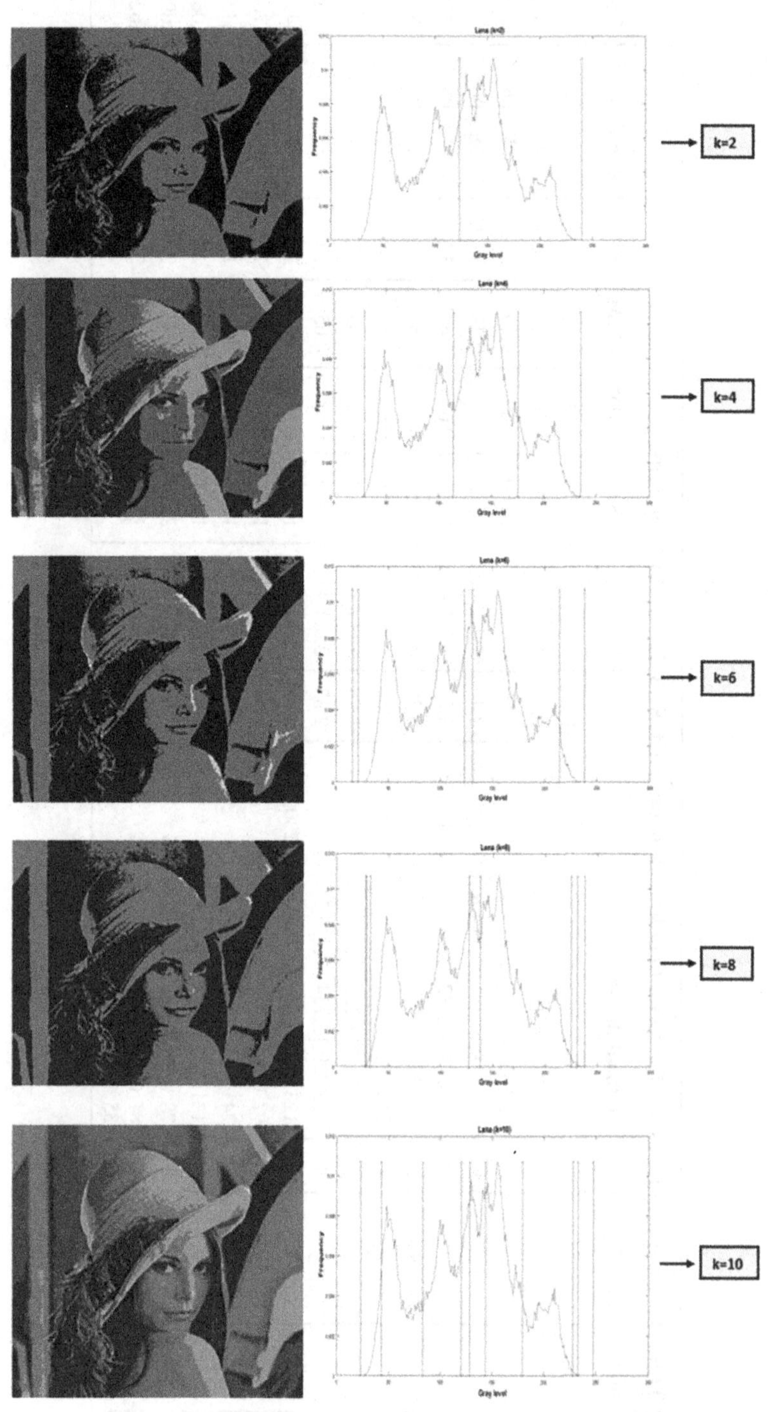

FIG. 8

LCGSA segmented images and histogram curves for Lena benchmark image at k = 2, 4, 6, 8, 10.

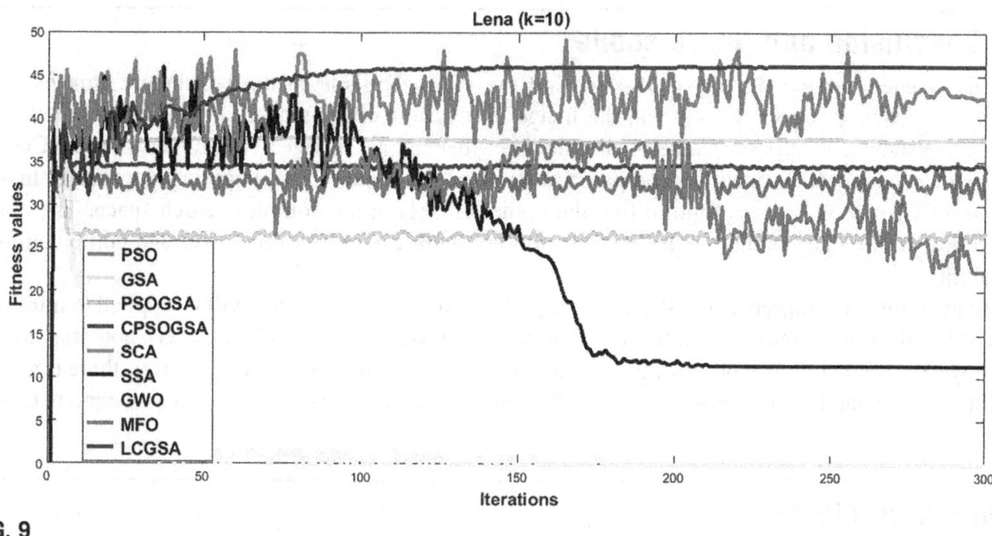

FIG. 9

Convergence curves of heuristic algorithms for Lena image.

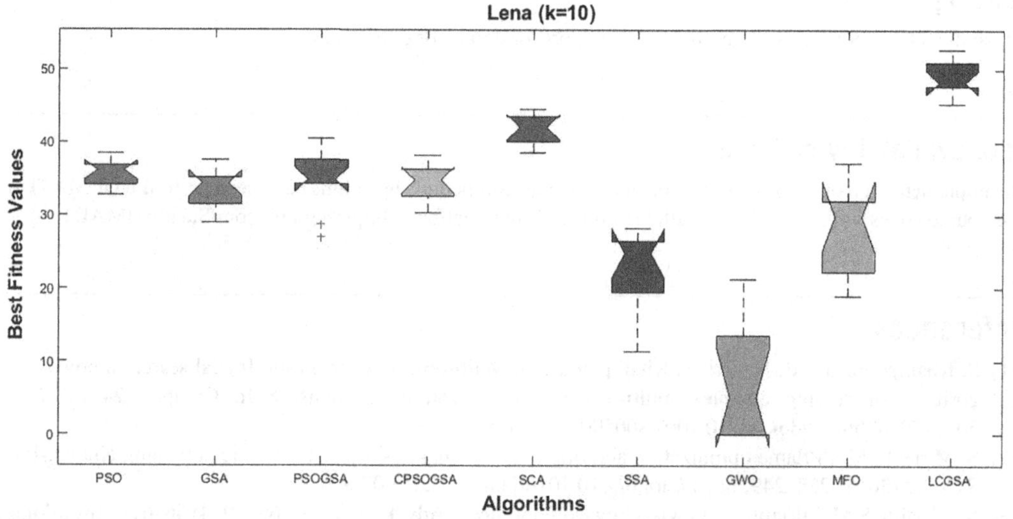

FIG. 10

Box plots of heuristic algorithms for Lena image.

7. Conclusion and future scope

In this chapter, we employed LCGSA to find the optimal pixels in the benchmark images. We employed Kapur's entropy to segment the image into various objects based on their pixel intensity. The experimental results on Cameraman and Lena benchmark images clearly show that LCGSA has better segmentation capability compared to standard GSA and other competing algorithms. In addition, LCGSA took less run time to find the optimal pixels in the complex search space. The main limitation of our study is that we have used a system with less processing power and computational capability.

In the future, a comprehensive study can be performed in which LCGSA will be applied to multiple USC-SIPI dataset images for multilevel thresholding tasks. Besides, LCGSA has potential to be employed in medical image processing, such as analysis of Covid-19 images. Moreover, there is a possibility for a separate study in which only GSA variants can be considered for image segmentation.

Conflict of interest

The authors clearly state that there is no conflict of interest whether financial or professional regarding the publication of the work.

Funding

The authors have not received any financial support for conducting the work.

Data availability statement

The implementation of the proposed algorithm and other competing algorithms have been performed in MATLAB. The source codes will be publicly available on the Github platform, https://github.com/SajadAHMAD1.

References

[1] S. Khalilpourazari, B. Naderi, S. Khalilpourazary, Multi-objective stochastic fractal search: a powerful algorithm for solving complex multi-objective optimization problems, Soft. Comput. 24 (4) (2020) 3037–3066, https://doi.org/10.1007/s00500-019-04080.

[2] S. Mirjalili, Moth-flame optimization algorithm: a novel nature-inspired heuristic paradigm, Knowl.-Based Syst. 89 (2015) 228–249, https://doi.org/10.1016/j.knosys.2015.07.006.

[3] S. Mirjalili, S.M. Mirjalili, A. Lewis, Grey wolf optimizer, Adv. Eng. Softw. 69 (2014) 46–61, https://doi.org/10.1016/j.advengsoft.2013.12.007.

[4] D.H. Wolpert, W.G. Macready, No free lunch theorems for optimization, IEEE Trans. Evol. Comput. 1 (1) (1997) 67–82, https://doi.org/10.1109/4235.585893.

[5] M. Alinaghian, E.B. Tirkolaee, Z.K. Dezaki, S.R. Hejazi, W. Ding, An augmented Tabu search algorithm for the green inventory-routing problem with time windows, Swarm Evol. Comput. 60 (2021) 100802, https://doi.org/10.1016/j.swevo.2020.100802 (November 2020).

[6] S. Khalilpourazari, H. Hashemi Doulabi, A. Özyüksel Çiftçioğlu, G.W. Weber, Gradient-based grey wolf optimizer with Gaussian walk: application in modelling and prediction of the COVID-19 pandemic, Expert Syst. Appl. 177 (2021), https://doi.org/10.1016/j.eswa.2021.114920.

[7] S. Khalilpourazari, S.H.R. Pasandideh, Modeling and optimization of multi-item multi-constrained EOQ model for growing items, Knowl.-Based Syst. 164 (2019) 150–162, https://doi.org/10.1016/j.knosys.2018.10.032.

[8] P. Bansal, S. Kumar, S. Pasrija, S. Singh, A hybrid grasshopper and new cat swarm optimization algorithm for feature selection and optimization of multi-layer perceptron, Soft. Comput. (2020), 0123456789, https://doi.org/10.1007/s00500-020-04877-w.

[9] S.A. Rather, P.S. Bala, Hybridization of constriction coefficient based particle swarm optimization and gravitational search algorithm for function optimization, SSRN Electron. J. (2020) 1–10, https://doi.org/10.2139/ssrn.3576489.

[10] E.B. Tirkolaee, A. Mardani, Z. Dashtian, M. Soltani, G.W. Weber, A novel hybrid method using fuzzy decision making and multi-objective programming for sustainable-reliable supplier selection in two-echelon supply chain design, J. Clean. Prod. 250 (2020), 119517, https://doi.org/10.1016/j.jclepro.2019.119517.

[11] S. Mirjalili, SCA: a sine cosine algorithm for solving optimization problems, Knowl.-Based Syst. 96 (2016) 120–133, https://doi.org/10.1016/j.knosys.2015.12.022.

[12] S. Mirjalili, A.H. Gandomi, S.Z. Mirjalili, S. Saremi, H. Faris, S.M. Mirjalili, Salp Swarm algorithm: a bio-inspired optimizer for engineering design problems, Adv. Eng. Softw. 114 (2017) 163–191, https://doi.org/10.1016/j.advengsoft.2017.07.002.

[13] F. Erdal, E. Dogan, M.P. Saka, Optimum design of cellular beams using harmony search and particle swarm optimizers, J. Constr. Steel Res. 67 (2) (2011) 237–247, https://doi.org/10.1016/j.jcsr.2010.07.014.

[14] D. Karaboga, An Idea Based on Honeybee Swarm for Numerical Optimization, Technical Report, Erciyes University, Turkey, 2005.

[15] J. Kennedy, R. Eberhart, Particle swarm optimization, in: IEEE International Conference on Neural Networks, 1995, pp. 1942–1948.

[16] R.A. Khurma, I. Aljarah, A. Sharieh, S. Mirjalili, EvoloPy-FS: An Open-Source Nature-Inspired Optimization Framework in Python for Feature Selection, 2020, pp. 131–173, https://doi.org/10.1007/978-981-32-9990-0_8.

[17] O.K. Erol, I. Eksin, A new optimization method: Big Bang–Big Crunch, Adv. Eng. Softw. 37 (2) (2006) 106–111, https://doi.org/10.1016/j.advengsoft.2005.04.005.

[18] J.L.J. Pereira, M.B. Francisco, C.A. Diniz, G. Antônio Oliver, S.S. Cunha, G.F. Gomes, Lichtenberg algorithm: a novel hybrid physics-based meta-heuristic for global optimization, Expert Syst. Appl. 170 (2021), https://doi.org/10.1016/j.eswa.2020.114522 (August 2020).

[19] E. Rashedi, H. Nezamabadi-pour, S. Saryazdi, GSA: a gravitational search algorithm, Inform. Sci. 179 (13) (2009) 2232–2248, https://doi.org/10.1016/j.ins.2009.03.004.

[20] J.H. Holland, Genetic algorithms, Sci. Am. 267 (1) (1992) 66–73. http://www.jstor.org/stable/24939139.

[21] H. Ma, D. Simon, Biogeography-based optimization with blended migration for constrained optimization problems, in: Proceedings of the 12th Annual Conference on Genetic and Evolutionary Computation, 2010, pp. 417–418, https://doi.org/10.1145/1830483.1830561.

[22] R. Storn, K. Price, Differential evolution—a simple and efficient heuristic for global optimization over continuous spaces, J. Glob. Optim. 11 (4) (1997) 341–359, https://doi.org/10.1023/A:1008202821328.

[23] P. Kandhway, A.K. Bhandari, A water cycle algorithm-based multilevel thresholding system for color image segmentation using Masi entropy, Circuits Systems Signal Process. 38 (7) (2019) 3058–3106, https://doi.org/10.1007/s00034-018-0993-3.

[24] Z. Tan, K. Li, Y. Wang, An improved cuckoo search algorithm for multilevel color image thresholding based on modified fuzzy entropy, J. Ambient. Intell. Humaniz. Comput. (2021), 0123456789, https://doi.org/10.1007/s12652-021-03001-6.

[25] J. Anitha, S. Immanuel Alex Pandian, S. Akila Agnes, An efficient multilevel color image thresholding based on modified whale optimization algorithm, Expert Syst. Appl. 178 (2021) 115003, https://doi.org/10.1016/j.eswa.2021.115003 (May 2020).
[26] L. He, S. Huang, An efficient krill herd algorithm for color image multilevel thresholding segmentation problem, Appl. Soft Aust. Comput. J. 89 (2020), 106063, https://doi.org/10.1016/j.asoc.2020.106063.
[27] A.K. Bhandari, A novel beta differential evolution algorithm-based fast multilevel thresholding for color image segmentation, in: Neural Computing and Applications, vol. 32, Springer, London, 2020, https://doi.org/10.1007/s00521-018-3771-z. Issue 9.
[28] Z. Xing, An improved emperor penguin optimization based multilevel thresholding for color image segmentation, Knowl.-Based Syst. 194 (2020), 105570, https://doi.org/10.1016/j.knosys.2020.105570.
[29] S. Borjigin, P.K. Sahoo, Color image segmentation based on multi-level Tsallis–Havrda–Charvát entropy and 2D histogram using PSO algorithms, Pattern Recogn. 92 (2019) 107–118, https://doi.org/10.1016/j.patcog.2019.03.011.
[30] E.H. Houssein, B.E.d. Helmy, D. Oliva, A.A. Elngar, H. Shaban, A novel black widow optimization algorithm for multilevel thresholding image segmentation, Expert Syst. Appl. 167 (2021), https://doi.org/10.1016/j.eswa.2020.114159 (May 2020).
[31] G. Ding, F. Dong, H. Zou, Fruit fly optimization algorithm based on a hybrid adaptive-cooperative learning and its application in multilevel image thresholding, Appl. Soft Aust. Comput. J. 84 (2019), 105704, https://doi.org/10.1016/j.asoc.2019.105704.
[32] K. Sowjanya, S.K. Injeti, Investigation of butterfly optimization and gases Brownian motion optimization algorithms for optimal multilevel image thresholding, Expert Syst. Appl. 182 (2021), 115286, https://doi.org/10.1016/j.eswa.2021.115286.
[33] B. Wu, J. Zhou, X. Ji, Y. Yin, X. Shen, An ameliorated teaching–learning-based optimization algorithm based study of image segmentation for multilevel thresholding using Kapur's entropy and Otsu's between class variance, Inform. Sci. 533 (2020) 72–107, https://doi.org/10.1016/j.ins.2020.05.033.
[34] J. Rahaman, M. Sing, An efficient multilevel thresholding based satellite image segmentation approach using a new adaptive cuckoo search algorithm, Expert Syst. Appl. 174 (2021), 114633, https://doi.org/10.1016/j.eswa.2021.114633.
[35] M. Abd Elaziz, S. Lu, S. He, A multi-leader whale optimization algorithm for global optimization and image segmentation, Expert Syst. Appl. 175 (2021) 114841, https://doi.org/10.1016/j.eswa.2021.114841 (2020).
[36] A.K. Bhandari, K. Rahul, A context sensitive Masi entropy for multilevel image segmentation using moth swarm algorithm, Infrared Phys. Technol. 98 (2019) 132–154, https://doi.org/10.1016/j.infrared.2019.03.010.
[37] X. Yue, H. Zhang, Modified hybrid bat algorithm with genetic crossover operation and smart inertia weight for multilevel image segmentation, Appl. Soft Aust. Comput. J. 90 (2020), 106157, https://doi.org/10.1016/j.asoc.2020.106157.
[38] M. Ahmadi, K. Kazemi, A. Aarabi, T. Niknam, M.S. Helfroush, Image segmentation using multilevel thresholding based on modified bird mating optimization, Multimed. Tools Appl. 78 (16) (2019) 23003–23027, https://doi.org/10.1007/s11042-019-7515-6.
[39] S.J. Mousavirad, H. Ebrahimpour-Komleh, Human mental search-based multilevel thresholding for image segmentation, Appl. Soft Comput. 97 (572086) (2020), 105427, https://doi.org/10.1016/j.asoc.2019.04.002.
[40] A.K.M. Khairuzzaman, S. Chaudhury, Multilevel thresholding using grey wolf optimizer for image segmentation, Expert Syst. Appl. 86 (2017) 64–76, https://doi.org/10.1016/j.eswa.2017.04.029.
[41] A.K. Bhandari, A. Kumar, S. Chaudhary, G.K. Singh, A novel color image multilevel thresholding based segmentation using nature inspired optimization algorithms, Expert Syst. Appl. 63 (2016) 112–133, https://doi.org/10.1016/j.eswa.2016.06.044.

[42] A. Raj, G. Gautam, S.N. Huda Sheikh Abdullah, A. Salimi Zaini, S. Mukhopadhyay, Multi-level thresholding based on differential evolution and Tsallis Fuzzy entropy, Image Vis. Comput. 91 (2019), 103792, https://doi.org/10.1016/j.imavis.2019.07.004.
[43] P. Upadhyay, J.K. Chhabra, Kapur's entropy based optimal multilevel image segmentation using crow search algorithm, Appl. Soft Comput. 97 (2020), 105522, https://doi.org/10.1016/j.asoc.2019.105522.
[44] R. Kalyani, P.D. Sathya, V.P. Sakthivel, Medical image segmentation using exchange market algorithm, Alex. Eng. J. 60 (6) (2021) 5039–5063, https://doi.org/10.1016/j.aej.2021.04.054.
[45] S. Zhao, P. Wang, A.A. Heidari, H. Chen, H. Turabieh, M. Mafarja, C. Li, Multilevel threshold image segmentation with diffusion association slime mould algorithm and Renyi's entropy for chronic obstructive pulmonary disease, Comput. Biol. Med. 134 (2021), 104427, https://doi.org/10.1016/j.compbiomed.2021.104427.
[46] B.B. Mandelbrot, The Fractal Geometry of Nature, W.H. Freeman, San Francisco, 1982.
[47] X.-S. Yang, Chapter 3—Random walks and optimization, in: X.-S. Yang (Ed.), Nature-Inspired Optimization Algorithms, Elsevier, 2014, pp. 45–65, https://doi.org/10.1016/B978-0-12-416743-8.00003-8.
[48] M. Gutowski, Levy Flights as an Underlying Mechanism for Global Optimization Algorithms, 2001. http://arxiv.org/abs/math-ph/0106003.
[49] I. Pavlyukevich, Levy flights, non-local search and simulated annealing, J. Comput. Phys. 226 (2) (2007) 1830–1844, https://doi.org/10.1016/j.jcp.2007.06.008.
[50] G. Ramos-Fernández, J.L. Mateos, O. Miramontes, G. Cocho, H. Larralde, B. Ayala-Orozco, Levy walk patterns in the foraging movements of spider monkeys (Ateles geoffroyi), Behav. Ecol. Sociobiol. 55 (3) (2004) 223–230, https://doi.org/10.1007/s00265-003-0700-6.
[51] S. Mirjalili, A.H. Gandomi, Chaotic gravitational constants for the gravitational search algorithm, Appl. Soft Aust. Comput. J. 53 (2017) 407–419, https://doi.org/10.1016/j.asoc.2017.01.008.
[52] B. Alatas, Chaotic bee colony algorithms for global numerical optimization, Expert Syst. Appl. 37 (8) (2010) 5682–5687, https://doi.org/10.1016/j.eswa.2010.02.042.
[53] C. Li, J. Zhou, J. Xiao, H. Xiao, Parameters identification of chaotic system by chaotic gravitational search algorithm, Chaos, Solitons Fractals 45 (4) (2012) 539–547, https://doi.org/10.1016/j.chaos.2012.02.005.
[54] N. Wang, L. Liu, L. Liu, Genetic algorithm in chaos, OR Trans. 5 (2001) 1–10.
[55] Y. Li-Jiang, C. Tian-Lun, Application of Chaos in genetic algorithms, Commun. Theor. Phys. 38 (2) (2002) 168–172, https://doi.org/10.1088/0253-6102/38/2/168.
[56] V. Jothiprakash, R. Arunkumar, Optimization of hydropower reservoir using evolutionary algorithms coupled with Chaos, Water Resour. Manag. 27 (7) (2013) 1963–1979, https://doi.org/10.1007/s11269-013-0265-8.
[57] G. Zhenyu, C. Bo, Y. Min, C. Binggang, in: L. Jiao, L. Wang, X. Gao, J. Liu, F. Wu (Eds.), Self-Adaptive Chaos Differential Evolution BT—Advances in Natural Computation, Springer, Berlin Heidelberg, 2006, pp. 972–975.
[58] S. Saremi, S.M. Mirjalili, S. Mirjalili, Chaotic krill herd optimization algorithm, Procedia Technol. 12 (2014) 180–185, https://doi.org/10.1016/j.protcy.2013.12.473.
[59] G.-G. Wang, L. Guo, A.H. Gandomi, G.-S. Hao, H. Wang, Chaotic krill herd algorithm, Inform. Sci. 274 (2014) 17–34, https://doi.org/10.1016/j.ins.2014.02.123.
[60] H. Peitgen, H. Jurgens, D. Saupes, Chaos and Fractals, Springer-Verlag, 1992.
[61] Y. Li, S. Deng, D. Xiao, A novel Hash algorithm construction based on chaotic neural network, Neural Comput. Applic. 20 (1) (2011) 133–141, https://doi.org/10.1007/s00521-010-0432-2.
[62] E. Ott, Chaos in Dynamical Systems, Cambridge University Press, 2002.
[63] J. Mingjun, T. Huanwen, Application of chaos in simulated annealing, Chaos, Solitons Fractals 21 (4) (2004) 933–941, https://doi.org/10.1016/j.chaos.2003.12.032.
[64] A.H. Gandomi, X.-S. Yang, S. Talatahari, A.H. Alavi, Firefly algorithm with chaos, Commun. Nonlinear Sci. Numer. Simul. 18 (1) (2013) 89–98, https://doi.org/10.1016/j.cnsns.2012.06.009.
[65] S.A. Rather, P.S. Bala, Swarm-based chaotic gravitational search algorithm for solving mechanical engineering design problems, World J. Eng. 17 (1) (2020) 97–114, https://doi.org/10.1108/WJE-09-2019-0254.

CHAPTER

Metaheuristics for optimal feature selection in high-dimensional datasets

13

Davies Segera, Mwangi Mbuthia, and Abraham Nyete
Electrical and Information Engineering, University of Nairobi, Nairobi, Kenya

1. Introduction

We live in an era where the complexity and volume of available data in the domain of machine learning is increasing. This has elevated the need for efficient machine learning approaches to help extract useful information from the existing large amounts of meaningless data. Among the many existing machine learning approaches, feature selection is considered one of the most important preprocessing tasks prior to classification of high-dimensional datasets. The process of feature selection aims at deriving features that can allow a given problem to be defined clearly, apart from the redundant and irrelevant ones [1–3].

Several feature selection techniques have been proposed to deal with the high dimensionality of existing data. Traditionally, these approaches are broadly categorized as wrappers, filters, and embedded. Though slower compared to filters, wrappers normally achieve better results than both filters and embedded types, a reason why they are currently widely adopted in selecting informative features in high-dimensional datasets. Wrappers depend on modeling algorithms (in most cases metaheuristics) whereby every potential informative subset is derived and evaluated.

This chapter presents an overview of recent metaheuristic-based feature selectors (wrappers) proposed to handle high-dimensional data and an analysis of their performance in various conditions. Section 2 presents characteristics of high-dimensional data. Section 3 details the process of feature selection using high-dimensional datasets. Section 4 provides an overview of metaheuristic approaches. Section 5 presents the experimental results of the most recent and significant wrappers on high-dimensional datasets. Finally, Section 6 concludes and identifies areas for further study.

2. Characteristics of high-dimensional data

To date, data continues to be pertinent in scientific research domains as well as industrial production processes. In literature, many approaches exist to obtain data. Some of the utilized approaches might be expensive, thus limiting the number of samples collected. Alternatively, some approaches might end up

Comprehensive Metaheuristics. https://doi.org/10.1016/B978-0-323-91781-0.00013-2

adding noise to the obtained data [1–3]. Thus, it is especially important to fully understand the data in question. This section presents some inherent characteristics of high-dimensional data.

2.1 High dimensionality

The dimensions of generated datasets continue to grow considerably. For instance, the prostate cancer dataset, which is a binary microarray dataset, contains 10,509 genes. In addition, the 11_tumour dataset has 11 classes with 12,533 genes. Though several datasets contain many features, most of the features are noninformative to the classes. Moreover, for an effective machine learning process, only the relevant and nonredundant attributes are considered [4].

Considering a complex model where the number of attributes is much more than the available samples, the overfitting problem will arise. Thus, in situations where it is not possible to increase the number of samples, reducing the dataset dimensions is crucial for an effective classification process [5].

Thus, feature selection is a process of shrinking the dimensions of data by effectively deriving a subset of informative features from the original large number of features, which is required in the classification phase [2,3,6]. The process of feature selection should be optimal such that the derived informative subsets should contain sufficient information from the original datasets [7,8].

Generally, the possible number of informative subsets of features increases exponentially with the dimensionality of the dataset (i.e., 2^Y where Y is the dimensionality of the dataset). Thus, deriving an optimal feature subset is a challenging task, a reason why feature selection problems are regarded as nondeterministic polynomial-time (NP)-hard type [3].

2.2 Limited sample size

For a classification challenge with Y dimensionality and W classes, a training subset with $10 \times Y \times W$ samples is needed [9]. For example, a colon cancer dataset with two classes and 2000 genes requires a training subset with 40,000 samples. However, this dataset has 62 samples only. The small sample size problem arises when there are a few S samples in a large V dimensionality (i.e.,) $S < V$. Such datasets are known as high dimension, low sample size (HDLSS). The HDLSS datasets are common in a number of domains such as DNA microarray data, text recognition, medical imaging, and face recognition [10]. As mentioned, a limited sample size increases the chances of overfitting.

2.3 Class imbalance

The problem of class imbalance occurs when within a given dataset a given class has a higher or lower number of samples in comparison with the other classes. The class with few samples is the minority class. In a dataset with two classes, the positive class is normally the minority class, while the negative class is the majority class.

A number of existing classifiers assume classes with balanced distributions. Hence, when subjected to classes with imbalanced distributions, they will be trained mostly on samples from the majority classes. Finally, we will have trained classifiers with poor predictions on samples from the minority classes.

The ratio of imbalance (IR) is the ratio between the sample size of the majority class (Maj_{ss}) and the sample size of the minority class (Min_{ss}), as given by Eq. (1).

$$IR = \frac{Maj_{ss}}{\mathrm{Min}_{ss}} \tag{1}$$

The main problem associated with imbalanced datasets is how to correctly classify samples within the minority class. Moreover, the minority class is of greater importance and thus inappropriate classification of samples within the minority class elevates the risk [3,7,9]. For example, the number of positive cases in the cancer disease diagnosis is far less in comparison to the negative cases. However, identification of the positive cases is of greater importance compared to that of the negative cases.

Table 1 presents the inherent imbalance ratio of the 11 commonly utilized high-dimensional datasets.

Thus, in trying to improve the performance of classifiers when subjected to these kinds of datasets, a number of techniques have been proposed [3,10–14]. These techniques can be categorized as follows [3,9]:

(a) *Classifier-independent processing techniques*

In these techniques, two different methods are adopted to rebalance distribution of samples from various classes within the training data subset. The first method is *oversampling,* whereby the sample size of the minority class is increased by adding more samples to it. In *undersampling*, the second approach, the sample size of the minority class is reduced by removing some samples from it. It is evident that both methods try to balance the classes. In [9], an undersampling approach called Hyperheuristic Training Set Selection (HTSS) is proposed.

(b) *Modification of classifiers*

These approaches modify and improve a given classifier with the aim of matching it with the imbalanced data. In [3,15], a formulation of an approach adopting ensemble techniques is proposed to improve the classifier performance in handling high-dimensional data.

(c) *Ensemble learning techniques*

These approaches utilize results achieved by a number of classifiers. For example, in [16], an ensemble technique is proposed. In this approach, the imbalanced data is rebalanced and then subjected to several classifiers. Finally, the results from the individual classifiers are combined.

Table 1 Inherent imbalance ratio of 11 commonly used high-dimensional datasets.

Name of dataset	Type of dataset	Inherent imbalance ratio
Brain_Tumors1	DNA microarray	15.0
CNS	DNA microarray	1.857
Leukemia	DNA microarray	1.780
SRBCT	DNA Microarray	2.64
Ovarian	DNA microarray	1.78
Gli	DNA microarray	2.27
Breast cancer	DNA microarray	1.10
Colon	DNA microarray	1.818
Prostate cancer	DNA microarray	1.04
Lung cancer	DNA microarray	23.17
RELATHE	Text	1.2

2.4 Label noise

Nowadays, existence of noise in datasets derived from real-world applications using common techniques is rampant. This noise may arise due to the use of defective measuring equipment or transmission irregularities. Data that is full of noise can adversely affect the performance of classifiers. That is, the quality of the training subset greatly influences the accuracy of the classifier.

Mislabeling may arise for different reasons. Foremost of which is if the available data is not enough for a reliable labeling [7,15,17,18]. Data with a significantly low quality is an example [19,20]. In most cases, an expert is involved in both the collection and labeling of data. The labeling phase may be prone to human errors. In addition, the process of labeling largely depends on the expert's opinion; this implies that two experts might label the same data differently.

As mentioned, a noisy data greatly degrades the performance of a classifier. Moreover, label noise can cause a number of samples in a class to change. This is a recurrent problem in the field of medicine. This is important because determining the contingency of a disease in a given population is a key objective. Again, normally the population size is not always large, thus a label noise might lead to biased measurements.

Label noise can largely affect feature selection techniques adopting ranking methods. This is because these techniques might discard an informative feature or select an irrelevant one.

Many techniques have been proposed to tackle the label noise challenge. These techniques are categorized as follows [3,21]:

(a) *Label noise robust approaches*

These approaches try to solve the label noise challenge by reducing the risks of overfitting. Boosting and bagging techniques are some examples [3,7].

(b) *Data cleansing approaches*

These techniques discard samples that seem mislabeled. Outlier detectors are some examples [3,22].

(c) *Learning techniques that are tolerant to label noise*

In these techniques, the label noise issue is managed by the modeling step. As an example, the loss function is modified to handle the label noise issue in [23].

2.5 Inherent attributes of microarray data

Gene expression is a fundamental issue in genetics. Genes are atomic elements of genetic inheritance in a genome that hold information pertaining all corporal characteristics of a person. A gene expression in DNA can either transfer or reject a certain property to an individual [24].

In the field of bioinformatics, much emphasis has been given to the study of gene expression. This is because information related to gene expressions helps in the prediction and diagnosis of various diseases like cancer [7].

In bioinformatics, DNA microarrays are useful tools in cancer detection. They are obtained from cells and tissues taking into consideration gene variations that might be helpful in diagnosing diseases and tumors. Though useful, these datasets have features (genes) that far exceeds the number of samples. This characteristic of microarrays hinders the process of selecting informative gene subsets. Moreover, they possess the risk of overfitting due to their limited sample size. As an example, the breast cancer dataset contains 24,481 genes and just 60 samples [3].

The *curse of dimensionality* problem is a major problem associated with microarray data. This problem occurs when feature vector far exceeds the number of available samples. This degrades the efficiency and extensibility of classifiers. In addition, microarray data suffer the class imbalance problem. As mentioned, many existing classifiers assume balanced classes within a given dataset. Thus, subjecting these classifiers to microarrays will normally not yield acceptable results. This is because the classifier in use will be predominantly trained on samples from the majority class, hence yielding errors in trying to classify samples from the minority class [5,24].

Among the many approaches employed in machine learning, the process of feature selection plays a critical role, particularly in data classification. In the field of medicine, optimal gene selection can improve the prediction and diagnosis of cancer. After effectively selecting an informative gene subset, a given classifier is employed to distinguish healthy people from cancer patients based on gene expression levels [2,25].

3. Feature selection in high-dimensional datasets

The feature selection process deals with irrelevant and redundant features. Its aim is to derive informative feature subsets from the original datasets [2,26]. With the ever-increasing data dimensions, feature selection is deemed the most important and challenging task in machine learning. Feature selection is applicable in a number of fields such as text mining [27], biomedical problems [28], image analysis [29], and so on. A feature selection task can be mathematically formulated as follows:

Consider a dataset A with e number of attributes. The aim of feature selection is to derive an informative feature subset from the e number of features. Given dataset $A = \{a_1, a_2, a_3, \ldots, a_e\}$, the aim is to derive an informative feature subset $I = \{a_1, a_2, a_3, \ldots, a_d\}$ whereby $d < e$ and a_1, a_2, a_3, ..., a_e are the individual features of any given dataset. Fig. 1 depicts the overall feature selection process.

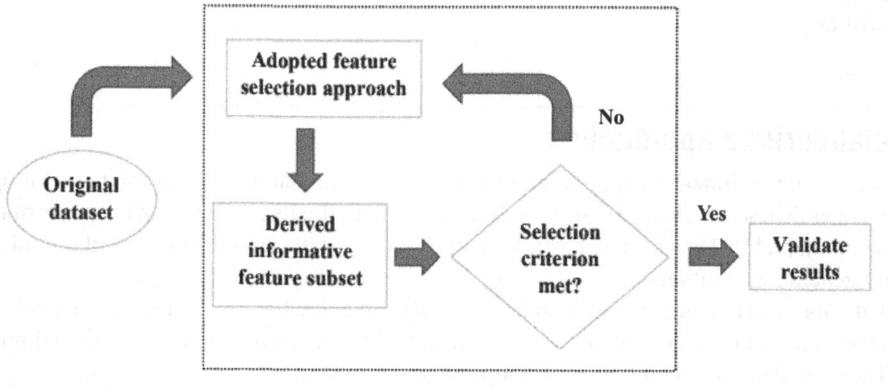

FIG. 1

The feature selection process.

As shown, the feature selection process consists of five major components: the original dataset, the adopted feature selection process, the derived informative feature subset, the selection criterion, and validation of the obtained results.

There are several approaches proposed in the literature to derive informative feature subsets. These approaches are broadly categorized as filters, embedded approaches, and wrappers [2,30–33]. Filter approaches normally focus on the general properties of datasets and are independent of classifiers [2,34]. Alternatively, wrappers always incorporate and interact with classifiers. Though computationally expensive, wrappers will always achieve more accurate results compared to filters. The embedded approaches combine filters and wrappers in the feature selection process. With embedded approaches, the feature selection process is incorporated within the training process, which is normally carried out by classifiers. Since the embedded approaches utilize classifiers in the feature selection process, they are normally categorized as wrappers [35].

Wrappers typically achieve better results than filters. However, they are slow. These wrappers depend on modeling algorithms whereby every potential informative subset is derived and evaluated. Derivation of these potential subsets is based of different search techniques. Jović et al. [36] categorize these search techniques into three types: randomized selection technique, sequential, and exponential.

In exponential techniques, the evaluated attributes increase exponentially with the number of presented attributes. Though they attain accurate results, they are not practically viable due to their high computational complexity. Branch-and-bound approach and exhaustive search techniques are examples of exponential approaches [37,38].

With sequential techniques, features are added or removed successively. Once a given feature is added to or removed from a potential subset, further changes cannot be made to it. This leads to local stagnation. The linear forward selection technique, floating backward or forward selection approach, and best first technique are some examples of sequential approaches [2].

Randomized selection approaches adopt randomness in exploring the search space. This prevents them from being trapped in the local optima. Randomized selection techniques are commonly referred to as population-based techniques. Metaheuristics, simulated annealing, and scatter-search are examples of population-based approaches [39]. Fig. 2 presents the existing categorization of feature selection approaches. A much-detailed account of this categorization can be found in [40]. In Fig. 2, the dashed boxes depict the chronology followed in this chapter to arrive at the metaheuristics.

4. Metaheuristic approaches

Metaheuristics are optimization approaches that attain near optimal or optimal solutions for optimization challenges. These are derivative-free techniques that are flexible, simple, and capable of avoiding the local optima [2,41]. They are stochastic by nature and normally start the process of optimization by randomly generating solutions.

They do not need to compute the search space derivative like the gradient search approaches. The metaheuristics are straightforward and simple due to their associated simple concepts and implementation. They can be easily modified to suit a particular problem.

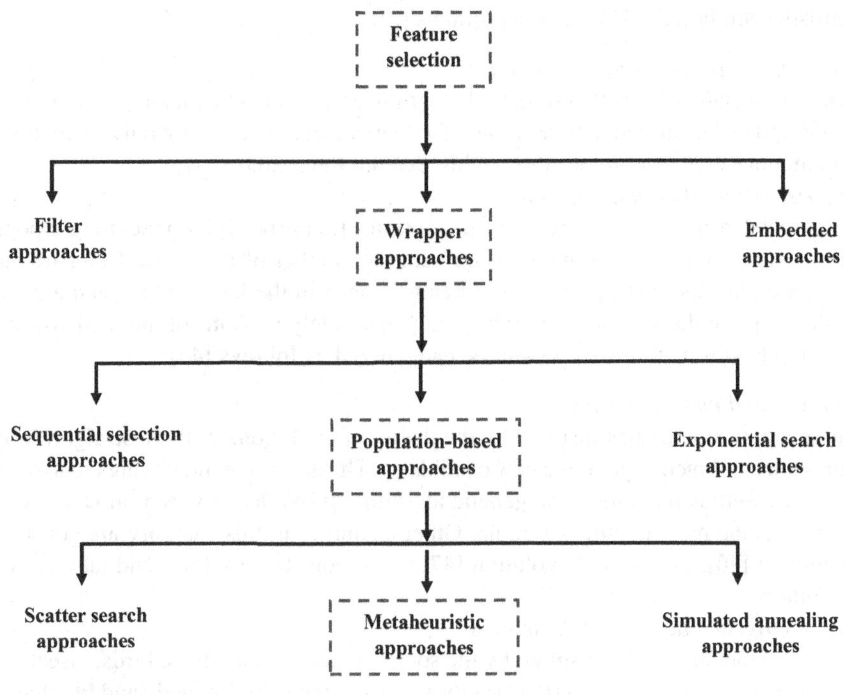

FIG. 2

Taxonomy of feature selection approaches.

Metaheuristics have an ability avoid premature convergence. This is the most attractive feature of these techniques. Due to their stochastic nature, metaheuristics work as a black box, avoiding being stuck in the local optima and efficiently exploring the search space [2].

The metaheuristic approaches try to strike a balance between their two critical aspects: exploitation and exploration. When exploring, these approaches thoroughly investigate the potential search space. In the exploitation phase, they locally search for the promising solutions in the identified potential search space [42,43].

To date, metaheuristic approaches have successfully attained promising results in various fields of engineering and science. For instance, in electrical engineering they are widely used to determine the optimal solution for generation of power. In industrial fields, they are used to schedule jobs, determine optimal routes for vehicles, and identify optimal facility locations. In civil engineering, they are adopted in designing bridges and buildings. They are widely used in designing radars and optimizing networks in the field of communication. In data mining, they are applied in clustering, classification, system modeling, and prediction [2].

Metaheuristics are broadly classified as follows [2]:

(a) *Metaheuristics based on single solutions*

These are metaheuristics that initiate the optimization process by a single solution. Then, this solution is updated in subsequent iterations. These metaheuristics are not only easily trapped in the local optima but they also do not thoroughly explore the search space.

(b) *Metaheuristics based on multiple solutions*

These metaheuristic approaches start the optimization process by generating a population of solutions. These solutions then get updated with the number of iterations. Compared to single solution-based metaheuristics, they avoid being trapped in the local optima and are able to thoroughly explore the search space. Thus, they are widely used in solving real-world solutions.

Based on their behavior, metaheuristics can be categorized as follows [44]:

(i) *Evolution-based metaheuristics*

These are metaheuristics inspired by the theory of evolution. With these algorithms, the best solutions are combined to generate new candidates. These new candidates are created by adopting crossover as well as mutation. The genetic algorithm (GA) that is based on Darwin's evolution theory [45] is the most popular example. Other examples in this category are genetic programming [46], differential evolution [47], evolution strategy [48], and tabu search [49], among others.

(ii) *Swarm-intelligence-based metaheuristics*

These are metaheuristics inspired by the social behavior of animals, birds, insects, and so on. The particle swarm optimization (PSO) algorithm formulated by Kennedy and Eberhart [50] is the most popular in this category. Ant colony optimization (ACO) [51], gray wolf optimization (GWO) [41], whale optimization algorithm (WOA) [52], salp optimization algorithm (SOA) [53], and honey bee optimization algorithm (HBOA) [54] are some of the other examples in this category.

(iii) *Physics-based metaheuristics*

These metaheuristics are inspired by rules of physics available in the universe. Harmony search [55] and simulated annealing [56] are two examples.

(iv) *Human behavior-based metaheuristics*

These are inspired by human behavior. Every person has a way of carrying out activities, which normally affects their performance. The teaching learning-based optimizer (TLO) [57] and the league championship algorithm (LCA) [58] are examples.

It is important to note that there are several metaheuristic approaches that have been developed so far. A detailed account on this can be found in [2]. Fig. 3 illustrates the broad categorization of metaheuristics.

4.1 Major metaheuristics as feature selectors in high-dimensional data

From the reviews presented in [59–61], it is evident that the relatively new swarm-intelligence-based metaheuristics are gaining much attention in solving NP-hard computational problems like feature selection. The five major swarm optimization techniques adopted in feature selection tasks include particle swarm optimization (PSO), invasive weed optimization (IWO), gray wolf optimization (GWO), bat algorithm (BA), and artificial bee colony (ABC) algorithm [1]. We discuss the details of these algorithms in the sections that follow.

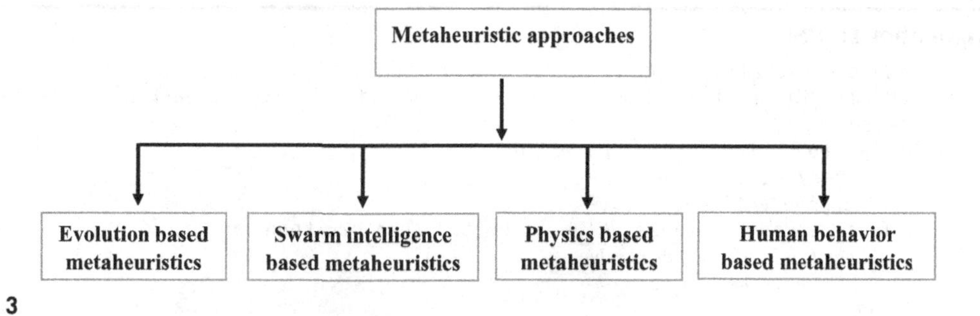

FIG. 3

Categorization of metaheuristic approaches.

4.1.1 Particle swarm optimization

PSO was introduced in 1995 by Kennedy and Eberhart [50]. Each particle in a swarm comprises three A-dimensional vectors where A is the search space of the dimensionality. The three vectors are the particle's current position x_i, its velocity v_i, and previous best position p_i. In addition, each particle in the swarm knows the current global best solution g. Each particle's velocity is iteratively modified to ensure all the particles hover around g and p_i positions. Ultimately, the whole swarm will closely move to an optimum of the considered fitness function [62]. Algorithm 1 depicts this process.

4.1.2 Artificial bee colony

ABC, proposed by Karaboga, is inspired by the unique foraging behavior of honey bees [54]. These insects normally collect nectar from various places around their hive and relay information among themselves via dances. The information relayed includes the distance, direction, and the quality of food sources [63]. In a hive there are three types of honeybees, namely, *scouts*, *onlookers*, and *employees*. The scouts adopt random flying patterns in search of new food sources to replace the abandoned ones [64]. The employees fly to the identified food sources to collect honey. On arrival at the hive, they do a dance that helps the onlookers identify and select the best food sources brought in by the employees. Algorithm 2 lists the detailed steps of the ABC algorithm.

4.1.3 Invasive weed optimization

IWO [65] is inspired by the colonization exhibited by invasive weeds in nature. Foremost, a certain number of invasive weeds is laid out in a field. The most adaptable weeds will utilize unused resources, grow rapidly, and yield more seeds. These newly yielded seeds are spread randomly over the parent weed and soon will grow into flowering weeds. Due to limited resources, when a certain maximum number of invasive weeds is attained in that field, only the fittest weeds are allowed to survive and yield other new weeds [66]. The unfit weeds are eliminated. This survival for the fittest scenario forces the weeds be more adaptable over time. Algorithm 3 shows this process.

Algorithm 1: PSO

Results: Global best position, optimal fitness value
Randomly initialize A-dimensioned position and velocity of each particle in the swarm
While *stop criterion is not met* **do**
  For every particle **do**
    Determine a fitness value
    If determined fitness value is better than $pbest_i$ **then**
      $pbest_i = determined\ fitness\ value$
      $p_i = current\ location\ x_i$
    End
    If $pbest_i$ is better than global best ($gbest_i$) **then**
      $gbest_i = pbest_i$
      $g = p_i$
    End
    Update the velocity and position of this particle as follows:

$$v_i = (\omega \times v_i) + \cup(0, a_1) \times (p_i - x_i) \times c_1 + \cup(0, a_2) \times (g - x_i) \times c_2 \quad (2)$$

$$x_i = x_i + v_i \quad (3)$$

  End
End
/* Notes
$\cup(0, a_i)$: is a vector of random numbers uniformly distributed in $[0, a_i]$ that is randomly generated for every particle during each iteration.
$\times$: represents component-wise multiplication
ω, c_1 *and* c_2: are parameters that are used to control exploitation and exploration processes. Normally, ω linearly decreases with iterations.
*/

Algorithm 2: ABC

Results: Global best position, optimal fitness value
Randomly initialize $\frac{n}{2}$ food sources
Set $\frac{1}{2}$ of the population as onlookers and the remaining $\frac{1}{2}$ as employees
While *stop criterion is not met* **do**
  For every *employee* **do**
    Randomly select a food source y
    Randomly select a dimension j
    Update the position of this employee bee as follows:

$$x\_new_j = x\_old_j + \left(x_{old\,j} - y_j\right) \times \varphi \quad (4)$$

    Compute a new fitness value
    If computed fitness value is better than the previous one **then**
      Accept new position
    End
End
For every *food source* **do**

Algorithm 2: ABC—cont'd

```
Determine the probability for it to be selected as follows:
```

$$P_{Source} = \frac{f_i}{f_s} \tag{5}$$

```
End
For every onlooker do
 Select a single food source(h) to exploit based on its
 associated probability
 Randomly select a food source y
 Randomly select a dimension j
 Update the position of this onlooker bee as follows:
```

$$x.new_j = h_j + (h_j - y_j) \times \varphi \tag{6}$$

```
 Compute a new fitness value
 If computed fitness value is better than the previous one then
  Accept new position
 End
End
For every food source do ()
 If was visited more than t times then
  Abandon it, create a random new now and evaluate fitness value
  End
 End
Select best bee so far
End
/* Notes
f_i is the current employee bee's fitness value while f_s is the cumulative sum of
all the employee bees' fitness values
φ is a random number uniformly distributed in [-1, 1]
*/
```

4.1.4 Bat algorithm

BA mimics the echolocational behavior portrayed by micro-bats. All existing bats produce loud and short sound pulses, and by being able to detect returning echoes these bats can determine the vicinity of an object. In some way, these creatures can differentiate between objects and their prey, which enables them to hunt in darkness.

These bats can fly randomly at position x_i with velocity v_i. They normally do this at a frequency f and loudness A in search of a prey. Moreover, they can easily adjust both the frequency of emitted pulses and the rate of pulse emission in accordance to the proximity of their target [67]. It is important to note that each bat carefully listens to the voices of nearby bats and is able to approach the prey. Algorithm 4 shows this process.

4.1.5 Gray wolf optimization

GWO [41] is inspired by the social intelligence portrayed by a pack comprised of 5–12 gray wolves. To mimic the social hierarchy of the pack, four distinct levels are considered in the GWO algorithm: alpha (α) wolf, beta (β) wolf, delta (δ) wolf, and omega (ω) wolf [68]. Foremost, the whole pack is involved in

Algorithm 3: IWO

Results: Global best position, optimal fitness value
An initial number of weeds is laid out over a A-dimensioned problem space
Evaluation of random positions and fitness values is conducted
While *stop criterion is not met* do
For every *weed* do
Compute the number of seeds this weed will produce as follows

$$S = (\min_{seeds} + (\max_{seeds} - \min_{seeds}) \times ratio) \tag{7}$$

Derive S new seeds whose positions are random numbers that are normally distributed. The mean of these positions is equal to the position of the parent weed while their standard deviation linearly decreases with the number of generations (iterations).
Evaluate their associated fitness values and add them to the population
End
If the population size has hit its limit M then
Only retain M best weeds, eliminate the rest
End
Determine fit and unfit weeds
End
/* Notes
$ratio = \frac{fit_i - fit_w}{fit_b - fit_w}$, where fit_b and fit_w are the current fittest and worst values respectively
max_{seeds} and min_{seeds} are the maximum and minimum number of seeds that a given weed can yield respectively
*/

searching for and tracking prey. Next, the α wolf leads the other wolves to encircle the prey. Then, the α wolf directs the β and δ wolves to launch an attack. If the prey tries to escape, the remaining ω wolves will continue attacking it until it is seized. In the GWO optimization algorithm, best location of the prey is estimated by the α, β and δ wolves, as depicted in Algorithm 5.

4.2 Improved metaheuristics as feature selectors in high-dimensional data

As a feature selector, the recent GWO metaheuristic has proved to offer better performance in comparison to the state-of-the-art metaheuristics [69]. However, the creation of new positions for the wolves is based on the experience of the three leaders of the pack: alpha (α), beta (β), and delta (δ). Moreover, attaining a proper balance between global search (diversification) and local search (intensification) remains a challenge in GWO [70].

In trying to overcome the aforementioned challenges of the existing GWO, an excited binary gray wolf optimizer (EBGWO) is proposed in [68]. Foremost, to overcome the limitation of the of GWO's position updating equation, which is good at exploitation (intensification) but poor in exploitation (diversification), a new criterion adopting the fitness values of vectors $\vec{X}_1$,$\vec{X}_2$, and $\vec{X}_3$ is proposed to create new wolves. Moreover, inspired by the widely known concept of the complete current response of a direct current (DC) excited resistor-capacitor (RC) circuit, a novel nonlinear criterion for parameter $\vec{a}$ is proposed. This criterion tries to make full use of and strike a proper balance between intensification and diversification of the existing GWO-based feature selector.

Algorithm 4: IWO

Results: Global best position, optimal fitness value
For each bat initialize loudness A_i, velocity v_i, pulse rate r_i, frequency f_i and position x_i
Evaluate their fitness values
While *stop criterion is not met* **do**
For every *bat* do
Generate a new solution by modifying its frequency, velocity, position as follows:

$$F_i = F_{min} + randNum_1 \times (F_{max} - F_{\min}) \tag{8}$$

$$v_i = v_i + (x_i - x_{best}) \times F_i \tag{9}$$

$$x_i = x_i + v_i \tag{10}$$

If $randNum_2 > r_i$ then
Derive a local solution that is close to best solution

$$x_i = x_i + A_{av} \times randNum_1 \tag{11}$$

End
If *new solution is better than existing* And $randNum_4 < A_i$ then
Accept it

$$A_i = 0.9 \times A_i \tag{12}$$

$$r_i = r_i \times (1 - \exp(-0.05 \times t)) \tag{13}$$

End
End
Select best bat
End
/* Notes
–$randNum_1$, $randNum_2$, $randNum_4$ are random values uniformly distributed in [0,1] while, $randNum_3$ is uniformly distributed in [-1.1]
-A_{av} is the average loudness of whole group of bats
- t is the actual iteration number
*/

4.3 Hybrid metaheuristics as feature selectors in in high-dimensional data

While formulating or using a metaheuristic, intensification (exploitation) and diversification (exploration) are two antagonistic principles that must be effectively balanced in order for the metaheuristic to perform well [71]. In this regard, developing a memetic metaheuristic that integrates two or more algorithms to improve the overall performance is a promising choice.

Motivated by this, a number of hybrid metaheuristics have recently been proposed to solve various optimization and feature selection tasks. However, for these hybrid metaheuristics to attain enhanced diversification (exploration) and intensification (exploitation), the basic constituent metaheuristics need to be fine-tuned prior to hybridization [2,25].

This implies that there are several constituent metaheuristics lying within the hybrid metaheuristics that need investigation. First, the approach of hybridizing one or more metaheuristics needs to be determined. Second, the criterion of ascertaining how many metaheuristics need to be integrated must be

Algorithm 5: GWO

Results: Global best position, optimal fitness value

Randomly initialize the positions of a pack of n gray wolves

While *stop criterion is not met* **do**

Determine the best (α), 2nd best (β) and 3rd best (δ) wolves based on the pack's fitness values.

Set coefficient a as follows:

$$a = 2 \times \left(1 - \frac{t}{T}\right) \tag{14}$$

For every *wolf* **do**

Update its position X_i as follows:

$$D_\alpha = |C_1 \times X_\alpha - X_i| \tag{15}$$

$$D_\beta = |C_2 \times X_\beta - X_i| \tag{16}$$

$$D_\delta = |C_2 \times X_\delta - X_i| \tag{17}$$

$$X_1 = X_\alpha - A_1 \times D_\alpha \tag{18}$$

$$X_2 = X_\beta - A_2 \times D_\beta \tag{19}$$

$$X_3 = X_\delta - A_3 \times D_\delta \tag{20}$$

$$X_3 = \frac{(X_1 + X_2 + X_3)}{3} \tag{21}$$

End

End

/* Notes

$$A = 2 \times a \times rand_1 - a$$

$$C = 2 \times rand_2$$

-$rand_1$ and $rand_2$ are random vectors in the scope of [0,1]

*/

accomplished. Third, an approach to determine the application domain of the proposed memetic needs to be developed. Finally, the method of applying the memetic in the identified domain has to be formulated [72].

Motivated by all these factors, in [73], a new memetic called Excited-Adaptive Cuckoo Search-Intensification Dedicated Gray Wolf Optimization (EACSIDGWO) is proposed to solve the feature selection task in the domain of biomedical science. In this memetic, the concept of both the complete current and voltage responses of a direct current (DC) excited resistor capacitor (RC) circuit is innovatively adopted to make the step size of the ACS and the nonlinear control criterion of parameter $\vec{a}$ of the IDGWO adaptive. Because the diversity of the population is greater during the initial stages of the EACSIDGWO algorithm, both IDGWO and ACS are jointly utilized to achieve accelerated convergence. Alternatively, to promote mature convergence while still striking an optimal balance between diversification (exploration) and intensification (exploitation) in latter stages, the task of ACS is switched to global exploration, while IDGWO is still left to carry out local exploitation.

5. Practical evaluation

This subsection describes in detail recent metaheuristics utilized in feature selection using high-dimensional datasets. Binary vectors are used to represent the informative features. For a given considered algorithm, a solution vector can be represented as (1100110000…), where the 1 implies that a given feature is selected and 0 implies that it is not selected.

5.1 Analysis of major metaheuristics as feature selectors in high-dimensional datasets

Taking into consideration experiments conducted in [1], binary versions of five metaheuristics, namely, PSO, ABC, IWO, BA, and GWO, are implemented and the decision tree classifier is used to evaluate their performance. Both the population size and the maximum number of iterations are set to 100.

5.1.1 Datasets

Seventeen datasets obtained from the UC Irvine (UCI) machine learning repository [74] are utilized. Table 2 presents a summary of these datasets. For effective comparisons, the researchers utilized datasets from different domains with varied numbers of both samples and features.

Table 2 A summary of datasets utilized.

ID	Name of dataset	Type of features	Samples	Features
1	Arrhythmia	Categorical, real	452	279
2	Chess (KR vs KP)	Categorical	3196	36
3	Crowdsourced mapping	Real	10,546	29
4	Hepatitis C virus (HCV) for Egyptian patients	Integer, real	1385	29
5	MEU-mobile KSD	Integer, real	2856	71
6	Phishing websites	Integer	2456	30
7	QSAR androgen receptor	Integer	1687	1024
8	DBWorld e-mails	Integer	64	4702
9	Musk (version 2)	Integer	6598	168
10	A study of Asian religious and biblical texts	Integer	590	8265
11	Ozone level detection	Real	2536	73
12	Diabetic RETINOPATHY	Integer, real	1151	20
13	Geographical Origin of music	Real	1059	68
14	Wall-following robot navigation	Real	5456	24
15	Insurance company benchmark	Categorical, integer	9000	85
16	Internet advertisements	Integer, real	3279	1558
17	Amazon commerce reviews	Real	1500	10,000

5.1.2 Data preprocessing

Generally, real-world data has noise and missing values. In addition, this data might be imbalanced or exist in a format that cannot be directly adopted by machine learning models. Thus, it is important to conduct data preprocessing to clean the data as well as make it fit for classifiers.

First, rows of data with missing values were discarded. Second, for numerical columns standard scaling was applied. This implies removing the mean and then scaling to unit variance. The standard score of sample a is computed using Eq. (4).

$$Q = \frac{(a-u)}{s} \tag{22}$$

For categorical columns, dummy encoding was applied. This implies creating a column for every categorical value to indicate that a given category is available in that instance. It is important to note that only those categories with a $\frac{1}{50}$ frequency were retained; the rest were regarded as infrequent. This kind of transformation creates a new set of features; thus, the feature selection process was conducted on an extended number of columns.

Lastly, 20% of the whole dataset was set as the test subset, while the remaining 80% was set as the training subset. Imbalanced datasets were split to ensure that the available classes were distributed proportionally between the test and train subsets.

5.1.3 Results

Tables 3 and 4 present most of the experimental results obtained. Table 3 presents the maximum accuracy attained by the considered metaheuristics for each dataset. Table 4 presents the optimal number of features reported by each algorithm. Each row for Tables 3 and 4 contains the results attained for a single dataset. The column labeled "Original" shows the values attained before optimization.

As shown in Table 3, most of the higher accuracy scores were attained after feature selection. This suggests good performance because almost all the feature selection processes were lossless in terms of the accuracy scores attained. Moreover, the number of features selected was significantly reduced to almost half of the original datasets. For example, a very good improvement is evident in dataset 8 in which all the considered metaheuristics attained an accuracy score of 1.0 by a reduced number of features.

Table 5 presents the average number of iterations required to attain the values in Tables 3 and 4. Since the population size was set to 100 search agents, in one iteration 100 models are successively trained. However, due to feature selection, the sizes of these models are greatly reduced in comparison with the size of the original model. Thus, the duration of computation for a single training phase is less as well.

Table 6 presents the ranking of the considered metaheuristics utilizing the results presented in Tables 3 and 4. For each dataset, the algorithms were ranked from position 1 to 5 based on the fitness value they attained. Rank 1 indicates an algorithm with the fittest value, while Rank 5 indicates an algorithm with the worst value. After awarding each algorithm a rank for every considered dataset, a cumulative sum for all the ranks was computed and tabulated for each algorithm, as shown in Table 6.

As shown, the GWO algorithm outperformed the other metaheuristics. The IWO and ABC algorithms did not perform well. This is because ABC is best suited for datasets with fewer dimensions. When the ABC algorithm is used to search for fittest solutions, one bee only changes one dimension of its location, which consequently changes a single feature. For example, considering datasets 4 and

Table 3 Accuracy scores achieved before and after feature selection.

ID	Original	Binary PSO	Binary IWO	Binary ABC	Binary BA	Binary GWO
1	0.357	0.643	0.714	0.643	0.643	0.643
2	0.997	0.992	0.997	0.994	0.997	0.997
3	0.874	0.895	0.886	0.890	0.892	0.896
4	0.274	0.343	0.368	0.365	0.357	0.357
5	0.642	0.710	0.682	0.682	0.685	0.699
6	0.961	0.964	0.961	0.961	0.962	0.969
7	0.875	0.931	0.920	0.917	0.937	0.931
8	0.727	1.0	1.0	1.0	1.0	1.0
9	0.965	0.989	0.994	0.986	0.993	0.992
10	0.712	0.890	0.847	0.831	0.864	0.890
11	0.915	0.949	0.947	0.945	0.949	0.954
12	0.621	0.709	0.691	0.730	0.739	0.739
13	0.207	0.349	0.325	0.312	0.382	0.373
14	0.996	0.996	0.994	0.997	0.995	0.995
15	0.880	0.936	0.923	0.920	0.926	0.931
16	0.968	0.983	0.977	0.980	0.980	0.979
17	0.337	0.497	0.460	0.413	0.483	0.473

Table 4 Number of features reported before and after feature selection.

ID	Original	Binary PSO	Binary IWO	Binary ABC	Binary BA	Binary GWO
1	279	118	126	140	102	77
2	73	36	42	43	41	28
3	28	15	17	18	12	19
4	28	13	12	14	11	13
5	71	36	39	36	38	39
6	30	22	22	20	20	21
7	1024	493	512	515	503	511
8	4702	2131	2271	2311	1971	2034
9	168	84	83	104	89	93
10	8266	4085	4118	4235	4101	4063
11	73	39	32	40	31	38
12	19	9	11	12	8	7
13	68	38	32	36	38	28
14	24	11	12	16	9	7
15	85	28	43	31	34	33
16	1558	732	784	776	681	751
17	10,000	4960	5012	4969	4958	4008

Table 5 Average number of iterations to attain best solution.

Algorithm	Average number of iterations
Binary PSO	41.5
Binary IWO	27.5
Binary ABC	52.2
Binary BA	49.6
Binary GWO	30.8

Table 6 Ranking of the considered algorithms based on obtained results.

Algorithm	Total sum of ranks	Final rank
Binary PSO	45	3
Binary IWO	63	4
Binary ABC	71	5
Binary BA	40	2
Binary GWO	36	1

14, the ABC algorithm outperformed the other metaheuristics. The challenge with the IWO algorithm lies with the criterion used to decrease its step size λ. With the current criterion, despite using a higher value for λ, after a few iterations the resultant value for λ is too small to effect changes in solutions. This is clear from Table 5, which shows that IWO required the least number of iterations. To solve this challenge, one can use higher values for the λ such as 0.9.

The utilized version of PSO is good at exploration compared to exploitation. This gives it the upper hand in learning the search space. However, the results show that in most cases the algorithm could not find the best solutions. This necessitates a better balance between exploration and exploitation.

The BA and GWO algorithms attained the best performance in terms of accuracy scores as well as the number of features selected. The effectiveness of the GWO algorithm lies in its logic to advance in the direction of three points as compared to one or two like the other metaheuristics. This logic enables it to actively move in different directions. For the BA algorithm, an agent to facilitate greater exploration does not always accept the fittest solution. This is why BA takes longer to attain best solutions. Thus, the GWO is recommended for selecting features in high-dimensional datasets.

5.2 Analysis of improved metaheuristics as feature selectors in high-dimensional datasets

In [68], researchers implemented a classifier improved version of GWO called EBGWO alongside five other algorithms: two versions of binary gray wolf optimization (BGWO1 and BGWO2), binary differential evolution (BDE), binary particle swarm optimization algorithm (BPSO), and binary genetic algorithm (BGA). The researchers used K-Nearest Neighbor (KNN) to evaluate performance. Moreover, for each algorithm, the population size was set to 10, while the maximum number of iterations was set to 100.

To ensure statistical significance and the stability of the attained results, all the considered metaheuristics were repeated over 10 independent runs. Moreover, the 10-fold cross-validation scheme was used to split the datasets into training and testing subsets [75]. A wrapper approach utilizing the KNN (where $k = 5$) classifier was adopted for feature selection.

5.2.1 Datasets

Table 7 lists the seven microarray datasets utilized. The datasets have varying numbers of samples, genes, and classes.

5.2.2 Results

Tables 8–14 presents the performance of all the considered algorithms in selecting informative gene subsets for the seven microarray datasets (see Table 7).

Table 7 A summary of microarray datasets utilized.

Microarray dataset	Number of samples	Number of genes	Number of classes
Brain Tumor 1	90	5920	5
Brain Tumor 2	50	10,367	4
CNS	60	7129	2
DLBCL	77	5469	2
Leukemia	72	7129	2
Colon	62	2000	2
Lung Cancer	203	12,600	4

Table 8 Accuracy attained and number of genes selected using Brain Tumor 1 dataset.

	Accuracy			Number of genes selected		
Algorithm	*Max_Acc*	*Min_Acc*	*Avg_Acc*	*Max_Nfeat*	*Min_Nfeat*	*Avg_Nfeat*
EBGWO	0.933	0.911	0.919	673	440	501.9
BGWO1	0.889	0.856	0.871	3831	2952	3356.9
BGWO2	0.911	0.878	0.894	1656	1094	1343.3
BPSO	0.854	0.823	0.843	2972	2763	2863.9
BDE	0.864	0.834	0.854	3017	2737	2937.6
BGA	0.869	0.844	0.859	2950	2840	2889.4

Table 9 Accuracy attained and number of genes selected using Brain Tumor 2 dataset.

	Accuracy			Number of genes selected		
Algorithm	*Max_Acc*	*Min_Acc*	*Avg_Acc*	*Max_Nfeat*	*Min_Nfeat*	*Avg_Nfeat*
EBGWO	0.920	0.840	0.884	2811	712	1151.5
BGWO1	0.840	0.820	0.838	7415	6103	6813.4
BGWO2	0.880	0.820	0.846	4019	2528	3083.8
BPSO	0.800	0.780	0.798	5126	5090	5122.4
BDE	0.728	0.713	0.714	5198	5076	5172.3
BGA	0.764	0.753	0.752	5139	5039	5089.5

Table 10 Accuracy attained and number of genes selected using CNS dataset.

	Accuracy			Number of genes selected		
Algorithm	*Max_Acc*	*Min_Acc*	*Avg_Acc*	*Max_Nfeat*	*Min_Nfeat*	*Avg_Nfeat*
EBGWO	0.850	0.800	0.827	1020	564	710.3
BGWO1	0.783	0.750	0.760	4942	4217	4606.4
BGWO2	0.800	0.750	0.780	2502	1842	2175.8
BPSO	0.767	0.733	0.737	3502	3486	3487.6
BDE	0.693	0.663	0.683	3530	3478	3521.9
BGA	0.727	0.707	0.717	3528	3428	3501.7

Table 11 Accuracy attained and number of genes selected using DLBCL dataset.

	Accuracy			Number of genes selected		
Algorithm	*Max_Acc*	*Min_Acc*	*Avg_Acc*	*Max_Nfeat*	*Min_Nfeat*	*Avg_Nfeat*
EBGWO	1.000	0.987	0.997	534	333	426.7
BGWO1	0.987	0.961	0.971	3706	2826	3343.4
BGWO2	1.000	0.948	0.986	1700	1002	1408.3
BPSO	0.919	0.891	0.901	2703	2672	2675.1
BDE	0.885	0.869	0.882	2732	2687	2721.4
BGA	0.906	0.883	0.896	2709	2699	2685.1

Tables 8–14 include the following information:

(i) ***Max_Acc***: the maximum accuracy value attained by a given algorithm over the 10 independent runs
(ii) ***Min_Acc***: the minimum accuracy value attained by a given algorithm over the 10 independent runs
(iii) ***Avg_Acc***: the average accuracy value attained by a given algorithm over the 10 independent runs
(iv) ***Max_Nfeat***: the highest number of features reported by a considered algorithm over the 10 independent runs

Table 12 Accuracy attained and number of genes selected using Leukemia dataset.

	Accuracy			Number of genes selected		
Algorithm	*Max_Acc*	*Min_Acc*	*Avg_Acc*	*Max_Nfeat*	*Min_Nfeat*	*Avg_Nfeat*
EBGWO	0.931	0.889	0.903	913	524	649.8
BGWO1	0.861	0.833	0.849	5065	3897	4428.5
BGWO2	0.889	0.847	0.874	2141	1618	1805.5
BPSO	0.828	0.809	0.814	3516	3505	3514.9
BDE	0.782	0.751	0.784	3537	3527	3531.2
BGA	0.801	0.782	0.792	3501	3461	3481.8

Table 13 Accuracy attained and number of genes selected using Colon dataset.

	Accuracy			Number of genes selected		
Algorithm	*Max_Acc*	*Min_Acc*	*Avg_Acc*	*Max_Nfeat*	*Min_Nfeat*	*Avg_Nfeat*
EBGWO	0.935	0.903	0.919	220	103	143.4
BGWO1	0.887	0.855	0.865	1316	1096	1189.4
BGWO2	0.919	0.871	0.900	622	351	455.2
BPSO	0.849	0.829	0.839	986	931	936.5
BDE	0.810	0.780	0.794	995	955	965.3
BGA	0.881	0.875	0.878	990	984	987.3

Table 14 Accuracy attained and number of genes selected using Lung Cancer dataset.

	Accuracy			Number of genes selected		
Algorithm	*Max_Acc*	*Min_Acc*	*Avg_Acc*	*Max_Nfeat*	*Min_Nfeat*	*Avg_Nfeat*
EBGWO	0.985	0.970	0.977	1148	781	1005.5
BGWO1	0.966	0.941	0.951	7598	6621	7211
BGWO2	0.975	0.956	0.966	2672	2167	2413.2
BPSO	0.936	0.931	0.935	6196	6179	6180.7
BDE	0.931	0.921	0.924	6256	6218	6226.8
BGA	0.952	0.939	0.945	6235	6214	6218.2

(v) ***Min_Nfeat***: the least number of features reported by a considered algorithm over the 10 independent runs

(vi) ***Avg_Nfeat***: the average number of features reported by a considered algorithm over the 10 independent runs

As shown, EBGWO outperformed all its contenders in both the classification accuracy attained as well as the reported number of informative genes. This is largely attributed to the novel criteria adopted to improve both its exploration phase as well as strike a proper balance between its two antagonistic phases: exploration and exploitation.

5.3 Analysis of hybrid metaheuristics as feature selectors in high-dimensional datasets

In [73], the EACSIDGWO feature selector was tested on six standard biomedical datasets obtained from the university of UCI repository. The experimental results were compared with those of four state-of-the-art feature selection approaches, including extended binary cuckoo search algorithm (EBCS), binary genetic algorithm (BGA), binary ant colony optimization (BACO), and binary particle swarm optimization algorithm (BPSO).

For each algorithm, the population size was set to 30, and to ensure statistical significance and the stability of the attained results, all the considered metaheuristics were repeated over 10 independent runs. Moreover, each run was terminated when the fitness function was evaluated 10,000 times. A wrapper approach utilizing the support vector machine (SVM) classifier was adopted for feature selection.

5.3.1 Considered biomedical datasets

Six benchmark biomedical datasets obtained from the UCI repository were utilized. Each dataset contains two classes and the performance of each considered memetic was evaluated based on its ability to correctly classify samples of these classes. Table 15 gives the details of the considered datasets.

5.3.2 Results

As shown in Tables 16–21:

(i) The EACSIDGWO metaheuristic outperformed all the other considered metaheuristics in terms of the classification for the utilized datasets. It attained the highest classification on both the three high-dimensional datasets (i.e., CNS, Colon, and Ovarian) and the three small-sample-sized datasets. This attractive performance is due to the joint exploitation carried out by ACS and IDGWO metaheuristic components of the EACSIDGWO during early stages and the single-handed exploitation (intensification) and exploration (diversification) conducted by IDGWO and ACS, respectively, at latter stages.

(ii) For the utilized datasets (i.e., Heart, Colon, Ovarian, and CNS), EACSIDGWO attained an average classification accuracy (i.e., *Avg_Acc*) greater than the maximum classification accuracy (*Max_Acc*) attained by EBCS (the second-best metaheuristic). The EBCS algorithm is an

Table 15 Considered biomedical datasets.

Biomedical dataset	Sample size	Number of features
Breast cancer Wisconsin (prognosis)	198	33
Breast cancer Wisconsin (diagnostic)	569	30
SPECTF heart	267	44
Ovarian cancer	216	4000
CNS	60	7129
Colon	62	2000

Table 16 Experimental results attained for the Ovarian Cancer dataset.

	Accuracy			Number of features selected		
Algorithm	*Max_Acc*	*Min_Acc*	*Avg_Acc*	*Max_Nfeat*	*Min_Nfeat*	*Avg_Nfeat*
EACSIDGWO	1.000	1.000	1.000	292	264	247.8
EBCS	0.991	0.991	0.991	1855	1747	1811.6
BACO	0.991	0.986	0.990	1975	1912	1945.7
BGA	0.991	0.991	0.991	1830	1755	1887.3
BPSO	0.991	0.990	0.990	1913	1777	1857

Table 17 Experimental results attained for the Breast Cancer Wisconsin (Diagnosis) dataset.

	Accuracy			Number of features selected		
Algorithm	*Max_Acc*	*Min_Acc*	*Avg_Acc*	*Max_Nfeat*	*Min_Nfeat*	*Avg_Nfeat*
EACSIDGWO	0.977	0.974	0.975	3	3	3
EBCS	0.981	0.974	0.973	4	3	3,1
BACO	0.972	0.960	0.969	8	6	7
BGA	0.975	0.965	0.972	6	3	3,6
BPSO	0.981	0.963	0.974	8	3	5,4

Table 18 Experimental results attained for the Breast Cancer Wisconsin (Prognosis) dataset.

	Accuracy			Number of features selected		
Algorithm	*Max_Acc*	*Min_Acc*	*Avg_Acc*	*Max_Nfeat*	*Min_Nfeat*	*Avg_Nfeat*
EACSIDGWO	0.879	0.864	0.873	7	3	5,6
EBCS	0.874	0.828	0.856	8	4	6,2
BACO	0.818	0.768	0.794	12	5	8,4
BGA	0.874	0.793	0.843	10	4	6,5
BPSO	0.848	0.798	0.821	11	4	8,3

Table 19 Experimental results attained for the SPECTF Heart dataset.

	Accuracy			Number of features selected		
Algorithm	*Max_Acc*	*Min_Acc*	*Avg_Acc*	*Max_Nfeat*	*Min_Nfeat*	*Avg_Nfeat*
EACSIDGWO	0.884	0.861	0.875	6	3	4.5
EBCS	0.873	0.846	0.861	8	5	6.2
BACO	0.846	0.813	0.831	15	10	12.1
BGA	0.884	0.846	0.866	11	4	8.4
BPSO	0.865	0.846	0.854	15	9	10.9

Table 20 Experimental results attained for the CNS dataset.

	Accuracy			Number of features selected		
Algorithm	*Max_Acc*	*Min_Acc*	*Avg_Acc*	*Max_Nfeat*	*Min_Nfeat*	*Avg_Nfeat*
EACSIDGWO	0.767	0.700	0.718	1623	807	1208.1
EBCS	0.667	0.667	0.667	3490	3391	3446.1
BACO	0.667	0.650	0.660	3589	3432	3522.9
BGA	0.683	0.667	0.668	3566	3438	3489.7
BPSO	0.667	0.667	0.667	3547	3359	3474.3

Table 21 Experimental results attained for the Colon dataset.

	Accuracy			Number of features selected		
Algorithm	*Max_Acc*	*Min_Acc*	*Avg_Acc*	*Max_Nfeat*	*Min_Nfeat*	*Avg_Nfeat*
EACSIDGWO	0.919	0.887	0.905	637	397	538.5
EBCS	0.903	0.871	0.887	1016	961	988.7
BACO	0.903	0.871	0.881	1002	932	976
BGA	0.887	0.871	0.882	1003	944	962.8
BPSO	0.887	0.855	0.879	1003	933	971.2

improved version of the cuckoo search algorithm, which is a component of the EACSIDGWO algorithm. This attractive performance proves the superiority of the proposed EACSIDGWO algorithm to efficiently find the optimal solution within the search space.

(iii) In regard to average number of features selected (*Avg_Nfeat*), EACSIDGWO demonstrated promising results by deriving subsets with the least number of features in comparison with the other algorithms. EACSIDGWO performed better than the state-of-the-art metaheuristics for the considered datasets.

In comparison with the features of the original biomedical datasets, EACSIDGWO was able to reduce the size of derived informative feature subsets. For example, the original number of features in CNS, Ovarian Cancer, and Colon Cancer datasets is 7129, 4000, and 2000, respectively, whereas the number of features selected by EACSIDGWO are 1208.1, 274.8, and 538.5, respectively. This is a clear indication that EACSIDGWO can significantly reduce the number of selected features as well as locate the most informative feature subsets. This is attributed to its well-formulated algorithm, which enhances both its exploration and exploitation capabilities, enabling it to discard noninformative features and actively search within the high-performance regions of the considered search space.

Table 22 Utilizing Wilcoxon's rank-sum test at $p = 0.05$ to compare EACSIDGWO with other considered metaheuristics.

Dataset	Wilcoxon's rank-sum test	EBCS versus EACSIDGWO	BACO versus EACSIDGWO	BGA versus EACSIDGWO	BGA versus EACSIDGWO
Ovarian cancer	*p*-value	0.000181651	0.000181651	0.000182672	0.000181651
	h-value	1.000000000	1.000000000	1.000000000	1.000000000
	z-value	3.743255786	3.743225786	3.741848283	3.743225786
Breast cancer Wisconsin (diagnostic)	*p*-value	0.022591996	0.000146767	0.017044126	0.000582314
	h-value	1.000000000	1.000000000	1.000000000	1.000000000
	z-value	2.28026466	3.796476695	2.38575448	3.439721266
Breast cancer Wisconsin (prognosis)	*p*-value	0.000730466	0.0001707	0.00073729	0.000174624
	h-value	1.000000000	1.000000000	1.000000000	1.000000000
	z-value	3.377881495	3.758843896	3.375323463	3.753152986
SPECTF heart	*p*-value	0.000321376	0.000176611	0.000176611	0.000176611
	h-value	1.000000000	1.000000000	1.000000000	1.000000000
	z-value	3.597430949	3.750317207	3.750317207	3.748901726
CNS	*p*-value	0.000182672	0.000182672	0.000182672	0.000182672
	h-value	1.000000000	1.000000000	1.000000000	1.000000000
	z-value	3.741848283	3.741848283	3.741848283	3.741848283
Colon	*p*-value	0.000182672	0.000182672	0.000182672	0.000181651
	h-value	1.000000000	1.000000000	1.000000000	1.000000000
	z-value	3.741848283	3.741848283	3.741848283	3.743255786

The superiority of EACSIDGWO metaheuristic has also been verified by the Wilcoxon's rank-sum test (i.e., a nonparametric test with a significance level of 5%). Table 22 presents the results obtained for the pairwise comparison of the four groups. The statistical significance of the experimental results for all the considered datasets can be deduced from this table. EACSIDGWO exhibited good performance compared to the remaining four metaheuristics. Therefore, the overall statistical results of EACSIDGWO are highly significant from those of the remaining four metaheuristics for all the considered biomedical datasets.

Table 23 outlines a detailed ranking of all the considered metaheuristics with their associated comparative analysis. The ranking is based on minimum classification accuracy (*Min_Acc*), maximum classification accuracy (*Max_Acc*), average classification accuracy (*Avg_Acc*), minimum number of informative features selected (*Min_NFeat*), maximum number of informative features selected (*Max_NFeat*, and average number of informative features selected (*Avg_NFeat*). EACSIDGWO obtained the best values in all these ranking measures for all the considered datasets. Taking into consideration the final ranks, EACSIDGWO attained the most attractive performance with an overall rank value of 37. As such, EACSIDGWO is a good feature selector for high-dimensional datasets.

Table 23 Overall ranking of the considered metaheuristics.

Method	Criterion	Datasets						Sum of ranks	Overall rank	Total rank	Final rank
		Ovarian Cancer	Breast cancer Wisconsin (diagnostic)	Breast cancer Wisconsin (prognosis)	SPECTF Heart	CNS	Colon				
EACSIDGWO	*Max_Acc*	1	2	1	1	1	1	7	1	37	1
	Min_Acc	1	1	1	1	1	1	6	1		
	Avg_Acc	1	1	1	1	1	1	6	1		
	Max_Nfeat	1	1	1	1	1	1	6	1		
	Min_Nfeat	1	1	1	1	1	1	6	1		
	Avg_Nfeat	1	1	1	1	1	1	6	1		
EBCS	*Max_Acc*	2	1	2	3	3	2	13	2	84	2
	Min_Acc	2	1	2	2	2	2	11	2		
	Avg_Acc	2	2	2	3	3	2	14	2		
	Max_Nfeat	3	2	2	2	2	4	15	2		
	Min_Nfeat	2	1	2	3	3	5	16	2		
	Avg_Nfeat	2	2	2	2	2	5	15	2		
BACO	*Max_Acc*	2	4	4	4	2	2	18	4	138	5
	Min_Acc	3	4	5	3	3	2	20	4		
	Avg_Acc	3	5	5	5	4	4	26	5		
	Max_Nfeat	5	4	5	4	5	2	25	5		
	Min_Nfeat	5	2	3	5	3	2	20	3		
	Avg_Nfeat	5	5	5	5	5	4	29	5		
BGA	*Max_Acc*	2	3	2	1	2	3	13	2	95	3
	Min_Acc	2	2	5	2	2	2	15	3		
	Avg_Acc	2	4	3	2	2	3	16	3		
	Max_Nfeat	2	3	3	3	4	3	18	3		
	Min_Nfeat	3	1	2	2	4	4	16	2		
	Avg_Nfeat	3	3	3	3	3	2	17	3		
BPSO	*Max_Acc*	2	1	3	3	3	3	15	3	110	4
	Min_Acc	3	3	2	2	2	3	15	3		
	Avg_Acc	3	2	4	4	3	5	21	4		
	Max_Nfeat	4	4	4	4	3	3	22	4		
	Min_Nfeat	4	1	2	4	2	3	16	2		
	Avg_Nfeat	3	4	4	4	3	3	21	4		

6. Conclusion

In a time when the size of datasets is increasing, optimal feature selection in high-dimensional datasets is considered the most important yet challenging task in bioinformatics, computer vision, text mining, industrial applications, image processing, and more.

In the literature, metaheuristics have proved efficient and effective in deriving optimal informative feature subsets while maintaining the accuracy of the classifiers. As such, the number of new metaheuristics, variants of existing metaheuristics, and memetic metaheuristics is growing.

In this chapter, we described and evaluated state-of-the-art metaheuristics, their improved versions, and hybridized versions as optimal feature selectors for high-dimensional datasets. The GWO algorithm and its variants have proved to be superior feature selectors. Thus, the GWO metaheuristic and its variants deserve more attention and further studies in the domain of feature selection.

Choosing a good classifier significantly affects the quality of the informative features selected. The KNN classifier is a widely adopted classifier in metaheuristic-based feature selectors for datasets derived from the UCI repository. The SVM classifier has also been adopted in different applications such as pattern recognition, medical diagnosis, and image analysis.

Despite the attractive performance of metaheuristic-based feature selectors for high-dimensional datasets, the dimensions of real-world datasets are growing more demanding and thus less scalable. Metaheuristic-based feature selectors must incorporate a scalable classifier to handle high-dimensional datasets. Thus, scalability is an especially important factor to consider in developing metaheuristic-based feature selectors. Moreover, stability is another important factor to consider in developing metaheuristic-based feature selectors. A stable feature selector is one that derives the same informative feature subset for a different number of sample of datasets. Instability occurs where a dataset contains highly correlated features, which are discarded over generations in trying to obtain the best classification accuracy. Thus, further research is needed.

References

[1] G. Kicska, A. Kiss, Comparing swarm intelligence algorithms for dimension reduction in machine learning, Big Data Cogn. Comput. 5 (3) (2021), https://doi.org/10.3390/bdcc5030036. Art. no. 3.

[2] P. Agrawal, H.F. Abutarboush, T. Ganesh, A.W. Mohamed, Metaheuristic algorithms on feature selection: a survey of one decade of research (2009-2019), IEEE Access 9 (2021) 26766–26791, https://doi.org/10.1109/ACCESS.2021.3056407.

[3] A. Rouhi, H. Nezamabadi-Pour, Feature selection in high-dimensional data, in: M.H. Amini (Ed.), Optimization, Learning, and Control for Interdependent Complex Networks, Springer International Publishing, Cham, 2020, pp. 85–128, https://doi.org/10.1007/978-3-030-34094-0_5.

[4] "Pattern Classification, second ed. Wiley," *Wiley.com*, November, 2012. https://www.wiley.com/en-us/Pattern+Classification%2C+2nd+Edition-p-9780471056690 (Accessed 26 October 2020).

[5] P.M. Kitonyi, D.R. Segera, Hybrid gradient descent Grey wolf optimizer for optimal feature selection, Biomed. Res. Int. 2021 (2021), e2555622, https://doi.org/10.1155/2021/2555622.

[6] Q. Al-Tashi, S.J.A. Kadir, H.M. Rais, S. Mirjalili, H. Alhussian, Binary optimization using hybrid Grey wolf optimization for feature selection, IEEE Access 7 (2019) 39496–39508, https://doi.org/10.1109/ACCESS.2019.2906757.

[7] V. Bolón-Canedo, N. Sánchez-Maroño, A. Alonso-Betanzos, Feature Selection for High-Dimensional Data, first ed., Springer, 2015. 2015 edition.
[8] E. Momanyi, D. Segera, A master-slave binary Grey wolf optimizer for optimal feature selection in biomedical data classification, Biomed. Res. Int. 2021 (2021), e5556941, https://doi.org/10.1155/2021/5556941.
[9] B. Nikpour, H. Nezamabadi-pour, HTSS: a hyper-heuristic training set selection method for imbalanced data sets, Iran J. Comput. Sci. 1 (2) (2018), https://doi.org/10.1007/s42044-018-0009-2.
[10] K. Borowska, J. Stepaniuk, A rough-granular approach to the imbalanced data classification problem, Appl. Soft Comput. 83 (2019) 105607, https://doi.org/10.1016/j.asoc.2019.105607.
[11] A. Reyes-Nava, H. Cruz-Reyes, R. Alejo, E. Rendón-Lara, A.A. Flores-Fuentes, E.E. Granda-Gutiérrez, Using deep learning to classify class imbalanced gene-expression microarrays datasets, in: Progress in Pattern Recognition, Image Analysis, Computer Vision, and Applications, 2019, pp. 46–54, https://doi.org/10.1007/978-3-030-13469-3_6. Cham.
[12] P. Branco, L. Torgo, R.P. Ribeiro, A survey of predictive modeling on imbalanced domains, ACM Comput. Surv. 49 (2) (2016) 31:1–31:50, https://doi.org/10.1145/2907070.
[13] H. He, E.A. Garcia, Learning from imbalanced data, IEEE Trans. Knowl. Data Eng. 21 (9) (2009) 1263–1284, https://doi.org/10.1109/TKDE.2008.239.
[14] J. Blaszczynski, J. Stefanowski, Improving bagging ensembles for class imbalanced data by active learning, 2018, https://doi.org/10.1007/978-3-319-67588-6_3.
[15] R.J. Hickey, Noise modelling and evaluating learning from examples, Artif. Intell. 82 (1) (1996) 157–179, https://doi.org/10.1016/0004-3702(94)00094-8.
[16] Z. Sun, Q. Song, X. Zhu, H. Sun, B. Xu, Y. Zhou, A novel ensemble method for classifying imbalanced data, Pattern Recogn. 48 (5) (2015) 1623–1637, https://doi.org/10.1016/j.patcog.2014.11.014.
[17] C.E. Brodley, M.A. Friedl, Identifying mislabeled training data, J. Artif. Intell. Res. 11 (1) (1999) 131–167.
[18] B. Frénay and A. Kaban, "A comprehensive introduction to label noise," Proc. 2014 Eur. Symp. Artif. Neural Netw. Comput. Intell. Mach. Learn. ESANN 2014, 2014, Accessed 26 October 2020. [Online]. Available: https://researchportal.unamur.be/en/publications/a-comprehensive-introduction-to-label-noise-proceedings-of-the-20.
[19] F. Barani, M. Mirhosseini, H. Nezamabadi-Pour, Application of binary quantum-inspired gravitational search algorithm in feature subset selection, Appl. Intell. 47 (2) (2017) 304–318, https://doi.org/10.1007/s10489-017-0894-3.
[20] A.P. Dawid, A.M. Skene, Maximum likelihood estimation of observer error-rates using the EM algorithm, J. R. Stat. Soc. Ser. C. Appl. Stat. 28 (1) (1979) 20–28, https://doi.org/10.2307/2346806.
[21] T.R. Golub, et al., Molecular classification of cancer: class discovery and class prediction by gene expression monitoring, Science 286 (5439) (1999) 531–537, https://doi.org/10.1126/science.286.5439.531.
[22] I. Kamkar, S.K. Gupta, D. Phung, S. Venkatesh, Stable feature selection for clinical prediction: exploiting ICD tree structure using tree-lasso, J. Biomed. Inform. 53 (2015) 277–290, https://doi.org/10.1016/j.jbi.2014.11.013.
[23] A. Rouhi, H. Nezamabadi-pour, A hybrid feature selection approach based on ensemble method for high-dimensional data, in: 2017 2nd Conference on Swarm Intelligence and Evolutionary Computation (CSIEC), Mar. 2017, pp. 16–20, https://doi.org/10.1109/CSIEC.2017.7940163.
[24] D. Segera, M. Mbuthia, A. Nyete, Particle swarm optimized hybrid kernel-based multiclass support vector machine for microarray cancer data analysis, Biomed. Res. Int. (2019). https://www.hindawi.com/journals/bmri/2019/4085725/. (Accessed 4 November 2020).
[25] N. Almugren, H. Alshamlan, A survey on hybrid feature selection methods in microarray gene expression data for Cancer classification, IEEE Access 7 (2019) 78533–78548, https://doi.org/10.1109/ACCESS.2019.2922987.

[26] H. Liu, L. Yu, Toward integrating feature selection algorithms for classification and clustering, IEEE Trans. Knowl. Data Eng. 17 (4) (2005) 491–502, https://doi.org/10.1109/TKDE.2005.66.
[27] M.H. Aghdam, N. Ghasem-Aghaee, M.E. Basiri, Text feature selection using ant colony optimization, Expert Syst. Appl. 36 (3, Part 2) (2009) 6843–6853, https://doi.org/10.1016/j.eswa.2008.08.022.
[28] S. Ahmed, M. Zhang, L. Peng, Enhanced feature selection for biomarker discovery in LC-MS data using GP, in: 2013 IEEE Congress on Evolutionary Computation, Jun. 2013, pp. 584–591, https://doi.org/10.1109/CEC.2013.6557621.
[29] A. Ghosh, A. Datta, S. Ghosh, Self-adaptive differential evolution for feature selection in hyperspectral image data, Appl. Soft Comput. 13 (4) (2013) 1969–1977, https://doi.org/10.1016/j.asoc.2012.11.042.
[30] M. Dash, H. Liu, Feature selection for classification, Intell. Data Anal. 1 (1) (1997) 131–156, https://doi.org/10.1016/S1088-467X(97)00008-5.
[31] I. Guyon, A. Elisseeff, An introduction to variable and feature selection, J. Mach. Learn. Res. 3 (2003) 1157–1182. null.
[32] H. Liu, H. Motoda, R. Setiono, and Z. Zhao, "Feature selection: an ever evolving frontier in data mining," in Feature Selection in Data Mining, May 2010, pp. 4–13. Accessed 3 November 2020. [Online]. Available: http://proceedings.mlr.press/v10/liu10b.html.
[33] N. Hoque, D.K. Bhattacharyya, J.K. Kalita, MIFS-ND: a mutual information-based feature selection method, Expert Syst. Appl. 41 (14) (2014) 6371–6385, https://doi.org/10.1016/j.eswa.2014.04.019.
[34] Z. Xu, I. King, M.R.-T. Lyu, R. Jin, Discriminative semi-supervised feature selection via manifold regularization, IEEE Trans. Neural Netw. (2010), https://doi.org/10.1109/tnn.2010.2047114. 006.
[35] J. Tang, S. Alelyani, H. Liu, Feature selection for classification: a review, in: Data Classification, CRC Press, 2014, pp. 37–64, https://doi.org/10.1201/b17320.
[36] A. Jović, K. Brkić, N. Bogunović, A review of feature selection methods with applications, in: 2015 38th International Convention on Information and Communication Technology, Electronics and Microelectronics (MIPRO), May 2015, pp. 1200–1205, https://doi.org/10.1109/MIPRO.2015.7160458.
[37] Z. Sun, G. Bebis, R. Miller, Object detection using feature subset selection, Pattern Recogn. 37 (11) (2004) 2165–2176.
[38] A.K. Jain, R.P.W. Duin, J. Mao, Statistical pattern recognition: a review, IEEE Trans. Pattern Anal. Mach. Intell. 22 (1) (2000) 4–37, https://doi.org/10.1109/34.824819.
[39] H. Liu, H. Motoda (Eds.), Feature extraction, construction and selection: A Data Mining Perspective, Springer US, 1998, https://doi.org/10.1007/978-1-4615-5725-8.
[40] B. Xue, M. Zhang, W.N. Browne, X. Yao, A survey on evolutionary computation approaches to feature selection, IEEE Trans. Evol. Comput. 20 (4) (2016) 606–626, https://doi.org/10.1109/TEVC.2015.2504420.
[41] S. Mirjalili, S.M. Mirjalili, A. Lewis, Grey wolf optimizer, Adv. Eng. Softw. 69 (2014) 46–61, https://doi.org/10.1016/j.advengsoft.2013.12.007.
[42] O. Olorunda, A.P. Engelbrecht, Measuring exploration/exploitation in particle swarms using swarm diversity, in: 2008 IEEE Congress on Evolutionary Computation (IEEE World Congress on Computational Intelligence), Jun. 2008, pp. 1128–1134, https://doi.org/10.1109/CEC.2008.4630938.
[43] L. Lin, M. Gen, Auto-tuning strategy for evolutionary algorithms: balancing between exploration and exploitation, Soft. Comput. 13 (2) (2009) 157–168, https://doi.org/10.1007/s00500-008-0303-2.
[44] A.W. Mohamed, A.A. Hadi, A.K. Mohamed, Gaining-sharing knowledge based algorithm for solving optimization problems: a novel nature-inspired algorithm, Int. J. Mach. Learn. Cybern. 11 (7) (2020) 1501–1529, https://doi.org/10.1007/s13042-019-01053-x.
[45] J.H. Holland, Genetic algorithms, Sci. Am. 267 (1) (1992) 66–73.
[46] J.H. Holland, Adaptation in Natural and Artificial Systems: An Introductory Analysis with Applications to Biology, Control, and Artificial Intelligence, A Bradford Book, Cambridge, MA, 1992.

[47] R. Storn, K. Price, Differential evolution—a simple and efficient heuristic for global optimization over continuous spaces, J. Glob. Optim. 11 (4) (1997) 341–359, https://doi.org/10.1023/A:1008202821328.
[48] W. Kinnebrock, 4. Evolutionsstrategien, 4. Evolutionsstrategien, Oldenbourg Wissenschaftsverlag, 2018, pp. 95–100, https://doi.org/10.1515/9783486785203-006.
[49] F. Glover, Future paths for integer programming and links to artificial intelligence, Comput. Oper. Res. (1986), https://doi.org/10.1016/0305-0548(86)90048-1.
[50] J. Kennedy, R. Eberhart, Particle swarm optimization, in: Proceedings of ICNN'95—International Conference on Neural Networks, vol. 4, Nov. 1995, pp. 1942–1948, https://doi.org/10.1109/ICNN.1995.488968.
[51] M. Dorigo, V. Maniezzo, A. Colorni, Ant system: optimization by a colony of cooperating agents, IEEE Trans. Syst. Man Cybern. B Cybern. 26 (1) (1996) 29–41, https://doi.org/10.1109/3477.484436.
[52] S. Mirjalili, A. Lewis, The whale optimization algorithm, Adv. Eng. Softw. 95 (2016) 51–67, https://doi.org/10.1016/j.advengsoft.2016.01.008.
[53] S. Mirjalili, A.H. Gandomi, S.Z. Mirjalili, S. Saremi, H. Faris, S.M. Mirjalili, Salp swarm algorithm: a bio-inspired optimizer for engineering design problems, Adv. Eng. Softw. 114 (2017) 163–191, https://doi.org/10.1016/j.advengsoft.2017.07.002.
[54] D. Karaboga, An Idea Based on Honey Bee Swarm for Numerical Optimization, Technical Report-TR06, Department of Computer Engineering, Engineering Faculty, Erciyes University, 2005.
[55] Z.W. Geem, J.H. Kim, G.V. Loganathan, A new heuristic optimization algorithm: harmony search, SIMULATION 76 (2) (2001) 60–68, https://doi.org/10.1177/003754970107600201.
[56] S. Kirkpatrick, C.D. Gelatt, M.P. Vecchi, Optimization by simulated annealing, in: M.A. Fischler, O. Firschein (Eds.), Readings in Computer Vision, Morgan Kaufmann, San Francisco, CA, 1987, pp. 606–615, https://doi.org/10.1016/B978-0-08-051581-6.50059-3.
[57] R.V. Rao, V.J. Savsani, D.P. Vakharia, Teaching–learning-based optimization: an optimization method for continuous non-linear large scale problems, Inform. Sci. 183 (1) (2012) 1–15, https://doi.org/10.1016/j.ins.2011.08.006.
[58] A.H. Kashan, League championship algorithm: a new algorithm for numerical function optimization, in: 2009 International Conference of Soft Computing and Pattern Recognition, Dec 2009, pp. 43–48, https://doi.org/10.1109/SoCPaR.2009.21.
[59] L. Brezočnik, I. Fister, V. Podgorelec, Swarm intelligence algorithms for feature selection: a review, Appl. Sci 8 (9) (2018), https://doi.org/10.3390/app8091521. Art. no. 9.
[60] B.H. Nguyen, B. Xue, M. Zhang, A survey on swarm intelligence approaches to feature selection in data mining, Swarm Evol. Comput. 54 (2020) 100663, https://doi.org/10.1016/j.swevo.2020.100663.
[61] M. Rostami, K. Berahmand, E. Nasiri, S. Forouzandeh, Review of swarm intelligence-based feature selection methods, Eng. Appl. Artif. Intel. 100 (2021) 104210, https://doi.org/10.1016/j.engappai.2021.104210.
[62] R. Poli, J. Kennedy, T. Blackwell, Particle swarm optimization, Swarm Intell. 1 (1) (2007) 33–57, https://doi.org/10.1007/s11721-007-0002-0.
[63] J. Brownlee, Clever Algorithms: Nature-Inspired Programming Recipes, 2011, s.l Lulu.com.
[64] I. Fister Jr., X.-S. Yang, I. Fister, J. Brest, and D. Fister, "A brief review of nature-inspired algorithms for optimization," *ArXiv13074186 Cs*, 2013, Accessed 15 February 2022. [Online]. Available: http://arxiv.org/abs/1307.4186.
[65] A.R. Mehrabian, C. Lucas, A novel numerical optimization algorithm inspired from weed colonization, Eco. Inform. 1 (4) (2006) 355–366, https://doi.org/10.1016/j.ecoinf.2006.07.003.
[66] S. Karimkashi, A. Kishk, Invasive weed optimization and its features in electromagnetics, IEEE Trans. Antennas Propag. (2010), https://doi.org/10.1109/TAP.2010.2041163.
[67] X.-S. Yang, A new metaheuristic bat-inspired algorithm, in: J.R. González, D.A. Pelta, C. Cruz, G. Terrazas, N. Krasnogor (Eds.), Nature Inspired Cooperative Strategies for Optimization (NICSO 2010), Springer, Berlin, Heidelberg, 2010, pp. 65–74, https://doi.org/10.1007/978-3-642-12538-6_6.

[68] D. Segera, M. Mbuthia, and A. Nyete, "An Excited Binary Grey Wolf Optimizer for Feature Selection in Highly Dimensional Datasets," Feb. 2022, pp. 125–133. Accessed 15 February 2022. [Online]. Available: https://www.scitepress.org/Link.aspx?doi=10.5220/0009805101250133.

[69] E. Emary, H.M. Zawbaa, A.E. Hassanien, Binary grey wolf optimization approaches for feature selection, Neurocomputing 172 (2016) 371–381, https://doi.org/10.1016/j.neucom.2015.06.083.

[70] J. Too, A.R. Abdullah, N. Mohd Saad, N. Mohd Ali, W. Tee, A new competitive binary grey wolf optimizer to solve the feature selection problem in EMG signals classification, Computers 7 (4) (2018), https://doi.org/10.3390/computers7040058. Art. no. 4.

[71] E.-G. Talbi, Metaheuristics: from design to implementation, Wiley, 2009. Accessed 3 November 2020. New. [Online]. Available: http://gen.lib.rus.ec/book/index.php?md5=77cda47604699c0464cd5ce53c753e89.

[72] R. Sindhu, R. Ngadiran, Y.M. Yacob, N.A. Hanin Zahri, M. Hariharan, K. Polat, A Hybrid SCA inspired BBO for feature selection problems, Math. Probl. Eng. (2019). https://www.hindawi.com/journals/mpe/2019/9517568/. (Accessed 3 November 2020).

[73] D. Segera, M. Mbuthia, A. Nyete, An innovative excited-ACS-IDGWO algorithm for optimal biomedical data feature selection, Biomed. Res. Int. 2020 (2020), e8506365, https://doi.org/10.1155/2020/8506365.

[74] D. Dua, C. Graff, UCI Machine Learning Repository, 2019. https://archive.ics.uci.edu/ml/index.php. (Accessed 15 February 2022).

[75] S. Arlot, A. Celisse, A survey of cross-validation procedures for model selection, Stat. Surv. 4 (2010) 40–79, https://doi.org/10.1214/09-SS054.

CHAPTER

14

Optimal deployment of sensors for leakage detection in water distribution systems using metaheuristics

Ehsan Yousefi-Khoshqalb[a], Mohammad Reza Nikoo[b], and Amir H. Gandomi[c,d]

[a]Faculty of Civil, Water and Environmental Engineering, Shahid Beheshti University, Tehran, Iran, [b]Department of Civil and Architectural Engineering, College of Engineering, Sultan Qaboos University, Muscat, Oman, [c]Faculty of Engineering and Information Technology, University of Technology Sydney, Ultimo, NSW, Australia, [d]University Research and Innovation Center, Obuda University, Budapest, Hungary

1. Introduction

Water crises rank among the top ten global risks menacing economic growth, according to the World Economic Forum's Global Risk Reports [1]. Many factors can threaten water, such as rapid population growth, consumption changes, severe droughts, consumer competition, poor management, and unforeseen risks posed by climate change and environmental degradation [2]. Therefore, all countries must equally contribute to finding solutions to the world's urgent challenges. Although water is an indispensable resource for human activity, protecting water resources can be challenging since a significant portion of it is lost through leaks or stolen from water distribution systems (WDSs) [3]. Over the past few decades, leakage has been the leading reason for water loss in WDSs. Leakage causes problems such as water pollution, land subsidence, water waste, service interruption, water distribution inequity, increasing cost of operation and maintenance, and revenue and energy losses. Various types of damage, such as elliptical and circular holes, fractures, cracks, and excessive pressure differences, cause leakage in WDSs. Worldwide, nearly 184 billion dollars are spent annually to provide clean water; however, 9.6 billion dollars are lost due to leakages in WDSs [4]. The cost of leakage alone in the United States exceeds 2.8 billion dollars annually [5]. Hence, detection and management of leakages are imperative to determining how much water is lost and where it occurs.

There are three types of leakage detection strategies: passive, active, and proactive. The passive strategies detect leaks that are easily spotted from the ground and are frequently reported. The active strategies detect leaks directly by using acoustic equipment, thermography, leak-noise correlation, ground-penetrating radar, pig-mounted acoustic sensing, tracer gas, and video inspection. Even so, these strategies are more costly for water providers and, in many cases, impractical because they require special equipment and expertise to use as well as time and cost [4]. For example, a matched-field processing (MFP) approach was proposed by Wang and Ghidaoui [6] to detect leakage

Comprehensive Metaheuristics. https://doi.org/10.1016/B978-0-323-91781-0.00014-4

location and volume, but cannot be applied to large WDSs as a result of costly operations [7]; unless the optimal intervention is determined, namely, the economic level of leakage (ELL) [8–11]. In recent years, the affordability of telemetry equipment led to pilot projects that analyze real-time pressure and flow data [12–14]. Since conventional ways are expensive, arduous, and specialized, studies have examined proactive leakage detection approaches that utilize lower operational costs by comparing observed and estimated values of flow and pressure in a WDS. By doing so, nowadays, proactive strategies are more efficient due to the increasing availability of pressure and flow data, and water losses from leaks and bursts (events) in WDSs can be revealed by these data. Hence, sensor data analysis can yield useful information, and subsequently, the number, type, and location of sensors determine the usefulness of the collected data. This chapter discusses methods, developments, and applications of leakage detection using metaheuristic algorithms.

2. Background

The detection of a leak or burst event occurs when a water utility becomes aware of it. Isolation defines the distinction between two possible events, whereas localization provides a way to determine where the event is located [15,16]. The implementation of district metered areas (DMAs) allows monitoring of data flow throughout WDSs since each DMA is equipped with flow and pressure monitoring devices at the inlet [12]. The inlet flowmeter is typically the primary source of leak detection information. However, it is recommended to place the additional pressure measurement arbitrarily at the highest point (which is the most sensitive to low-pressure events) or at the farthest point from the inlet. In the standard approach, water loss per DMA is calculated using the nightline method, in which significant fluctuations over time indicate potential problems [12,17,18]. This may be useful in countries with limited access to technology or alternative methods. Although, an improvement of the current method for locating instruments in DMAs is needed since it relies on obtaining stable spatial coverage instead of detection [12,19]. Due to this, the presented points cannot reflect the overall condition of the network in real-time. Therefore, sensor deployment methods are developed, which typically include pressure sensors and in some cases, additional flow meters. Nevertheless, Farley et al. [12] noted that pressure is often viewed as a "poor cousin" of flow when it comes to detecting events because (1) pressure and flow have a squared relationship, as defined by Bernoulli; (2) measurements of flow are generally based on pulse counting; and (3) instant pressure data has noise and fluctuation, whereas smoothed values over a short period of time (15 min) are considered reliable. Flow meters are sensitive to downstream variations, but pressure sensors respond to variations in head loss on a given flow path. Despite this, its low cost has made the pressure sensor a valuable tool regardless of its low sensitivity.

Numerous studies have been published on model-based event detection, and localization since Pudar and Liggett [20] formulated the problem of event detection in terms of an inverse least-squares parameter estimation problem [15,20,21]. The event detection problem was subsequently proposed as an inverse problem method using pressure measurements and a sensitivity matrix [20]. This methodology analyzes the difference between measurement and estimation based on hydraulic models, residuals on the line, and considering a threshold to reflect uncertainty in the model. Whenever residuals break their threshold, they are compared to the leak sensitivity matrix to identify potential leaks. With this in mind that nodal demand uncertainty and measurement noise can significantly impact the performance of this approach [15]. For most studies, automatic event detection is used by simulating various leak scenarios in hydraulic models to estimate each monitoring station's sensitivity to simulated artificial faults. Yet, several approaches have been developed by experts in this area. Using these approaches, sensors can be deployed to detect leaks, isolate them, and locate them.

The deployment of sensors is evaluated using a variety of objectives. The main objective of each method is to use several sensors, which plays a significant role in determining detection efficiency. That is,to determine a monitoring strategy, the number of sensors must be weighed against the sensor layout cost. For the second objective, sensors deployment is evaluated using performance criteria, the most common of which is maximizing the number of detected events; the trade-offs between objectives can only be discernible when different criteria are used. Subsequently, a combination of sensors deployment and hydraulic models enables leak detection, narrows search area, and facilitates on-site investigations for optimal inspection by concentrating on leaky areas. Numerous studies have been conducted on leak detection by deploying sensors; here we discuss methods that use metaheuristic algorithms.

3. Related works

Farley et al. [22] took the first step in deploying sensors for leak detection by using a DMA simulation over a day-night period to create an $N \times M$ matrix (N is event points and M is monitor points). The sum of chi-squares between the pressures recorded under different leak scenarios and those registered under control states are placed in this matrix. Moreover, to distinguish events from one another, they suggest the matrix' mean as the threshold value. Accordingly, the column with the greatest number of events is considered the sensor location with the most sensitivity. This method has been used in two ideal case studies, finding that choosing the wrong threshold prevented them from detecting small leaks, and deploying two sensors optimally allowed them to detect a high number of artificial events. Then, Farley et al. [12] expanded the number of sensors to three. By comparing modeled data with field data, they validated the methodology using several different measurement locations, demonstrating successful operationalization. To determine the method's accuracy for pressure sensor deployment, five different leak scenarios were simulated by opening fire hydrants and inserting pressure sensors at various points in the DMA. The results indicated that installing pressure sensors at appropriate locations are crucial to determining sensitivity. Also, in this study, the best deployment is determined by the Max-Sum method adopted by Bush and Uber [23]. In the following, Farley et al. [24], using the Genetic Algorithm (GA) [25], examined a supplementary study on search efficiency for selecting the best sensor locations to detect events in different networks. Through this method, a DMA could be partitioned into detection subareas to localize events. In this study, using 14 real-life DMAs, the objective is to minimize fitness function and determine the best sensor combination to divide the DMA into equal zones (see Table 1).

As discussed in Farley et al. [22], Pérez et al. [26] proposed a similar sensor deployment method by optimizing residuals between pressure measurements. The residual can be thought of as an event's signature, measured against a threshold to distinguish events. They created a signature matrix from the normalized sensitive matrix, allowing them to set an identical threshold for all sensors. A binary threshold was then set for all sensors. For a fixed number of sensors, they employed a GA in three real-life case studies to identify the sensor deployment that minimized the maximum number of events containing unique signatures (see Table 1). In the same way as Pérez et al. [26], Wu and Song [28] proposed an optimal deployment method for pressure sensors based on a binary matrix with the GA optimization, in

Table 1 Review of related works on optimal deployment of sensors for leakage detection using metaheuristic algorithms.

Author (s)	Optimizer (s)	Objective(s)	Consideration (s)	Hydraulic simulator	Period (s)	Leak location	Uncertainty	Event(s) magnitude	Case study	Field validation
[26]	GA	▪ Minimizing the maximum number of events containing an identical signature	• Using a fixed number of sensors	PICCOLO	Single	Node	No	Single	▪ Real-life, Placa del Diamant (1600 nodes, 41 km of pipes), Barcelona, Spain	No
[12]	GA	▪ Max-Sum [23]	• Using a fixed number of sensors	AQUIS	Extended	Node	No	Multiple	▪ Real-life, DMA (17.8 km pipes, 28 service connections and 925 consumers)	Yes
[27]	GA	▪ Maximizing the leak isolability with a reasonable number of sensors	• Particular time step during the nighttime period for 15 days	PICCOLO	Extended	Node	Yes-demand uncertainty	Multiple	▪ Real-life, Placa del Diamant (1600 nodes, 41 km of pipes), Barcelona, Spain	Yes
[28]	GA	▪ Maximizing the number of events that can be detected	• Using a fixed number of sensors (5, 10, 15, 20, 25, 28 sensors)	WaterGEMS	NA	Node	No	Multiple (leaks simultaneously)	▪ Real-life, DMA (1321 pipes), UK ▪ Real-life, DMA (86 pipes), UAE	No

[24]	GA	▪ Minimizing fitness function and finding the best combination of sensors for dividing the DMA into equally sized zones	• Using a fixed number of sensors • There is an uncertainty band applied on either side of the threshold to penalize sensors	AQUIS	Extended	Node	Yes-measurement	Single	▪ 14 real-life, DMAs (204–1091 nodes, 6.3–36 km of pipes), UK	Yes
[15]	GA	▪ Minimizing nonisolable leakages, based on the introduced isolation criteria	• Using a fixed number of sensors • Using the angle method to analyze the sensitivity matrix	EPANET	Extended	Node	Yes-measurement	Multiple	▪ Real-life, Hanoi (31 nodes, 34 pipes), Vietnam ▪ Real-life, Limassol (197 nodes, 239 pipes), Cyprus	No
[29]	GA and PSO	▪ Minimizing overlapping signatures derived from primary leak signatures	• Using a fixed number of sensors Leak signature space method	EPANET	Extended	Node	Yes-measurement	Multiple	▪ Real-life, Hanoi (31 nodes, 34 pipes), Vietnam ▪ Real-life, Limassol (197 nodes, 239 pipes), Cyprus	No
[30]	GA	▪ Minimizing cost-benefit depending on the number of sensors	• Using a different number of sensors • Using a fixed penalty is used	EPANET and OOPNET	NA	Node	Yes -demand uncertainty	Single	▪ Real-life, DMA (392 nodes, 452 pipes, 37 km of pipes)	No

Continued

Table 1 Review of related works on optimal deployment of sensors for leakage detection using metaheuristic algorithms—cont'd

Author (s)	Optimizer (s)	Objective(s)	Consideration (s)	Hydraulic simulator	Period (s)	Leak location	Uncertainty	Event(s) magnitude	Case study	Field validation
[31]	GALAXY	▪ Maximizing the event localization performance using the inverse distance weighted geospatial interpolation	• Using a fixed number of sensors	EPANET	Single	Node	No	Single	▪ Synthetic, Bakryan (35 nodes, 58 pipes, 102 km of pipes)	No
[32]	ABC	▪ Minimizing the average mutual coherence	• Using a fixed number of sensors that is estimated roughly by a greedy algorithm	NA	NA	Node	No	Multiple	▪ Real-life, DMA (171 nodes, 239 pipes), China	No
[7]	NSGA-II	▪ Maximizing accuracy of identified leakage zone ▪ Minimizing number of sensors	• Identifying candidate flow and pressure sensors using k-means clustering	EPANET	Extended	Node	No	Multiple	▪ Synthetic, Mesopolis (1588 nodes, 2176 pipes)	No
[33]	NSGA-II	▪ Minimizing the number of pressure sensors ▪ Minimizing the detection time	• Iteratively, the pressure error threshold reduces and allows for the use of increasingly accurate sensors	EPANET	Extended	Node	Yes-measurement	Multiple	▪ Synthetic, C-town (388 nodes, 429 pipes)	No

[34]	GWO	▪ Minimizing the number of deployed sensors to leak detection with a minimum of time	• Based on the pressure and flow rate of the nodes, sensitivity and entropy values are derived	EPANET	Extended	Node	No	Multiple	▪ Real-life, DMA (883 nodes, 927 pipes, 17.4 km of pipes), Barcelona, Spain	No
[35]	DEA and PSO	▪ Minimizing error in the inverse analysis of pressure and flow data	• Using a fixed number of sensors	EPANET	Extended	Pipe	Yes-demand uncertainty	Multiple	▪ Synthetic, Apulian (23 nodes, 34 pipes)	No
[36]	NSGA-II	▪ Minimizing the number of pressure sensors ▪ Maximizing VOI ▪ Minimizing TE		EPANET	Extended	Node	No	Multiple	▪ Synthetic, C-town (388 nodes, 429 pipes)	No
[37]	NSGA-II	▪ Minimizing the number of pressure sensors ▪ Maximizing the sensor coverage ▪ Minimizing the extra information derived from pressure sensors	• Identifying candidate pressure sensors using k-means clustering	EPANET	Extended	Node	Yes-Measurement	Multiple	▪ Synthetic, C-town (388 nodes, 429 pipes)	No

which a fixed number of sensors can detect the maximum number of events. In contrast to Pérez et al. [26], they generated random events at one node or two nodes simultaneously by using the Monte Carlo Method (MCM) [38] and determined a binary matrix based on the residual matrix while the sensors' accuracy was used as the threshold. They optimized pressure sensors deployment using two real-life models, providing maximum coverage of events. Pérez et al. [27] examined the impact of several time steps and uncertainties in nodal demands (see Table 1). They found that because of the large variation in consumption over a given period of time, the performance of localization suffers from uncertainty in nodal demand, so sensitivity matrices must be calculated no matter whether there is uncertainty or not. In this study, a voting criterion was used to assign each event to a node group. Then an objective was set up to minimize the size of the maximum group, which is claimed to achieve success even in the face of uncertain demand.

Casillas et al. [15] developed a method for maximizing leak isolation through the deployment of sensors created by a nonbinarized event sensitivity matrix coupled with exhaustive search and GA, solving it as an integer optimization problem. This Isolation, as a user-independent method, relies on the analysis of the residual vector as well as the sensitivity vectors to recognize which node has the greatest likelihood of leaking. Using the angle method [15], they determined the angle between the actual residual vector and each column of the sensitivity matrix (each event node). Then, the angle mean and leak index were found for each time step, and the one that showed the smallest event index was identified as the event node. This study suggests that the angle method yields the best results for isolation tasks. And, to maximize isolation regarding events across all nodes, GA can find near-optimal solutions with less effort than an exhaustive search. Furthermore, Casillas et al. [29] proposed a new sensor deployment method using leak signature space (LSS) method [39], in which the residual vector is transformed into an original representation where event location can be assigned with a specific signature independent of event magnitude. To optimize sensor deployment, overlapping signature domains from the LSS representation should be minimized. Whereby the semi-exhaustive search method (SEM) is proposed for small networks, GA [25] or particle swarm optimization (PSO) [40] are separately proposed for larger networks. Subsequently, two networks were validated using the proposed methodology (see Table 1), under perfect models, but considering measurement uncertainties using Gaussian white noise. Results show that for certain parameters, PSO yields results faster than GA and is better suited to small networks or fewer sensors; GA tends to be better deployed and efficient for larger networks or more sensors, and time-horizon evaluation can notably enhance efficiency under complex scenarios. They also noted that PSO could be stuck in a local suboptimum; maybe it tends to explore the same configurations that led to past success. However, crossover operations in GAs are probably perform better when when moving between search spaces.

Steffelbauer and Fuchs-Hanusch [30] used Monte Carlo simulations to estimate pressure under multinodal demand scenarios to optimize sensor deployment based on demand uncertainty and different numbers of sensors. Several levels of uncertainty were considered, which resulted in wildly different optimal deployments than those without uncertainty; that is, without uncertainty, the method sought to deploy sensors in points of high demand uncertainty across the system and vice versa; that is, increasing uncertainties lead to a decrease in the number of sensors. Therefore, the sensitive points to leaks are also susceptible to variations in demand, which are nonideal places for sensors. Yet, for the purpose of determining how many sensors must be deployed to identify a given scenarios correctly, they developed a cost-benefit function to determine event localization accuracy based on the number of sensors, in which the derived function follows a power law, even when demand uncertainties are considered.

An inverse distance weighted (IDW) geospatial interpolation was used by Boatwright et al. [31] to integrate pressure sensor deployment and leak localization, which was optimized with the GALAXY multiobjective algorithm [41] through a fixed number of sensors. This study focused on a single period (time step) in which optimizing the sensor deployment was done by simulating leaks at all nodes and then creating a matrix of instantaneous chi-square values; this matrix served to compute interpolation surfaces for optimization, which maximizes locating performance among the sensors under different leak scenarios. With the optimally deployed sensors, spatially confined IDW interpolation can then be applied to estimate where a leak event can occur by measuring the real pressure at the sensors. Geostatistical methods can be used to reduce the deployed sensors in a network as they can estimate parameters at unmeasured points using the data from nearby sensors, thereby enabling more efficient leak detection performance with fewer sensors [42].

Using an inverse analysis of flow rates delivered by the reservoir and pressure measurements measured at specific points in the network, Righetti et al. [35] sought to locate the location of an imposed leakage in the well-known Apulian network [43]. An evolutionary algorithm called differential evolutionary particle swarm optimization (DEPSO) was used to identify the leaks into the WDS [44]. This algorithm combined the powerful exploration capabilities of DE and the fast convergence of PSO.

Xie et al. [32] proposed an optimal sensor deployment method aimed at meeting budget constraints and enabling reasonable diagnosis. They developed a diagnosability criterion for measurements using average mutual coherence to determine whether they contain enough information to detect the leak locations. Using compressed sensing theory, they determine leak sparse demands from measurements. Then, an optimization method using the enhanced binary artificial bee colony (ABC) algorithm [45,46], enhanced by genetic operators, including crossover and swap, was proposed for deploying sensors optimally by minimizing the average mutual coherence (maximizing the diagnostic degree), which is defined as a $\{0, 1\}$ binary problem. This method worked well in a real-life network (see Table 1) and provided a quasi-optimal sensor deployment.

Rayaroth and Sivaradje [34] proposed the Bivariate Correlation and Sensitivity Analysis-based Meta-Heuristic Grey Wolf Optimization (BCSA-MHGWO) method to detect leaks with a minimal number of pressure sensors. The Grey Wolf Optimization (GWO) [47] evaluates sensors' sensitivity, entropy, and correlation to deploy them optimally. They use pressure and flow data to determine sensitivity and entropy values and use bivariate correlation coefficients and pressure series to determine correlation coefficients. In this way, accuracy and time of leakage detection are improved, where the BCSA-MHGWO performance is measured in terms of leakage detection accuracy, leakage detection time, and incorrectly detected leakages. Comparing the results of this study with the comparative studies, the authors concluded that BCSA-MHGWO appears to perform well in terms of leakage detection accuracy and time.

Using a multiobjective approach, Raei et al. [7] proposed a two-step method to deploy pressure and flow sensors; Raei et al. [33], Khorshidi et al. [36], and Taravatrooy et al. [37] proposed various methods to deploy pressure sensors optimally, including (1) maximizing the identified leakage zone and minimizing the number of flow and pressure sensors [7]; (2) minimizing the number of pressure sensors and the detection time [33]; (3) minimizing the number of pressure sensors and transinformation entropy, and maximizing the value of information [36]; and (4) minimizing the number of pressure sensors and the extra information derived from pressure sensors, and maximizing the sensor coverage [37]. These methods enable the evaluation of nondominated solutions with a multiobjective optimization algorithm namely NSGA-II [48]. It is noteworthy that the reviewed studies also assumed that only one event could occur simultaneously. Also, different approaches can be used to select the optimal sensor deployment according to the constraints. Table 1 summarizes the characteristics of these studies. In the following sections, the methods presented in these studies are described, and the results are compared.

4. Methods

Leakage monitoring networks must meet two key performance criteria: (1) minimizing the time between leak initiation and detection and (2) minimizing the volume of leaked water. To achieve these criteria, it is first necessary to describe how to model leakage in a network and then specify objective functions. To cover the mentioned criteria directly or indirectly, this section discusses how to prepare the recently developed model and the objective functions. For calculating these criteria, when leakage occurs in the WDS, in any scenario, the absolute pressure difference between the node n at time-step i and the corresponding nodal pressure in the control state must exceed the tolerance pressure variation threshold. Also, by choosing the appropriate nodes, measurement errors are considered.

4.1 Leakage simulation

There are several modeling techniques that can be used to model events (leaks and bursts). For instance, all nodes can be modeled as possible events for each location, which has been done in numerous studies [7,12,22,33,36,37]. In contrast to the previous model, events can instead be modeled at middle points in pipes rather than at nodes [35,49]. Experimentally, longitudinal cracks, circumferential cracks, and round holes were considered leakage types [50,51]; however, when an orifice model is used, leakages can be modeled as round holes, and leakage flow can be simulated by:

$$q = cp^{\gamma} \tag{1}$$

where q is a leakage flow (lit/s), p is a pressure (Pa), c is a discharge or emitter coefficient (lit/s/Pa$^{0.5}$), and variable γ is the pressure exponent, which is assumed 0.5 [49,52].

In WDSs, hydraulic models can be used to simulate a wide range of leakage scenarios. This chapter refers to EPANET as a hydraulic model [52], frequently used in the literature. The EPANET model does not include emitter coefficients, so leakage scenarios cannot be accounted for; therefore, extra-nodal demand can be added to the existing nodal demand. The demand-driven simulation calculates a new demand for each time step based on the pressure at each node, which is expressed for each node as follows:

$$d_{new(n,i)} = d_{initial(n,i)} + \begin{cases} c\dfrac{p^{\gamma}_{(n,i)}}{d_c} & i \geq t_m \\ 0 & \text{otherwise} \end{cases}, \quad \forall i \in \{1, \ldots, (T/\text{time step})\} \text{ and } n \in \{1, \ldots, N\} \tag{2}$$

where i is a time step of the simulation, $d_{new(n,i)}$ and $d_{initial(n,i)}$ are new and initial demand for a simulated leak at node n, constant c is the emitter coefficient, $p_{(n,i)}$ is the pressure for a simulated leak at node n at time step i, d_c is the demand pattern coefficient, γ is the pressure exponent (0.5), t_m is the initial time step of leakage at node n, T is the total simulation time, and N is the total nodes in a network.

The demand pattern coefficient is chosen to resolve the issue that cannot add the emitter coefficient to EPANET for the hours of 6 am to 12 am.

It should be noted that a one-hour simulation of the leak-free (control) state model will generate a $T \times N$ matrix with nodal pressure values for one hour. In leakage-affected nodes, the pressure is different than expected from the leak-free state, and the same thing happens when a node receives added demand. Accordingly, pressure sensors at the leakage nodes produce various pressure profiles according to when and where leakage begins. Yet, it is necessary to determine nodal pressure to calculate an emitter coefficient for any given node that leaks at the specified rate. Furthermore, the ideal way to simulate leakage in a network is with small time steps, which can be a time-consuming process.

4.2 Candidate locations for sensor deployment

When a node leaks, it causes a pressure drop, which in turn creates a gradient in adjacent pipes, drawing flow into the node. As the flow drops in the main pipes adjacent to the leaky node, the secondary pipes will also be affected, and water may be allowed to enter the leaky node from other pipes in the loop [33]. Furthermore, the network flow rate varies with time compared to the leak-free state, and components of the network attempt to pay for the pressure drop. Subsequently, since the flow rate fluctuates in different pipes, it would be inappropriate to recommend a large number of pipes/nodes as candidate locations to place sensors in each leakage scenario for a few time steps. As follows, it might be better to select candidates based on a threshold:

$$\begin{cases} M[i,j] = M[i,j] & \text{if } abs(M[i,j]) \geq \text{threshold} \\ M[i,j] = 0 & \text{otherwise} \end{cases}, \quad \forall i \in \{o, \ldots, (T/\text{time step})\},\ j \in \{1, \ldots, N\} \tag{3}$$

where $M[i,j]$ is an element in column j and row i of the differentiation matrix for flow and pressure resulting from leakage at the target point.

4.3 Candidate pipes for flow sensor deployment

According to Raei et al. [7], the candidate selection should meet two objectives: (1) reducing the number of pipes showing flow divergence compared to a leak-free state that can be selected as candidate flow sensor locations; and (2) finding the number of pipes consistently showing flow divergence over successive time steps. Yet, Raei et al. [7] recommend that a large network can be divided into homogeneous zones first, tolerance thresholds and candidate pipes determined, then sensors ranked in descending order based on zones that are common to all sensors. Rank incrementally decreases from (Z-1) to 0, where Z is the number of zones, (Z-1) represents a sensor enveloping only one zone, and 0 represents a sensor enveloping all zones. Sensors with (Z-1) and (Z-2) ranks are recommended for flow sensors to minimize sensor overlap and operational costs, that is, sharing one flow sensor between two zones at most. Furthermore, it seems that sensors with rank "(Z-2)" indicate that more than one zone could be considered for the leakage node, whereas pressure sensors can be examined to enhance the accuracy of leakage zone detection. They also noted that the number of required flow sensors directly corresponds to the tolerance threshold; as the threshold increases, the number of deployed flow sensors has a minor impact on detection accuracy.

4.4 Candidate nodes for pressure sensor deployment

There are several ways to find candidate pressure nodes. Each leak state can be compared to a leak-free state by computing a matrix of pressure residuals, representing the absolute pressure differences. Specifically, the pressure residual matrix is determined as follows:

$$M_S = \begin{bmatrix} p_{11} - \hat{p}_{11} & \cdots & p_{1j} - \hat{p}_{1j} \\ \vdots & \ddots & \vdots \\ p_{i1} - \hat{p}_{i1} & \cdots & p_{ij} - \hat{p}_{ij} \end{bmatrix}, \ \forall i \text{ and } S \in \{1, \dots, N\}, \ j \in \{1, \dots, (T/\text{time step})\} \tag{4}$$

where M_S is the pressure residual matrix between the measured pressure (p_{ij}, leak state) and the estimated pressure ($\hat{p}_{ij}$, leak-free state), induced by leak scenario S at all nodes (N) (j, matrix columns) during the simulation time steps (i, matrix rows).

It should be noted that for each column in M_S, the lowest and highest values get ranks of $(N-1)$ and 0, respectively. When the maximum pressure residual for a node is below the error threshold, it is not considered a candidate node for detecting leakage and receives a rank of 0. By running all leak scenarios (S) for each simulation time step (ts), each node receives $(N-1)$ candidacy value, which is then summed as a node score (R_S^{ts}). The node score ranges from 0 to $(N-1)^2$, where the best score is $(N-1)^2$, and this is the location where a pressure sensor can detect leaks at other nodes. In each simulation time step, the top-ranking nodes are candidates for pressure sensors deployment. A node is scored as follows:

$$R = \sum \left[R_1^{ts}, \dots, R_S^{ts}\right], \ \forall S \in \{1, \dots, N\} \tag{5}$$

5. Objective functions

5.1 Minimizing the number of sensors

Decision variables in the optimization process can be defined as a binary vector ($\vec{d}$), which has the size of the number of nodes in WDS, and values of 1 and 0 for deployment and nondeployment of sensors at each node. The summation of decision vector elements can therefore be functionalized to minimize the number of sensors, as follows:

$$\begin{gathered} \text{Minimize}: \ F = \sum_{\forall i} d_i \\ \text{Constraint}: \ m \le F \le m_s \le N \end{gathered} \tag{6}$$

where m is the minimum number of required sensors and m_s is the maximum number of sensors.

5.2 Minimizing the leak detection time

Leak detection in the first hour of event occurrence has the least destructive effects in terms of water loss, infrastructure damage, and contamination of pipelines. Therefore, the following function can be used as the criterion for sensor placement to minimize leak detection time:

$$\text{Minimize}: \ F = \frac{\sum_i^N \sum_t^T t_{\text{det},i,t}(\overline{e}_s)}{N \times T} \tag{7}$$

where T is the total time steps in which a leakage starts at the ith node, N is the total number of nodes, $t_{\text{det},i,t}(\overline{e}_s)$ is the time when a leakage is detected by one of the deployed sensors, and $\overline{e}_s$ is detection time when one of the sensors measures a pressure residual larger than an error threshold.

5.3 Maximizing the value of information (VOI)

Grayson [53] introduced the value of information (VOI) theory, which has since been adopted by numerous researchers for designing various monitoring networks, whereas Khorshidi et al. [36] applied it for leakage monitoring. VOI theory states that each possible sensor could have different states and messages with a certain prior probability. The message received about the system's performance will either be valuable or detrimental to the decision-maker's opinion depending on the state of the system or the sensor. Therefore, updating the opinion is required to achieve a better decision at the node. In order to integrate the prior probability and the likelihood of an event, the Bayes theorem can be used.

$$P(s|\,m) = \frac{P(m|\,s)P(s)}{P(m)} \tag{8}$$

where $P(s\,|\,m)$ is a new opinion after receiving message m, $P(m\,|\,s)$ is the conditional probability of receiving message m when the system state is s, $P(s)$ is the prior probability of state s, and $P(m)$ is the probability of receiving message m.

Khorshidi et al. [36] interpreted that the state s of each node can be viewed as detection time. Then, the detection time matrix can be used to determine the vector $P(s)$ for all nodes. In this case, if a sensor is deployed at node i, and a leak event occurs at node j, message m represents the event warning received by the decision-maker. When a warning message w is sent with a different detection delay, the associated costs of node j in detection state s can be determined as $C(w,s)$. Additionally, the efficacy of warning messages in terms of whether a message is received (E_m) or not (E_n) can be determined using Eqs. (9) and (10). Based on a received message m, the value of the selected warning is determined by Eq. (11). Consequently, the VOI of node i for node state determination is determined by Eq. (12).

$$E_m = \sum_{DS} C(w,s)P(s|\,m) \tag{9}$$

$$E_n = \sum_{DS} C(w,s)P(s) \tag{10}$$

$$V_m = E_m - E_n \tag{11}$$

$$VOI_i(j) = \sum_{MT} P(m)\left[\max_w (E_m) - \max_w (E_n)\right] \tag{12}$$

where DS is the total detection states and MT is the total number of messages. For node i, calculating $VOI_i(j)$ provides a numerical VOI curve for all other nodes. The VOI for node i in a system state readout is the area under the VOI_i curve. It is noteworthy that for a set of nodes that is used to detect the state of node j, the union of their VOI curves will represent the VOI curve of all the nodes in the set.

The VOI of nodes i and j are equal to $\max\{VOI_i(gn), VOI_j(gn)\}$, where gn is a given node in the WDS. As previously noted, a set of nodes with the maximum area under their union VOI curves provides the most information to a decision-maker regarding the state of the WDS. The summation of the maximum of all selected rows of the VOI matrix for all nodes can be functionalized to maximize the normalized VOI as follows:

$$\text{Maximize}: \ F = \frac{1}{\max\{VOI\}} \sum_{\forall j} \max_i \{d_i \times VOI_i(j)\} \tag{13}$$

There are the same number of decision variables ($\vec{d}$) as there are nodes in WDSs, where a population of decision vectors is used to determine optimal nodes for deploying sensors in WDSs.

5.4 Minimizing the transinformation entropy (TE)

In information theory (entropy), which was introduced by Shannon [54], mutual information (or transinformation) quantifies how much information is obtained about one random variable by observing another. The concept has been applied to sensor deployment in WDS by [36,37]. As a result, to obtain the information transmitted between each pair of sensors at nodes x and y in the discrete form, the following equation can be used:

$$TE(x, y) = -\sum_{i=1}^{a} \sum_{j=1}^{b} P(x_i, y_j) \ \ln \left[\frac{P(x_i, y_j)}{P(x_i)P(y_j)} \right] \tag{14}$$

where a and b are the total number of pressure sensors in the x- and y-axis, i and j are sensor counters in the x- and y-axis, x_i is the pressure residual of ith sensor in the x-axis, y_j is the pressure residual of jth sensor in the y-axis, $P(x_i)$ is the probability of x_i, $P(y_j)$ is the probability of y_j, and $P(x_i, y_j)$ is the joint probability of (x_i, y_j). Khorshidi et al. [36] defined x_i as the detection state of ith sensor in x-axis and y_j as the detection state of jth sensor in the y-axis. Also, like VOI, for a set of nodes that is used to detect the state of node j, the union of their TE curves will represent the TE curve of all the nodes in the set.

It is possible to deduce VOI and TE from any pair of nodes (i, j) located in a network, using the detection time and leakage matrices derived from simulations and storing them in two matrices. However, detection states and costs are arbitrary parameters; they may vary based on decision-makers' priorities and the condition of the WDSs. The TE of pair nodes (i, j) is the maximum of their specific $TE(i,j)$. As noted in VOI, the TE of the set of nodes is the union of their TE curves, that is, a set of nodes with the minimum area under their union TE curves provides less mutual information to a decision-maker regarding the state of the WDS. The summation of the minimum of all selected rows of the TE matrix for all nodes can be functionalized to minimize normalized TE, as follows:

$$\text{Minimize}: \ F = \frac{1}{\max\{TE\}} \sum_{\forall j \neq i} \max_i \{d_i \times d_j \times TE(i,j)\} \tag{15}$$

There are the same number of decision variables ($\vec{d}$) as there are nodes in WDSs, where a population of decision vectors is used to determine optimal nodes for deploying sensors in WDSs.

5.5 Minimizing the extra information derived from pressure sensors

For each pair of sensors, the derived transinformation can be plotted against the distance between them, and an exponential *T-D* curve can be fitted to these data (Fig. 1), indicating that transinformation diminishes with distance until it is minimized at the optimal distance.

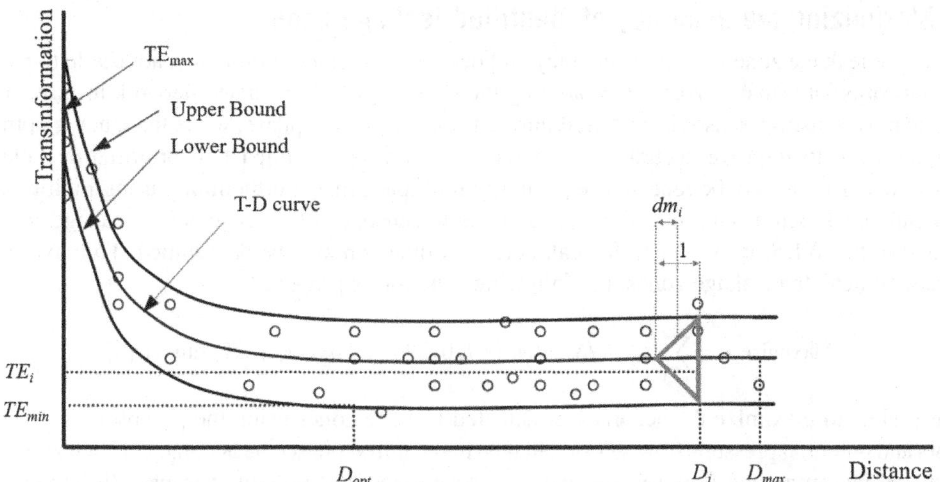

FIG. 1

A typical scheme of transinformation and distance (*T-D*) curve.

Adopted from F. Masoumi, R. Kerachian, Optimal redesign of groundwater quality monitoring networks: a case study, Environ. Monitor. Assess. 161 (1) (2010) 247–257.

$$\text{Minimize}: F = a_1 \sum_{i=1}^{cn} \left\{ d_i \frac{1 - dm_i}{cn} \right\} + a_2 \sum_{i=1}^{cn} \begin{Bmatrix} d_i \dfrac{TE_i - TE_{\min}}{TE_{\max} - TE_{\min}} / cn & D_i < D_{opt} \\ d_i / cn & D_i \geq D_{opt} \end{Bmatrix} \tag{16}$$

where a is the coefficient ($a_1 + a_2 = 1$), dm is the degree of membership for the ith pressure sensor considering the uncertainty in T-D curves, TE_i is the average of TE using D_i for pressure sensor i for all leakage scenarios, $TE_{\min}$ is the minimum average value of TE for all leakage scenarios, and $TE_{\max}$ is the maximum average value of TE for all leakage scenarios.

5.6 Maximizing the sensor coverage

There should be sufficient coverage of the sensors. In addition, the sensors deployment should be identified based on nodal pressure, hydraulic system, and flow direction [50], which Taravatrooy et al. [37] formulated as follows:

$$\text{Maximize}: F = \sum_{i=1}^{cn} \begin{Bmatrix} d_i \times \dfrac{D_{\max} - D_i}{D_{\max} - D_{opt}} / cn & D_i > D_{opt} \\ d_i / cn & D_i \leq D_{opt} \end{Bmatrix} \tag{17}$$

where $\vec{d}$ is a binary vector of decision variable, which is set to values of 1 and 0 for deployment and nondeployment of sensors at each node, cn is the number of candidate sensors, $D_{\max}$ is the maximum distance between each pair of pressure sensors (meters), D_k is the distance from ith pressure sensor to the closest pressure sensor (meters), and D_{opt} is the optimal distance (meter) between all pressure sensors using the $TE_{\min}$ values.

5.7 Maximizing the accuracy of identified leakage zone

To improve leakage zone detection accuracy and decrease calculation time, it is advisable to partition large networks into smaller zones. Considering the sensitivity of the studied network to flow or pressure, the most sensitive sensor is preferred during the optimization phase, while the other is optimized at a later stage to improve accuracy. However, to minimize the impact of prioritizing preference sensors, this process may be repeated sequentially multiple times. Furthermore, using multiobjective optimization, a Pareto front according to the number of sensors and the accuracy of leakage zones can be found in the WDS, from which the leak detection time can also be determined. To maximize the accuracy of detected leakage zones, the following function is proposed.

$$\text{Maximize } F = \sum_{n=1}^{N} d_n > o/N, \quad d_n \in \{-1, 0.5, 1\} \text{ and } o \in \{1, \ldots, (T/\text{time step})\} \tag{18}$$

where F aims to maximize the accuracy of detected leakage zones using the proposed flow/pressure sensors and potential pressure/flow sensors; that is, in the initial phase, the accuracy of each zone can be maximized by optimizing flow sensors and considering constant pressure sensors. By implementing this function in the next phase, the accuracy can be maximized by achieving optimized flow sensors and minimizing pressure sensors; N is the number of possible leaking nodes, n is the leaking node, d_n is the detection value (1 means that one zone that is detected properly, 0.5 means that two zones are detected and only one is correct, and -1 means that no zone is identified properly or more than two zones are detected for a node). Eq. (6) was used to minimize sensors.

It should be noted that a number of flow sensors (k_f) from possible flow sensors (n_{sf}) is selected during optimization. For each leakage scenario, the possibility of leak detection using the selected sensors is explored, and a matrix of size (T/time step) $\times K_f$ is mapped. The optimization algorithm seeks flow divergence that exceeds the tolerance threshold within determined time steps, in which flow sensors that meet this condition are selected. This step also selects all possible pressure sensors whose pressure divergence exceeds the threshold. Now that we have optimized flow sensor deployments and all possible pressure sensors, the corresponding zones for the flow sensors are specified. After that, an $N \times Z$ matrix is created using N nodes and Z zones. The rows of this matrix represent the frequency of detecting each zone using the combination of optimized flow sensors. The detection probability for each zone is calculated by dividing the total number of times that each zone is detected by the sum of the elements in each row. Rows with detection probabilities greater than 80% are selected. Depending on the leakage scenario, there could potentially be more than one zone selected for each node. Furthermore, pressure sensors are optimized to reduce the number of leakage zones detected per leakage node and to improve leakage zone detection accuracy. Afterward, the leakage zones detected by the flow sensors that coincide with those detected by the pressure sensors are selected; otherwise, flow sensors will determine the leakage zone. The detection zones are now compared to the defined zones for each node, based on the values of d_n. Finally, the sum of all detection values more than zero is divided by N.

6. Computational experiments

In this section, we compare the results of various methods applied to the same case study proposed by a collaborative team of researchers [7,33,36,37]. These proposed methodologies have been evaluated on the virtual city of C-Town, one of the most popular case studies, which is more complex than a typical

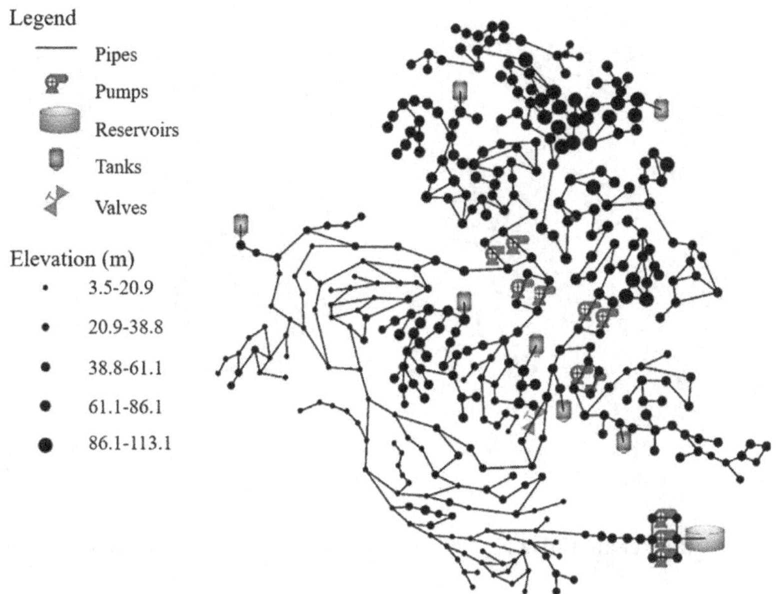

FIG. 2

C-Town water distribution system and its characteristics.

Adopted from M.S. Khorshidi, M.R. Nikoo, N. Taravatrooy, M. Sadegh, M. Al-Wardy, G.A. Al-Rawas, Pressure sensor placement in water distribution networks for leak detection using a hybrid information-entropy approach, Inform. Sci. 516 (2020) 56–71.

WDS. In this network, there are 388 nodes in five different elevations, 1 reservoir, 11 pumps, and 7 water tanks (Fig. 2), which are simulated by EPANET hydraulic simulation [52]. Similar assumptions are made throughout these studies, including: (1) leakage is modeled as an orifice; (2) leakage is modeled as nodes, not pipes; (3) nodal leakages are variable with pressure; (4) leakage is simulated using an emitter equation; (5) each leakage scenario only involves one leak at a time, and background leaks are not considered; (6) the nodes/pipes that are most sensitive to pressure or flow are selected as candidates for sensors; and (7) the simulation time is 97 hours, but four leakage intervals are defined by Khorshidi et al. [36], Taravatrooy et al. [37], and Raei et al. [33] to evaluate changes in all nodal demands and leaks, including 0–6, 6–12, 12–18, and 18–24. By introducing leakage intervals in hydraulic simulations, the computational effort is reduced, and a source of demand uncertainty that has been neglected is exposed.

These studies employed four multiobjective optimization methods for sensor deployment (see "Related Works"). These methods were implemented using NSGA-II, which considers a set of candidate pressure and flow sensors as decision variables. Fig. 3 shows the results of the sensor deployment layout using these methods.

As illustrated in Fig. 3, no sensor deployment is equally optimal under all algorithms. This difference is mostly due to the fact that each study considered a different aspect of sensor deployment. To compare these differences, Khorshidi et al. [36] found that a single pressure sensor can detect leaks in 50% of the scenarios, but five or more sensors would detect leaks in more than 98% of scenarios. They compared the values of VOI for the Pareto optimal layout of ten sensors proposed by Raei et al. [33] with similar ones on their own method and found that the VOI proposed by Raei et al. [33] is smaller.

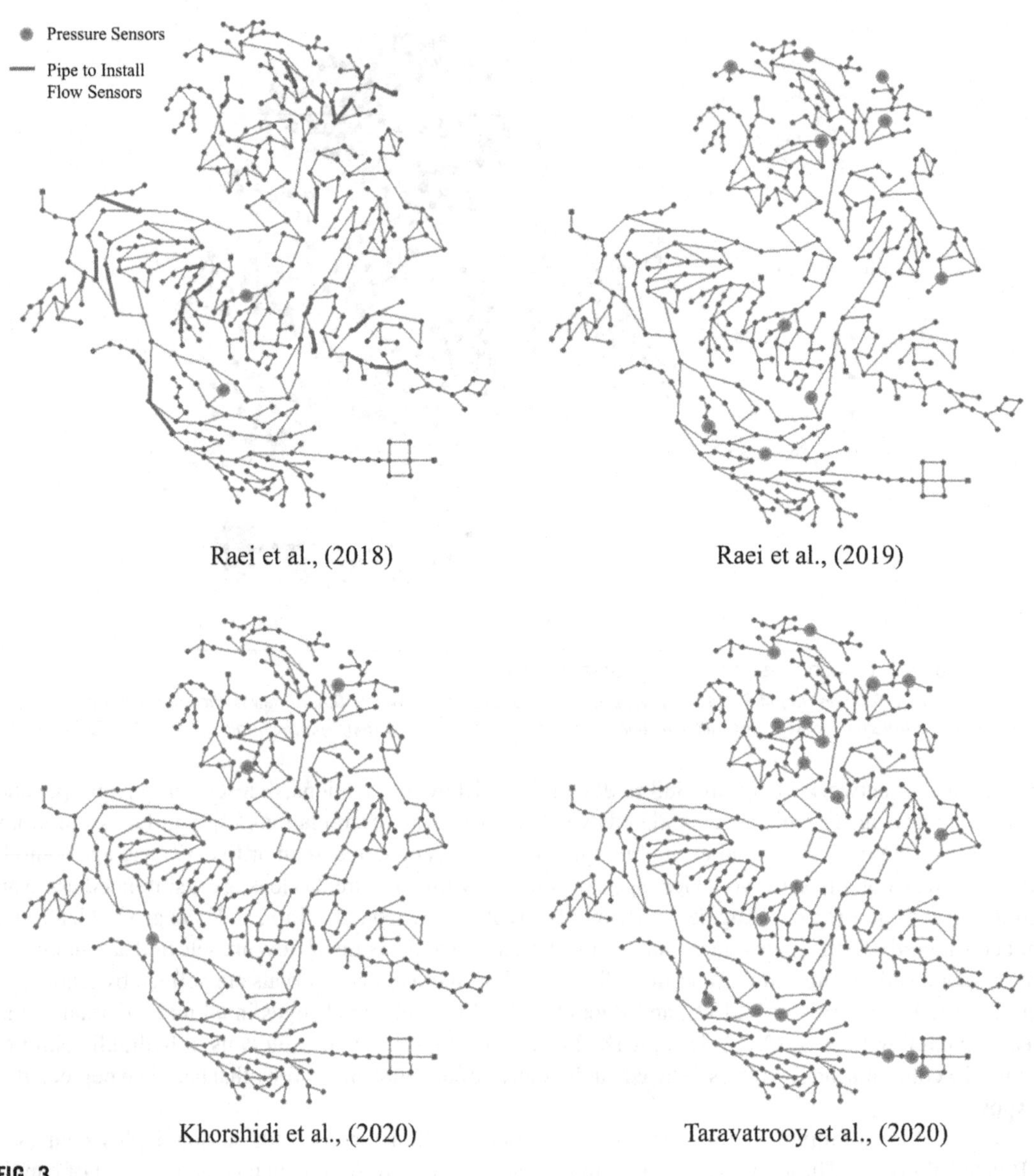

FIG. 3

The optimal sensor deployment layout in C-Town derived by Raei et al. [7,33], Khorshidi et al. [36], and Tarsavatrooy et al. [37].

This suggests that the layout of the sensors they proposed provides less information for decision-maker than the layout of the similar sensors by their own method. The proposed layout by Raei et al. [33] resulted in TE at rates equal to or greater than those of Khorshidi et al. [36], indicating that sensor deployment would yield more redundant information to the decision-maker. Furthermore, the probability of detecting leakage scenarios provided by Raei et al. [33] is similar to those provided by Khorshidi et al. [36], but the average detection time and the average volume of water leaked are significantly different. According to Khorshidi et al. [36], the average detection time is between 1206 and 1237 minutes, while Raei et al. [33] reported that it is between 1221 and 1578 minutes. Subsequently, Khorshidi et al. [36] estimated the volume of the leaked water to be between 93.85 and 96.14 m^3, whereas Raei et al. [33] estimated it to be between 94.95 and 121.75 m^3. However, Khorshidi et al. [36] found that deploying more than three sensors helps detect leak events precisely and promptly. As shown in Fig. 3, the sensor layout by Raei et al. [33] covers the network well, but the layout proposed by Khorshidi et al. [36] may provide more effective and economic information to decision-makers with fewer sensors. As such, it is simply not feasible to decide the coverage of a network unilaterally.

In the instance of sensor coverage, Taravatrooy et al. [37] examined a study (see Fig. 3) in which the authors asserted that the sensors are sufficient for covering the network under uncertainty so that extra information is minimized as pressure sensors are reduced. This study, however, divides the candidate nodes into three clusters with a k-means clustering algorithm using homogeneity and calculates the layout of each cluster simultaneously, which may affect the results. In this study, the leak detection time has not been investigated, and therefore it cannot be compared with the two studies by Raei et al. [33] and Khorshidi et al. [36]. Still, it seems that the leak detection time is acceptable due to the way the sensors are covered. However, the number of pressure sensors used in this study is greater than in other studies, which makes it unlikely that they are economically feasible. There seems to be a need to carefully examine the methods proposed by Taravatrooy et al. [37] and Khorshidi et al. [36], as it has already been argued that the method proposed by Khorshidi et al. [36] produces better results than Raei et al. [33], whereas Khorshidi et al. [36] have significantly fewer sensors. It is noteworthy to mention that Raei et al. [33] and Taravatrooy et al. [37] discussed the uncertainty in the analysis of error thresholds for pressure sensors that could affect the precision of leak detection, in which triangle fuzzy memberships (with a fuzzy membership degree of 1 were utilized to resolve this uncertainty. The results showed that detection times increased as the sensors' accuracy decreased, and in the C-Town case, sensor uncertainties did not seem to have a significant impact on sensor deployment.

In comparison to pressure sensors layout, flow sensors layout must be examined since a network may be more sensitive to flow variations than the pressure, as suggested by Raei et al. [7] in two virtual cities: Mesopolis and C-Town (this section focuses on C-Town). They first used multiobjective optimization to determine the optimal deployment of flow sensors, and then determined the optimal placement of pressure sensors using the optimal layout of flow sensors. To reduce search space and computation time, they dived the network into five zones with similar sensor locations using a k-means clustering algorithm before the optimal deployment. They noted that as the number of zones increases, the number of sensors required for leakage zone detection rises, but as more sensors are added to the network, the accuracy of leakage zone detection decreases, which may be a result of network circumstances, whereby five zones are chosen. Furthermore, the leakage detection time will increase with an increase in zones.

As shown in Fig. 3, in the layout proposed by Raei et al. [7], flow sensors are deployed more frequently than pressure sensors, in which the effects of flow variations may be shown in deciding how sensors are deployed. According to this study, the layout of optimized flow and pressure sensors in C-Town can detect more than 80% of leaks. However, it does not mention the leakage detection time in C-Town. Leakage detection time results are presented for Mesopolis (a large network), however, there are similar zones with 1588 leakage scenarios for 1588 nodes. In this case, a leak begins at hour 96 and lasts 360 hours in hourly time steps. They detect leaks in Mesopolis for four detection states of 0–14, 14–58, 58–119, and 119–219 hours. It is notable that in these studies [33,36,37], C-Town consists of 388 nodes and four leakage intervals that involve 1552 ($388 \times 4 = 1552$) leakage scenarios. Subsequently, an economic evaluation of this sensor deployment should be performed in comparison to other methods provided to determine its effectiveness and implementation.

7. Conclusions

Nowadays, water utilities are increasingly utilizing pressure and flow sensors in their networks to detect and localize leak events in near real-time due to the importance of preventing leaks as well as the availability of sensors. However, due to budget constraints, only a limited number of sensors can be used, and therefore it is very important to use the fewest sensors that can provide as much network information as possible in the shortest time frame. Some locations within a network are likely to be more capable of measuring pressure and flow than others. Subsequently, various methods have been proposed for assessing the optimal deployment of several sensors to detect an event wherever it occurs and for determining the approximate location of the event based on this information. In light of reviewing relevant studies, it appears that solving this complex problem will require further research. We can draw several conclusions by taking these study findings into account:

- The hydraulic models for optimal sensor deployment are valuable tools and simulating near realistic models may provide a better understanding of the problem. An optimal method is also needed if no hydraulic model exists or if one exists but has been poorly calibrated, which is the case in most developing countries.
- Since the network complexity makes solving the optimal deployment of sensors more complex, it is necessary to reduce the search space in the optimization problem to speed up computing by including another type of multiobjective algorithm, hybrid optimization algorithms, clustering algorithms, data mining methods, and so on.
- There are studies that use residuals or a sensitivity matrix in the presence or absence of leaks, which may have uncertain performances in the case of simultaneous leaks, or may have different outcomes for each leakage scenario. It seems using artificial intelligence methods for network training under different scenarios can help in better results.
- There are uncertainties in determining the optimal placement of sensors using existing methods, which may appear differently depending on the different scenarios. Examining the uncertainty in both the model and the measurements is important.
- Quantitative studies have combined the simultaneous use of pressure and flow sensors, which may have variable deployment depending upon the network model's sensitivity to these two sensors. Further research is needed to capture the concurrent effect of these sensors; using transinformation entropy might reduce uncertainty in leak detection calculations and improve method performance.
- The studies assume that sensors collect data without malfunction, yet the sensors can be out of order for a while. As such, more research is needed to assess the risk of sensor malfunction.

- Sensor deployment depends on multiple criteria, which makes this a multiobjective problem that should be optimally satisfied simultaneously. The combination of criteria can lead to a trade-off of objectives and provide more options for decision-making. In addition, utility priorities can be considered when choosing sensor layouts with the multicriteria decision-making model.

References

[1] M. McLennan, The Global Risks Report 2021, World Economic Forum, 2021.

[2] M. Salehi, Global water shortage and potable water safety; today's concern and tomorrow's crisis, Environ. Int. 158 (2022), 106936, https://doi.org/10.1016/j.envint.2021.106936.

[3] T. Al-Washali, S. Sharma, R. Lupoja, A.N. Fadhl, M. Haidera, M. Kennedy, Assessment of water losses in distribution networks: methods, applications, uncertainties, and implications in intermittent supply, Resour. Conserv. Recycl. 152 (2020), 104515, https://doi.org/10.1016/j.resconrec.2019.104515.

[4] G. Moser, S.G. Paal, I.F.C. Smith, Performance comparison of reduced models for leak detection in water distribution networks, Adv. Eng. Inform. 29 (3) (2015) 714–726, https://doi.org/10.1016/j.aei.2015.07.003.

[5] W. Gong, M.A. Suresh, L. Smith, A. Ostfeld, R. Stoleru, A. Rasekh, M.K. Banks, Mobile sensor networks for optimal leak and backflow detection and localization in municipal water networks, Environ. Model. Softw. 80 (2016) 306–321, https://doi.org/10.1016/j.envsoft.2016.02.001.

[6] X. Wang, M.S. Ghidaoui, Pipeline leak detection using the matched-field processing method, J. Hydraul. Eng. 144 (6) (2018) 04018030, https://doi.org/10.1061/(ASCE)HY.1943-7900.0001476.

[7] E. Raei, M.R. Nikoo, S. Pourshahabi, M. Sadegh, Optimal joint deployment of flow and pressure sensors for leak identification in water distribution networks, Urban Water J. 15 (9) (2018) 837–846, https://doi.org/10.1080/1573062X.2018.1561915.

[8] H. Haider, I.S. Al-Salamah, Y.M. Ghazaw, R.H. Abdel-Maguid, M. Shafiquzzaman, A.R. Ghumman, Framework to establish economic level of leakage for intermittent water supplies in arid environments, J. Water Resour. Plan. Manag. 145 (2) (2019) 05018018, https://doi.org/10.1061/(ASCE)WR.1943-5452.0001027.

[9] M.S. Islam, M.S. Babel, Economic analysis of leakage in the Bangkok water distribution system, J. Water Resour. Plan. Manag. 139 (2) (2013) 209–216, https://doi.org/10.1061/(ASCE)WR.1943-5452.0000235.

[10] I. Moslehi, M.J. Ghazizadeh, E. Yousefi-Khoshqalb, An economic valuation model for alternative pressure management schemes in water distribution networks, Util. Policy 67 (2020), 101129, https://doi.org/10.1016/j.jup.2020.101129.

[11] I. Moslehi, M. Jalili-Ghazizadeh, E. Yousefi-Khoshqalb, Developing a framework for leakage target setting in water distribution networks from an economic perspective, Struct. Infrastruct. Eng. 17 (6) (2021) 821–837, https://doi.org/10.1080/15732479.2020.1777568.

[12] B. Farley, S.R. Mounce, J.B. Boxall, Field testing of an optimal sensor placement methodology for event detection in an urban water distribution network, Urban Water J. 7 (6) (2010) 345–356, https://doi.org/10.1080/1573062X.2010.526230.

[13] D. Misiunas, J. Vítkovský, G. Olsson, M. Lambert, A. Simpson, Failure monitoring in water distribution networks, Water Sci. Technol. 53 (4–5) (2006) 503–511, https://doi.org/10.2166/wst.2006.154.

[14] S.R. Mounce, J. Boxall, J. Machell, Development and verification of an online artificial intelligence system for burst detection in water distribution systems, ASCE J. Water Resour. Plan. Manag. 136 (3) (2010) 309–318, https://doi.org/10.1061/(ASCE)WR.1943-5452.0000030.

[15] M.V. Casillas, V. Puig, L.E. Garza-Castanon, A. Rosich, Optimal sensor placement for leak location in water distribution networks using genetic algorithms, Sensors 13 (11) (2013) 14984–15005, https://doi.org/10.1016/j.ifacol.2015.09.530.

[16] R. Puust, Z. Kapelan, D.A. Savic, T. Koppel, A review of methods for leakage management in pipe networks, Urban Water J. 7 (1) (2010) 25–45, https://doi.org/10.1080/15730621003610878.

[17] M. Farley, S. Trow, Losses in Water Distribution Networks—A Practitioner's Guide to Assessment, Monitoring and Control, IWA Publishing, 2003.
[18] UK Water Industry, The Managing Leakage Series of Reports, Report A: Summary Report, 1994.
[19] R. Janke, R. Murray, J. Uber, T. Taxon, Comparison of physical sampling and real-time monitoring strategies for designing a contamination warning system in a drinking water distribution system, J. Water Resour. Plan. Manag. 132 (4) (2006) 310–313, https://doi.org/10.1061/(ASCE)0733-9496(2006)132:4(310).
[20] R.S. Pudar, J.A. Liggett, Leaks in pipe networks, J. Hydraul. Eng. 118 (7) (1992) 1031–1046.
[21] A.F. Colombo, P. Lee, B.W. Karney, A selective literature review of transient-based leak detection methods, J. Hydro-Environ. Res. 2 (4) (2009) 212–227, https://doi.org/10.1016/j.jher.2009.02.003.
[22] B. Farley, J.B. Boxall, S.R. Mounce, Optimal locations of pressure meter for burst detection, in: Proc., 10th Water Distribution System Analysis Symp., ASCE, Reston, VA, 2008, pp. 1–11.
[23] C.A. Bush, J.G. Uber, Sampling design methods for water distribution model calibration, J. Water Resour. Plan. Manag. 124 (6) (1998) 334–344, https://doi.org/10.1061/(ASCE)0733-9496(1998)124:6(334).
[24] B. Farley, S.R. Mounce, J.B. Boxall, Development and field validation of a burst localization methodology, J. Water Resour. Plan. Manag. 139 (6) (2013) 604–613, https://doi.org/10.1061/(asce)wr.1943-5452.0000290.
[25] D.E. Goldberg, Genetic Algorithms in Search, Optimization, and Machine Learning, Addison-Wesley, Reading, MA, 1989.
[26] R. Pérez, V. Puig, J. Pascual, A. Peralta, E. Landeros, L. Jordanas, Pressure sensor distribution for leak detection in Barcelona water distribution network, Water Sci. Technol. Water Supply 9 (6) (2009) 715–721, https://doi.org/10.2166/ws.2009.372.
[27] R. Pérez, V. Puig, J. Pascual, J. Quevedo, E. Landeros, A. Peralta, Leakage isolation using pressure sensitivity analysis in water distribution networks: application to the Barcelona case study, IFAC Proc. Vol. 43 (8) (2010) 578–584, https://doi.org/10.3182/20100712-3-FR-2020.00094.
[28] Z.Y. Wu, Y. Song, Optimizing pressure logger placement for leakage detection and model calibration, in: Proceedings of the 14th Water Distribution Systems Analysis Conference, Adelaide, Australia, 2012, September, pp. 24–27.
[29] M.V. Casillas, L.E. Garza-Castañón, V. Puig, Optimal sensor placement for leak location in water distribution networks using evolutionary algorithms, Water 7 (11) (2015) 6496–6515, https://doi.org/10.3390/w7116496.
[30] D. Steffelbauer, D. Fuchs-Hanusch, Efficient sensor placement for leak localization considering uncertainties, Water Resour. Manag. 30 (14) (2016) 5517–5533, https://doi.org/10.1007/s11269-016-1504-6.
[31] S. Boatwright, M. Romano, S. Mounce, K. Woodward, J. Boxall, in: Optimal sensor placement and leak/burst localisation in a water distribution system using spatially-constrained inverse-distance weighted interpolation, Proceedings of the 13th International Conference on Hydroinformatics, Palermo, Italy, 2018.
[32] X. Xie, Q. Zhou, D. Hou, H. Zhang, Compressed sensing based optimal sensor placement for leak localization in water distribution networks, J. Hydroinform. 20 (6) (2018) 1286–1295, https://doi.org/10.2166/hydro.2017.145.
[33] E. Raei, M.E. Shafiee, M.R. Nikoo, E. Berglund, Placing an ensemble of pressure sensors for leak detection in water distribution networks under measurement uncertainty, J. Hydroinform. 21 (2) (2019) 223–239, https://doi.org/10.2166/hydro.2018.032.
[34] R. Rayaroth, G. Sivaradje, Grey wolf optimization based sensor placement for leakage detection in water distribution system, Int. J. Recent Technol. Eng. 7 (5) (2019) 180–188.
[35] M. Righetti, C.M.G. Bort, M. Bottazzi, A. Menapace, A. Zanfei, Optimal selection and monitoring of nodes aimed at supporting leakages identification in WDS, Water 11 (3) (2019) 629, https://doi.org/10.3390/w11030629.
[36] M.S. Khorshidi, M.R. Nikoo, N. Taravatrooy, M. Sadegh, M. Al-Wardy, G.A. Al-Rawas, Pressure sensor placement in water distribution networks for leak detection using a hybrid information-entropy approach, Inform. Sci. 516 (2020) 56–71, https://doi.org/10.1016/j.ins.2019.12.043.

[37] N. Taravatrooy, M.R. Nikoo, S. Hobbi, M. Sadegh, A. Izady, A novel hybrid entropy-clustering approach for optimal placement of pressure sensors for leakage detection in water distribution systems under uncertainty, Urban Water J. 17 (3) (2020) 185–198, https://doi.org/10.1080/1573062X.2020.1758162.

[38] N. Metropolis, S. Ulam, The Monte Carlo method, J. Am. Stat. Assoc. 44 (247) (1949) 335–341, https://doi.org/10.1080/01621459.1949.10483310.

[39] M.V. Casillas, L.E. Garza-Castañón, V. Puig, A. Vargas-Martinez, Leak signature space: an original representation for robust leak location in water distribution networks, Water 7 (3) (2015) 1129–1148, https://doi.org/10.3390/w7031129.

[40] J. Kennedy, R. Eberhart, Particle swarm optimization, in: Proceedings of ICNN'95-International Conference on Neural Networks, vol. 4, IEEE, 1995, November, pp. 1942–1948.

[41] Q. Wang, D.A. Savić, Z. Kapelan, GALAXY: a new hybrid MOEA for the optimal design of water distribution systems, Water Resour. Res. 53 (3) (2017) 1997–2015, https://doi.org/10.1002/2016WR019854.

[42] M. Romano, Review of techniques for optimal placement of pressure and flow sensors for leak/burst detection and localisation in water distribution systems, in: ICT for Smart Water Systems: Measurements and Data Science, Springer, Cham, 2019, pp. 27–63.

[43] O. Giustolisi, D. Savic, Z. Kapelan, Pressure-driven demand and leakage simulation for water distribution networks, J. Hydraul. Eng. 134 (2008) 626–635, https://doi.org/10.1061/(ASCE)0733-9429(2008)134:5(626).

[44] W.-J. Zhang, X.-F. Xie, DEPSO: hybrid particle swarm with differential evolution operator, in: Proceedings of the 2003 IEEE International Conference on Systems, Man and Cybernetics, Washington, DC, USA, 5–8 October 2003, vol. 4, 2003, pp. 3816–3821.

[45] D. Karaboga, B. Basturk, A powerful and efficient algorithm for numerical function optimization: artificial bee colony (ABC) algorithm, J. Global Optim. 39 (3) (2007) 459–471, https://doi.org/10.1007/s10898-007-9149-x.

[46] C. Ozturk, E. Hancer, D. Karaboga, A novel binary artificial bee colony algorithm based on genetic operators, Inform. Sci. 297 (2015) 154–170, https://doi.org/10.1016/j.ins.2014.10.060.

[47] S. Mirjalili, S.M. Mirjalili, A. Lewis, Grey wolf optimizer, Adv. Eng. Softw. 69 (2014) 46–61, https://doi.org/10.1016/j.advengsoft.2013.12.007.

[48] K. Deb, A. Pratap, S. Agarwal, T.A.M.T. Meyarivan, A fast and elitist multiobjective genetic algorithm: NSGA-II, IEEE Trans. Evol. Comput. 6 (2) (2002) 182–197, https://doi.org/10.1109/4235.996017.

[49] E. Forconi, Z. Kapelan, M. Ferrante, H. Mahmoud, C. Capponi, Risk-based sensor placement methods for burst/leak detection in water distribution systems, Water Sci. Technol.: Water Supply 17 (6) (2017) 1663–1672, https://doi.org/10.2166/ws.2017.069.

[50] B. Greyvenstein, J.E. Van Zyl, An experimental investigation into the pressure-leakage relationship of some failed water pipes, J. Water Supply: Res. Technol.—AQUA 56 (2) (2007) 117–124, https://doi.org/10.2166/aqua.2007.065.

[51] M.E. Shafiee, A. Berglund, E.Z. Berglund, E.D. Brill Jr., G. Mahinthakumar, 2016. Parallel evolutionary algorithm for designing water distribution networks to minimize background leakage. J. Water Resour. Plan. Manag., 142(5), C4015007. https://doi.org/10.1061/(ASCE)WR.1943-5452.0000601.

[52] L.A. Rossman, EPANET 2: Users Manual, Water Supply and Water Resources Division, National Risk Management Research Laboratory, Cincinnati, OH, 2000.

[53] C.J. Grayson, Decisions Under Uncertainty: Drilling Decisions by Oil and Gas Operators, Ayer, Harvard Business School, 1960.

[54] C.E. Shannon, A mathematical theory of communication, II and I, Bell Syst. Tech. J. 27 (1948) (1948) 379–423, https://doi.org/10.1002/j.1538-7305.1948.tb01338.x.

CHAPTER

15 Metaheuristic-based automatic generation controller in interconnected power systems with renewable energy sources

Özay Can[a], Hasan Eroğlu[b], and Ali Öztürk[c]

[a]*Department of Electronics and Automation, Technical Sciences Vocational School, Recep Tayyip Erdogan University, Rize, Turkey,* [b]*Department of Electrical-Electronics Engineering, Faculty of Engineering and Architecture, Recep Tayyip Erdogan University, Rize, Turkey,* [c]*Department of Electrical-Electronics Engineering, Faculty of Engineering, Duzce University, Duzce, Turkey*

1. Introduction

In networks where alternative electricity is used, the balance of the network frequency is very important in terms of compliance with power quality standards. However, it becomes difficult to keep the frequency constant as the network loads fluctuate in different conditions and time periods. In case of an unbalanced difference between the demanded power and the generated power, the system frequency differs from the desired value and these frequency fluctuations cause significant network problems. Therefore, a control process called Automatic Generation Control (AGC) or Load Frequency Control (LFC) is required to keep the system frequency and tie-line power variation within the specified values.

The integration of renewable energy sources (RESs) such as wind turbines (WTs) and photovoltaic (PV) panels into the conventional energy networks is increasing daily in line with environmental and sustainable energy targets. Although RESs are plentiful, accessible, cost competitive and environmentally friendly, their variability and stochastic power generation are disadvantages in terms of frequency control. Thus, it is also very difficult to keep the frequency constant in cases where there is a demand increase and/or RES generation in certain time periods on the network side. The fact that the network frequency is below or above the desired level causes power quality problems such as malfunctioning, shortening of the life of the devices fed from the network, and even system crashes.

Many studies on AGC have been carried out in the literature [1,2]. With the classic, inflexible AGC models, it has become very difficult to meet the system frequency requirements. To solve this fundamental problem, flexible AGC models such as Genetic Algorithms (GAs), Fuzzy Logic, Artificial Neural Networks (ANNs), and Particle Swarm Optimization (PSO) have been developed. The common aspect of these models is the fact that they use modern, intelligent, flexible control methods. On the

Comprehensive Metaheuristics. https://doi.org/10.1016/B978-0-323-91781-0.00015-6

other hand, the integration of renewable energy systems into the grid, load frequency control has gained a new dimension.

In recent years, different strategies with different tools have been developed to solve the frequency problem. Since the use of fixed parameter control methods in power systems does not give optimal performance of the system, a new fuzzy proportional–integral–derivative (PID) controller is introduced to AGC for interconnected multiarea power systems in the presence of external interferences in [3]. The scaling factors and modal parameters of the input and output membership functions in the PID controller are arranged using the Tribe-DE (TDE) algorithm. Conventional LFC design techniques can become inaccurate and unstable, as control signals transmitted from remote terminal units (RTUs) to the control center are complicated by the inherent delay of communication channels. Therefore, a PID controller is proposed by [4] using a stability boundary locus (SBL) approach for interval single-area power system. An event-triggered load frequency controller has been developed [5] for an isolated hybrid power system with wind, diesel, and battery storage to minimize the use of network resources, as the wind speed is intermittent and high sampling rate consumes more network bandwidth.

Heuristic optimization algorithms including Teaching-Learning-Based Optimization [6], Equilibrium Optimization Algorithm [7], Salp Swarm Algorithm [8], Bacterial Foraging Algorithm [9], Marine Predator Algorithm [10], Particle Swarm Optimization [11], Sine Cosine Algorithm [12], Modified Sine Cosine Algorithm [13], Manta Ray Foraging Optimization [14], Gray Wolf Optimization [15], Crow Search Algorithm [16], Firefly Algorithm [17], Moth Flame-Generalized Hopfield Neural Network [18], Sine adapted improved Whale Optimization Algorithm [19], Jaya Algorithm [20], Lightning Flash Algorithm [21], Grasshopper Optimization Algorithm [22], and Imperialist Competitive Algorithm [23] are frequently employed to obtain more efficient gain and control parameters for AGC in power systems.

In solving the AGC problem with classical PID controllers, the system has some disadvantages such as stability [24], peak deviation [25], and transient response [26]. Therefore, improved classical PID controllers need to be developed [3,27]. With the integration of RESs into conventional grids, new power systems have become more complex. Power systems containing different RESs are connected to each other by tie-lines. Many complex nonlinear models have been developed for modeling these large-scale power systems [28]. However, simplified models are generally used in AGC studies [29].

The most important motivation for this study is the search for a new and novel framework with improved parameters such as maximum overshoot (MO^+), minimum overshoot (MO^-), and settling time (t_s). In this study, we propose a novel cascaded controller for AGC in a two-area nonreheat thermal power system with RESs such as PV panels and WTs. The parameters of the controller are optimized by different techniques such as Gorilla Troops Optimizer (GTO), African Vulture Optimization Algorithm (AVOA), and Honey Badger Algorithm (HBA). The efficiency of the proposed controller is tested under different situations such as random load change, RES generation, and a sensitivity analysis with different system parameters.

The main contributions of the study are summarized as follows:

- A two-area nonreheat thermal power system with RES such as WTs and PV panels has been modeled in MATLAB/SIMULINK environment to keep the system frequency and tie-line power variation within the determined values.

- A cascade controller named PID-(1+I) has been proposed for AGC in the studied power system.
- The parameters of the proposed controller have been optimally adjusted by recent and effective optimization techniques such as GTO, AVOA, and HBA, which have not been applied in the AGC process before.
- The performances of the proposed controller and optimization techniques have been tested under different case conditions such as random load variation, RES fluctuations, and system parameters variation.
- The performances of GTO, AVOA, and HBA algorithms have been found very close to each other.
- It has been seen that the proposed controller gives remarkable results in terms of settling time and overshoot values of frequency and tie-line power change in AGC using the mentioned algorithms.

The remainder of the chapter is organized as follows: Section 2 highlights the components of the model in the power system in which AGC is performed. Section 3 focuses on the controller method used for AGC. Section 4 gives information about the optimization techniques. Section 5 presents the simulation results obtained for different scenarios. Finally, Section 6 concludes.

2. Renewable energy sources integrated power systems

The power system where AGC is performed consists of two areas connected by alternating current (AC) tie-line, as shown in Fig. 1. Area-1 consists of a nonreheat thermal power system, wind power system, and loads. In contrast, area-2 consists of a nonreheat thermal power system, PV power system, and loads. Sudden load changes due to consumer demands or reasons originating from RES such as solar radiation and wind speed changes cause system instability. Therefore, each area is controlled with a controller called cascade PID-(1+I) to minimize the frequency variation (ΔF_1, ΔF_2) and the tie-line power variation (ΔP_{tie}).

Determining the parameters of the controllers used in the control process is important to improve system performance under variable conditions. In this study, the parameters of the proposed controller are optimally tuned using the swarm-based metaheuristic optimization techniques of GTO, AVOA, and HBA. Fig. 2 shows the transfer function block diagram of a two-area interconnected by AC tie-line power system.

For simplicity, each component in the studied power system can be considered linear and modeled using a first-order transfer function on MATLAB/SIMULINK. The generation units in the two areas consist of a traditional thermal power system, wind power system, and PV generator. The transfer functions of the turbine and the governor, which form the nonreheat thermal power system, are expressed by Eqs. (1) and (2).

$$G_T(s) = \frac{K_T}{sT_T + 1} \tag{1}$$

$$G_G(s) = \frac{K_G}{sT_G + 1} \tag{2}$$

where K_T and K_G are the turbine and governor gain coefficients, and T_T and T_G are the turbine and generator time constants.

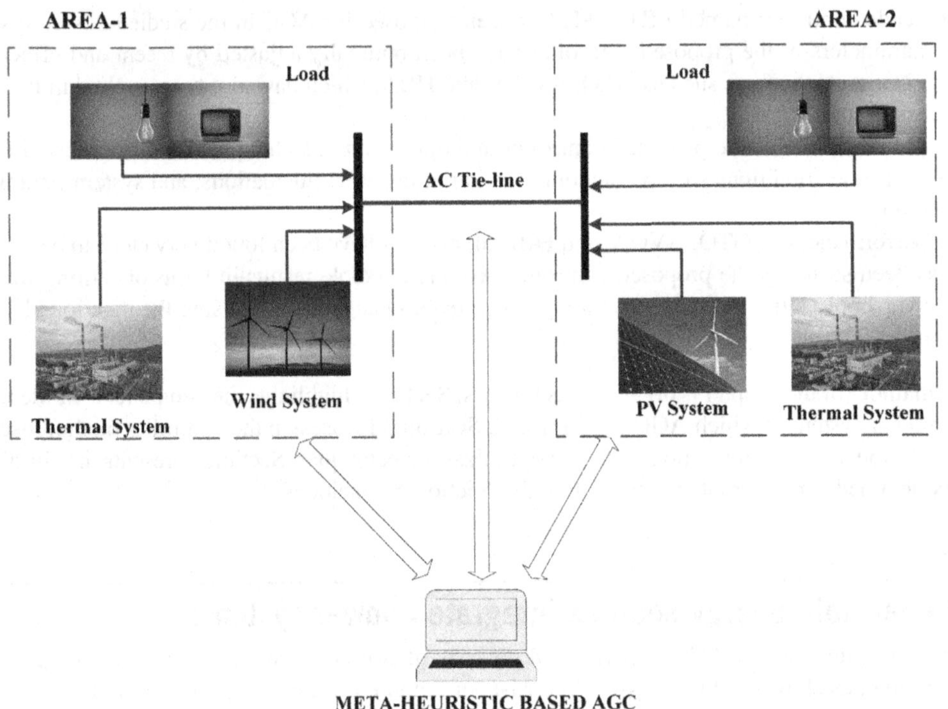

FIG. 1

The framework of the power system.

The simplified transfer function model of wind power systems is shown in Eq. (3). Similarly, the first-order transfer function model of the PV generation system is expressed by Eq. (4).

$$G_{WT}(s) = \frac{K_{WT}}{sT_{WT} + 1} \tag{3}$$

$$G_{PV}(s) = \frac{K_{PV}}{sT_{PV} + 1} \tag{4}$$

where K_{WT} and K_{PV} represent the wind generation system and PV generation systems gain coefficients, while T_{WT} and T_{PV} represent the wind generation system and PV generation systems time constants.

The power flow from area-1 to area-2 over the tie-line is calculated by Eq. (5).

$$G_{AC}(s) = \Delta P_{tie} = \frac{2\pi T_{12}}{s}(\Delta F_1 - \Delta F_2) \tag{5}$$

The transfer function of the power system is given in Eq. (6).

$$G_{PS}(s) = \frac{K_{PS}}{sT_{PS} + 1} \tag{6}$$

where K_{PS} is the gain coefficient of the power system, and T_{PS} is the time constant of the power system.

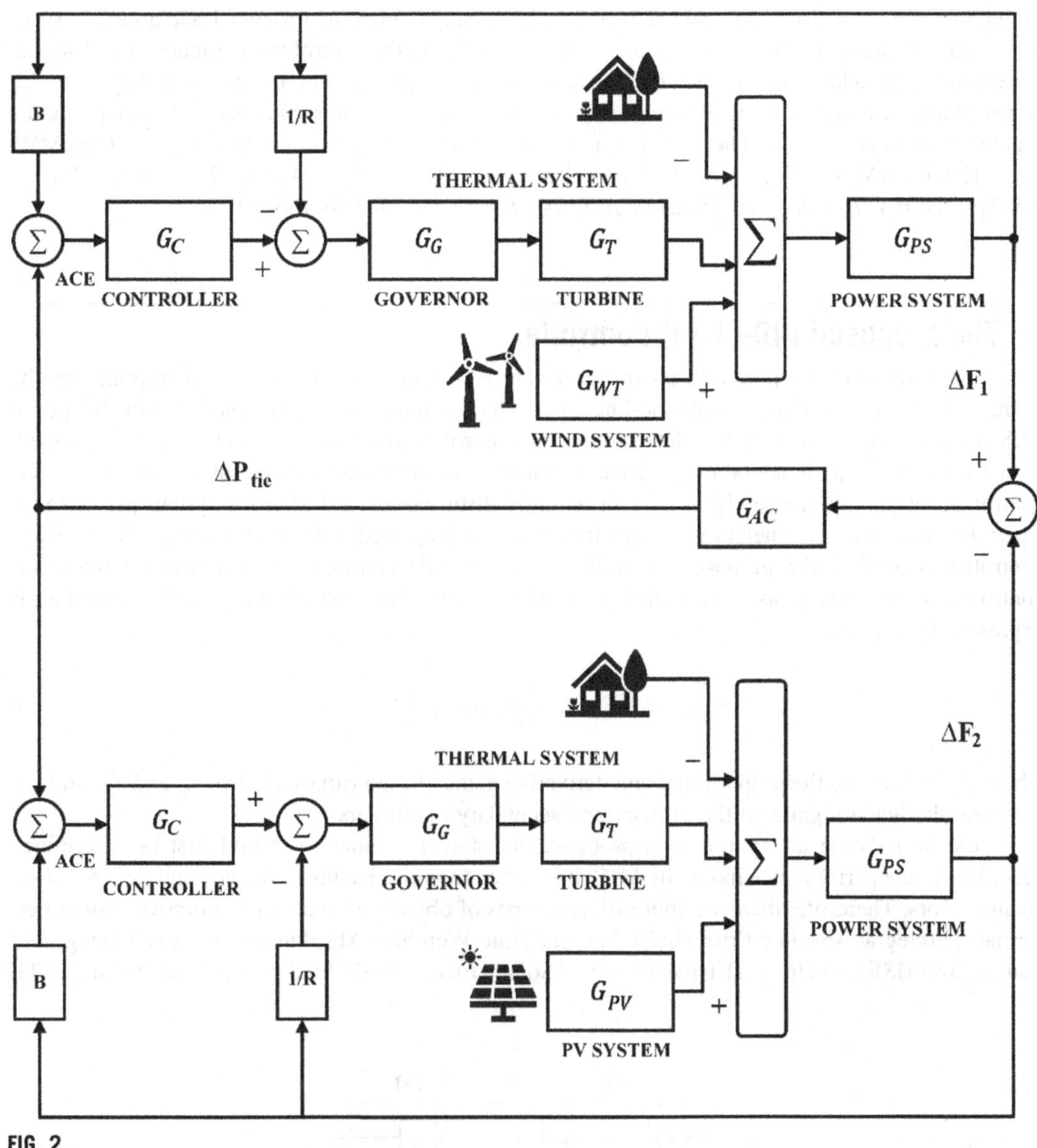

FIG. 2

The transfer function model of the system.

The Area Control Error (ACE), the controller input signal of each area in the power system, defines the unbalance between the generator and the load demands. It is calculated by Eqs. (7) and (8) for area-1 and area-2.

$$ACE_1 = B_1 \cdot \Delta F_1 + \Delta P_{tie} \tag{7}$$

$$ACE_2 = B_2 \cdot \Delta F_2 + \Delta P_{tie} \tag{8}$$

In the preceding equations, ΔF_1 and ΔF_2 are the frequency deviations for area-1 and area-2, respectively; ΔP_{tie} is the tie-line power variation; ACE_1 and ACE_2 are the control errors for area-1 and area-2, respectively. In addition, R_1 and R_2 are the regulation parameters of the nonreheat thermal power plant; B_1 and B_2 are frequency bias parameters; a_{12} is the area capacity rate; and T_{12} is the synchronization coefficient. The system parameters are set as: $K_{WT}=K_{PV}=K_T=K_G=1$ Hz/pu MW, $K_{PS}=120$ Hz/pu MW, $T_{WT}=1.5$ s, $T_{PV}=1.3$ s, $T_T=0.3$ s, $T_G=0.08$ s, $T_{PS}=20$ s, $T_{12}=0.545$ pu MW/Hz, $B_1=B_2=0.425$ pu MW/Hz, $R_1=R_2=2.4$ Hz/pu MW [30–32].

3. The proposed PID-(1 + I) controller

The main purpose of the proposed controller design is to reduce the value to zero or minimize the changes in the system frequency and tie-line power change quickly in case of sudden load changes or RES fluctuations. For this reason, determining the controller to be used and the optimal adjustment of the controller's parameters are of great importance in achieving successful results. A cascade controller design can efficiently operate in multiple disturbances and improve system performance since the number of parameters is greater than a single-loop feedback control design. A two-stage controller is used in the proposed controller, with the PID controller as primary and the (1+I) controller as secondary, as shown in Fig. 3. The transfer function of the cascade controller is expressed by Eq. (9).

$$G_{controller}(s)=\left(K_P+\frac{K_{I1}}{s}+K_D s\right)\cdot\left(1+\frac{K_{I2}}{s}\right) \tag{9}$$

where K_P and K_D are the proportional and derivative gains of the primary controller, and K_{I1} and K_{I2} represent the integral gains of the primary and secondary controllers.

While formulating an optimization problem, the objective function should first be determined according to the performance index. In AGC, it is aimed to minimize the frequency and tie-line power change errors. Therefore, there are four different types of objective functions to improve system performance: Integral Absolute Error (IAE), Integral Time-Weighted Absolute Error (IATE), Integral of Square Error (ISE), and Integral Time-Weighted Square Error (ITSE). Studies show that ISE and IATE

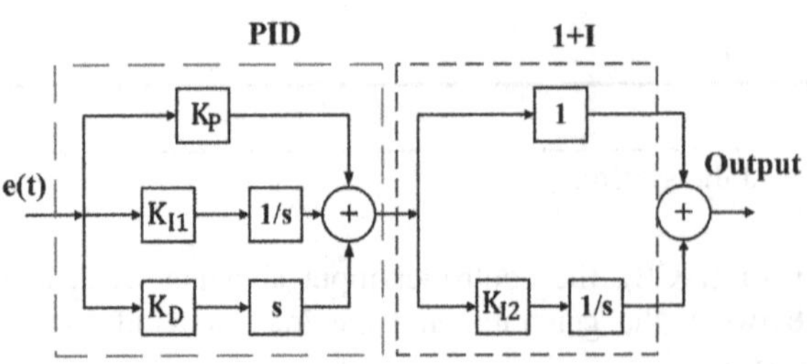

FIG. 3

The block diagram of the proposed controller.

are the most used objective functions to achieve better results in AGC studies [31]. The ISE objective function results in increased settling time while providing lower overshoot. In addition, the IATE objective function provides a faster settling time than ISE. ISE is used as the objective function applied to the optimization problem, and this function is expressed by Eq. (10).

$$ISE = \int_0^\infty \left\{ (\Delta F_1)^2 + (\Delta F_2)^2 + (\Delta P_{tie})^2 \right\} \cdot dt \tag{10}$$

The following constraints should be considered in optimizing the controller parameters: the lower limits are −0.5 and the upper limits are 0.5.

$$K_{P,\min} \le K_P \le K_{P,\max}$$

$$K_{I1,\min} \le K_{I1} \le K_{I1,\max}$$

$$K_{I2,\min} \le K_{I2} \le K_{I2,\max}$$

$$K_{D,\min} \le K_D \le K_{D,\max}$$

4. Metaheuristic optimization techniques

The controller and the optimization method to optimally adjust the parameters of the controller play a major role in the success of the AGC process in power systems. Fluctuations in system frequency and tie-line power changes should be quickly kept within the determined values in AGC. Therefore, different optimization techniques should improve the time spent for setting the controller parameters. The parameters of the proposed PID-(1+I) controller for AGC in the two-area power system were determined through newly developed swarm-based meta-heuristic optimization techniques such as GTO, AVOA, and HBA.

4.1 Gorilla troops optimizer (GTO)

Gorillas, like other apes, think about the past and the future, form strong family ties, make/use tools, and have feelings. Some researchers claim that they also have inner feelings or religious beliefs. Gorillas generally spend their days resting, traveling, and eating food. They live in groups called troops consisting of one adult male or a silverback gorilla group and several adult female gorillas and their offspring.

GTO is a recent swarm-based metaheuristic optimization technique based on gorillas' group behaviors. Gorillas hunt in groups and continue their lives under the leadership of the silverback, the group leader. The silverback, named after the silvery hair that grows on its back during adolescence, makes all decisions, directs fights, determines group movements, directs gorillas to food sources, and takes responsibility for the group's welfare [31].

In GTO, the silverback is considered the best candidate, and other candidate gorillas tend to follow them. Certain mathematical mechanisms explain two different phases of exploration and exploitation.

Algorithm 1. Pseudocode of the GTO.

```
Inputs: The population size N and maximum number of iterations T and parameters β
and p
  Outputs: The location of Gorilla and its fitness value
  Initialize randomly population
  Calculate the fitness values
  while (stopping criteria is not met) do
    Update the C
    Update the L
    % Exploration
    for (each Gorilla) do
    Update the location Gorilla
    end for
    % Create group
    Calculate the fitness values of Gorilla
    if new is better than old, replace them
    Set X_silverback as the silverback location
    % Exploitation
    for (each Gorilla) do
      if ( |C| ≥ 1 ) then
      Update the location Gorilla
      Else
      Update the location Gorilla
      End if
    end for
    % Create group
    Calculate the fitness values of Gorilla
    if new solutions are better than previous solutions, replace them
    Set X_silverback as the location of silverback
  end while
  Return X_BestGorilla
```

Each gorilla is evaluated as a candidate for the best candidate in each iteration in the exploration phase; the best candidate is the silverback. There are three different operators at this stage: migration to an unknown location, movement to other gorillas, and migration to a known location. Two different strategies are applied in the exploitation phase, such as following the silverback and competing for adult females. The pseudocode of GTO is given in Algorithm 1. More detailed information, such as the mathematical model of the algorithm, can be found in [33,34].

4.2 African vulture optimization algorithm (AVOA)

AVOA is a swarm-based metaheuristic optimization technique inspired by the hunting behavior of vultures. Vultures, a bird species found worldwide except in Antarctica and Australia, are useful animals, especially in tropical areas, because they prevent carrion from rotting. These birds feed on carrion containing diseases and infections such as cholera and anthrax that kill living things. The most interesting situation in vultures is that they travel far to find food, and when they find food, they fight among themselves for access to it.

Based on the four assumptions presented to simulate AVOA and the basic concepts of vultures, the proposed algorithm is formulated in four separate steps.

In the first phase, the best vulture in the group is determined. After the initial population is created, all candidates' fitness is calculated, and the best candidate is the best vulture of the first group. The second-best candidate is the best vulture of the second group. Other candidates move towards the best candidate of the first and second groups.

In the second phase, the vultures hunger rates are determined. Frequent foraging vultures have high energy and travel long distances searching for food when they are full. However, vultures do not have enough energy to fly and forage for food when they are hungry, making them aggressive. These behaviors are modeled in this phase of the algorithm.

The third phase is exploration. Vultures have a high vision ability to find food and detect dying animals in their natural environment. However, it can be difficult for them to find food. Therefore, vultures carefully inspect their habitat and travel long distances searching for food. These distances need to be modeled mathematically in AVOA.

The fourth phase is exploitation. When food is available, other vultures compete for food by coming to the food's location. This stage models those behaviors. The pseudocode of AVOA is given in Algorithm 2. More detailed information, such as the mathematical model of the algorithm, can be found in [35].

Algorithm 2. Pseudocode of the AVOA.

```
Inputs: The population size N and maximum number of iterations T
  Outputs: The location of vulture and its fitness value
  Initialize randomly population
  while (stopping criteria is not met) do
   Calculate the fitness values
   Set P_BestVulture1 as the location of vulture (First best location)
   Set P_BestVulture2 as the location of vulture (Second best location)
   for (each Vulture) do
   Select R(i)
    Update the F
    if ( |F| ≥ 1 ) then
       if (P1≥rand_P1) then
          Update the location vulture
       else
    Update the location vulture
    if ( |F| < 1 ) then
        if ( |F| ≥ 0.5 ) then
           if (P2≥rand_P2) then
              Update the location vulture
           else
              Update the location vulture
        else
           if (P3≥rand_P3) then
              Update the location vulture
           else
              Update the location vulture
  Return P_BestVulture1
```

4.3 Honey badger algorithm (HBA)

The honey badger is a mammal with black and white fluffy fur found in rainforests and semi-deserts in Africa, Southwest Asia, and the Indian subcontinent. This fearless dog-sized hunter hunts 60 different species, including dangerous snakes. With the ability to use tools, the honey badger is an intelligent animal that loves honey. It prefers to be alone in the holes it digs and meets other badgers for mating purposes.

A honey badger finds its prey by walking slowly using its olfactory abilities. The honey badger determines the approximate location of its prey and eventually catches it by digging. They love honey but are not skilled at finding honey hives. In contrast, a bird named a honeyguide can locate the hives but are unsuccessful in extracting the honey. These two situations cause cooperation between these two species; the bird helps the badger find the hives and opens the hives with the help of the badger's claws.

The HBA is a metaheuristic optimization technique developed by mimicking the foraging behavior of honey badgers. The honey badger either sniffs and digs or follows the bird to get the food. The first stage is called the digging mode. The second stage is called the honey mode. The honey badger uses its olfactory ability to locate prey in the first phase. When it reaches the prey, it moves around the prey to choose the appropriate place to dig and catch the prey. The badger follows the bird's instructions to reach the hives in the other phase.

While constructing the mathematical model of the algorithm, the two stages mentioned previously are considered. The number of honey badger populations and possible candidates' positions in the population is determined in the initial step. The second step is to create a mathematical model for determining the intensity of the odor. If the odor intensity is high, the honey badger will move fast; if it is low, it will move slowly. The next steps identify the honey badger with the best position. The pseudocode of HBA is given in Algorithm 3. More detailed information, such as the mathematical model of the algorithm, can be found in [36].

5. Simulation results and discussions

In this section, we model the proposed two-area nonreheat thermal power system in the MATLAB/SIMULINK environment. Three algorithms, namely GTO, AVOA, and HBA, are created in the *m.file* file. The behavior of the power system examined under various scenarios such as random load variation, RES generation, and system parameters variation has been tested. In each case study, the parameters of the proposed controller are determined using metaheuristic algorithms such as GTO, AVOA, and HBA, and the results are compared.

5.1 Case study I: Random load change in Area-1

This scenario investigates system performance under load demand in area-1. RES fluctuations are not considered in this scenario. The parameters of the proposed controller are optimally adjusted with GTO, AVOA, and HBA to control the system frequency and tie-line power variation under random load fluctuations (RLFs) varying between −0.02 and 0.03 pu for 100 s in area-1. The numerical data

Table 3 The system performance values for case study III.

	ΔF_1 (Hz)			ΔF_2 (Hz)			ΔP_{tie} (pu MW)		
Method	**MO⁺**	**MO⁻**	t_s (s)	**MO⁺**	**MO⁻**	t_s (s)	**MO⁺**	**MO⁻**	t_s (s)
GTO	0.09	−0.167	16.48	0.033	−0.045	19.71	0.034	−0.0764	16.52
AVOA	0.064	−0.15	16.24	0.0325	−0.0262	19.57	0.0235	−0.051	15.99
HBA	0.063	−0.15	20.82	0.0317	−0.0298	21.24	0.023	−0.0483	16.66

Table 4 The optimized controller gains for case study III.

	Gain	**GTO**	**AVOA**	**HBA**
AREA-1	K_P	0.4201	−0.5	−0.5
	K_{I1}	−0.4854	−0.4268	−0.44372
	K_{I2}	0.0468	0.1463	0.10845
	K_D	−0.4982	−0.5	−0.5
AREA-2	K_P	−0.4855	−0.5	−0.5
	K_{I1}	−0.4433	−0.5	−0.5
	K_{I2}	0.5	0.4996	0.5
	K_D	−0.2124	−0.2933	−0.5

Table 5 The system performance values for case study IV.

		ΔF_1 (Hz) (10^{-3})		ΔF_2 (Hz) (10^{-3})		ΔP_{tie} (pu MW) (10^{-3})	
Method	**Parameter change**	**MO⁺**	**MO⁻**	**MO⁺**	**MO⁻**	**MO⁺**	**MO⁻**
GTO	B_1 +25%	2.46	−10	2.25	−5.9	0.933	−2.13
	B_2 −25%	1.72	−15.2	1.76	−10.3	0.35	−3.79
	R_1 +25%	2.79	−11.15	2.58	−6.98	1.08	−2.57
	R_2 −25%	2.08	−11.05	2.47	−6.73	1.25	−2.7
AVOA	B_1 +25%	2.45	−9.89	2.25	−5.85	0.9	−2.12
	B_2 −25%	1.8	−11.2	1.75	−7.84	0.61	−2.56
	R_1 +25%	0	−13.27	0	−9.34	0	−3.27
	R_2 −25%	1.9	−11.15	1.74	−6.38	0.86	−2.63
HBA	B_1 +25%	2.42	−9.94	2.25	−5.58	0.95	−2.17
	B_2 −25%	1.28	−11.34	1.16	−4.82	1.77	−2.93
	R_1 +25%	2.81	−11.16	2.59	−6.76	1.08	−2.48
	R_2 −25%	2.14	−10.99	2.18	−5.97	1.21	−2.44

References

[1] S. Arunagirinathan, P. Muthukumar, S.S. Yu, H. Trinh, A novel observer-based approach to delay-dependent LFC of power systems with actuator faults and uncertain communications conditions, Int. J. Electr. Power Energy Syst. 131 (2021) 106957.

[2] M. Shouran, F. Anayi, M. Packianather, Design of sliding mode control optimised by the Bees algorithm for LFC in the Great Britain power system, Mater. Today: Proc. 52 (2021) 937–943.

[3] N. Jalali, H. Razmi, H. Doagou-Mojarrad, Optimized fuzzy self-tuning PID controller design based on Tribe-DE optimization algorithm and rule weight adjustment method for load frequency control of interconnected multi-area power systems, Appl. Soft Comput. J. 93 (2020) 106424.

[4] J. Sharma, Y.V. Hote, R. Prasad, PID controller design for interval load frequency control system with communication time delay, Control. Eng. Pract. 89 (2019) 154–168.

[5] P. Dahiya, P. Mukhija, A.R. Saxena, Design of sampled data and event-triggered load frequency controller for isolated hybrid power system, Int. J. Electr. Power Energy Syst. 100 (2018) 331–349.

[6] A.K. Barisal, Comparative performance analysis of teaching learning based optimization for automatic load frequency control of multi-source power systems, Int. J. Electr. Power Energy Syst. 66 (2015) 67–77.

[7] A.M. Agwa, Equilibrium optimization algorithm for automatic generation control of interconnected power systems, Przegląd Elektrotechniczny R. 96 (9) (2020) 143–148.

[8] H.M. Hasanien, A.A. El-Fergany, Salp swarm algorithm-based optimal load frequency control of hybrid renewable power systems with communication delay and excitation cross-coupling effect, Electr. Power Syst. Res. 176 (2019) 105938.

[9] S.S. Dhillon, J.S. Lather, S. Marwaha, Multi objective load frequency control using hybrid bacterial foraging and particle swarm optimized PI controller, Int. J. Electr. Power Energy Syst. 79 (2016) 196–209.

[10] A.H. Yakout, M.A. Attia, H. Kotb, Marine predator algorithm based cascaded PIDA load frequency controller for electric power systems with wave energy conversion systems, Alex. Eng. J. 60 (2021) 4213–4222.

[11] M.S. Alam, F.S. Al-Ismail, M.A. Abido, PV/wind-integrated low-inertia system frequency control: PSO-optimized fractional-order PI-based SMES approach, Sustainability 13 (2021) 7622.

[12] B. Khokhar, S. Dahiya, K.P.S. Parmar, Load frequency control of a microgrid employing a 2D sine logistic map based chaotic sine cosine algorithm, Appl. Soft Comput. 109 (2021) 107564.

[13] P.C. Sahu, R.C. Prusty, B.K. Sahoo, Modified sine cosine algorithm-based fuzzy-aided PID controller for automatic generation control of multiarea power systems, Soft. Comput. 24 (2020) 12919–12936.

[14] E.A. Mohamed, E.M. Ahmed, A. Elmelegi, M. Aly, O. Elbaksawi, A.A.A. Mohamed, An optimized hybrid fractional order controller for frequency regulation in multi-area power systems, IEEE Access 8 (2020) 213899–213915.

[15] T. Shanmuga Priya, M. Sarojini Devi, A. Usha, P. Radha, A. Jamuna, Controller parameters tuning for multi area automatic generation control using modified Grey Wolf optimization algorithm, Mater. Today: Proc. (2021).

[16] J.R. Nayak, B. Shaw, B.K. Sahu, K.A. Naidu, Application of optimized adaptive crow search algorithm based two degree of freedom optimal fuzzy PID controller for AGC system, Eng. Sci. Technol. Int. J. 32 (2021) 1–14.

[17] R.K. Sahu, S. Panda, P.C. Pradhan, Design and analysis of hybrid firefly algorithm-pattern search based fuzzy PID controller for LFC of multi area power systems, Int. J. Electr. Power Energy Syst. 69 (2015) 200–212.

[18] R. Ramachandran, J. Satheesh Kumar, B. Madasamy, V. Veerasamy, A hybrid MFO-GHNN tuned self-adaptive FOPID controller for ALFC of renewable energy integrated hybrid power system, IET Renew. Power Gener. 15 (2021) 1582–1595.

[19] D. Mohanty, S. Panda, Frequency control of hybrid power system by sine function adapted improved whale optimisation technique, Int. J. Ambient Energy (2020), https://doi.org/10.1080/01430750.2020.1839550.

[20] C. Pradhan, C.N. Bhende, Online load frequency control in wind integrated power systems using modified Jaya optimization, Eng. Appl. Artif. Intell. 77 (2019) 212–228.

[21] M. Kheshti, L. Ding, H. Askarian-Abyaneh, A.R. Singh, S. Zare, V. Terzija, Improving frequency regulation of wind-integrated multi-area systems using LFA-fuzzy PID control, Int. Trans. Electr. Energy Syst. 31 (2021), e12802.

[22] S.M. Nosratabadi, M. Bornapour, M.A. Gharaei, Grasshopper optimization algorithm for optimal load frequency control considering predictive functional modified PID controller in restructured multi-resource multi-area power system with redox flow battery units, Control. Eng. Pract. 89 (2019) 204–227.

[23] S.A. Taher, M. Hajiakbari Fini, S. Falahati Aliabadi, Fractional order PID controller design for LFC in electric power systems using imperialist competitive algorithm, Ain Shams Eng. J. 5 (2014) 121–135.

[24] A. Boudia, S. Messalti, A. Harrag, M. Boukhnifer, New hybrid photovoltaic system connected to superconducting magnetic energy storage controlled by PID-fuzzy controller, Energy Convers. Manag. 244 (2021), 114435.

[25] N. Sinha, L.L. Loi, V.G. Rao, GA optimized PID controllers for automatic generation control of two area reheat thermal systems under deregulated environment, in: 3rd International Conference on Deregulation and Restructuring and Power Technologies, DRPT, 2008, pp. 1186–1191.

[26] T.A. Jumani, M.W. Mustafa, Z. Hussain, M. Md. Rasid, M.S. Saeed, M.M. Memon, I. Khan, K.S. Nisar, Jaya optimization algorithm for transient response and stability enhancement of a fractional-order PID based automatic voltage regulator system, Alex. Eng. J. 59 (2020) 2429–2440.

[27] M.Z. Bernard, T.H. Mohamed, Y.S. Qudaih, Y. Mitani, Decentralized load frequency control in an interconnected power system using coefficient diagram method, Int. J. Electr. Power Energy Syst. 63 (2014) 165–172.

[28] T.S. Tsay, Load–frequency control of interconnected power system with governor backlash nonlinearities, Int. J. Electr. Power Energy Syst. 33 (2011) 1542–1549.

[29] H. Shayeghi, H.A. Shayanfar, Power system load frequency control using RBF neural networks based on μ-synthesis theory, in: 2004 IEEE Conference on Cybernetics and Intelligent Systems, 2004, pp. 93–98.

[30] E.M. Ahmed, A. Elmelegi, A. Shawky, M. Aly, W. Alhosaini, E.A. Mohamed, Frequency regulation of electric vehicle-penetrated power system using MPA-tuned new combined fractional order controllers, IEEE Access 9 (2021) 107548–107565.

[31] M. Khamies, G. Magdy, A. Selim, S. Kamel, An improved Rao algorithm for frequency stability enhancement of nonlinear power system interconnected by AC/DC links with high renewables penetration, in: Neural Computing and Applications 0123456789, 2021.

[32] E. Sahin, Design of an optimized fractional high order differential feedback controller for load frequency control of a multi-area multi-source power system with nonlinearity, IEEE Access 8 (2020) 12327–12342.

[33] B. Abdollahzadeh, F. Soleimanian Gharehchopogh, S. Mirjalili, Artificial gorilla troops optimizer: a new nature-inspired metaheuristic algorithm for global optimization problems, Int. J. Intell. Syst. 36 (2021) 5887–5958.

[34] M. Ali, H. Kotb, K.M. Aboras, N.H. Abbasy, Design of cascaded pi-fractional order PID controller for improving the frequency response of hybrid microgrid system using gorilla troops optimizer, IEEE Access 9 (2021) 150715–150732.

[35] B. Abdollahzadeh, F.S. Gharehchopogh, S. Mirjalili, African vultures optimization algorithm: a new nature-inspired metaheuristic algorithm for global optimization problems, Comput. Ind. Eng. 158 (2021), 107408.

[36] F.A. Hashim, E.H. Houssein, K. Hussain, M.S. Mabrouk, W. Al-Atabany, Honey badger algorithm: new metaheuristic algorithm for solving optimization problems, Math. Comput. Simul. 192 (2022) 84–110.

CHAPTER

Route optimization in MANET using swarm intelligence algorithm 16

Anita R. Patil and Gautam M. Borkar

Department of Information Technology, Ramrao Adik Institute of Technology, D Y Patil Deemed To Be University, Navi Mumbai, Maharashtra, India

1. Introduction

A mobile ad hoc network (MANET) is a self-organized network that has several mobile nodes with dynamic movement. It provides an infrastructure-less communication system with available resources, like the mobile node itself acting as a router. With an increase in the number of users and new applications, MANETs are progressively taking up space in domestic and corporate organizations. Different insects and living things (ants, grey wolves, lions, dolphins, fungi, and many others) show the same behavior as a MANET with capacity for self-organization of adaptive, resilient, and efficient networks in transporting nutrients, fungal colonies, and so on [1]. This type of bio-inspiration is used in MANETs to design new, efficient routing protocols. The mobility, power, battery, storage, and communication capabilities of MANET nodes are all constrained. Some examples of MANET applications include wireless sensor networks, body and health monitoring systems, military applications, emergency operations, search and rescue with unmanned aerial vehicles (UAVs), multiagent exploration of unknown areas, vehicular networks, network connectivity, interplanetary Internet, rural area networks, and more.

MANETs are a fast-growing and on-demand networking technology. In a MANET, when a new node enters the network, it always sends an arrival beacon, and other nodes make an entry into the routing table. Routing protocols play an important role in finding the path to the destination and maintaining the route between source and destination node [2]. Routing protocols decide which route will be available. The nodes are not familiar with dynamic topology, which is also considered during designing of the routing protocol. In such cases, routing algorithms play a vital role in finding the best path and maintaining connectivity between source and destination node. Routing protocol mainly based routing information which consist of past and predictable routing information also combination of both (hybrid routing). Routing protocols for a MANET are categorized as [3]:

- Proactive routing protocol: These protocols exchange topological data among every one of the nodes in the network. They vary in the method for trading and refreshing routing data, the quantity of routing tables utilized, and the kind of data stored in the table. Every node in the network may easily establish the connection from the routing information, which is available in the table. When there is a failure in the link, the reconstruction process is very slow. In addition, a node for route construction and maintenance requires much data. Destination-Sequenced Distance Vector

Comprehensive Metaheuristics. https://doi.org/10.1016/B978-0-323-91781-0.00016-8

(DSDV), Optimized Link State Routing (OLSR), and Wireless Routing Protocol (WRP) are examples of proactive protocols.

- Reactive routing protocol: These are the most preferred for MANETs. The routing information is available only when it is required, thus these algorithms require minimized traffic overhead and have efficient bandwidth usage. These methods waste a large amount of time in finding routes, as well as cause flooding that results in network clog. Dynamic Source Routing (DSR), Ad Hoc On-demand Distance Vector (AODV), and Temporally Ordered Routing Algorithm (TORA) have been efficiently used as reactive routing protocols.
- Hybrid routing protocol: Hybrid routing is a combination of benefits of both table-driven and on-demand routing protocols. The Zone Routing Protocol (ZRP) is an example of hybrid routing that is adaptive in nature to maintain the source and destination information according to zone. If source and destination nodes are occupied in the same zone or location, a table-driven approach is used to deliver the data packet. If source and destination nodes are in different zones, on-demand routing is used.

Performance of a MANET is affected by design and configuration issues in the routing protocol. One of the most challenging issues in a MANET is providing an optimal path and maintaining quality of service (QoS) in the network.

2. Related work

Ronaldo et al. [1] proposed the HyphaNet system, combining a fungi dynamics optimization technique with a routing algorithm for optimal route discovery in a MANET. The routes found in HyphaNet are built similarly to the fungal mycelium in which multiple paths are initiated but only the hyphae of the best routes receive biomass for reinforcement and thickening of their walls over time. This is resulting in the best routes remaining and presenting greater flow attractiveness to find optimal routes. The routing mechanism is based on the attraction principle, which states that data flows tend to move through nodes and links with greater concentration of immobile biomass, indicating lower costs and better resource availability. Best optimal path is calculated using route heuristic values.

Shivakumar et al. [4] proposed a cross-layer routing protocol with Particle Swarm Optimization (PSO) algorithm to improve QoS in a MANET. Design factors of energy-efficient routing protocols are available current energy level of node, traffic overhead, and channel conflict. The proposed protocol had an efficient energy route and maintained stability of the network. In this system, packet delivery rate, mobility of nodes, and available energy level are measured from network layer, and the network contention is evaluated from the MAC layer. The contention window (CW) is adjusted based on network contention and available remaining energy. When compared to existing systems, this system effectively improved packet delivery ratio while reducing energy usage and overhead.

Jamali et al. [5] applied Binary Particle Swarm Optimization (BPSO) on a TORA protocol to efficient use of energy level of node. The proposed combination assumes path cost and route energy to evaluate the best fitted path. Optimization with the TORA protocol improved network life and total delivery time of the packet.

Zhang et al. [6] proposed a system based on Ant Colony Optimization (ACO) and Physarum Autonomic Optimization (PAO) called Bio-inspired Hybrid Trusted Routing Protocol (B-iHTRP). In the proposed system, the MANET is divided into zones. Among these zones, ants are sent to reactively find routes to destinations. ACO was used to evaluate the optimal route and PAO was used to optimize the path in multipath domain. Compared to other algorithms, B-iHTRP achieved better performance with respect to minimum delay and control overhead. It also maximized packet delivery ratio.

Sabree et al. [7] developed the Lion Optimization Routing Protocol (LORP), which uses a Lion Optimization Algorithm (LOA). The authors used an LOA maximization technique to choose the best path based on three primary metrics: energy conservation, throughput, and packet delivery ratio. They also used an LOA minimization technique to choose the optimal route based on delay and short path metrics.

Guerriero et al. [8] proposed a bicriteria optimization approach that allows for the complete integration of energy consumption and mobile node link stability. To test the validity of the presented model, the authors devised a greedy approach. Results show that a shorter route results in a more stable route but increases energy consumption. In contrast, longer routes enhance route fragility while lowering average energy consumption.

Path-finding problems in MANETs are also significant prospects for bio-inspired techniques. ACO, for example, is inspired by ant foraging operations and uses a positive feedback mechanism to achieve an optimal solution. Some routing protocols based on ACO include ant-colony-based routing algorithm (ARA), AntHocNet, HOPNET, Ant-based Dynamic Zone Routing Protocol (AD-ZRP), and Hybrid ACO Routing (HACOR).

3. Workflow of MANETs

Fig. 1 shows the working principles of a MANET and how nodes are synchronized with each other and select the route for data transmission. This process is entirely managed by a routing protocol. Dynamic topology will affect the route discovery that is also considered in routing. Source nodes send broadcast messages to all neighbors and neighboring nodes forwarded to their neighboring nodes like that request message (RREQ) received by the destination node. Destination nodes send reply (RREP) messages with sequence numbers. Source nodes decide the path based on hop count or highest sequence number to start communication [9]. Route discovery and route maintenance are major tasks of all routing algorithms.

3.1 Limitations of MANETs

MANETs are highly adaptive ad hoc networks, but they do have some limitations, as discussed in the following [9].

3.1.1 Hop count

Data can be sent directly through neighboring nodes within transmission range. The data sent through multiple hops and the intermediary node act as a router. When the number of hops is increased, throughput drops quickly.

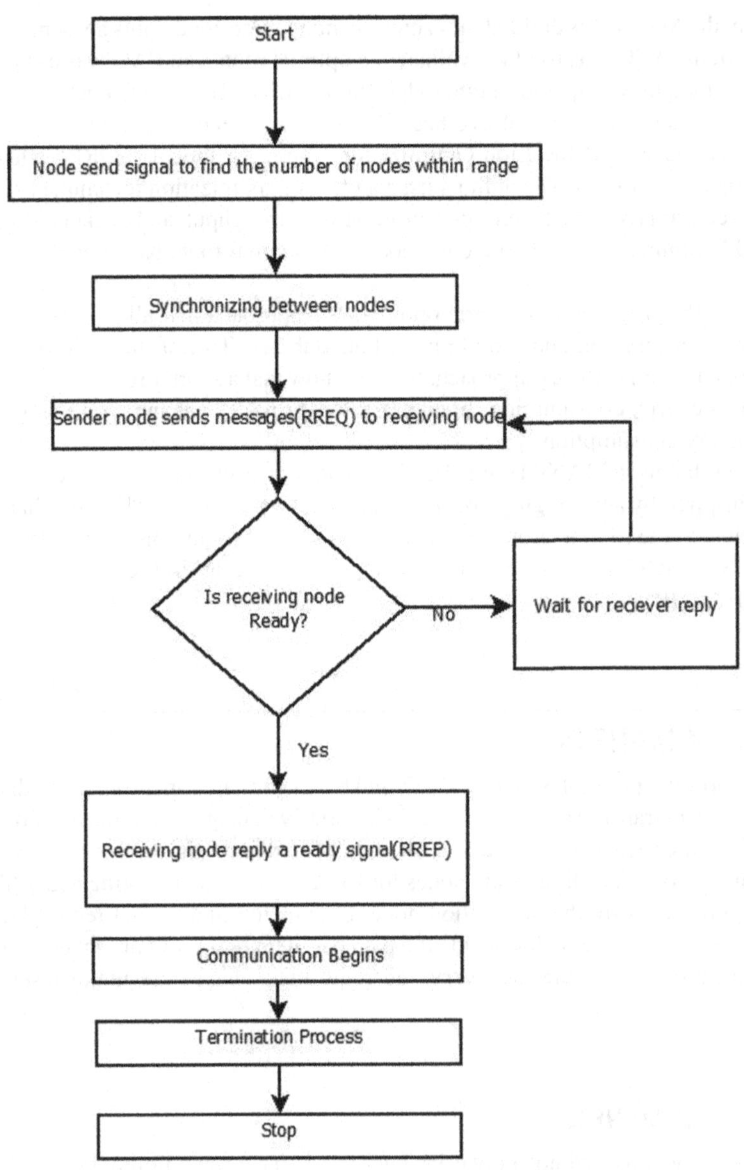

FIG. 1

Working principles of MANET.

3.1.2 Node mobility

Due to frequent topology changes, highly mobile nodes will incur higher overheads. Because of the requirement to determine alternative routes following route failures, the number of routing packet broadcasts increases. The routing table stores a record of the destination nodes and number of hops required to reach each one. Any modifications to the routing table are broadcast to all other nodes.

The overall network will be burdened with more overhead because of this. When the overhead is high, a smaller portion of the packet is dedicated to data transmission, resulting in a reduced throughput.

3.1.3 Delay

The average time it takes for a packet to travel from the time it leaves a source to the time it arrives at its destination is called delay. To improve throughput, it is necessary to keep the nodes busy with packet transmission and reception. As a result, the queue of each node will never be empty, resulting in a longer latency.

4. Routing challenges in MANETs

The routing table maintains and shares routing information between neighboring nodes. This information includes hop length, energy consumption, broken link information, and bandwidth utilization. With respect to these parameters, routing protocols face several challenges in achieving the best performance in a MANET [10]

- Mobility of node: Throughput measurement in a MANET is affected during a variety of mobility models, including global mobility model, Brownian mobility model, random walk model, and so on. Most of these studies assume global mobility, which means that each node maintain throughput. As a result, there is a significant disparity between static networks with low throughput and low delay and mobile networks with high throughput and high delay.
- Loop-free routing: When an error occurs in the operation of the routing algorithm, the path to a certain destination forms a loop in a group of nodes. Intermediate nodes record multipath links during the route request phase, which are used to generate multiple paths during the route discovery phase. To avoid routing loop concerns, each path is given its own flow identification.
- Minimum route acquisition delay: In on-demand routing protocols, route acquisition is majorly considered to minimize end-to-end delay and find the route when it is requested.
- Control overhead: This is the ratio of the control data transmission to the number of packets sent and number of received packets at each node in a MANET. It always maintains to minimum control overhead.
- Fast route reconfiguration: If link failure occurs, then immediately search for a new route to avoid end-to-end delay and overall impact on network performance.
- QoS Provisioning: Network performance is measured according to what facilities or services are provided to the end user. Due to lack of central coordination and limited resource access in MANETs, it is difficult to deploy QoS. QoS-aware routing consists of different parameters such as throughput, delay, packet loss rate, bit error rate, jitter, packet delivery ratio, pathloss, and so on.
- Energy awareness in node: Mobile nodes have limited battery capacity. If the energy level is decreased, the performance of the entire network is affected. If energy awareness is done properly then other alternative will be available before node discontinuity.

5. Routing issues resolved by optimization

Different optimization algorithms have been used by researchers to solve routing challenges in MANETs. In the sections that follow, we consider several routing challenges and discuss the optimization techniques used to optimize the route from source to destination node. These challenges include:

- route optimization
- energy-aware routing
- loop-free routing
- link failure and route rediscovery

As per research, optimization techniques play an important role in improving MANET performance and increasing the reliability of ad hoc networks in real-time applications.

5.1 Genetic algorithm (GA)

Holland proposed the Genetic Algorithm (GA) [11] in 1975. A GA is based on natural selection principles. Among the existing optimization techniques, GA is the most effective. Function optimizers are GAs that perform well in optimization. The algorithm is initialized in this population of solutions called a chromosome. Each chromosome's fitness is assessed using the relevant fitness function. The best chromosomes are chosen and undergo crossover and mutation to produce superior progeny. When the search space is big, complex, or poorly known, GA is beneficial and efficient.

By considering strength of GA, Multi Objective function used to enhance QoS Routing in MANET [12]. In this technique, available routes are needed to satisfy jitter, delay, packet loss rate, packet delivery ratio, bandwidth utilization, and so on. Evaluated fitness function as shown in Eq. (1) is calculated based on number of packets generated (λ_{out}) and number of data packets transmitted from node to other (λ_{in}).

$$f(t) = \sqrt{\frac{2J}{\lambda_{out}} e^{-\frac{\lambda_g^2}{2\lambda_{in}} J} PDR} \tag{1}$$

where J represents maximum jitter, λ_{in} represents packet received at node, λ_g represents the number of packets generated by node, $\lambda_{out} = \lambda_{in} + \lambda_g$, and PDR is the packet delivery ratio.

5.2 Particle swam optimization (PSO)

Kennedy and Eberhart proposed PSO [13], a population-based stochastic optimization technique, in 1995. PSO is inspired by the flocking behavior of birds and swimming behavior of fish. Each member of the PSO is represented as a particle with its own velocity and position. The highest fitness value is used to determine the particle's optimum position.

In a MANET, PSO specifies the possible solutions to every node with optimized N-dimensional space [14]. Every node particle contains location and velocity. These parameters are used to calculate link life, node power life, and available bandwidth of the network. Fig. 2 shows different variations of PSO combined with routing algorithms applied to enhance QoS. Fitness function as calculated by Eq. (2) is defined as follows:

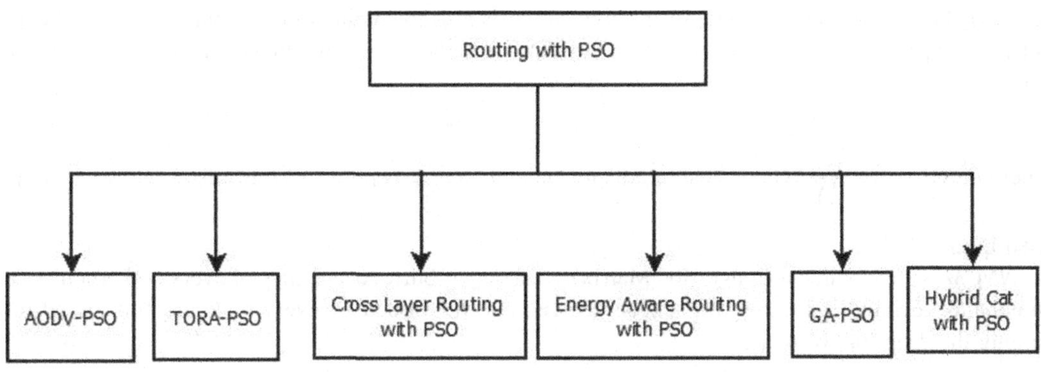

FIG. 2

Different variations of PSO for QoS improvement in MANETs.

$$F_i = (\alpha_1 * LTM_i) + (\alpha_2 * LTN_i) + (\alpha_3 * \beta_i) \tag{2}$$

where α_1, α_2, and α_3 represent weight, LTM_i represents link life, LTN_i represents node life, and β_i represents available bandwidth.

5.3 Ant Colony Optimization (ACO)

The ACO technique is a metaheuristic strategy inspired by ant foraging behavior. Dorgio and Dicario proposed this optimization technique in 1999 [15]. Its three main functions are to: (1) construct an ant solution (artificial ants move between nearby states of the problem); (2) update pheromone trails (pheromone trails are updated after the solution is totally developed); and (3) apply the added pheromone to the best solution in this daemon operation.

For discovery of optimal route and energy awareness purposes, ACO is usually implemented with different routing protocols [16] such as AODV, DSR, DSDV, and others, as shown in Fig. 3. In enhanced DSR, route discovery is based on the stored link path in the received strength signal matrix

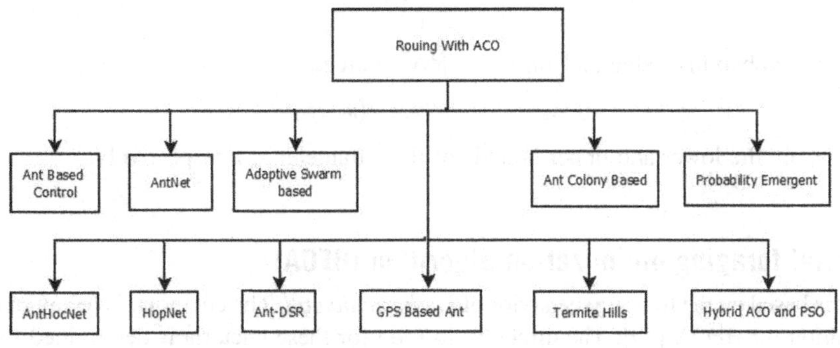

FIG. 3

Different variations of ACO for QoS improvement in MANETs.

cache (RSSM) at the source node. Pheromone count is also maintained in the routing table and considered for path selection. Route selection consists of highest ratio of fitness function, as shown in Eq. (3):

$$f = RSSM/NNCM \tag{3}$$

where $RSSM_i = \frac{G_e * G_t * S_t}{\left(4\pi * \frac{x}{\lambda}\right)2}$, G_t is transmitted antenna gain, and S_t is power of transmission λ wavelength used in the MANET.

NNCM is the Nonlinear Congestion Matrix calculated as buffer occupancy of every node such as λt. In enhance technique, number of packets received and forwarded by every node are considered for finding the drop trail of node.

5.4 Artificial Bee Colony (ABC)

Various available swarm intelligence algorithms are based on the natural behavior of bees. Basturk and Karaboga proposed the Artificial Bee Colony (ABC) method [17], which is based on honeybee swarm foraging behavior. There are three types of bees in the ABC algorithm: onlooker, employed, and scout bees. A bee waiting to decide about a food source is referred to as an onlooker. As it travels to the food source that it has already visited, it is referred to as an employed bee. The bee conducts scouts, which are random searches.

In a MANET, ABC used with AODV and TORA routing protocols resulted in fast route discovery and enhanced energy saving of the node [18]. To evaluate fitness value of employed, onlooker, and scout bees, we use Eqs. (4)–(6):

$$f_m(x_m) = \left\{ \begin{array}{ll} \frac{1}{1+f_m(x_m)} & \text{if} f_m(x_m) \geq 0 \\ 1+abs(f_m(x_m)) & \text{if} f_m(x_m) \geq 0 \end{array} \right\} \tag{4}$$

where x_m is food source and $f_m(x_m)$ is the objective of the employed bee.

$$p_m = \frac{f_m(x_m)}{\sum_{m=1}^{SN} f_m(x_m)} \tag{5}$$

where *pm* is the probability value and *xm* is the food source.

$$x_{mi} = l_i + rand(0, 1) * (u_i - l_i) \tag{6}$$

where l_i and u_i are the lower and upper bounds of the parameter x_{mi}, respectively.

5.5 Bacterial foraging optimization algorithm (BFOA)

This method is based on the foraging behavior of *Escherichia coli*. The chemotaxis behavior of bacteria is the inspiration for BFOA [19]. The direction to food for these bacteria is determined by chemical gradients. A set of operations is used to achieve the information processing approach. Chemotaxis is the movement of cells along a surface one by one. The best strain of bacteria is chosen to contribute to the next generation. Cells are eliminated and new samples are added throughout the elimination and dispersal processes.

Kodavayur et al. [20] proposed the protocol based on cross-layer energy routing protocol and BFOA. This protocol presents initial route with minimum transmission energy and maximum signal-to-noise ratio. A fitness function is calculated by using node energy level, node degree, and remaining energy metric. To further optimize energy consumption, BFOA is applied throughout the chosen path, and nodes with a better fitness function are selected as forwarding or relay nodes.

5.6 Bat optimization

Yang proposed the Bat Algorithm (BA) [21] in 2010. It is used to solve optimization issues by replicating the behavior of bats, which rely on echolocation to update their position and velocity. This algorithm works iteratively, and the position of each bat is updated based on the velocity and hertz number of the sound wave in each iteration. The parameters of the problem to be solved are represented by the position vectors of bats in network optimization.

To enhance route optimization and reduce energy consumption, BA can be combined with PSO in a MANET routing protocol.

5.7 Gray Wolf Optimization

Mirjalili et al. [22] introduced Grey Wolf Optimization (GWO). This algorithm represents the social behavior of gray wolves, meaning it follows wolves' leadership hierarchy and attacking strategy. To simulate the leadership hierarchy in a MANET, four types of gray wolves are identified: alpha, beta, delta, and omega [23]. The GWO selects the optimal value of nodes based on their location. This proposed technique proved to efficiently increase network performance, including increased network lifetime, decreased delay, and minimized energy consumption. Fitness function is shown in Eq. (7):

$$f_{min}(X_i) = \alpha\left[d(e+1)/e^{\frac{L}{Q}+1}\right] + [(1-\alpha)/PDR] \tag{7}$$

where d is the end-to-end delay, L is the link quality between two different nodes, Q is maximum link quality in the network, PDR is the packet delivery ratio, and α is defined as a constant with value of 0.5.

5.8 Dolphin optimization

Dolphins have many noteworthy biological characteristics and living habits such as echolocation, information exchanges, cooperation, and division of labor. Combining these biological characteristics and living habits of dolphin with swarm intelligence and bringing them into optimization problems [24].

A dolphin optimizer is used to design and configure a secure routing algorithm. In this technique, trust value is computed based on echolocation level on dolphin [25].

6. Comparative analysis

According to analysis of various combinations of routing protocols with swarm intelligence algorithms, MANET performance is improved with respect to different QoS parameters such as maximized throughput, reduced end-to-end delay, increased packet delivery ration, decreased packet loss, less energy consumption, maintained overhead, and so on. To design and configure optimized routing protocols, routing table parameters such as destination sequence number, hop count, fitness value, node energy, path length, mobility, speed of node, forwarding status of node, trust level of node, and so on should be updated accordingly [12,14,16,18,20,23,25,26]. Table 1 shows a comparative analysis of different optimized routing protocols.

Table 1 Comparative analysis of different optimized routing protocol.

Applied approach	Operators	Applied on routing algorithm	Parameter from routing table	Performance of optimized routing protocol
Genetic Algorithm (GA)	Crossover Mutation Selection Inversion	CGSR, AODV	Sequence number	Routing information Overhead
Particle Swarm Optimization (PSO)	Initialize Updater Evaluator	AODV, TORA, Cross layer protocol	Route/path length, route energy	Network life and delivery ratio increased Faster route rediscovery
Ant Colony Optimization (ACO)	Pheromone update and measure, Trail evaporation	AODV, AOMDV, DSDV, DSR	Sequence number and hop count, pheromone count, lifetime of node	Improved packet delivery rate, throughput-minimum energy consumption and packet delay Link failure detection
ACO and PSO	Pheromone update and multi path	AODV, ZRP	Sequence number, mobility, hop count	Routing overhead Route optimality Energy efficiency
Artificial Bee Colony (ABC)	Reproduction Replacement of bee, Selection	AODV	Mobility, hop count	Route optimality Less energy consumption
Bacterial Foraging Optimization (BFO)	Reproduction, Chemotaxis, Dispersion, Elimination	Cluster routing algorithm	Fuzzy clustering algorithm, n trust levels	Secure routing
Gray Wolf Optimizer (GWO)	Gray wolf hunting mechanism, optimize hunting, search for prey, encircle prey and attacking prey	AODV, GWO-GPR	Sequence number, mobility, hop count, location of node	Increased network Lifetime Reduced delay and energy consumption

Table 1 Comparative analysis of different optimized routing protocol—cont'd

Applied approach	Operators	Applied on routing algorithm	Parameter from routing table	Performance of optimized routing protocol
Fungi dynamics	Hyphae of the best routes receive biomass for reinforcement, attractiveness of route	Bio-inspired routing algorithm	Hop count, route forwarding status	Heuristic route provided
Bat Algorithm (BA)	Echolocation	AODV, DSR	Mobility of node and hop count	Increased packet delivery ratio Reduced delay and drop packets

7. Conclusion

This chapter discussed routing protocols based on different mechanisms, including routing information, route discovery, route maintenance, link breakage, and so on. It also compared the performance of MANETs based on packet loss, jitter, delay, congestion, transmission speed, throughput, drop rate, mean opinion score, latency, and so on. The essential criterion for determining the best and most optimal solution from all conceivable outcomes is optimization. There are a variety of optimization approaches available that can be used to achieve better results. This chapter presented comparative analysis of MANET performance based on optimized routing protocols, including AODV, DSDV, DSR, TORA, and others. The focus is on bio-inspired optimization strategies like ACO, PSO, GWO, dolphin, bacteria, fungi, and so on that make the network efficient, prevent data transfer from losing its original link, and improve QoS parameters.

References

[1] C.R. da Costa Bento, E.C.G. Wille, Bio-inspired routing algorithm for MANETs based on fungi networks, Ad Hoc Netw. 1570-8705, 107 (2020) 102248.

[2] M. Elshaikh, M.F.M. Fadzil, C.M.N.C. Isa, N. Kamel, L. Hasnawi, SNR-based dynamic sequence distance vector routing protocol metric for mobile ad-hoc networks, in: International Conference on Computer & Information Science (ICCIS), 2012, pp. 643–648.

[3] J. Khan, S.I. Hayder, A comprehensive performance comparison of on-demand routing protocols in mobile ad-hoc networks, in: T. Kim; T. Vasilakos, K. Sakurai, Y. Xiao, G. Zhao, D. Ślęzak (Eds.), Communication and Networking. FGCN 2010, Communications in Computer and Information Science, vol. 120, Springer, Berlin, Heidelberg, 2010.

[4] K.S. Shivakumar, V.C. Patil, An optimal energy efficient cross-layer routing in MANETs, Sustain. Comput. Inform. Syst. 2210-5379, 28 (2020) 100458.

[5] S. Jamali, L. Rezaei, S.J. Gudakahriz, An energy-efficient routing protocol for MANETs: a particle swarm optimization approach, J. Appl. Res. Technol. 1665-6423, 11 (6) (2013) 803–812.

[6] M. Zhang, M. Yang, Q. Wu, R. Zheng, J. Zhu, Smart perception and autonomic optimization: a novel bio-inspired hybrid routing protocol for MANETs, Futur. Gener. Comput. Syst. 0167-739X, 81 (2018) 505–513.

[7] A.S. Awad, K.S. Alheety, A.N. Rashid, An optimal path selection using lion optimization routing protocol for mobile ad-hoc network, Period. Eng. Nat. Sci. 8 (4) (2020) 2346–2357.
[8] F. Guerriero, F. De Rango, S. Marano, E. Bruno, A biobjective optimization model for routing in mobile ad hoc networks, Appl. Math. Model. 33 (3) (2009) 1493–1512.
[9] N. Taing, S. Thipchaksurat, R. Varakulsiripunth, H. Ishii, Routing scheme for multimedia services in mobile ad hoc network, in: 5th International Conference on Information Communications & Signal Processing, 2005, 2005, pp. 11–15.
[10] S. Singh, A. Pise, O. Alfarraj, A. Tolba, B. Yoon, A cryptographic approach to prevent network incursion for enhancement of QoS in sustainable smart city using MANET, Sustain. Cities Soc. 2210-6707, (2021) 103483.
[11] J.H. Holland, Genetic algorithms, Sci. Am. 267 (1) (1992) 66–73.
[12] C. Rajan, N. Shanthi, Genetic based optimization for multicast routing algorithm for MANET, Sadhana 40 (8) (2015) 2341–2352.
[13] R. Eberhart, A. James Kennedy, New optimizer using particle swarm theory, in: Sixth International Symposium on Micro Machine and Human Science, IEEE, 1995. 0-7803-2676-8/95.
[14] R. Kalaiarasi, D. Sridharan, Performance improvement of mobile ad hoc network using particle swarm optimization, J. Comput. Inf. Syst. 9 (2013) 4213–4221.
[15] M. Dorigo, G. Di Caro, Ant colony optimization: a new meta-heuristic, in: Proceedings of the 1999 Congress on Evolutionary Computation, CEC 99, vol. 2, IEEE, 1999, pp. 1470–1477.
[16] D. Sinwar, N. Sharma, S.K. Maakar, S. Kumar, Analysis and comparison of ant colony optimization algorithm with DSDV, AODV, and AOMDV based on shortest path in MANET, J. Inf. Optim. Sci. 41 (2) (2020) 621–632.
[17] D. Karaboga, Artificial bee colony algorithm, Scholarpedia 5 (3) (2010) 6915.
[18] N.K. Vandana, An energy efficient enhanced hybrid routing protocol for MANET utilizing artificial bee colony, Int. J. Innov. Technol. Explor. Eng. 2278-3075, 8 (10) (2019).
[19] S. Das, A. Biswas, S. Dasgupta, A. Abraham, Bacterial foraging optimization Algorithm: theoretical foundations, analysis, and applications, in: A. Abraham, A.E. Hassanien, P. Siarry, A. Engelbrecht (Eds.), Foundations of Computational Intelligence Volume 3, Studies in Computational Intelligence, vol. 203, Springer, Berlin, Heidelberg, 2009.
[20] V.C. Patil, K.S. Shivakumar, Cross-layer based energy optimized routing using bacterial foraging optimization algorithm in MANET, Des. Eng. (2021) 3313–3331.
[21] X.-S. Yang, A. Gandomi, Bat algorithm: a novel approach for global engineering optimization, J. Eng. Comput. (2012).
[22] S. Mirjalili, S.M. Mirjalili, A. Lewis, Grey Wolf Optimizer, Adv. Eng. Softw. 0965-9978, 69 (2014) 46–61.
[23] S. Vimalnath, G. Ravi, Optimized geographic routing in mobile ad hoc network using Gray Wolf Optimization, EAI Endorsed Trans. Energy Web (28) (2020), e6.
[24] T.-q. Wu, M. Yao, J.-h. Yang, Dolphin swarm algorithm, Front. Inf. Technol. Electron. Eng. 2095-9184, 17 (8) (2016) 717–729.
[25] G.M. Borkar, A.R. Mahajan, P. Jawade, Trust based secure optimal route selection and enhancing QoS in mobile ad-hoc networks using AOMDV-SAPTV and DE-algorithm, IOSR J. Comput. Eng. (2017) 55–62. e-ISSN: 2278-0661, p-ISSN: 2278-8727.
[26] M.E. Manaa, S.D. Shamsi, Improved MANET routing protocols performance using optimization methods, Int. J. Eng. Technol. 7 (4.19) (2018) 642–648.

CHAPTER 17

The promise of metaheuristic algorithms for efficient operation of a highly complex power system

Davut Izci[a] and Serdar Ekinci[b]

[a]*Department of Electronics & Automation, Batman University, Batman, Turkey,* [b]*Department of Computer Engineering, Batman University, Batman, Turkey*

1. Introduction

There is a growing demand for energy worldwide, which is difficult to address by keeping power systems in isolated forms [1]. Therefore, interconnecting power systems is necessary. Doing this makes it feasible to observe highly interconnected, widely distributed power networks, which make real-world power system problems complex and highly nonlinear [2]. In an example of an interconnected power system, each of the synchronous machines rotates at the same speed for a steady-state operation and the electrical output power is equal to the feeding mechanical power [3]. However, such a configuration is prone to disturbances occurring because of variations or faults in the load. The disturbances cause a mismatch between the input mechanical and output electrical powers leading to low-frequency oscillations with frequency in the range of 0.2–3 Hz [4]. Low-frequency oscillations generally start with a small load variation in the system and then unexpectedly increase and cause the system to lose synchronism [5]. If such oscillations are not damped appropriately, they will force the system to become unstable and lead to blackouts, as were observed previously in the United States and Canada [1]. Therefore, it is necessary to appropriately control the low-frequency oscillations to achieve system stability.

Transient stability and electromechanical modes' damping are two key requirements for appropriately designing large interconnected multimachine power systems. An automatic voltage regulator in the generator excitation system can be used for improving the damping characteristics of the low-frequency oscillations and increasing the system stability, however, it is not sufficient alone to achieve adequate performance [1]. Therefore, a power system stabilizer (PSS) is used as an efficient and cost-effective solution to appropriately overcome the issues related to low-frequency oscillations [6]. As a result of the latter, conventional PSSs have widely been employed as essential components to damp low-frequency oscillations and improve the stability of the system.

It is worth noting that the conventional PSS considers a specific operation point, thus it has fixed parameters that are tuned based on trial-and-error method [7]. That means the low-frequency oscillations may not be damped out efficiently, as the performance of the system is limited to specific operating conditions. Different design approaches such as eigenvalue method, robust control method, numerical programing method, and gradient method have been proposed in the literature to design

Comprehensive Metaheuristics. https://doi.org/10.1016/B978-0-323-91781-0.00017-X

the parameters of a conventional PSS [1,3]. However, those numerical methods are not convenient to design the parameters of a PSS employed for interconnected multimachine power systems, as the latter systems operate on different conditions and configurations. Besides, the employment of numerical methods involves greater computation load [8]. In addition, it is also difficult to reach the global solution using those methods due to the complexity and dynamical nature of interconnected multimachine power systems [9].

Metaheuristic algorithms have recently attracted significant attention as intelligent optimization techniques since they have been demonstrated to effectively solve complex nonlinear, nonconvex, nondifferentiable, high-dimensional, and multimodal real-world problems [10–14]. Different metaheuristic optimization algorithms such as bacteria foraging optimization algorithm [15], cultural algorithm [16], genetic algorithm [17], honeybee mating optimization [18], artificial bee colony algorithm [19], gray wolf optimization algorithm [20], bat algorithm [21], salp swarm algorithm [22], sine-cosine algorithm [23], particle swarm optimization [24], whale optimization algorithm [25], and Harris hawks optimizer [26] have so far been adapted for PSS design by considering the limitations of the numerical methods. These methods have been demonstrated to provide good performance in this regard. However, achieving further enhancement of the PSS design is still feasible with novel algorithms despite the previously presented promise of the existing approaches just mentioned [4]. The latter case has also been well explained via the well-known no free lunch theorem in terms of the algorithmic point of view [27].

In this chapter, we use a recently developed metaheuristic algorithm called the reptile search algorithm (RSA) [28] to design the parameters of a PSS employed in a well-known Western System Coordinating Council (WSCC) three-machine, nine-bus power system [5,7].

We then provide the Simulink model of the employed power system and analyze its performance without a PSS controller. Then, we determine the optimal location to place the PSS for efficient operation. We use the performance index calculated as integral of time multiplied absolute error (ITAE) [29] as an objective function to convert the design of the PSS into an optimization problem. Initially, we compare the statistical performance of the RSA-based PSS design with particle swarm optimization (PSO)- [30], sine-cosine algorithm (SCA)- [31], and whale optimization algorithm (WOA)-based PSS designs [32]. Lastly, we compare the system stability for linear and nonlinear models. All performed analyses demonstrate the greater capability of the proposed method for designing a PSS employed in an interconnected multimachine power system.

2. Reptile search algorithm

The social behavior and hunting mechanisms of crocodiles are the main inspiration for the development of RSA as a metaheuristic optimizer [28]. Therefore, the RSA mathematically models the behavior of crocodiles to perform optimization tasks. The RSA consists of three phases: initialization, exploration (global search), and exploitation (local search).

In the initialization, an $N \times n$ matrix is stochastically generated. Here, N and n denote the number of candidate solutions and the dimension size of the problem, respectively. Each element of the matrix represents a candidate solution and created as given in Eq. (1), where x_{ij} is the jth position of the ith solution, $i = 1, 2, \ldots, N$ and $j = 1, 2, \ldots, n$.

$$x_{ij} = rand \times \left(UB_j - LB_j\right) + LB_j \tag{1}$$

In Eq. (1), LB_j and UB_j are the jth lower and upper bounds of the problem, respectively, whereas *rand* is a random number.

To perform the global search phase, the RSA mathematically models the encircling behavior of the crocodiles; they perform two types of movements: high walking and belly walking. They do not approach the prey directly when performing this behavior because they do not want to cause any disturbance so that they can perform an exploratory task. In the global search, the RSA mimics the high walking strategy for $t \leq \frac{T}{4}$, where t is the current iteration and T is the maximum iteration number. The belly walking strategy is performed for $t \leq 2\frac{T}{4}$ and $t > \frac{T}{4}$. Thus, the position update for the global search is provided as given in Eq. (2) where the current best-obtained solution's jth position is denoted by $Best_j(t)$.

$$x_{ij}(t+1) = \begin{cases} Best_j(t) - \eta_{ij}(t) \times \beta - R_{ij}(t) \times rand; t \leq \dfrac{T}{4} \\ Best_j(t) \times x_{r_1j}(t) \times ES(t) \times rand; t \leq 2\dfrac{T}{4} \text{ and } t > \dfrac{T}{4} \end{cases} \tag{2}$$

Here, β represents a sensitivity parameter controlling the accuracy of the exploration and has a fixed value of 0.1 [28] and r_1 represents a random number within $[1, N]$ where N is the number of candidate solutions and x_{r_1j} stands for a random position in ith solution. The hunting operator for ith solution's jth position is represented by η_{ij}, which is calculated as given in Eq. (3), where P_{ij} is the percentage difference between jth positions of the best-obtained and the current solutions.

$$\eta_{ij} = Best_j(t) \times P_{ij} \tag{3}$$

The term P_{ij} is calculated using Eq. (4), where $M(x_i)$ is the average positions of ith solution and ϵ is a small value.

$$P_{ij} = \alpha + \frac{x_{ij} - M(x_i)}{Best_j(t) \times (UB_j - LB_j) + \epsilon} \tag{4}$$

Through iterations, the accuracy of the exploration is controlled by the term α, which basically considers the difference between candidate solutions [28]. The function of R_{ij} given in Eq. (2) reduces the search space, which is calculated as given in Eq. (5).

$$R_{ij} = \frac{Best_j(t) - x_{r_2j}}{Best_j(t) + \epsilon} \tag{5}$$

Here, r_2 represents a random number within $[1, N]$ where N is the number of candidate solutions and x_{r_2j} is a random position in ith solution. $ES(t)$, given in Eq. (2), is a probability ratio and calculated by Eq. (6), where r_3 represents a random integer number between -1 and 1.

$$ES(t) = 2 \times r_3 \times \left(1 - \frac{1}{T}\right) \tag{6}$$

To perform the local search, the RSA mathematically models the hunting behavior of the crocodiles. Like the global search phase of the RSA, the local search phase also has two stages: hunting coordination and hunting cooperation. The crocodiles approach the prey quickly with hunting coordination and cooperation. The local search strategy of the RSA is defined by Eq. (7).

$$x_{ij}(t+1) = \begin{cases} Best_j(t) \times P_{ij}(t) \times \text{rand}; t \leq 3\frac{T}{4} \text{ and } t > 2\frac{T}{4} \\ Best_j(t) - \eta_{ij}(t) \times \epsilon - R_{ij}(t) \times \text{rand}; t \leq T \text{ and } t > 3\frac{T}{4} \end{cases} \tag{7}$$

The RSA performs the hunting coordination stage for $t \leq 3\frac{T}{4}$ and $t > 2\frac{T}{4}$, whereas it performs the hunting cooperation for $t \leq T$ and $t > 3\frac{T}{4}$, as can be observed from the latter equation. The flowchart in Fig. 1 provides a detailed illustration of the stages of the RSA.

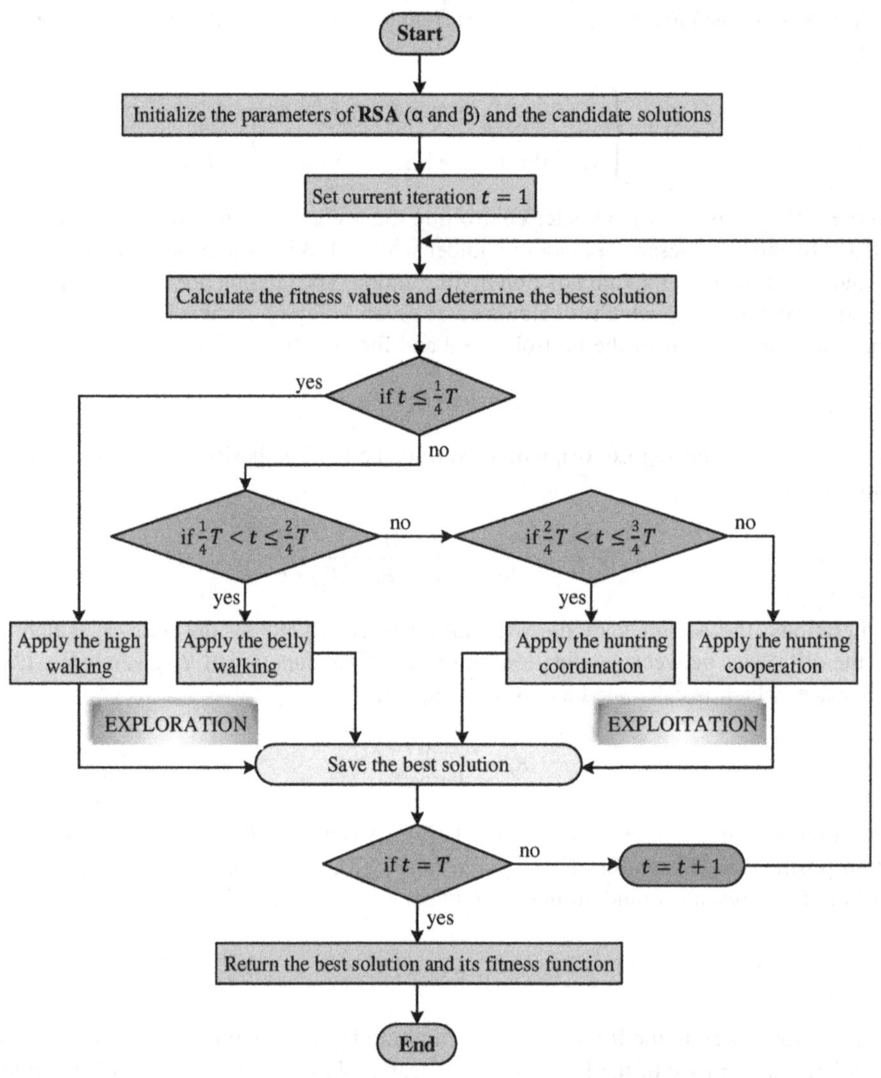

FIG. 1

Flowchart of the RSA.

3. Problem statement

3.1 Nonlinear model of multimachine power system

A set of nonlinear differential equations can be employed to describe the model of a power system. The following third-order model of the ith machine is used in this chapter since it is widely considered to be sufficient for accurately analyzing electromechanical dynamics [33,34]:

$$\frac{d\delta_i}{dt} = \omega_s(\omega_i - 1) \tag{8}$$

$$\frac{d\omega_i}{dt} = \frac{T_{mi} - T_{ei} - D_i(\omega_i - 1)}{M_i} \tag{9}$$

$$\frac{dE'_{qi}}{dt} = \frac{-E'_{qi} + E_{fdi} - \left(x_{di} - x'_{di}\right)i_{di}}{T'_{doi}} \tag{10}$$

$$T_{ei} = E'_{qi}i_{qi} + \left(x_{qi} - x'_{di}\right)i_{di}i_{qi} \tag{11}$$

where δ and ω are the rotor angle and the speed of the ith machine, respectively; M is the machine's inertia constant; D is the damping coefficient; ω_s is the synchronous speed; $E_q{}'$ is the internal voltage behind $x_d{}'$; i_d and i_q are the stator currents in d-axis and q-axis circuits, respectively; x_d and x_q are the synchronous reactances of d-axis and q-axis, respectively; $x_d{}'$ is the d-axis transient reactance; E_{fd} is exciter voltage; $T_{do}{}'$ is the d-axis open-circuit transient time constant; and T_e and T_m are electric torque and mechanical input torque, respectively. In this chapter, we consider the IEEE Type-ST1 excitation system [35]. This system is shown in Fig. 2 and can be described as given in Eq. (12):

$$\frac{dE_{fdi}}{dt} = \frac{-E_{fdi} + K_{Ai}\left(V_{refi} - V_i + V_{pssi}\right)}{T_{Ai}} \tag{12}$$

where the regulator gain and time constant are denoted by K_A and T_A, respectively. The reference and the terminal voltages are represented by V_{ref} and V, respectively, whereas V_{pss} denotes the PSS output signal.

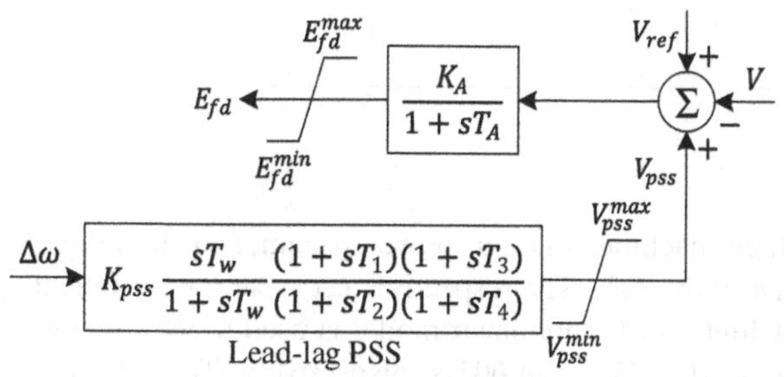

FIG. 2

Static exciter model with PSS.

3.2 Power system stabilizer structure

A PSS is basically a feedback controller that injects an additional signal to a fast-acting static automatic voltage regulator to stabilize the excitation system. The generator excitation is modulated by a PSS such that the electrical torque matches with the speed of the rotor, which consequently reduces the low-frequency oscillations [36]. As stated previously, this chapter considers the IEEE Type-ST1 excitation system with lead–lag PSS. Therefore, the structure of the PSS can be described as given in Eq. (13):

$$V_{pssi}(s) = K_{pssi}\frac{sT_w}{1+sT_w}\frac{1+sT_{1i}}{1+sT_{2i}}\frac{1+sT_{3i}}{1+sT_{4i}}\Delta\omega_i(s) \tag{13}$$

where $\Delta\omega$ stands for the speed deviation of the generator, T_w is the time constant of the washout, and K_{pss} is the dynamic gain. The terms T_1 and T_3 denote the phase-lead time constants, whereas T_2 and T_4 are the phase-lag time constants.

3.3 Linearized system model

The state equation provided in Eq. (14) can be obtained from the linearized multimachine power system model [37] where A is the state variables matrix and B is the input matrix.

$$\frac{d\Delta x}{dt} = A\Delta x + B\Delta u \tag{14}$$

The vectors of Δx and Δu represent the state and the input variables, respectively. In this chapter, Δx is taken to be equal to $[\Delta\delta \;\; \Delta\omega \;\; \Delta E_q' \;\; \Delta E_{fd}]^T$ and Δu is considered as the output signals of the PSS. The matrix of A can be used to evaluate all eigenvalues of the system using the following form where $i = 1, 2, \ldots, k$ and k is the total number of eigenvalues.

$$\lambda_i = \sigma_i \pm j\omega_i \tag{15}$$

In the latter definition, σ_i and ω_i represent the ith eigenvalue's real and imaginary parts, respectively. Thus, the eigenvalues may be real or complex. The damping ratio of the ith eigenvalue is represented by ζ_i, which is defined with Eq. (16).

$$\zeta_i = \frac{-\sigma_i}{\sqrt{\sigma_i^2 + \omega_i^2}} \tag{16}$$

4. Case study

4.1 Test system

As stated, the WSCC three-machine, nine-bus power system, shown in Fig. 3, is used as the multimachine test system for this study [5,7]. The respective test system is sufficiently large to be close to an actual power system. A fourth-order nonlinear model is used to represent each machine. The system base and the frequency are 100 MVA and 60 Hz, respectively. Table 1 lists the related generator and exciter data. The exciter is assumed to be IEEE Type-ST1, which is identical for all machines. Table 2 provides the converged load-flow result using Newton's method [7].

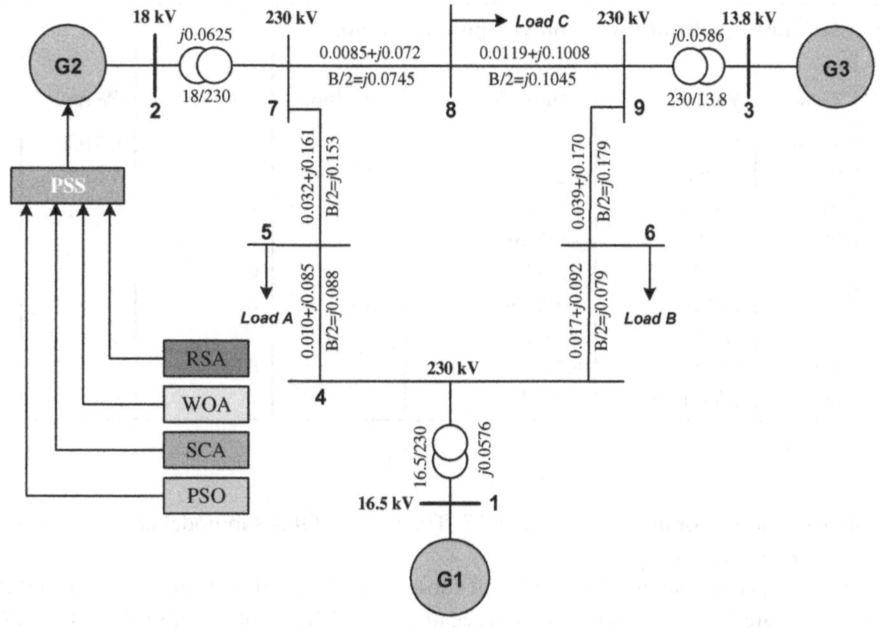

FIG. 3

Multimachine test system.

Table 1 Generator and exciter data of the employed power system.

Generator	1	2	3
H (s)	23.64	6.4	3.01
x_d (pu)	0.146	0.8958	1.3125
x_d' (pu)	0.0608	0.1198	0.1813
x_q (pu)	0.0969	0.8645	1.2578
x_q' (pu)	0.0969	0.1969	0.25
T_{do}' (s)	8.96	6.0	5.89
T_{qo}' (s)	0.31	0.535	0.6
K_A	100	100	100
T_A (s)	0.05	0.05	0.05

4.2 Developed Simulink model

Fig. 4 provides the complete Simulink model of the interconnected multimachine power system that is used in the nonlinear simulations in this chapter. As shown, the complete model is formed from three main submodels: differential equations, stator algebraic equations, and network equations. The differential equations submodel consists of subsystems 1, 2, and 3, shown in Fig. 4, and is used to calculate

Table 2 Load-flow results of the employed power system.

Bus no.	Type	Voltage (pu)	Angle (degrees)	$-P_L$ (pu)	$-Q_L$ (pu)	P_G (pu)	Q_G (pu)
1	Swing	1.0400	0	–	–	0.7164	0.2705
2	PV	1.0250	9.2800	–	–	1.63	0.0665
3	PV	1.0250	4.6648	–	–	0.85	−0.1086
4	PQ	1.0258	−2.2168	–	–	–	–
5	PQ	0.9956	−3.9888	1.25	0.5	–	–
6	PQ	1.0127	−3.6874	0.9	0.3	–	–
7	PQ	1.0258	3.7197	–	–	–	–
8	PQ	1.0159	0.7275	1	0.35	–	–
9	PQ	1.0324	1.9667	–	–	–	–

the differential equations for machines 1, 2, and 3. The inputs of this submodel are T_e, V, i_d, and i_q, and the outputs are δ, ω, E_q', and E_{fd}.

The algebraic equations submodel consists of subsystems 4, 5, and 6, shown in Fig. 4, and is used to compute the stator algebraic equations for all three machines. The inputs of this submodel are δ, E_q', i_d, and i_q, and the outputs are T_e, V, and θ. The network equations submodel consists of subsystems 7, 8, and 9, shown in Fig. 4, and is used to compute the values of electrical current outputs for different machines. The inputs of this submodel are the terminal voltages of V_1, V_2, and V_3 and their angles of θ_1, θ_2, and θ_3, whereas the outputs are the electrical currents of I_1, I_2, and I_3. The constructed Simulink model allows the simulation parameters of start and stop times, types of solvers, step size, tolerance, and output options to be facilitated as desired. In this chapter, load flow, fault clearance time, initial values of parameters, and fault-related changes in the network are controlled through an m-file developed in MATLAB software.

4.3 System analysis without PSS and optimal PSS location

To analyze the test system for transient stability without using any stabilizer, a three-phase fault is assumed to occur at $t = 1$ s near bus 7 at the end of line 5–7, which is a representation of a large disturbance. The fault continues for six cycles (100 ms) and is removed without line tripping. Then, the original system is restored. In transient stability simulation, for numerical integration of the differential equations, the fourth-order Runge-Kutta method is used with a step integration of $\Delta t = 0.01$ s. Then, the behavior of the system is evaluated for 10 s. Figs. 5 and 6, respectively, demonstrate the responses of the rotor angles and rotor speeds of generator 2 ($G2$) and generator 3 ($G3$) with respect to generator 1 ($G1$).

As shown, the power system oscillations are insufficiently damped without a stabilizer although the system is stable. Therefore, employment of a PSS would be required for quickly damping out the low-frequency oscillations.

The most important step in designing a stabilizer is to determine the most effective generators for locating the PSS. In this regard, the participation factor method is widely used in power systems to find the optimum PSS locations [23]. Therefore, for the system without PSS, we performed the linearization

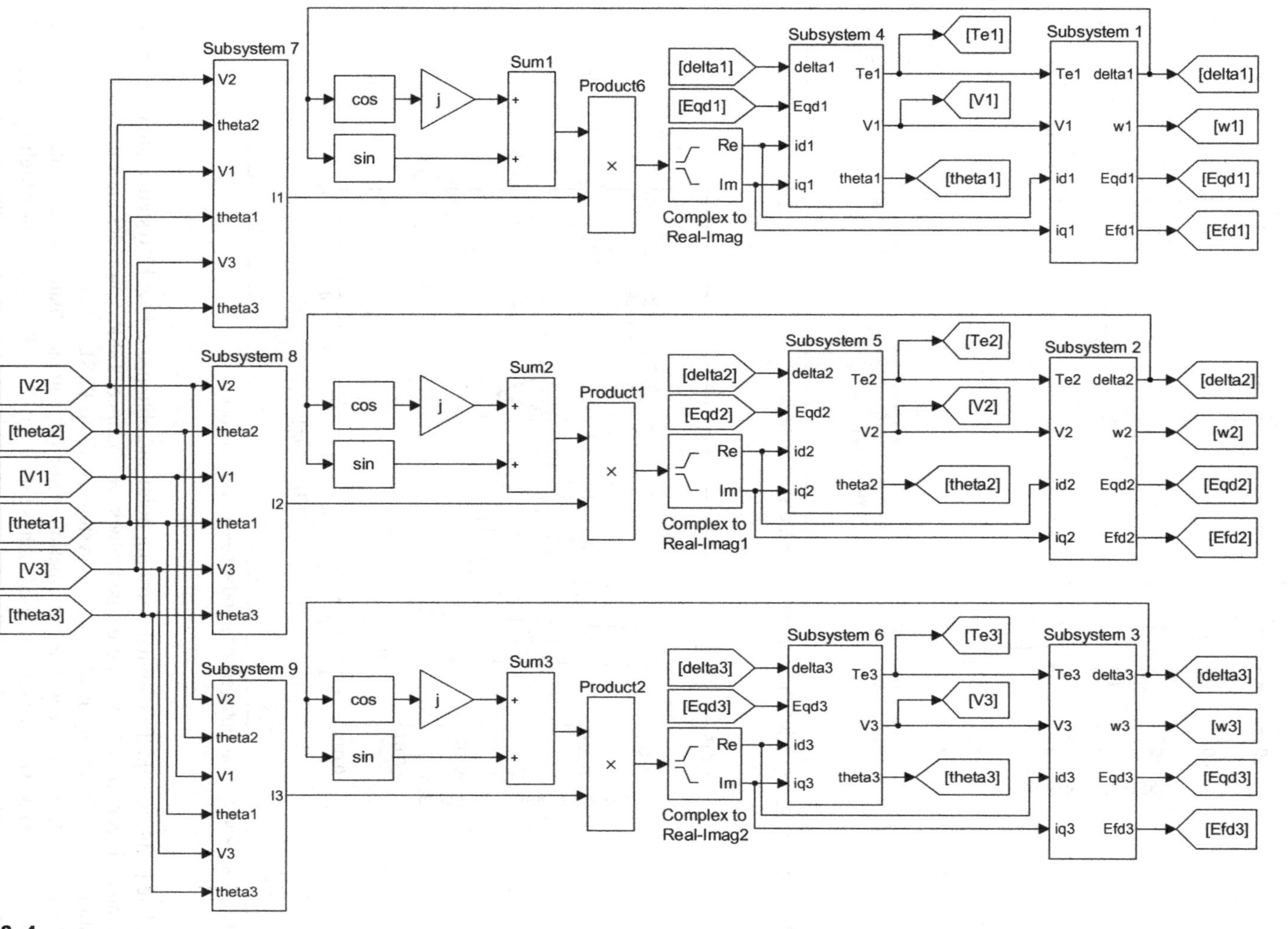

FIG. 4

The constructed Simulink model of the three-machine, nine-bus power system.

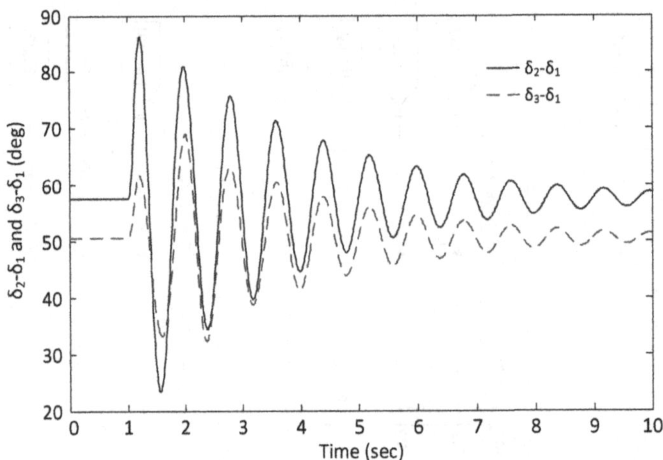

FIG. 5

Plot of $\delta_2-\delta_1$ and $\delta_3-\delta_1$ rotor angles under severe disturbance.

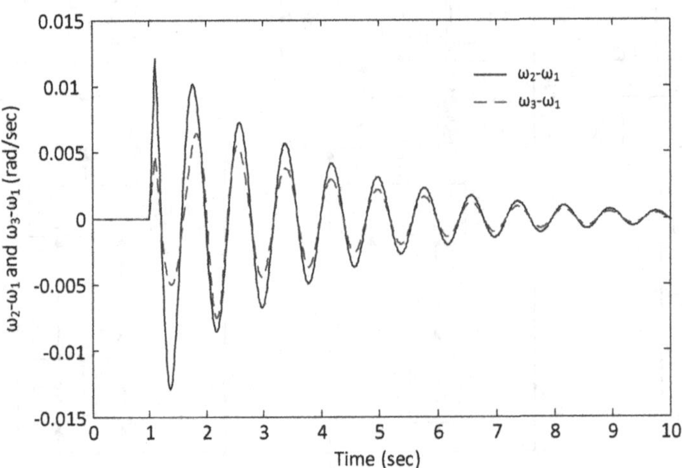

FIG. 6

Plot of $\omega_2-\omega_1$ and $\omega_3-\omega_1$ rotor speeds under severe disturbance.

procedure [7,34] around the nominal operating point and obtained a 12×12 system matrix (A). Since the damping is zero ($D=0$), two zero eigenvalues are obtained. Table 3 provides the values of the participation factors for all eigenvalues of the system (without PSS).

The table lists only those with a participation factor value greater than 0.2. The table also provides the state variables and the machines related to those state variables. The eigenvalues highlighted in bold indicate the electromechanical oscillation modes. As can be seen from the eigenvalues, the damping

Table 3 All eigenvalues and their frequencies, damping ratios, and participation factors.

Eigenvalue	Frequency	Damping ratio	Machine no.	Machine variable	Participation factor
$-9.8638 \pm 13.6643i$	2.6822	58.53%	1	E_q', E_{fd}	1.0000, 0.9886
			2	E_q', E_{fd}	0.8865, 0.8633
			3	E_q', E_{fd}	0.7463, 0.7257
$-1.3738 \pm 11.7499i$	1.8828	11.61%	3	δ, ω	1.0000, 1.0000
			2	δ, ω	0.2388, 0.2388
$-0.3831 \pm 7.8846i$	1.2563	**4.85%**	**2**	δ, ω	1.0000, 1.0000
			1	δ, ω	0.4634, 0.4634
-12.7012	2.0215	100%	3	E_{fd}, E_q'	1.0000, 0.5978
			2	E_{fd}, E_q'	0.4775, 0.2855
$-9.9194 \pm 6.4142i$	1.8800	83.97%	1	E_q', E_{fd}	1.0000, 0.9765
			2	E_q', E_{fd}	0.5898, 0.5657
-5.5006	0.8754	100%	3	E_q', E_{fd}	1.0000, 0.3602
			2	E_q'	0.3897
0	–	–	1	δ, ω	1.0000, 1.0000
			2	δ, ω	0.2783, 0.2783
0	–	–	1	δ, ω	1.0000, 1.0000
			2	δ, ω	0.2783, 0.2783

ratio of the system is significantly small, thus the system exhibits a highly oscillating response. Therefore, one or more PSSs should be deployed to improve the dynamic performance of the system. Considering results of the participation factor analyses given in Table 3, placement of a PSS only to the generator 2 ($G2$) would be sufficient to damp out the local modes of the oscillations because the damping ratio of the eigenvalue of $-0.3831 \pm 7.8846i$ is smaller than 5%.

4.4 Objective function and PSS tuning via RSA

The ITAE given in Eq. (17) is utilized as an objective function to design the PSS controller because it allows the parameters of the PSS to be chosen for minimization.

$$ITAE = \int_{t=t_{pre}}^{t=T_{sim}} (t - t_{pre})|\omega_2 - \omega_1| dt + \int_{t=t_{pre}}^{t=T_{sim}} (t - t_{pre})|\omega_3 - \omega_1| dt \tag{17}$$

In Eq. (17), T_{sim} stands for the simulation time and is chosen as 50 s for this work. The term t_{pre} represents the time where the fault occurs, whereas ω_1, ω_2, and ω_3 represent the rotor speeds for $G1$, $G2$, and $G3$, respectively. The advantage of the employed objective function is that it only requires minimal dynamic plant information. In this chapter, the design problem is performed by considering the criteria of $0.1 \leq K_{pss} \leq 100$ and $0.01 \leq T_i \leq 1$ where $i = 1, 2, 3, 4$.

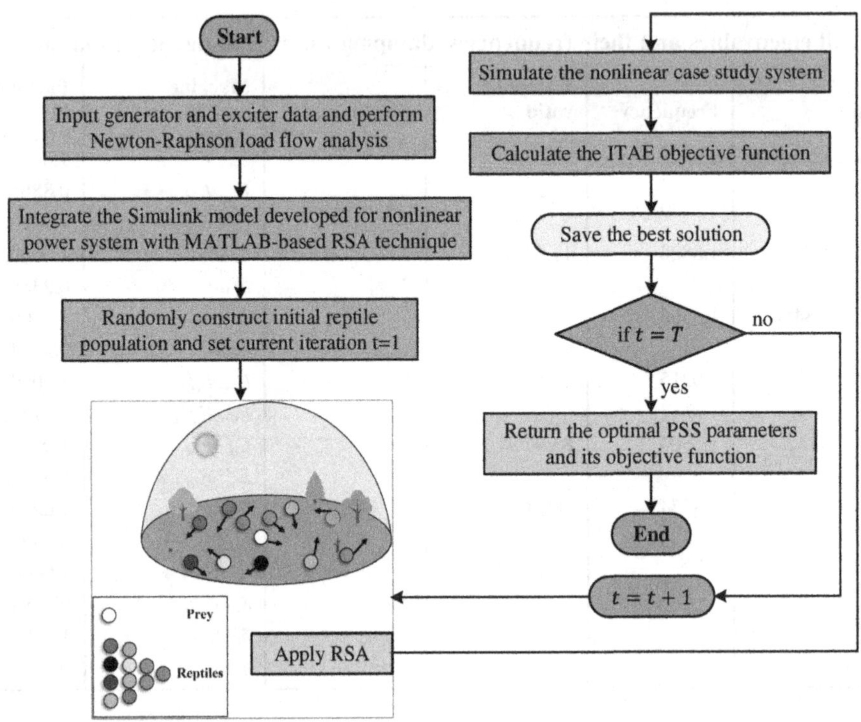

FIG. 7

Flowchart of RSA-based PSS design for multimachine power system.

Fig. 7 shows the implementation steps of RSA-based PSS design for an interconnected multimachine power system. The developed "Newton–Raphson load flow program" is run by using the generator and the exciter input data. This is followed by the integration of the developed Simulink model for a nonlinear power system with the MATLAB-based RSA technique. For the optimization problem of the respective PSS design, the population size (number of crocodiles) is determined to be 40 and the total number of iterations is set to 50. Then, the RSA performs the optimizations for 30 individual runs by simulating the nonlinear test system with the stated configuration, and the objective function is computed.

4.5 Statistical performance of RSA

In this chapter, we also use PSO [30], SCA [31], and WOA [32] to assess the proposed method. Table 4 lists the adopted parameters of these algorithms.

Table 5 provides the comparative average computational times of the adopted algorithms for PSS design. The reason for the longer computational times is the complexity of the system and the employment of a nonlinear stabilizer design instead of a linear stabilizer design, which is commonly used in the literature. Nevertheless, the RSA-based PSS design has less computational times compared to WOA-, SCA-, and PSO-based PSS designs.

Table 4 Parameter settings of RSA, WOA, SCA, and PSO algorithms.

Algorithm	Parameter	Value
RSA	α	0.1
	β	0.1
WOA	a_1	Decreases linearly from 2 to 0
	a_2	Decreases linearly from −1 to −2
SCA	A	2
PSO	c_1	2
	c_2	2
	Inertia weight	Decreases linearly from 0.9 to 0.1

Table 5 Average computation times of RSA, WOA, SCA, and PSO algorithms per run.

Algorithm	RSA	WOA	SCA	PSO
Time (s)	1626.97	2119.15	1823.32	2463.89

Fig. 8 further illustrates the ability of the proposed method in terms of obtained ITAE objective function values with respect to number of runs. Like with computational times, the proposed method achieved a minimum objective function value for each run compared to the other methods.

We also performed a statistical test for mean, standard deviation, best, and rank. Fig. 9 provides these results. As shown, the proposed RSA-based PSS design method is ranked first among the other methods as well as reaches the lowest values for other statistical parameters.

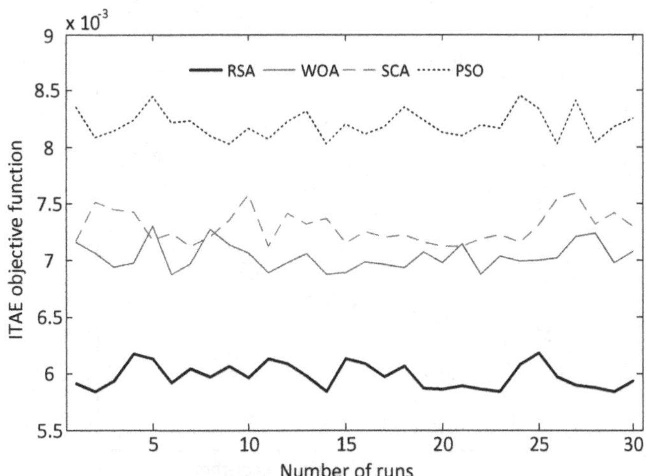

FIG. 8

Obtained ITAE objective function values from 30 runs.

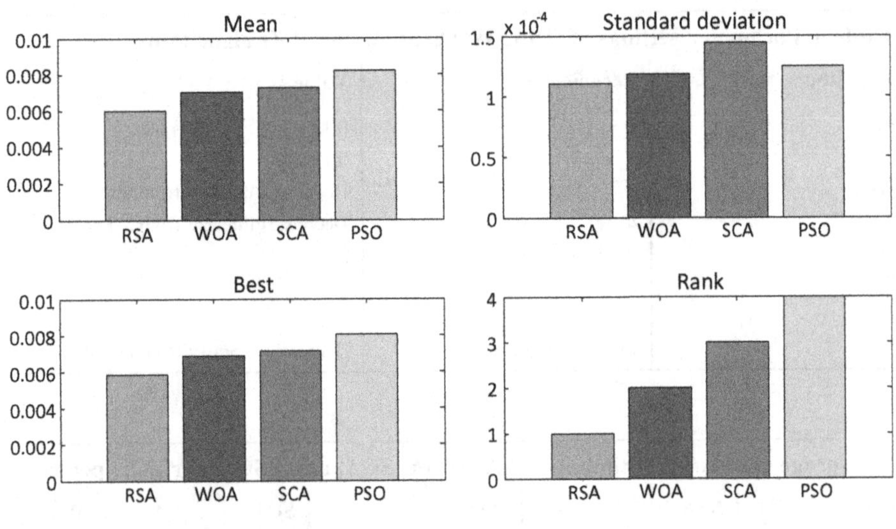

FIG. 9

Statistical values of ITAE objective function in terms of mean, standard deviation, best and rank.

To further visualize the statistical performance of the proposed method, the boxplot analysis is also performed comparatively as shown in Fig. 10. As illustrated in the latter figure, even the worst objective function value achieved by the RSA-based design method is well below the best values obtained by other compared methods further depicting the superior ability of the proposed method.

Fig. 11 demonstrates the convergence profiles of RSA, WOA, SCA, and PSO algorithms comparatively for their best runs. The RSA-based method reached lower objective function values later than

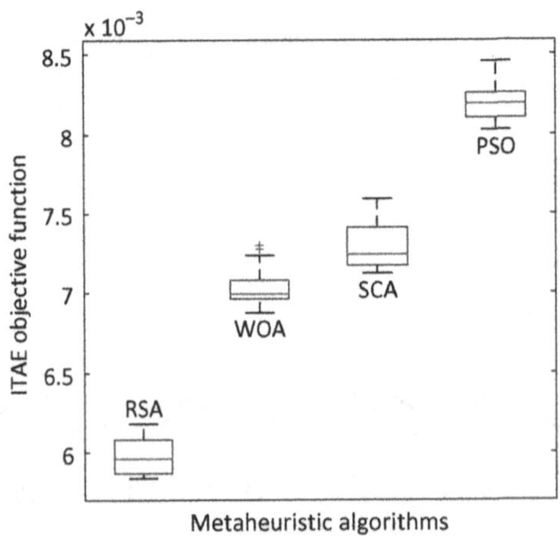

FIG. 10

Comparative boxplot analysis results for RSA, WOA, SCA, and PSO.

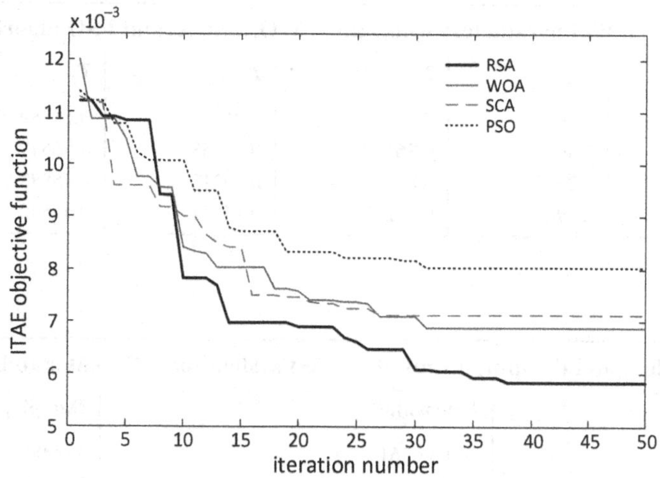

FIG. 11

Convergence profiles for RSA, WOA, SCA, and PSO algorithms.

any other compared algorithm-based approach since it does not stagnate into the local minimum. Therefore, it is more capable than WOA-, SCA-, and PSO-based methods in terms of escaping local minimum and converging to the global solution.

Lastly, we performed the Wilcoxon signed-rank test to demonstrate that the superior ability of the proposed method did not occur by chance. Therefore, we compared the proposed RSA-based method with WOA-, SCA-, and PSO-based methods. Table 6 provides the related *p*-values along with the statistical significance of the proposed method. As shown, the proposed RSA-based design method demonstrates a statistically significant difference, that is, it performed well compared to other methods, which did not occur by chance.

Table 7 lists the obtained optimal PSS parameters using different design methods (RSA, WOA, SCA, and PSO). The listed values for the respective methods are used in this chapter to perform stability analyses for the system both in linear and nonlinear forms.

4.6 System stability in linear model

Table 8 provides all the eigenvalues ($\lambda=\sigma\pm j\omega$) and damping ratios ($\zeta$) of the PSS-based systems designed with different methods. As shown, the eigenvalues for all methods are in the left side of the s-plane. However, the proposed RSA-based PSS design method provides the greater damping ratio

Table 6 Nonparametric statistical tests.

Objective function	RSA vs WOA		RSA vs SCA		RSA vs PSO	
	p-value	Significant	*p*-value	Significant	*p*-value	Significant
ITAE	1.7344E−06	+	1.7344E−06	+	1.7344E−06	+

Table 7 The optimal PSS parameters using RSA, WOA, SCA, and PSO algorithms.

Stabilizer type	K_{pss}	T_1	T_2	T_3	T_4
RSA-based PSS	5.7623	0.91085	0.01000	0.74389	0.60274
WOA-based PSS	7.8679	0.36193	0.02958	0.56647	0.30460
SCA-based PSS	6.2933	0.87199	0.55212	0.45883	0.03562
PSO-based PSS	2.4983	0.72262	0.14671	0.44540	0.05158

Table 8 Eigenvalues and damping ratios of the test system for different stabilizer types.

Stabilizer type	Eigenvalue	Damping ratio
RSA-based PSS	−105.6361	1.0000
	$-6.3376 \pm 25.6761i$	**0.2396**
	$-8.6874 \pm 10.6862i$	0.6308
	$-2.6033 \pm 10.4274i$	0.2422
	$-8.4830 \pm 3.1842i$	0.9362
	$-1.6629 \pm 4.3407i$	0.3577
	−1.6222	1.0000
	−0.2344	1.0000
	0	–
	0	–
WOA-based PSS	−44.6461	1.0000
	$-3.4156 \pm 21.6874i$	**0.1556**
	$-8.4120 \pm 10.8303i$	0.6134
	$-2.7962 \pm 10.2916i$	0.2622
	$-8.3913 \pm 3.1829i$	0.9350
	$-2.1882 \pm 4.5867i$	0.4306
	−0.3012	1.0000
	−3.1179	1.0000
	0	–
	0	–
SCA-based PSS	−38.8059	1.0000
	$-3.2641 \pm 20.0026i$	**0.1611**
	$-8.1950 \pm 10.8861i$	0.6014
	$-2.9621 \pm 10.2040i$	0.2788
	$-8.2753 \pm 3.1100i$	0.9361
	$-2.4784 \pm 4.7279i$	0.4643
	−1.8551	1.0000
	−0.2565	1.0000
	0	–
	0	–

Table 8 Eigenvalues and damping ratios of the test system for different stabilizer types—cont'd

Stabilizer type	Eigenvalue	Damping ratio
PSO-based PSS	−33.8817	1.0000
	$-3.3326 \pm 20.1436i$	**0.1632**
	$-8.5463 \pm 10.9427i$	0.6155
	$-2.8115 \pm 10.4565i$	0.2597
	$-8.7471 \pm 3.5028i$	0.9283
	$-1.3568 \pm 5.3380i$	0.2463
	−3.9510	1.0000
	−0.1642	1.0000
	0	–
	0	–

The bold ones signify the best obtained values.

(ζ_{min}=23.96%) compared to other PSS design methods based on WOA (ζ_{min}=15.56%), SCA (ζ_{min}=16.11%), and PSO (ζ_{min}=16.32%). Therefore, the RSA-based PSS design greatly improves the small signal stability of the interconnected multimachine power system and the damping characteristics of electromechanical modes.

4.7 System stability in nonlinear model

In this chapter, the system stability is also performed using the nonlinear model of the plant such that the ability of the proposed RSA-based PSS design method can be demonstrated for interconnected multimachine power system from a real-world complex system perspective. Therefore, a six-cycle, three-phase fault is applied on bus 7 at the end of line 5–7 at time $t=1$ s to further validate the performance of the RSA-based PSS design. The related fault is removed without line tripping. After the clearance of the fault, the original system is restored.

Figs. 12 and 13 respectively demonstrate the system rotor angle response of the $G2$ with respect to $G1$ ($\delta_{21}=\delta_2-\delta_1$) and $G3$ with respect to $G1$ ($\delta_{31}=\delta_3-\delta_1$) for the related severe disturbance. Similarly, Figs. 14 and 15 respectively demonstrate the variations of the rotor speed response of $G2$ with respect to $G1$ ($\omega_{21}=\omega_2-\omega_1$) and $G3$ with respect to $G1$ ($\omega_{31}=\omega_3-\omega_1$). As mentioned earlier, the power system is stable without a PSS, however, the oscillations of the power system are damped inadequately. Therefore, employment of a PSS helps maintain the stability of the power system and suppresses the oscillations effectively, which can be observed for all PSS design methods shown in Figs. 12–15. Meanwhile, compared to the other design methods, the RSA-based PSS design method suppresses the oscillations more quickly and helps the system to reach stability earlier.

The superior ability of the proposed RSA-based PSS design method can clearly be observed from all nonlinear simulation results, which indicate the method's better damping of low-frequency oscillations. Therefore, the dynamic stability of the system increases significantly with the proposed RSA-based PSS design for the stated severe disturbance compared to other design approaches.

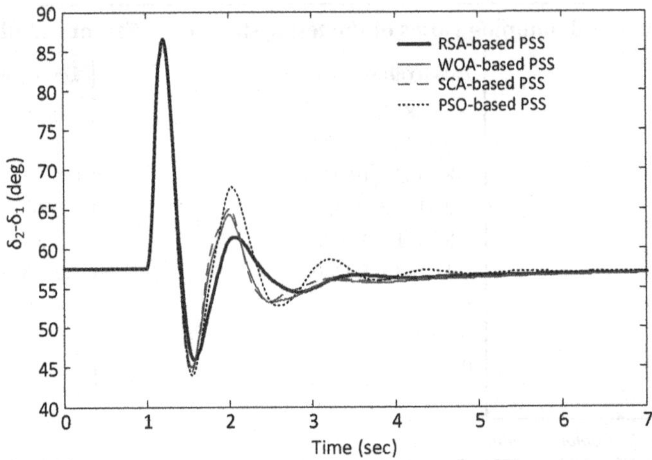

FIG. 12

$\delta_2 - \delta_1$ rotor angle response for various stabilizers.

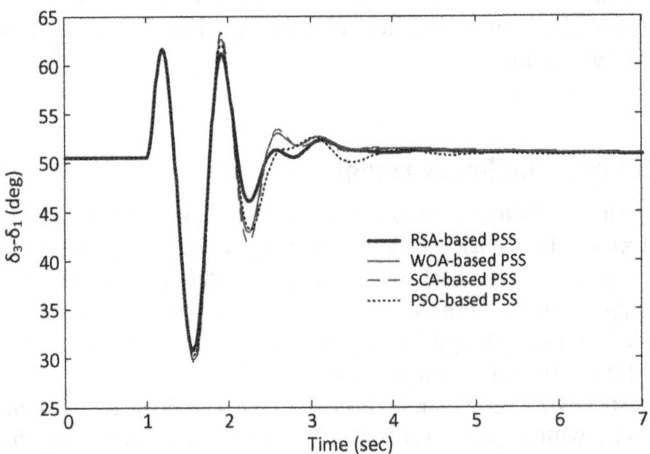

FIG. 13

$\delta_3 - \delta_1$ rotor angle response for various stabilizers.

5. Conclusions

In this chapter, we discussed the optimum tuning procedure of a PSS design for an interconnected multimachine power system. In this regard, we used the RSA optimizer as an efficient nature-inspired metaheuristic optimizer to tune the parameters of the PSS. We employed a well-known WSCC three-machine, nine-bus power system as a highly complex and real-world power system engineering

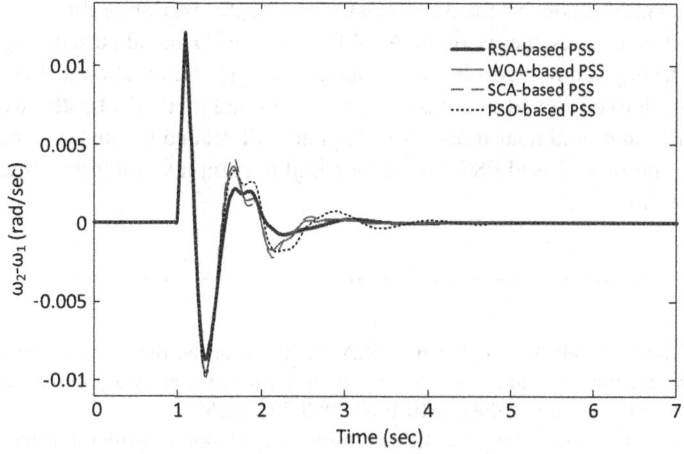

FIG. 14

$\omega_2 - \omega_1$ rotor speed response for various stabilizers.

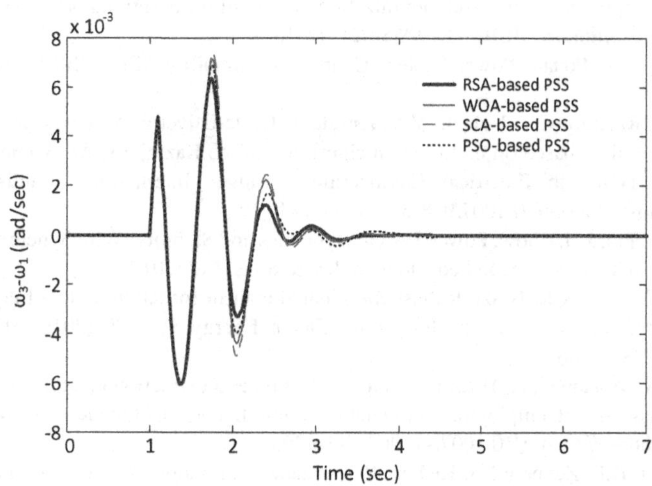

FIG. 15

$\omega_3 - \omega_1$ rotor speed response for various stabilizers.

design problem. We demonstrated the powerful capability of the RSA metaheuristic method to improve performance in difficult real-world engineering design problems. We also discussed the nonlinear and linearized models of the employed power system along with the structure of the PSS. We provided a detailed Simulink model of the employed nonlinear interconnected multimachine power system. Initially, we analyzed the system without PSS and determined the optimal PSS locations. We used the

ITAE performance index to convert the design task into an optimization problem. We followed this by performing comparative assessments of SCA, WOA, and PSO metaheuristic algorithms. We performed statistical testing; computed objective function value, computational time, and convergence rate; and used the Wilcoxon signed-rank test to rank the studied methods. Lastly, we analyzed the system using both linear and nonlinear models of the plant. All related evaluations indicate the superior ability of the RSA optimizer-based PSS design for a highly complex, nonlinear, interconnected multimachine power system.

References

[1] A. Sabo, N.I. Abdul Wahab, M.L. Othman, M.Z.A. Mohd Jaffar, H. Beiranvand, Optimal design of power system stabilizer for multimachine power system using farmland fertility algorithm, Int. Trans. Electr. Energy Syst. (2020) e12657, https://doi.org/10.1002/2050-7038.12657.

[2] B.M. Alshammari, T. Guesmi, New chaotic sunflower optimization algorithm for optimal tuning of power system stabilizers, J. Electr. Eng. Technol. 15 (2020) 1985–1997, https://doi.org/10.1007/s42835-020-00470-1.

[3] M. Singh, R.N. Patel, D.D. Neema, Robust tuning of excitation controller for stability enhancement using multi-objective metaheuristic Firefly algorithm, Swarm Evol. Comput. 44 (2019) 136–147, https://doi.org/10.1016/j.swevo.2018.01.010.

[4] D. Izci, A novel improved atom search optimization algorithm for designing power system stabilizer, Evol. Intell. (2021), https://doi.org/10.1007/s12065-021-00615-9.

[5] P.M. Anderson, A.A. Fouad, Power System Control and Stability, IEEE, 2002, https://doi.org/10.1109/9780470545577.

[6] N. Razmjooy, S. Razmjooy, Z. Vahedi, V.V. Estrela, G.G. de Oliveira, A new design for robust control of power system stabilizer based on moth search algorithm, in: N. Razmjooy, M. Ashourian, Z. Foroozandeh (Eds.), Lecture Notes in Electrical Engineering, Springer International Publishing, Cham, 2021, pp. 187–202, https://doi.org/10.1007/978-3-030-56689-0_10.

[7] P.W. Sauer, M.A. Pai, J.H. Chow, Power System Dynamics and Stability: With Synchrophasor Measurement and Power System Toolbox, second ed., John Wiley & Sons, Ltd, 2017.

[8] A.K. Gupta, K. Verma, K.R. Niazi, Robust coordinated control for damping low frequency oscillations in high wind penetration power system, Int. Trans. Electr. Energy Syst. 29 (2019) e12006, https://doi.org/10.1002/2050-7038.12006.

[9] N.I.A. Wahab, A. Mohamed, A. Hussain, Feature selection and extraction methods for power systems transient stability assessment employing computational intelligence techniques, Neural. Process. Lett. 35 (2012) 81–102, https://doi.org/10.1007/s11063-011-9205-x.

[10] D. Izci, S. Ekinci, H.L. Zeynelgil, J. Hedley, Performance evaluation of a novel improved slime mould algorithm for direct current motor and automatic voltage regulator systems, Trans. Inst. Meas. Control 44 (2022) 435–456, https://doi.org/10.1177/01423312211037967.

[11] D. Izci, An enhanced slime mould algorithm for function optimization, in: 2021 3rd International Congress on Human-Computer Interaction, Optimization and Robotic Applications (HORA), IEEE, 2021, pp. 1–5, https://doi.org/10.1109/HORA52670.2021.9461325.

[12] D. Izci, S. Ekinci, Comparative performance analysis of Slime mould algorithm for efficient design of proportional–integral–derivative controller, Electrica 21 (2021) 151–159, https://doi.org/10.5152/electrica.2021.20077.

[13] D. Izci, S. Ekinci, S. Orenc, A. Demiroren, Improved artificial electric field algorithm using Nelder-Mead simplex method for optimization problems, in: 2020 4th International Symposium on Multidisciplinary

Studies and Innovative Technologies (ISMSIT), IEEE, 2020, pp. 1–5, https://doi.org/10.1109/ISMSIT50672.2020.9255255.

[14] D. Izci, S. Ekinci, E. Eker, M. Kayri, Improved manta ray foraging optimization using opposition-based learning for optimization problems, in: 2020 International Congress on Human-Computer Interaction, Optimization and Robotic Applications (HORA), IEEE, 2020, pp. 1–6, https://doi.org/10.1109/HORA49412.2020.9152925.

[15] S.M. Abd-Elazim, E.S. Ali, Power system stability enhancement via bacteria foraging optimization algorithm, Arab. J. Sci. Eng. 38 (2013) 599–611, https://doi.org/10.1007/s13369-012-0423-y.

[16] A. Khodabakhshian, R. Hemmati, Multi-machine power system stabilizer design by using cultural algorithms, Int. J. Electr. Power Energy Syst. 44 (2013) 571–580, https://doi.org/10.1016/j.ijepes.2012.07.049.

[17] L.H. Hassan, M. Moghavvemi, H.A.F. Almurib, K.M. Muttaqi, V.G. Ganapathy, Optimization of power system stabilizers using participation factor and genetic algorithm, Int. J. Electr. Power Energy Syst. 55 (2014) 668–679, https://doi.org/10.1016/j.ijepes.2013.10.026.

[18] M. Mohammadi, N. Ghadimi, Optimal location and optimized parameters for robust power system stabilizer using honeybee mating optimization, Complexity 21 (2015) 242–258, https://doi.org/10.1002/cplx.21560.

[19] S. Ekinci, A. Demirören, Modeling, simulation, and optimal design of power system stabilizers using ABC algorithm, Turk. J. Electr. Eng. Comput. Sci. 24 (2016) 1532–1546, https://doi.org/10.3906/elk-1311-208.

[20] M.R. Shakarami, I. Faraji Davoudkhani, Wide-area power system stabilizer design based on Grey Wolf Optimization algorithm considering the time delay, Electr. Pow. Syst. Res. 133 (2016) 149–159, https://doi.org/10.1016/j.epsr.2015.12.019.

[21] L. Chaib, A. Choucha, S. Arif, Optimal design and tuning of novel fractional order PID power system stabilizer using a new metaheuristic Bat algorithm, Ain Shams Eng. J. 8 (2017) 113–125, https://doi.org/10.1016/j.asej.2015.08.003.

[22] S. Ekinci, B. Hekimoglu, Parameter optimization of power system stabilizer via Salp Swarm algorithm, in: 2018 5th International Conference on Electrical and Electronic Engineering (ICEEE), IEEE, 2018, pp. 143–147, https://doi.org/10.1109/ICEEE2.2018.8391318.

[23] S. Ekinci, Optimal design of power system stabilizer using sine cosine algorithm, J. Fac. Eng. Archit. Gazi Univ. 34 (2019) 1330–1350, https://doi.org/10.17341/gazimmfd.460529.

[24] B. Dasu, M. Siva Kumar, R. Srinivasa Rao, Design of robust modified power system stabilizer for dynamic stability improvement using Particle Swarm Optimization technique, Ain Shams Eng. J. 10 (2019) 769–783,-https://doi.org/10.1016/j.asej.2019.07.002.

[25] B. Dasu, M. Sivakumar, R. Srinivasarao, Interconnected multi-machine power system stabilizer design using whale optimization algorithm, Prot. Control Mod. Power Syst. 4 (2019) 2, https://doi.org/10.1186/s41601-019-0116-6.

[26] L. Chaib, A. Choucha, S. Arif, H.G. Zaini, A. El-Fergany, S.S.M. Ghoneim, Robust design of power system stabilizers using improved Harris hawk optimizer for interconnected power system, Sustainability 13 (2021) 11776, https://doi.org/10.3390/su132111776.

[27] D.H. Wolpert, W.G. Macready, No free lunch theorems for optimization, IEEE Trans. Evol. Comput. 1 (1997) 67–82, https://doi.org/10.1109/4235.585893.

[28] L. Abualigah, M.A. Elaziz, P. Sumari, Z.W. Geem, A.H. Gandomi, Reptile Search Algorithm (RSA): a nature-inspired meta-heuristic optimizer, Expert Syst. Appl. (2021) 116158, https://doi.org/10.1016/j.eswa.2021.116158.

[29] D. Izci, Design and application of an optimally tuned PID controller for DC motor speed regulation via a novel hybrid Lévy flight distribution and Nelder–Mead algorithm, Trans. Inst. Meas. Control 43 (2021) 3195–3211, https://doi.org/10.1177/01423312211019633.

[30] J. Kennedy, R. Eberhart, Particle swarm optimization, in: Proceedings of ICNN'95—International Conference on Neural Networks, IEEE, 1995, pp. 1942–1948.

[31] S. Mirjalili, SCA: a sine cosine algorithm for solving optimization problems, Knowl. Based Syst. 96 (2016) 120–133, https://doi.org/10.1016/j.knosys.2015.12.022.

[32] S. Mirjalili, A. Lewis, The whale optimization algorithm, Adv. Eng. Softw. 95 (2016) 51–67, https://doi.org/10.1016/j.advengsoft.2016.01.008.

[33] M.A. Abido, Y.L. Abdel-Magid, Eigenvalue assignments in multimachine power systems using tabu search algorithm, Comput. Electr. Eng. 28 (2002) 527–545, https://doi.org/10.1016/S0045-7906(01)00005-2.

[34] D. Mondal, A. Chakrabarti, A. Sengupta, Power System Small Signal Stability Analysis and Control, Elsevier, 2020, https://doi.org/10.1016/C2018-0-02439-1.

[35] M.A. Abido, Y.L. Abdel-Magid, Optimal design of power system stabilizers using evolutionary programming, IEEE Trans. Energy Convers. 17 (2002) 429–436, https://doi.org/10.1109/TEC.2002.805179.

[36] D. Chitara, K.R. Niazi, A. Swarnkar, N. Gupta, Cuckoo search optimization algorithm for designing of a multimachine power system stabilizer, IEEE Trans. Ind. Appl. 54 (2018) 3056–3065, https://doi.org/10.1109/TIA.2018.2811725.

[37] S. Ekinci, A. Demiroren, B. Hekimoglu, Parameter optimization of power system stabilizers via kidney-inspired algorithm, Trans. Inst. Meas. Control 41 (2019) 1405–1417, https://doi.org/10.1177/0142331218780947.

CHAPTER

Genome sequence assembly using metaheuristics

18

Sehej Jain and Kusum Kumari Bharti
Department of Computer Science and Engineering, PDPM Indian Institute of Information Technology, Design and Manufacturing, Jabalpur, India

1. Introduction

Genome sequencing is a process through which an organism's genome can be decoded [1]. It has immense applications in modern science, including personalized medicine, clan genomics, vaccine development, and identifying and developing cures for various diseases. The entire genome sequencing process consists of biological processes paired with sequencing algorithms through which genome sequences, which can have billions of protein bases, are decoded.

The human genome contains more than six billion base pairs. However, even a sequence with just 600 base pairs cannot be read with complete certainty. Therefore, the entire genome is broken down into smaller fragments. This process is a part of *shotgun sequencing.* Reading these small fragments is a simple task; however, reassembling the original genome from these fragments is difficult. This process is often referred to as DNA fragment assembly. The most commonly accepted method for this is the overlap-layout-consensus (OLC) approach [2]. In this approach, the first step entails the creation of an overlap graph between the fragments. The second phase is the layout phase, which bundles the fragments into *contigs*. The final phase is the consensus phase in which the DNA sequence is finally derived. The optimal placement of these fragments helps to reconstruct the original sequence.

As a consequence, it helps in the correct interpretation of the sequence. As such, this problem has been modeled as an optimization problem. The most optimal solution would be the one containing the shortest number of contigs. This is a combinatorial optimization problem that can be solved using metaheuristic algorithms such as Particle Swarm Optimization (PSO), Gray Wolf Optimizer (GWO), and Genetic Algorithms (GAs).

Metaheuristic algorithms are designed to find a heuristic that provides a sufficiently good solution to a problem with reduced computational resources and lesser time. Assembling genomes takes significant time. The first-ever genome sequencing of a human being took 13 years to complete as a part of the "Human Genome Project" in 2003. In this chapter, we examine the application of metaheuristic algorithms to solve the genome sequencing problem. The chapter presents a detailed description of the problem formulation, describes the metaheuristic algorithms used, discusses preprocessing, and presents the proposed model and its results.

Comprehensive Metaheuristics. https://doi.org/10.1016/B978-0-323-91781-0.00018-1

2. Past works

Parsons et al. [3] first proposed a GA-based approach to solve the DNA sequence assembly problem in 1994. It was done as part of the Human Genome Project [4], which was started in 1990. The project aimed to determine the base pairs that make up human DNA and identify, map, and sequence all the nucleotides of the human genome. The project was declared complete in 2003.

Over the past two decades, several fragment assembly techniques have been proposed. Express Sequence Tab (EST) Assemblers were created with the objective of assembling individual genes instead of whole genomes [5]. De-novo Assemblers [6,7] gained popularity since they do not require any reference genomes. De-novo can be either Greedy based or Graph based. Unlike De-novo Assemblers, Mapping Assemblers [8] make use of an existing backbone sequence to build a similar sequence that is not identical to the backbone sequence. The TIGR assembler [9] was developed for whole-genome shotgun sequencing. It overcame many obstacles that traditional assemblers had not. The EULER path [10] was an alternate approach to solve this problem instead of the OLC model. It succeeded in places where the OLC model failed to show better results.

Recently, many heuristic methods have been explored in the literature to solve this problem. GA-based approaches [3,11–13] were explored first since GA can be directly used to solve this problem. Since then, PSO [14], CS [15,16], Artificial Bee Colony (ABC) [17,18], and other metaheuristic algorithms [19,20] have been investigated to solve this problem. In this chapter, we give a detailed description of how PSO, GWO, and Cuckoo Search (CS) can be used to solve the genome sequencing problem.

3. Genome sequencing

DNA of any being consists of three primary components:

1. a nitrogenous base or nucleotide
2. a phosphate group
3. a sugar molecule

This nitrogenous base can be either a purine (adenine or guanine) or a pyrimidine (cytosine or thymine). These bases have a very specific binding. Adenine only bonds with thymine, and cytosine will only bond with guanine. DNA consists of two parallel strands held together by the bonds between these nucleotides.

Genome sequencing is a process through which the order of these nitrogenous bases is determined. After the biological phase of this process is completed, fragments of the genome being processed are obtained. These fragments are random and overlapping and can often be repetitive. The computational problem now becomes to find out the order of these fragments in which they can be combined to give the original genome. One of the most popular approaches to this computational problem is OLC approach [2]. It consists of three stages: overlap, layout, and consensus.

1. **Overlap stage:** In this stage, a semi-global alignment algorithm is used to evaluate the similarity score. This involves comparing all possible pairs among the fragments. This technique is used to

determine the overlapping fragments. It mainly uses semi-global alignments like the Smith-Waterman algorithm [21].

2. **Layout stage:** In this stage, the exact order of fragments is found based on calculated similarity scores in this step. This h proven to be an NP-hard problem [22].
3. **Consensus stage:** This stage helps derive the DNA sequence from the layout stage. Traditionally, the majority rule is used to build consensus.

The OLC model for genome sequence assembly has received much merit since being proposed in 1980 [23,24]. Fig. 1 is an example of the OLC approach.

4. Combinatorial optimization

The objective of the proposed approach is to optimally arrange the genome fragments so that the final genome is as close to the original genome as possible. The best solution would have the minimum number of contigs and the maximum overlap between the fragments. This kind of problem comes under the category of permutation optimization problems, which are combinatorial optimization [25] problems. In these problems, the set of feasible solutions is discrete or can be reduced to a discrete set. Similar problems in this category include the traveling salesman problem [26] and the knapsack problem [27].

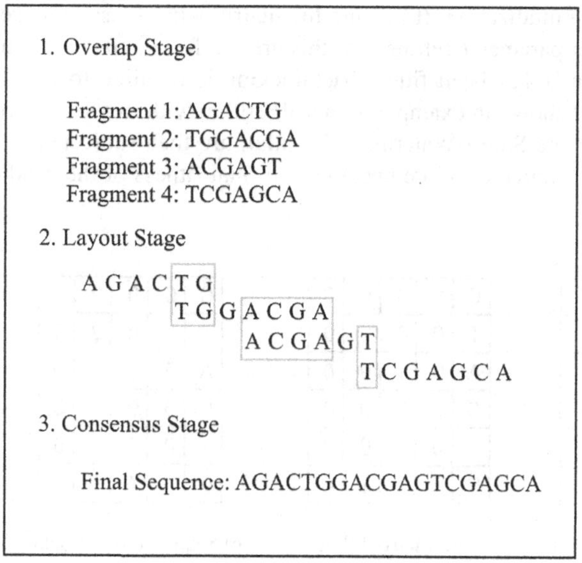

1. Overlap Stage

Fragment 1: AGACTG
Fragment 2: TGGACGA
Fragment 3: ACGAGT
Fragment 4: TCGAGCA

2. Layout Stage

```
A G A C T G
        T G G A C G A
              A C G A G T
                        T C G A G C A
```

3. Consensus Stage

Final Sequence: AGACTGGACGAGTCGAGCA

FIG. 1

Example of OLC model.

4.1 Fitness function

Researchers have given different fitness functions [11,28] in the past few decades for this problem. Previous works in related literature have either one or both of the following objectives: (a) reducing the number of contigs and/or (b) maximizing the sum of overlapping scores. We discuss two main methods of calculating the fitness score for a candidate solution in the sections that follow.

4.1.1 Overlap-based fitness

Overlap-based fitness can be calculated based on the summation of pairwise overlap score in a particular permutation of the given fragments in order. Scores are calculated for each pair starting from the first fragment to the last fragment [12,28].

The mathematical formulation of the simple overlap-based fitness is given by:

$$score_{i,i+1} = \begin{cases} 0 & \text{if } n \equiv 0 \\ 1 & \text{if } n > 0 \end{cases}$$

$$fitness_i = \sum_{j=0}^{D-1} score_{j,j+1}$$

where D is the number of fragments in the solution.

4.1.2 Smith-Waterman algorithm

This is a dynamic program [21] based Local Sequence Alignment algorithm. This algorithm contains three steps: (a) matrix initialization, (b) filling the matrix with the suitable scores, and (c) sequence backtracking. Common parameter settings for this are −2 for a gap, 1 for a match, and −3 for mismatches. Once the matrix has been filled, backtracking is required to find the maximum value in the entire matrix. Fig. 2 shows an example of this algorithm with two sample fragments. The pairwise scores generated using the Smith-Waterman algorithm are then subjected to equations discussed in Section 4.1.1, and the summation score becomes the final fitness of the candidate solution.

C	0			
T	0			
A	0			
G	0			
-	0			
0	-	A	T	C

C	0	0	0	3
T	0	0	2	0
A	0	1	0	0
G	0	0	0	0
-	0	0	0	0
0	-	A	T	C

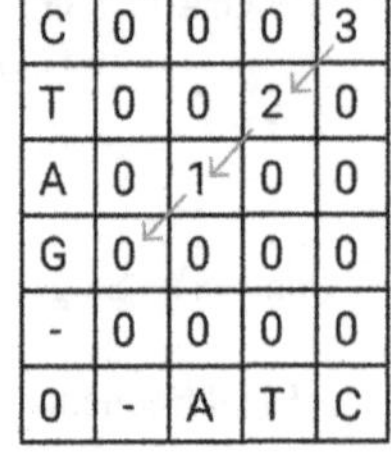

C	0	0	0	3
T	0	0	2	0
A	0	1	0	0
G	0	0	0	0
-	0	0	0	0
0	-	A	T	C

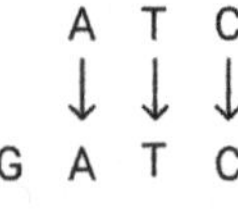

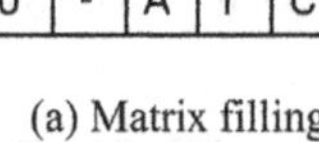

(a) Matrix Initialisation (a) Matrix filling (c) Sequence Backtracking (d) Alignment

FIG. 2

Smith-Waterman algorithm for local alignment.

4.2 Shortest position value (SPV) rule

Genome sequencing is a permutation optimization problem. Many algorithms have been proposed in the literature to solve this optimization problem. However, most of the algorithms were proposed to solve problems that are continuous in nature. PSO, CS, and GWO are examples of popular metaheuristics that work with continuous optimization problems since they operate using operations on real numbers [29–31]. A simple way to solve this problem would be to use algorithms such as GAs, which are well known to solve combinatorial optimization problems. Hence, they can be used to solve the considered problem directly. However, the drawback of a GA is that it is prone to getting stuck at a local optimum solution [11].

To use any metaheuristic algorithm in this context, the Shortest Position Value (SPV) rule can be used. The SPV rule helps convert any problem with a discrete set of feasible solutions into a continuous one.

Using the SPV fule, a particular solution can be represented as $X=\{x_1,x_2,x_3...x_D\}$, where D is the number of fragments and $x_1, x_2, x_3...x_D$ are in a fixed domain. Based on this, a new sequence vector can be generated.

For example, let the position vector of a particular solution be given by $X=\{3,5,8,7,0\}$ and the given fragments represented by the vector $F=\{f_1,f_2,f_3,f_4,f_5\}$. Next, the assignment of fragments to the sequence vector will be performed in the ascending order of their positions. The position vector in consideration represents the permutation $S = \{f_2, f_3, f_5, f_4, f_1\}$.

Using the SPV rule, this problem has been converted into a form that can be solved with any optimization algorithms without significantly modifying the algorithm. The candidate solutions generated by the metaheuristics in each iteration are the vectors generated by the SPV rule. Fig. 3 represents an example of the SPV rule.

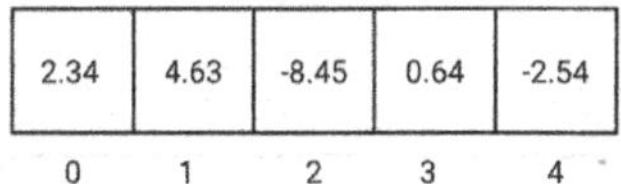

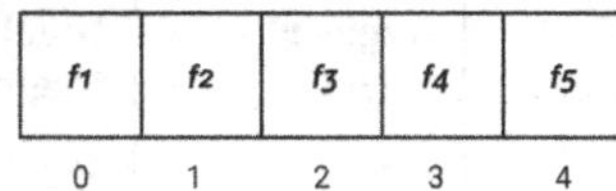

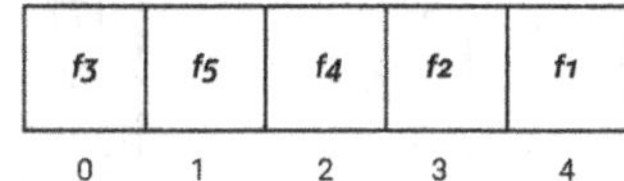

FIG. 3

Example of SPV rule.

5. Experiments

Three popular algorithms have been implemented and used to solve the genome sequencing problem using eight benchmark datasets created by Mallén-Fullerton et al. [32]. PSO [29] is an extremely popular metaheuristic that has been used to solve a multitude of problems in the past decade. Along with this, CS [30] and GWO [31] have been considered due to their popularity and excellence in solving optimization problems.

5.1 Datasets

There are two sources of genome sequencing data that are used by researchers. First, Mallén-Fullerton et al. [32] created a repository (http://chac.sis.uia.mx/fragbench/index1.php) with benchmark datasets for the DNA fragment assembly problem, which contains small datasets that have a lesser number of fragments and real datasets that are significantly bigger.

Second is the NCBI Sequence Read Archive (https://www.ncbi.nlm.nih.gov/sra). This is the largest publicly available repository of high-throughput sequencing data. The datasets available in this repository are massive and contain millions or even billions of fragments depending on the organism whose genome is being studied.

For our purposes, we chose eight datasets from the GenFrag database [32]. Table 1 presents the details of these datasets, including their coverage, mean fragment length, number of fragments, and their original sequence length.

5.2 Parameters

For all the algorithms, parameters were selected and fixed as proposed in the original publications for the experiment. For an unbiased analysis, the number of iterations and the population for all three algorithms were kept the same. The values for other parameters and their relevant algorithms are given in Table 2.

Table 1 Details of GenFrag datasets.

Benchmark	Coverage	Mean fragment length	Number of fragments	Original sequence length
j02459 7	7	405	352	20,000
m15421 5	5	398	127	10,089
m15421 6	6	350	173	
m15421 7	7	383	177	
x60189 4	4	395	39	3835
x60189 5	5	286	48	
x60189 6	6	343	66	
x60189 7	7	387	68	

Table 2 Parameters opted for the considered algorithms.

Parameter	Algorithm	Value	Reference
Population	PSO, CS, GWO	50	–
Iterations	PSO, CS, GWO	1000	–
$c1$, $c2$	PSO	2	Kennedy et al. [29]
Pa	CS	0.4	Yang et al. [30]
Alpha	CS	0.5	
Lambda	CS	1.5	
a	GWO	Decreases linearly from 2 to 0	Mirjalili et al. [31]

For PSO, the two main parameters that need to be considered are $c1$ and $c2$, which govern the social and cognitive components of the movement of particles. For CS, the main parameters are Pa and α, where α is the step size scaling factor, and Pa is the switching parameter. Table 2 summarizes these parameters.

5.3 Results

Eight GenFrag datasets were used for comparative analysis of the considered algorithm. GenFrag takes a known sequence (parent sequence) and generates random fragments from it based on given criteria. These criteria usually are mean fragment length and coverage of the parent strand.

The experiments were conducted using Python 3.7.9 and performed on a 3.1 GHz Dual-Core Intel Core i5 with 8GB of memory running Mac OS 11.2. The algorithms were run 20 times each with different random seeds, and their mean, standard deviation, best, and worst values were recorded.

Table 3 reports on the performance of the three metaheuristic algorithms used on the eight datasets. The table depicts the mean, standard deviation, and best and worst values of the conducted experiments.

It is evident from Table 3 that GWO gives the best mean solution compared to the other methods on all eight datasets. It shows that GWO is a reliable algorithm due to its consistency in giving higher fitness solutions. CS performs better than PSO but seldom gives results close to the results of GWO. A possible reason for this behavior lies in the algorithms that have been used. GWO offers faster convergence due to continuous reduction of search space and is proficient in avoiding local optima. CS fares better than PSO due to its random walks and Levy flights, but due to the vast search space of the problem, it gets stuck in local optimum solutions.

The fitness function considered for this experiment aims at maximizing the overlap score. The scores will differ if the fitness function aims to optimize a different objective from those discussed in Section 5.1.

Table 3 Comparative analysis of PSO, CS, and GWO algorithms on the considered datasets.

	Algorithm	Mean	Std Dev	Best	Worst
j02459 7	PSO	3.62E+03	2.91E+02	4.08E+03	3.24E+03
	CS	3.71E+03	3.50E+02	**4.44E+03**	3.14E+03
	GWO	**3.88E+03**	**2.25E+02**	4.30E+03	**3.46E+03**
m15421 5	PSO	2.74E+03	**2.39E+02**	3.21E+03	2.47E+03
	CS	3.04E+03	2.69E+02	3.33E+03	**2.58E+03**
	GWO	**3.17E+03**	5.45E+02	**3.94E+03**	2.43E+03
m15421 6	PSO	2.76E+03	**1.61E+02**	3.08E+03	2.57E+03
	CS	3.03E+03	2.80E+02	3.53E+03	**2.72E+03**
	GWO	**3.05E+03**	6.29E+02	**4.72E+03**	2.46E+03
m15421 7	PSO	3.48E+03	3.41E+02	4.14E+03	3.11E+03
	CS	3.32E+03	**1.89E+02**	3.59E+03	3.07E+03
	GWO	**3.92E+03**	7.89E+02	**5.35E+03**	**3.12E+03**
x60189 4	PSO	2.54E+03	**1.81E+02**	2.97E+03	2.29E+03
	CS	2.65E+03	3.12E+02	3.29E+03	**2.33E+03**
	GWO	**2.92E+03**	5.72E+02	**3.69E+03**	2.12E+03
x60189 5	PSO	2.74E+03	4.68E+02	3.56E+03	2.18E+03
	CS	2.67E+03	**2.25E+02**	3.18E+03	**2.43E+03**
	GWO	**3.69E+03**	8.85E+02	**5.11E+03**	2.22E+03
x60189 6	PSO	2.68E+03	**1.37E+02**	2.83E+03	2.44E+03
	CS	2.81E+03	2.64E+02	3.31E+03	2.43E+03
	GWO	**3.73E+03**	7.98E+02	**5.34E+03**	**2.72E+03**
x60189 7	PSO	3.36E+03	3.28E+02	3.98E+03	**3.07E+03**
	CS	3.29E+03	**1.96E+02**	3.65E+03	3.05E+03
	GWO	**4.44E+03**	9.79E+02	**6.00E+03**	2.93E+03

Here, bold values indicate the best results recorded with the respective algorithm on the considered dataset.

5.4 Convergence analysis

Exploration and exploitation are important aspects of any metaheuristic algorithm. To analyze these aspects in this problem, we performed a convergence analysis for the conducted experiments. Fig. 4 represents the convergence graph of the considered algorithms with the eight datasets. A flat line in a convergence graph means that there is no improvement to the best fitness found by the algorithm within the span of those iterations. This means that the optimizer is stuck in a local optimum. As shown, PSO (blue) gets stuck in local optima very frequently and for long periods. In contrast, GWO shows better convergence capabilities. It improves consistently and does not get stuck in local optima for long periods.

5.5 Statistical significance

It is essential to compare the statistical significance to assess if two population means differ. We performed T-tests on the experimental results to test their significance. In this problem, T-tests have been conducted with a significance value, $\alpha=0.05$. P-values have been calculated for each pair of

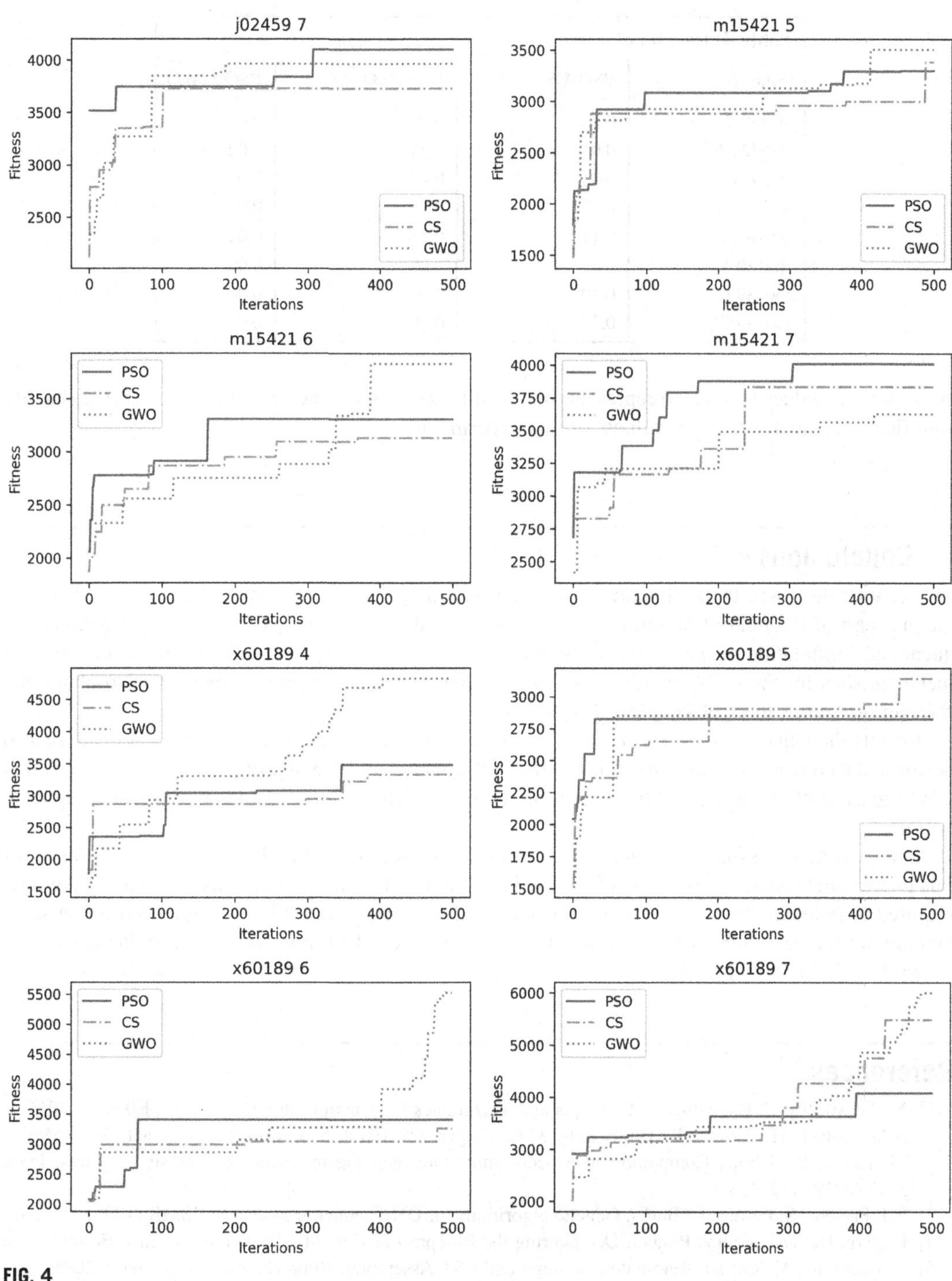

FIG. 4

Convergence graph of the considered algorithms.

Table 4 Results of *T*-tests.

Dataset	PSO-CS	CS-GWO	PSO-GWO
j02459 7	0.26	0.10	0.02
m15421 5	0.01	0.25	0.02
m15421 6	0.01	0.45	0.09
m15421 7	0.10	0.02	0.07
x60189 4	0.17	0.11	0.04
x60189 5	0.35	0.00	0.00
x60189 6	0.10	0.00	0.00
x60189 7	0.29	0.00	0.00

algorithms. P-values less than α represent that the difference of the means of the groups is statistically significant. Table 4 lists the P-values for this experiment.

6. Conclusions

This chapter discussed the application of metaheuristic algorithms in genome sequencing, which is an integral part of the field of bioinformatics. We presented in detail the basic concepts of genome sequencing, formulation of the problem statement, preprocessing steps, and finally, the application of metaheuristics for the problem. We chose three popular nature-inspired metaheuristics to obtain the optimum solutions to eight sample datasets.

Results show that GWO performed well in terms of obtaining consistently better solutions. We also performed a convergence analysis to judge the performance of these algorithms. It was observed that GWO consistently finds higher fitness solutions among the algorithms considered while avoiding local optima.

Future work in this direction can include designing efficient strategies that reduce the time taken for this process and improving the algorithms used to obtain higher fitness solutions. Decreasing the time required or reducing the time complexity would lead to more genomes being sequenced in the same amount of time. Obtaining better fitness solutions would mean that more valuable conclusions can be drawn from studying a genome.

References

[1] N. Saraswathy, P. Ramalingam, Concepts and Techniques in Genomics and Proteomics, Elsevier, 2011.

[2] A.M. Bolger, H. Poorter, K. Dumschott, M.E. Bolger, D. Arend, S. Osorio, H. Gundlach, K.F. Mayer, M. Lange, U. Scholz, Computational aspects underlying genome to phenome analysis in plants, Plant J. 97 (2019) 182–198.

[3] R.J. Parsons, S. Forrest, C. Burks, Genetic algorithms for DNA sequence assembly, ISMB (1993) 310–318.

[4] P. Berg, Human Genome Project: Deciphering the Blueprint of Heredity, University Science Books, 1994.

[5] B. Chevreux, MIRA: An Automated Genome and EST Assembler, Ruprecht-Karls-University, 2005.

[6] X. Huang, A contig assembly program based on sensitive detection of fragment overlaps, Genomics 14 (1992) 18–25.
[7] H. Peltola, H. Söderlund, E. Ukkonen, SEQAID: a DNA sequence assembling program based on a mathematical model, Nucleic Acids Res. 12 (1984) 307–321, https://doi.org/10.1093/nar/12.1Part1.307.
[8] T.D. Otto, ERRATUM: From Sequence Mapping to Genome Assemblies, in: Parasite Genomics Protoc, Springer, 2015, pp. E1–E4.
[9] G.G. Sutton, O. White, M.D. Adams, A.R. Kerlavage, T.I.G.R. Assembler, A new tool for assembling large shotgun sequencing projects, genome, Sci. Technol. 1 (1995) 9–19.
[10] P.A. Pevzner, H. Tang, M.S. Waterman, An Eulerian path approach to DNA fragment assembly, Proc. Natl. Acad. Sci. 98 (2001) 9748–9753.
[11] Z. Halim, Optimizing the DNA fragment assembly using metaheuristic-based overlap layout consensus approach, Appl. Soft Comput. 92 (2020), 106256.
[12] R.J. Parsons, S. Forrest, C. Burks, Genetic algorithms, operators, and DNA fragment assembly, Mach. Learn. 21 (1995) 11–33, https://doi.org/10.1023/A:1022613513712.
[13] A.J. Nebro, G. Luque, F. Luna, E. Alba, DNA fragment assembly using a grid-based genetic algorithm, Comput. Oper. Res. 35 (2008) 2776–2790.
[14] A.B. Ali, G. Luque, E. Alba, An efficient discrete PSO coupled with a fast local search heuristic for the DNA fragment assembly problem, Inf. Sci. 512 (2020) 880–908.
[15] R. Indumathy, S.U. Maheswari, G. Subashini, Nature-inspired novel cuckoo search algorithm for genome sequence assembly, Sadhana 40 (2015) 1–14.
[16] W. Kartous, S. Chikhi, Improved cuckoo search algorithm for dna fragment assembly problem, Netw. Adv. Syst. (2015) 117.
[17] J.S. Firoz, M.S. Rahman, T.K. Saha, Bee algorithms for solving DNA fragment assembly problem with noisy and noiseless data, in: Proc. 14th Annu. Conf. Genet. Evol. Comput, 2012, pp. 201–208.
[18] W. Wetcharaporn, N. Chaiyaratana, S. Tongsima, DNA fragment assembly by ant colony and nearest neighbour heuristics, in: Int. Conf. Artif. Intell. Soft Comput, Springer, 2006, pp. 1008–1017.
[19] P. Meksangsouy, N. Chaiyaratana, DNA fragment assembly using an ant colony system algorithm, in: 2003 Congr. Evol. Comput. 2003 CEC03, IEEE, 2003, pp. 1756–1763.
[20] P.J. Vidal, A.C. Olivera, Solving the DNA fragment assembly problem with a parallel discrete firefly algorithm implemented on GPU, Comput. Sci. Inf. Syst. 15 (2018) 273–293.
[21] T.F. Smith, M.S. Waterman, Identification of common molecular subsequences, J. Mol. Biol. 147 (1981) 195–197, https://doi.org/10.1016/0022-2836(81)90087-5.
[22] P. Pevzner, Computational Molecular Biology: An Algorithmic Approach, MIT Press, 2000.
[23] F. Sanger, A.R. Coulson, G.F. Hong, D.F. Hill, G.B. Petersen, Nucleotide sequence of bacteriophage λ DNA, J. Mol. Biol. 162 (1982) 729–773, https://doi.org/10.1016/0022-2836(82)90546-0.
[24] R. Staden, A new computer method for the storage and manipulation of DNA gel reading data, Nucleic Acids Res. 8 (1980) 3673–3694, https://doi.org/10.1093/nar/8.16.3673.
[25] G.C. Onwubolu, D. Davendra, Differential Evolution: A Handbook for Global Permutation-Based Combinatorial Optimization, Springer Science & Business Media, 2009.
[26] I.M. Ross, R.J. Proulx, M. Karpenko, An Optimal Control Theory for the Traveling Salesman Problem and its Variants, ArXiv200503186 Cs Math, 2020. http://arxiv.org/abs/2005.03186. (Accessed 28 January 2022).
[27] R. Lahyani, K. Chebil, M. Khemakhem, L.C. Coelho, Matheuristics for solving the multiple knapsack problem with setup, Comput. Ind. Eng. 129 (2019) 76–89.
[28] R. Shankar Verma, V. Singh, S. Kumar, DNA sequence assembly using particle swarm optimization, Int. J. Comput. Appl. 28 (2011) 33–38, https://doi.org/10.5120/3425-4777.
[29] J. Kennedy, R. Eberhart, Particle swarm optimization, in: Proc. ICNN95—Int. Conf. Neural Netw, vol. 4, 1995, pp. 1942–1948, https://doi.org/10.1109/ICNN.1995.488968.

[30] X.-S. Yang, S. Deb, Engineering optimisation by cuckoo search, Int. J. Math. Model. Numer. Optim. 1 (2010) 330–343, https://doi.org/10.1504/IJMMNO.2010.03543.
[31] S. Mirjalili, S.M. Mirjalili, A. Lewis, Grey wolf optimizer, Adv. Eng. Softw. 69 (2014) 46–61, https://doi.org/10.1016/j.advengsoft.2013.12.007.
[32] G.M. Mallén-Fullerton, J.A. Hughes, S. Houghten, G. Fernández-Anaya, Benchmark datasets for the DNA fragment assembly problem, Int. J. Bio-Inspired Comput. 5 (2013) 384–394, https://doi.org/10.1504/IJBIC.2013.058912.

CHAPTER

Metaheuristics for optimizing weights in neural networks

19

Mohammed A. Awadallah[a,b], Iyad Abu-Doush[c,d], Mohammed Azmi Al-Betar[e,f], and Malik Shehadeh Braik[g]

[a]*Department of Computer Science, Al-Aqsa University, Gaza, Palestine,* [b]*Artificial Intelligence Research Center (AIRC), Ajman University, Ajman, United Arab Emirates,* [c]*Department of Computing, College of Engineering and Applied Sciences, American University of Kuwait, Salmiya, Kuwait,* [d]*Computer Science Department, Yarmouk University, Irbid, Jordan,* [e]*Artificial Intelligence Research Center (AIRC), College of Engineering and Information Technology, Ajman University, Ajman, United Arab Emirates,* [f]*Department of Information Technology, Al-Huson University College, Al-Balqa Applied University, Al-Huson, Irbid, Jordan,* [g]*Department of Computer Science, Al-Balqa Applied University, Jordan*

1. Introduction

The biological neurons in the human brain naturally control the behavior of the entire body. The biggest achievement in the field of machine learning is innovation of neural networks (NNs) inspired by the behavior of biological neurons and modeled mathematically [1]. Due to impressive characteristics such as simplicity, efficiency, and reasonable computation cost, NNs are utilized for a wide range of problems such as prediction, classifications, function approximation, feature extractions, clustering, and so on [2–4]. There are several types of NNs such as feedforward neural network (FNN), backpropagation neural network (BNN), convolutional neural network (CNN), and others [5–7]. The most attractive and simplest one is FNN, which accepts the input data, and then trains the model on unidirection to connect the neurons in various layers [6]. FNN versions have two main types: single layer perceptron (SLP) [1] and multilayer perceptron (MLP) [6]. SLP is suitable for linear problems because it has only a single perceptron, while MLP is suitable for nonlinear problems because it has more than one perceptron on each layer.

The learning mechanism used in the FNN has the ability to learn from experience. The strength and power of the learning mechanism distinguishes the capability of MLP performance. This is the main part of any type of NN aiming to minimize the cost function, referred to as mean square error (MSE). The MSE evaluates the neuron weights modified between the actual and targeted output [8]. To optimize neuron weights and minimize the MSE, the supervised learning MLP uses a gradient-based algorithm represented by a backpropagation (BP) algorithm [9] or improved BP [10]. However, this gradient-based algorithm has two chronic dilemmas related to the slow convergence rate and local minima trap [11]. Therefore, heuristic-based approaches are being developed to overcome the dilemmas of gradient-based algorithms.

Comprehensive Metaheuristics. https://doi.org/10.1016/B978-0-323-91781-0.00005-3

In heuristic-based learning mechanisms, several areas in the search space are explored at the same time through several population members. Nowadays, the metaheuristic-based approaches have attracted considerable attention from machine learning researchers to serve as efficient training algorithms for FNN [12, 13]. Metaheuristics are generic optimization models applicable for a wide range of optimization problems such as training tasks in MLP. The metaheuristics used as a trainer in MLP can be divided into four different categories: Evolutionary algorithms include genetic algorithm (GA) [14], differential evolution (DE) [15, 16], biogeography-based optimization [17], and organisms search algorithm [18]. Swarm-based algorithms include particle swarm optimization [19, 20], grey wolf optimizer [8, 21], glowworm swarm optimization [22], salp swarm algorithm [23], artificial bee colony [24, 25], grasshopper optimization algorithm [26, 27], chimp optimization algorithm [28], monarch butterfly optimization [29], ant colony optimization [30], bat algorithm (BA) [25, 31], and butterfly optimization algorithm [32]. Human-based algorithms include harmony search (HS) [33] and fireworks algorithm [34]. Finally, physical-based algorithms include multiverse optimizer [35], sine cosine algorithm (SCA) [36], lightning search algorithm [37], and gravitational search algorithm [38].

Although the training-based metaheuristic algorithms have great performance to empower the convergence rate and to avoid the local minima that is caused by gradient-descent algorithms, in the optimization domain, the exact solution cannot be guaranteed by any metaheuristic algorithm. Furthermore, the No Free Lunch (NFL) theorem stated that there is no single outstanding metaheuristic algorithm that can tackle all optimization problems, even with different instances [39]. The slow convergence with local minima trap dilemmas in the gradient-descent algorithm and the NFL theorem motivate this work to investigate an optimization algorithm called horse herd optimization algorithm (HOA) as a training algorithm in MLP to improve FNN performance. The HOA is a new swarm-based algorithm that imitates the behavior of horses in grazing and surviving [40]. It has several advantages such as ease of use, few parameter settings, no requirement for derivative information in the initial search, flexible and dynamic behavior, and soundness and completeness. Although HOA is quite new, it has been successfully tailored to a wide range of optimization applications such as health problems [41], fuel-constrained day-ahead scheduling of isolated nanogrids [42], feature selection [43], high-dimensional benchmark functions [40], and standard benchmark functions [40].

In this chapter, HOA is modeled to work as a training algorithm for MLP. HOA-MLP is utilized to optimize the solution vector (i.e., weight and biases vector) to be suitable for the MLP. The main measurement used for the proposed HOA-MLP method is the MSE that is embedded to calculate the accuracy. The performance of the proposed HOA-MLP are evaluated using 15 classification test beds of different size and complexity. For comparative evaluation, the accuracy results obtained by HOA-MLP is evaluated against those obtained by five state-of-the-art methods using the same test beds. These are BA [31], HS [33], flower pollination algorithm (FPA) [44], SCA [36], and JAYA algorithm [45]. Interestingly, the HOA-MLP outperforms other comparative methods in 5 out of 15 test beds. Therefore, the proposed HOA-MLP is an efficient training algorithm for MLP that can be adopted to improve the performance of FNN for implementation in other classification applications.

This chapter is organized as follows: Section 2 gives background about FNNs. Section 3 describes the methodology of HOA-MLP. Section 4 presents the results of the proposed HOA-MLP and compares it to other algorithms. Finally, Section 5 concludes and presents possible future research directions.

2. Feedforward neural networks

An FNN is a machine learning algorithm based on the concept of human neurons [20]. It is a computational model inspired by biological concepts that consists of a pool of processing elements called *neurons* and connections between neurons with weights (i.e., coefficients) linked with the connections [25]. As artificial neural networks (ANNs) are inspired by the structure of the human brain, it is significant to identify the key elements under which neurons, cell bodies, dendrites, and axons work. The neurons are spread over a predefined number of amassed layers in which each layer is completely united with the following layer. The first layer is called the input, where the input variables are mapped to the network by this layer. The number of neurons in this layer is identical to the size of the input feature vector. The last layer of this network architecture is called the output layer, which is composed of the output neurons, which are aligned with the labels of the predicted classes. The layers between the input and output layers are called hidden layers [46].

MLP is the most widely used and common FNN. In MLP, neurons are connected in a unidirectional manner, in which information flows in one direction only. The connections are expressed by a set of weights, which are represented by real numbers that lie in the range $[-1, 1]$. Fig. 1 shows a general schematic of the structure of an MLP network with input, output, and one middle layer.

The MLP's mathematical model is based on three related main elements: input datasets, biases, and weights. The output of each node in the MLP network is computed in three successive steps, described as follows:

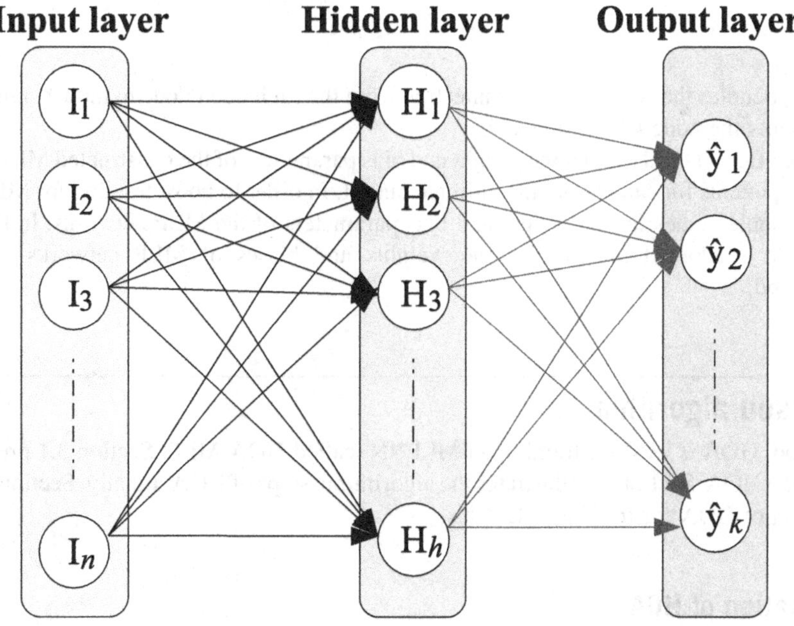

FIG. 1

Multilayer perceptron NN structure with one middle layer.

1. At the outset, the weighted pooling of the inputs is computed using Eq. (1).

$$S_j = \sum_{i=1}^{n} w_{ij} \cdot X_i + \beta_j \tag{1}$$

where $j = 1,2,\ldots,h$, h denotes the number of hidden nodes, n denotes the total number of input nodes in the MLP network architecture, w_{ij} denotes the association weight that connects the input node X_i to the hidden node j, X_i identifies the ith input node, and β_j denotes the bias term of the jth hidden neuron.

2. Then, an activation function is applied to activate the output of neurons on the basis of the value of the aggregation function defined in Eq. (1). There are various kinds of activation functions that could be used in the MLP network. At this point, the sigmoid function is utilized, which is typically used for MLP network structures by previous works. This function is embraced to move the output that is weighted in the hidden layer to the following layer. Node j output in the hidden layer can be computed using Eq. (2).

$$f_j = \frac{1}{1 + e^{-S_j}} \tag{2}$$

where f_j represents the sigmoid activation function of the node j in the middle layer and S_j is the weighted summation of the input nodes.

3. The last step after computing each neuron's output in the middle layer is to find the MLP network's output, which can be determined using Eq. (3).

$$\hat{y}_k = \sum_{i=1}^{m} w_{kj} f_i + \beta_k \tag{3}$$

where w_{jk} denotes the weight of the connection from the jth hidden node to the kth output node and β_k represents the node k bias outcome.

It is clear from Eqs. (1) and (3) that the weight and bias parameters of the constructed MLP network are essential components for calculating the ultimate output. In order to boost training for MLP networks, this needs suitable values for the weight and bias parameters of the MLP networks. In the following section, HOA is adopted to optimize the weights and biases of MLP networks and act as a training method.

3. Proposed algorithm

In this section, HOA is used for training an MLPNN, called HOA-MLP. Section 3.1 presents the inspiration of the HOA. Section 3.2 illustrates the algorithmic steps of HOA. Finally, Section 3.3 explains the adaptation of HOA for training MLPNNs.

3.1 Inspiration of HOA

HOA is a nature-based inspired optimization algorithm proposed by MiarNaeimi et al. in 2021 [40]. It is based on the social behavior of horses at different ages and is divided into six features.

Grazing feature: Horses are grazing animals and they prefer to stay where there is water, plants, and grasses provide. Horses spend most of their time grazing in pastures. This feature is innate in horses from birth to death.

Hierarchy feature: Horses by their nature are subject to the laws of hierarchy; the strongest horse is the leader and the rest of the horses follow the leader. The horses aged between 5 and 15 years follow the law of hierarchy according to McDonnell [47].

Sociability feature: Horses are social animals. They prefer to live in groups to survive and avoid predators. Furthermore, horses can live with other pet animals like sheep, cattle, chickens, dogs, and so on.

Imitation feature: Horses imitate each other's behavior. In general, young horses learn from older horses how to find an appropriate pasture, how to defend themselves, and so on.

Defense feature: When approached by predators, horses will flee. When confronted with limited resources such as food and water, horses may fight among themselves.

Roam feature: Horses move from one pasture to another during their different life stages to search for the best sources of food and water.

The social behavior of horses in nature is formulated in an optimization context for training MLPs in the following section.

3.2 Algorithmic steps of HOA

This section describes the steps of the HOA.

Step 1: Initialize parameters of HOA. In this step, the parameters of the HOA algorithm should be initialized before the HOA is executed. The parameters of the HOA algorithm include the size of the population (N) and the maximum number of generations/iterations (Itr_{max}).

Step 2: Construct the initial population. In this step, the initial positions of horses (i.e., solutions) are constructed and stored in **HP**, as shown in Eq. (4). **HP** is a two-dimensional matrix of size $N \times D$.

$$\mathbf{HP} = \begin{bmatrix} X_{1,1} & X_{1,2} & \cdots & X_{1,D-1} & X_{1,D} \\ X_{2,1} & X_{2,2} & \cdots & X_{2,D-1} & X_{2,D} \\ \vdots & \vdots & \cdots & \vdots & \vdots \\ X_{N,1} & X_{N,2} & \cdots & X_{N,D-1} & X_{N,D} \end{bmatrix} \tag{4}$$

Each row in **HP**, $X_k = (X_{k,1}, X_{k,2}, \ldots, X_{k,D})$, represents the horse at the kth position. Each decision variable $X_{k,r}$ from the solution X_k is assigned a random value using Eq. (5).

$$X_{k,r} = X_r^{min} + rand \times (X_r^{max} - X_r^{min}) \tag{5}$$

where X_r^{min} and X_r^{max} are the lower and upper bounds of the rth position and $rand$ is a random value between 0 and 1.

Step 3: Fitness evaluation. In this step, the fitness value (i.e., quality) of the position of each horse in **HP** is evaluated using the objective function as follows:

$$f(X_k), \quad \forall k = 1, 2, \ldots, N \tag{6}$$

Step 4: Horse age determination. In this step, the horses in the population are classified according to their age into four categories: δ refers to a horse with an age between 0 and 5 years, γ refers to a horse with an age between 5 and 10 years, β refers to a horse with an age between 10 and 15 years, and α refers to a horse with an age older than 15 years. The positions of the horses stored in **HP** are sorted from the fittest to the worst. The age category of the top 10% of the horses in **HP** is set to α. The age category of the next 20% of the horses in **HP** is assigned as β. The next 30% of the horses age category in the population is set as γ, while the age category of the remaining horses in the population is set as δ.

Step 5: Velocity update. At each iteration, the velocity of each horse in the population is updated according to its age. From Eqs. (7) to (10), it can be seen that each horse velocity is computed using its social skills (i.e., sociability, grazing, imitation, roam, hierarchy, and defense).

$$\vec{V}_k^{t,\alpha} = \vec{G}_k^{t,\alpha} + \vec{D}_k^{t,\alpha} \tag{7}$$

where $\vec{V}_k^{t,\alpha}$ is the horse velocity in the kth position at the t iteration. α is the horse age. The horse at this age has the characteristics of grazing ($\vec{G}_k^{t,\alpha}$) and self-defense ($\vec{D}_k^{t,\alpha}$).

$$\vec{V}_k^{t,\beta} = \vec{G}_k^{t,\beta} + \vec{H}_k^{t,\beta} + \vec{S}_k^{t,\beta} + \vec{D}_k^{t,\beta} \tag{8}$$

where $\vec{V}_k^{t,\beta}$ is the horse velocity in the kth position at the t iteration. The horse age is specified as β. The horse at this age has the features of grazing ($vecG_k^{t,\beta}$), hierarchy ($\vec{H}_k^{t,\beta}$), sociability ($\vec{S}_k^{t,\beta}$), and defense ($\vec{D}_k^{t,\beta}$).

$$\vec{V}_k^{t,\gamma} = \vec{G}_k^{t,\gamma} + \vec{H}_k^{t,\gamma} + \vec{S}_k^{t,\gamma} + \vec{I}_k^{t,\gamma} + \vec{D}_k^{t,\gamma} + \vec{R}_k^{t,\gamma} \tag{9}$$

where $\vec{V}_k^{t,\gamma}$ is the horse velocity in the kth position at the t iteration. The horse age is γ. The horse in this age has all social features as follows: grazing ($\vec{G}_k^{t,\gamma}$), hierarchy ($\vec{H}_k^{t,\gamma}$), sociability ($\vec{S}_k^{t,\gamma}$), imitation ($\vec{I}_k^{t,\gamma}$), defense ($\vec{D}_k^{t,\gamma}$), and roam ($\vec{R}_k^{t,\gamma}$).

$$\vec{V}_k^{t,\delta} = \vec{G}_k^{t,\delta} + \vec{I}_k^{t,\delta} + \vec{R}_k^{t,\delta} \tag{10}$$

where $\vec{V}_k^{t,\delta}$ is the horse velocity in the kth position at the t iteration. The horse age is specified as δ. The horse in this age has the characteristics of grazing ($\vec{G}_k^{t,\delta}$), imitation ($\vec{I}_k^{t,\delta}$), and roam ($\vec{R}_k^{t,\delta}$).

The six features of horses are formulated as follows:

1. **Grazing:** A horse has the ability to eat grass throughout its life from birth to death. Thus, the horses aged α, or β, or δ, or γ are grassing.

$$\vec{G}_k^{t,AGE} = \vec{g}_k^{t,AGE}\left(\breve{u} + p\breve{l}\right)\left[X_k^{t-1}\right], \quad AGE = \alpha,\beta,\delta,\gamma \tag{11}$$

$$\vec{g}_k^{t,AGE} = \vec{g}_k^{(t-1),AGE} \times w_g \tag{12}$$

where $\vec{G}_k^{t,AGE}$ is the motion vector of the horse at the kth position and $\breve{u}$ and $\breve{l}$ are the lower and upper bounds of grassing area. Notably, $\breve{u} = 1.05$ and $\breve{l} = 0.95$ are recommended in MiarNaeimi et al. [40]. The parameter p is assigned value from $U(0, 1)$. The value of the $\vec{G}_k^{t,AGE}$ is updated according to the value of $\vec{g}_k^{t,AGE}$ at each iteration. $\vec{g}_k^{t,AGE}$ is set to 1.5 at the beginning of search process, while this value is decreased according to the value of w_g, as shown in Eq. (12).

2. **Hierarchy:** The horses with age equal to β, α, and γ are following the rules of hierarchy. This concept is mathematically formulated as follows:

$$\vec{H}_k^{t,AGE} = \vec{h}_k^{t,AGE}\left[X_{best}^{t-1} - X_k^{t-1}\right], \quad AGE = \alpha,\beta,\gamma \tag{13}$$

$$\vec{h}_k^{t,AGE} = \vec{h}_k^{(t-1),AGE} \times w_h \tag{14}$$

where $\vec{H}_k^{t,AGE}$ is the hierarchy level of the horse at the kth position, X_{best}^{t-1} is the best horse location at the previous iteration $(t-1)$, and X_k^{t-1} is the location of the current horse at the previous iteration $(t-1)$. The value of $\vec{h}_k^{t,AGE}$ is set to 1.5, 0.9, and 0.5 for horses with age equal to α, β, and γ, respectively. The value of $\vec{h}_k^{t,AGE}$ decreases at each iteration according to value of w_h.

3. **Sociability:** Horses in nature tend to live in groups. This social behavior is formed as a moving toward the central position of the other horses in the population, as shown using Eq. (15).

$$\vec{S}_k^{t,AGE} = \vec{s}_k^{t,AGE}\left[\left(\frac{1}{N}\sum_{c=1}^{N} X_c^{t-1}\right) - X_k^{t-1}\right], \quad AGE=\beta,\gamma \tag{15}$$

$$\vec{s}_k^{t,AGE} = \vec{s}_k^{(t-1),AGE} \times w_s \tag{16}$$

where $\vec{S}_k^{t,AGE}$ is the motion vector of the horse at the kth position. The horses with ages β, and γ have this social behavior. The $\vec{s}_k^{t,AGE}$ represents the horse's move into the herd at iteration t. The value of $\vec{s}_k^{t,AGE}$ decreases based on the value of w_s. The value of $\vec{s}_k^{t,AGE}$ is set to 0.2 and 0.1 for horses with age β and γ, respectively.

4. **Imitation:** The horse seeks to imitate the behavior of other horses in the herd. This behavior is formulated as follows:

$$\vec{I}_k^{t,AGE} = \vec{i}_k^{t,AGE}\left[\left(\frac{1}{pN}\sum_{c=1}^{pN} \hat{X}_c^{(t-1)}\right) - X_k^{(t-1)}\right], \quad AGE=\gamma \tag{17}$$

$$\vec{i}_k^{t,AGE} = \vec{i}_k^{(t-1),AGE} \times w_i \tag{18}$$

where $\vec{I}_k^{t,AGE}$ is the horse motion vector at the kth position toward the average of the best horses at the population and p^N is the count of horses with the best locations in the population. The value of p^N is equal to $0.1 \times N$, recommended in MiarNaeimi et al. [40]. Iteratively, the value of $\vec{i}_m^{t,AGE}$ reduces according to the value of w_i.

5. **Defense:** The defense scheme of horses is expressed by running away from the inappropriate positions that are far from optimal. The defense schema is formulated as a negative coefficient to avoid the horse with the worst positions, as shown in the following equations:

$$\vec{D}_k^{t,AGE} = -\vec{d}_k^{t,AGE}\left[\left(\frac{1}{pN}\sum_{c=1}^{pN} \check{X}_c^{(t-1)}\right) - X_k^{(t-1)}\right], \quad AGE=\alpha,\beta,\gamma \tag{19}$$

$$\vec{d}_k^{t,AGE} = \vec{d}_k^{(t-1),AGE} \times w_d \tag{20}$$

where $\vec{D}_k^{t,AGE}$ is the horse escape vectors at the kth position to avoid the horses with the worst positions and p^N is the number of horses with the worst positions in the population. The value of p^N is equal to $0.2 \times N$, recommended in MiarNaeimi et al. [40]. In Eq. (20), the value of $\vec{d}_k^{t,AGE}$ is updated at each iteration according to the value of the reduction faction (w_d). It should be noted that the horses aged with α, β, or γ have the self-defense mechanism.

6. **Roam:** The horse is moving from one place to another searching for the source of food. This social behavior is expressed as a random movement formulated in the following equations:

$$\vec{R}_k^{t,AGE} = \vec{r}_k^{t,AGE} P X_k^{(t-1)}, \quad AGE=\delta,\gamma \tag{21}$$

$$\vec{r}_k^{t,AGE} = \vec{r}_k^{(t-1),AGE} \times w_r \tag{22}$$

where $\vec{R}_k^{t,AGE}$ is the horse vector of random velocity at the kth position. This is to avoid premature convergence problem. P is a random value uniformly distributed between 0 and 1. The value of the reduction factor w_r is used to update the value of $\vec{r}_k^{t,AGE}$.

[Step 6: Position update.] Each horse location in **HP** is updated using Eq. (23). In Eq. (23), $X_k^{t,AGE}$ is the new position of the horse at the kth position, $X_k^{(t-1),AGE}$ is the old position of the same horse at previous iteration, and $\vec{V}_k^{t,AGE}$ is the velocity of the same horse at the same kth position. Notably, the velocity and the age of each horse are calculated in the previous two steps:

$$X_k^{t,AGE} = \vec{V}_k^{t,AGE} + X_k^{(t-1),AGE} \quad AGE = \gamma, \delta, \alpha, \beta \tag{23}$$

[Step 7: Stop condition.] The maximum number of iterations is the main termination condition in this research. Therefore, Steps 3 to 6 are repeated until Itr_{max} is reached.pt

3.3 Proposed algorithm

Fig. 2 shows the flowchart of the proposed HOA for training the MLP. This algorithm is called HOA-MLP. It should be noted that the problem of training MLPs should be formulated with appropriate form in order to be solved by metaheuristic algorithms.

The settings of the MLP parameters are taken from the datasets. These parameters include the number of neurons in the hidden layer of neural networks (h), the number of features in the dataset (f), the number of weights (w), the number of outputs from the MLP neural networks (o), and the number of biases (b). The number of weights and the number of biases are calculated using the following equations:

$$h = 2 \times f + 1 \tag{24}$$

$$w = f \times h + h \times o \tag{25}$$

$$b = h + o \tag{26}$$

Each solution in the population represents the possible weights and biases inputs for the MLPNN. In this research, the solution $X = (X_1, X_2, \ldots, X_D)$ in MLP is represented as a vector $X = (w_1, w_2, \ldots, w_n, b_1, b_2, \ldots, b_m)$. Apparently, the variables of the solution X are divided into consecutive groups: weights $\boldsymbol{w} = (w_1, \ldots, w_n)$ and biases $\boldsymbol{b} = (b_1, \ldots, b_m)$. Again, n and m are the number of weights and the number of biases, respectively. The dimension of the solution (i.e., D) is equal to $n + m$.

After defining the solution representation in MLP, the fitness function is used to evaluate the quality of the solution. We use MSE to compute the variation between the predicted values and the actual values produced by the MLP-based trainer (i.e., metaheuristic algorithms) for all training samples, as shown in Eq. (27).

$$\overline{MSE} = \sum_{t=1}^{T} \frac{\sum_{i=1}^{o} (y_i^t - \hat{y}_i^t)^2}{T} \tag{27}$$

where T refers to the total number of used instances in the training dataset, t is the training instance, y is the actual classification value, $\hat{y}$ is the predicated classification value, and o is the number of outputs. The desired fitness function is formulated in Eq. (28).

$$\min f(x) = \overline{MSE} \tag{28}$$

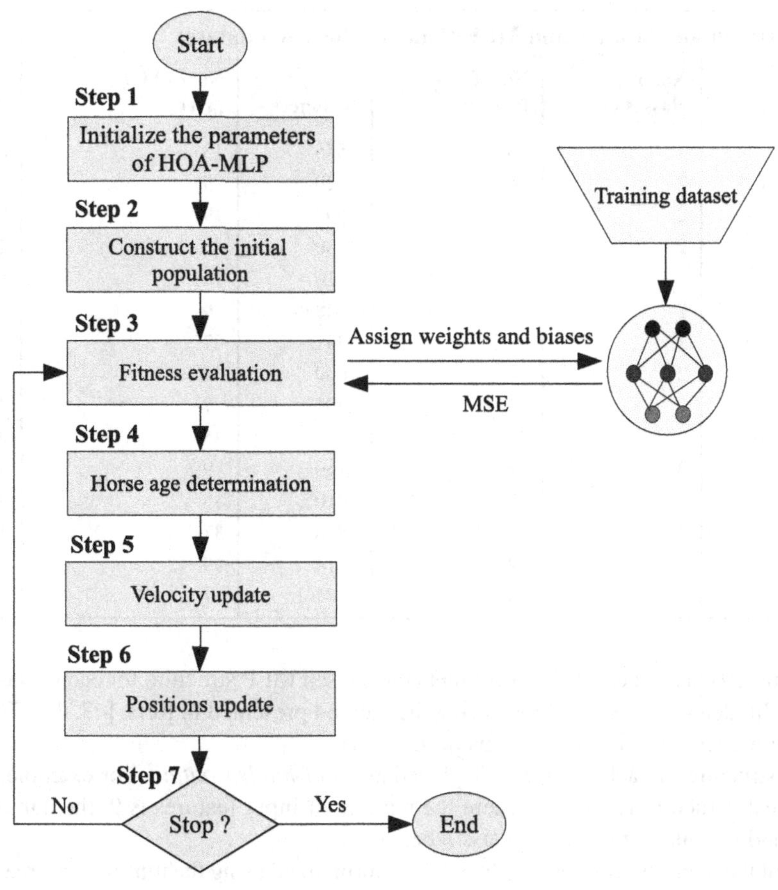

FIG. 2

Flowchart of the HOA-MLP algorithm.

4. Experiments and results

This section evaluates the proposed HOA-MLP algorithm using 15 datasets against five swarm-based metaheuristic algorithms. Section 4.1 describes the attributes of the datasets. Section 4.2 presents the setup of the experiment. Section 4.3 gives the classification accuracy and MSE to analyze the efficiency of the proposed algorithm when compared against comparable methods.

4.1 Test datasets

The proposed HOA-MLP competency is evaluated using 15 benchmark classification problems with different complexities (https://archive.ics.uci.edu/ml/index.php). Table 1 displays the number of classes, features, and instances. The selected benchmarks cover a wide range of classification complexities

Table 1 Classification datasets and MLP structure for each dataset.

No.	Dataset	No. of classes	No. of features	Instances	No. of hidden layers	MLP structure
1	Monk	2	6	556	13	6–13–2
2	Balloon	2	4	20	9	4–9–2
3	Cancer	2	9	699	19	9–19–2
4	Heart	2	22	80	45	22–45–2
5	Vertebral	2	6	310	13	6–13–2
6	Blood	2	4	748	9	4–9–2
7	Ionosphere	2	33	351	67	33–67–2
8	German	2	24	1000	49	24–49–2
9	Titanic	2	3	2201	7	3–7–2
10	Parkinson	2	22	195	45	22–45–2
11	Iris	3	4	150	9	4–9–3
12	Seeds	3	7	210	15	7–15–3
13	Vehicle	4	18	846	37	18–37–4
14	Glass	6	9	214	19	9–19–6
15	Yeast	10	8	1484	17	8–17–10

with 2, 4, 6, and 10 class labels. Table 1 presents the chosen MLP structure for each dataset. Note that the number of hidden layers is developed using the method presented in Refs. [48, 49]. The number of neurons in the hidden layer is computed using Eq. (24).

The MLP structure of each dataset is displayed as *input-hidden-output*. For example, in the Glass dataset, the MLP structure is 9–19–6 where the number of input features is 9, the number of hidden layers is 19, and the number of output class labels is 6.

The scale of the datasets features is reduced. It is normalized using the min-max normalization technique as follows:

$$x' = \frac{x_i - min_F}{max_F - min_F} \tag{29}$$

where x' is the normalized value of x in the range $[min_F, max_F]$.

Note that 30% of the dataset is used for testing and 70% for training. To keep a representative number of classes when splitting to train and test, a **stratified method** is utilized to split the dataset [50]. This can help in increasing the presence of the minority classes in both training and testing datasets.

4.2 Experimental settings

The proposed algorithm is evaluated on the same datasets against the other five swarm algorithms. We used a Microsoft Azure server with MATLAB version 9.7.0 on a PC with a Windows operating system, Intel R Xeon Silver 1.8 GHz CPU, and 6 GB of RAM to conduct the experiments. Each experiment is replicated 30 times with 250 iterations. All comparative algorithms set the populations size to 70.

Table 2 Initial parameters of the comparative algorithms.

Algorithm	Parameter	Settings
BA-MLP [31]	Population size	70
	Loudness (A_0)	0.5
	Pulse emission rate (r)	[0, 1]
	Minimum frequency (f_{min})	0
	Maximum frequency (f_{max})	2
HS-MLP [33]	Harmony memory size (HMS)	70
	Pitch adjustment rate (PAR)	0.1
	Memory consideration rate (HMCR)	0.95
	Bandwidth (bw)	0.02
FPA-MLP [44]	Population size	70
	Switch probability (P)	0.8
SCA-MLP [36]	Population size	70
JAYA-MLP [45]	Population size	70

Table 2 demonstrates the parameter settings of all comparative methods, which are based on the original paper's recommendation.

4.3 Comparison with other swarm-based optimization algorithms

This section presents the efficiency of the proposed HOA-MLP algorithm when compared against five swarm-based algorithms. The comparable methods are BA-MLP [31], HS-MLP [33], FPA-MLP [44], SCA-MLP [36], and JAYA-MLP [45]. The MLP trainer using flower pollination algorithm (FPA) is developed for this study.

Table 3 presents all the results of the proposed algorithm when compared with the comparable methods. The table presents the mean, standard deviation, and best classification accuracy captured.

Table 3 The accuracy results of the proposed HOA-MLP in comparison with other algorithms.

Dataset	Measures	HOA-MLP	BA-MLP	HS-MLP	FPA-MLP	SCA-MLP	JAYA-MLP
Monk	Mean	66.73	65.02	61.29	66.22	**69.64**	68.25
	STD	5.08	5.21	4.62	3.73	4.13	3.49
	Best	76.51	74.70	73.49	75.30	78.31	74.70
Balloon	Mean	**100.0**	87.78	81.11	94.44	95.56	91.67
	STD	0.0	14.47	16.22	12.63	10.66	12.18
	Best	100.0	100.0	100.0	100.0	100.0	100.0
Iris	Mean	87.78	66.15	65.48	87.11	87.04	**89.93**
	STD	9.30	14.99	7.82	6.94	6.78	3.81
	Best	97.78	93.33	86.67	97.78	97.78	97.78

Continued

Table 3 The accuracy results of the proposed HOA-MLP in comparison with other algorithms—cont'd

Dataset	Measures	HOA-MLP	BA-MLP	HS-MLP	FPA-MLP	SCA-MLP	JAYA-MLP
Cancer	Mean	**98.15**	95.84	94.21	94.82	95.30	97.07
	STD	0.64	0.68	1.37	0.85	0.79	0.78
	Best	99.04	97.13	96.65	96.17	96.17	98.56
Heart	Mean	**69.86**	68.06	63.33	64.58	63.75	62.50
	STD	7.31	10.17	8.85	9.65	9.23	8.19
	Best	79.17	87.50	75.0	79.17	79.17	79.17
Vertebral	Mean	**82.04**	80.04	73.33	78.85	79.32	81.83
	STD	3.64	4.33	4.44	2.19	2.58	2.94
	Best	89.25	86.02	81.72	83.87	82.80	87.10
Blood	Mean	78.08	76.61	76.65	**78.11**	76.90	77.98
	STD	0.49	1.28	1.80	0.60	1.04	1.11
	Best	79.02	78.57	79.91	79.02	79.02	79.91
Seeds	Mean	84.97	66.14	58.36	**87.99**	78.10	83.39
	STD	8.75	13.36	10.22	4.58	6.08	5.60
	Best	95.24	93.65	76.19	93.65	88.89	92.06
Glass	Mean	43.33	37.81	29.79	45.31	44.95	**47.03**
	STD	6.53	9.06	10.74	3.74	5.83	5.23
	Best	53.13	53.13	51.56	53.13	57.81	54.69
Ionosphere	Mean	83.59	79.05	72.92	**86.10**	78.13	81.33
	STD	4.35	5.58	4.39	3.66	3.60	4.90
	Best	94.29	88.57	81.90	92.38	85.71	87.62
German	Mean	**79.74**	73.71	68.66	79.40	77.31	77.81
	STD	4.01	3.91	2.43	2.48	2.25	2.54
	Best	84.67	81.0	74.67	83.67	81.67	80.67
Titanic	Mean	77.99	77.87	76.61	**78.99**	77.43	78.72
	STD	0.63	0.83	1.17	0.47	0.86	0.64
	Best	79.39	79.39	78.33	80.15	78.94	79.39
Vehicle	Mean	32.58	29.47	28.26	31.42	**35.18**	34.49
	STD	9.66	5.87	5.58	6.67	6.34	5.99
	Best	59.68	41.11	40.71	42.69	44.27	45.06
Parkinson	Mean	83.45	**87.30**	80.29	82.93	82.76	84.89
	STD	2.96	5.17	4.22	3.77	3.84	3.22
	Best	87.93	93.10	86.21	87.93	91.38	91.38
Yeast	Mean	32.92	24.99	21.87	**33.45**	31.78	31.26
	STD	3.55	8.39	10.77	2.77	2.98	3.01
	Best	39.78	32.58	34.61	37.53	37.98	36.85

Higher values of the two measurements (i.e., mean and Best) imply better performance. Note that best values are highlighted using bold font.

Table 3 summarizes the mean classification accuracy. The proposed HOA-MLP algorithm obtains the best mean results in five datasets (i.e., Balloon, Cancer, Heart, Vertebral, and German). This demonstrates the algorithm's ability in outperforming other comparative techniques when navigating the problem search space. The algorithm was able to escape from local optima because of its diversification ability. However, FPA-MLP was able to achieve the greatest accuracy in five datasets as well. JAYA-MLP and SCA-MLP each outperform the other comparative methods in two datasets, while the HS-MLP did not surpass any other method in any dataset.

Additionally, Table 3 presents the best classification accuracy results obtained by the comparative methods. The proposed HOA-MLP produces the best results in 12 datasets: Monk, Balloon, Iris, Cancer, Heart, Vertebral, Seeds, Ionosphere, German, Titanic, Vehicle, and Yeast. It is worth mentioning that HOA-MLP outperforms other comparative methods in Yeast (10 classes) and Vehicle (4 classes), which are two of the largest datasets. The algorithm's ability to balance between exploration and exploitation makes it reach new regions in the search space and thus provide a better solution.

Finally, the standard derivation (STD) results refer to robustness of the algorithm. Lower values of STD are better. Reading the standard derivation results of each method recorded in Table 3, it can be seen the HOA-MLP and FPA-MLP are ranked first because each obtained the minimum STD results in five datasets. SCA-MLP, JAYA-MLP, and HS-MLP obtained the minimum STD results in the 2, 2, and 1 datasets, respectively. This proves that the HOA-MLP and FPA-MLP are more robust than the remaining methods by achieving almost the same results over 30 runs.

4.3.1 Convergence analysis

Fig. 3 visualizes the comparative methods convergence behavior on all evaluated datasets. The figure plots the iteration number in the x-axis and the fitness values (i.e., MSE) in the y-axis.

Interestingly, the proposed algorithm shows faster convergence in obtaining better MSE. This is true in most of the evaluated datasets. HOA-MLP achieved the best results in 10 datasets: Monk, Iris, Cancer, Seeds, Vertebral, Glass, German, Parkinson, Vehicle, and Yeast. The algorithm operators enable reaching new regions in the search space, thus escaping being stuck in the local optima trap during the search.

4.3.2 Friedman's statistical test

In this section, we use the Friedman test and Holm's and Hochberg's procedures to statistically analyze the proposed method. The null hypothesis (H_0) assumes that we have equal behavior among the comparative methods when solving the problem. Additionally, the alternative hypothesis (H_1) assumes that there is inequality among the comparative methods when solving the problem. Fig. 4 shows Friedman's statistical test. The proposed HOA-MLP algorithm ranked first, whereas FPA-MLP and JAYA-MLP are ranked second.

We used Holm and Hochberg procedures as post hoc techniques to calculate adjusted ρ value to determine whether there are significant differences between the proposed algorithm and the comparative methods. The HOA-MLP is the controlled algorithm because it is ranked first, as determined by the Friedman test. The null hypothesis H_0 is not accepted when the ρ-value ≤ 0.01667, and the null hypothesis H_0 is abandoned using Hochberg's procedure when the ρ-value ≤ 0.0125. Table 4 illustrates that there is a significant difference between the proposed algorithm and HS-MLP and BA-MLP.

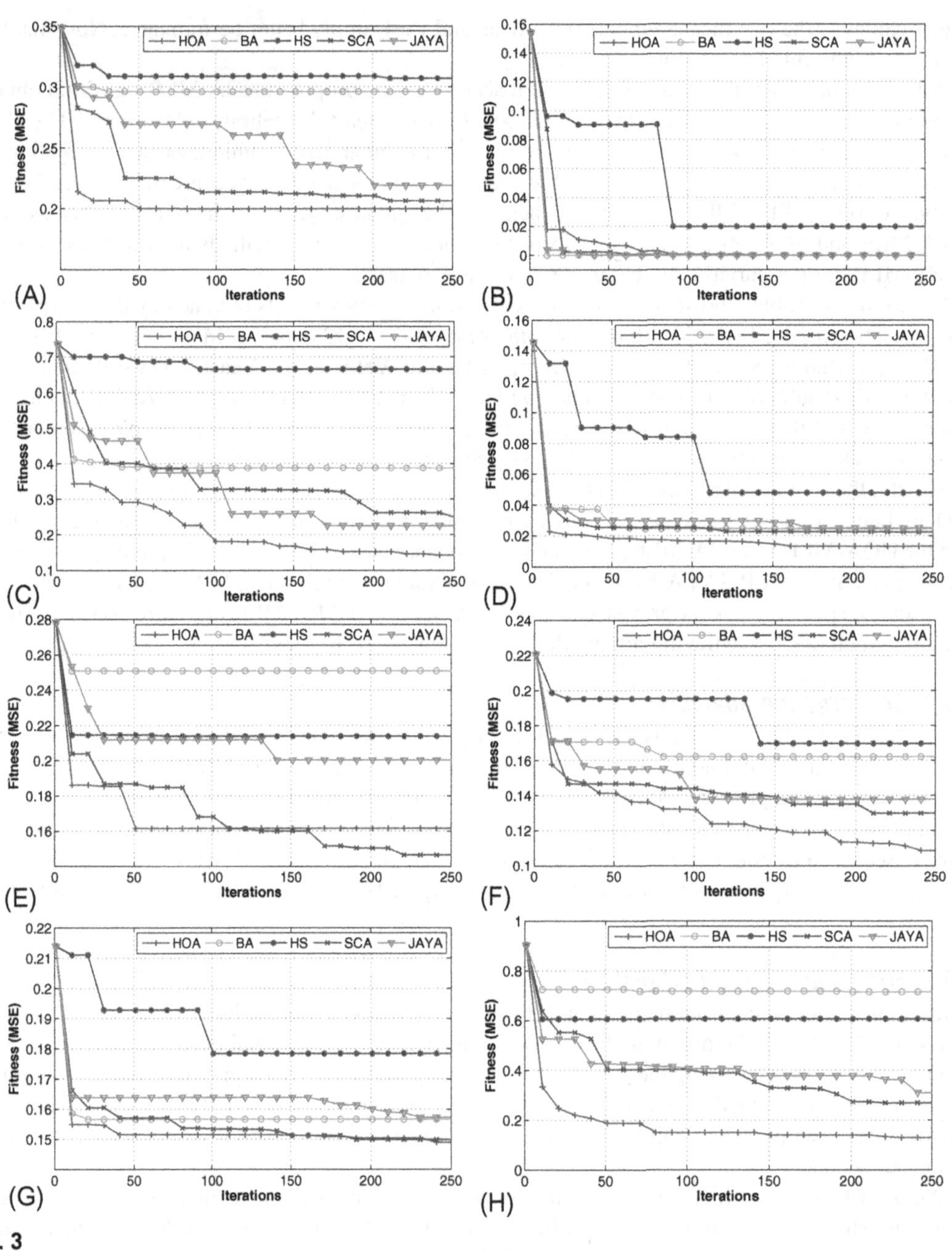

FIG. 3

The convergence behavior of the proposed HOA against BA, HS, SCA, and JAYA algorithms. (A) Monk, (B) Balloon, (C) Iris, (D) Cancer, (E) Heart, (F) Vertebral, (G) Blood, (H) Seeds,

(Continued)

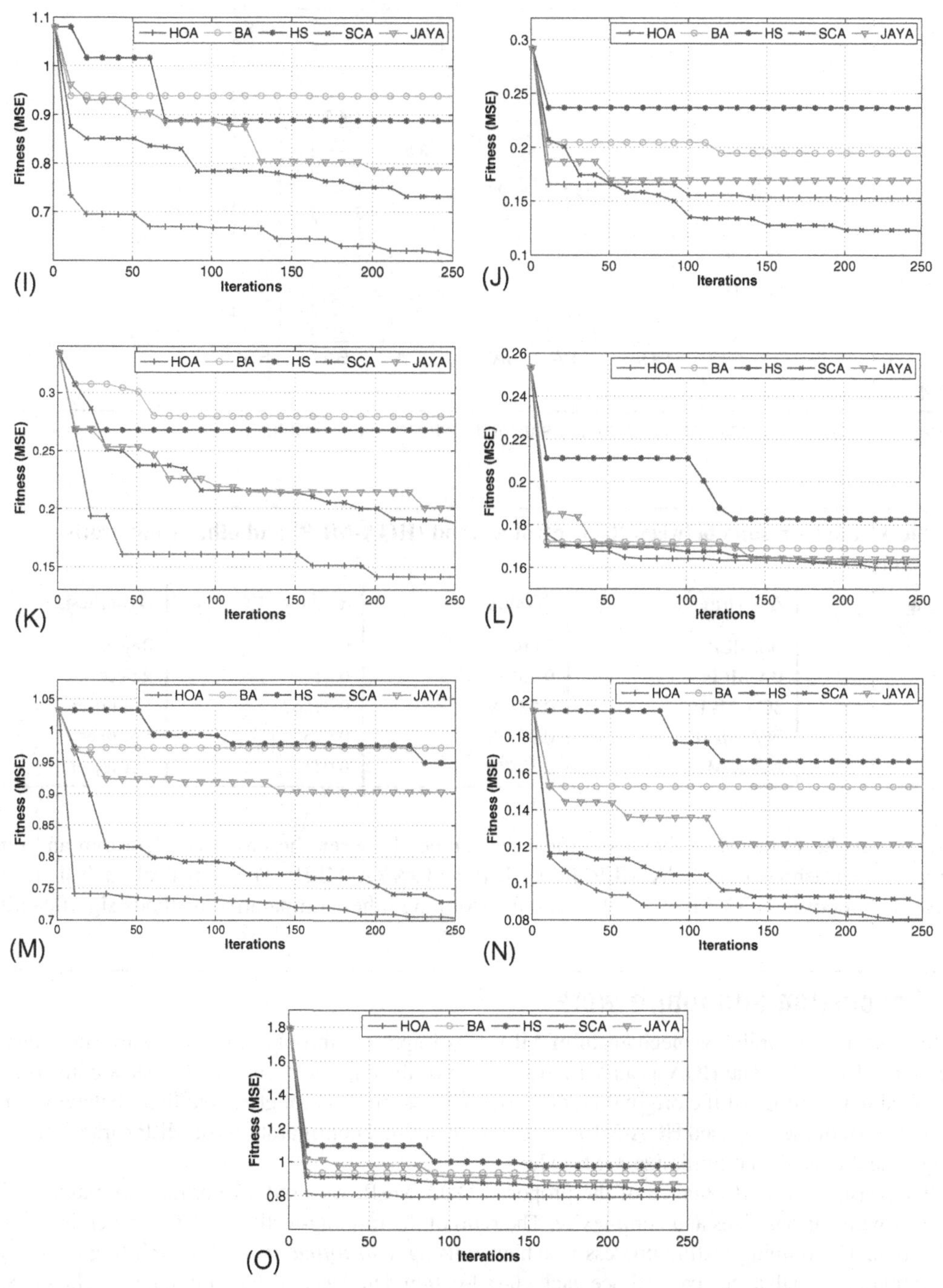

FIG. 3, CONT'D

(I) Glass, (J) Ionosphere, (K) German, (L) Titanic, (M) Vehicle, (N) Parkinson, and (O) Yeast.

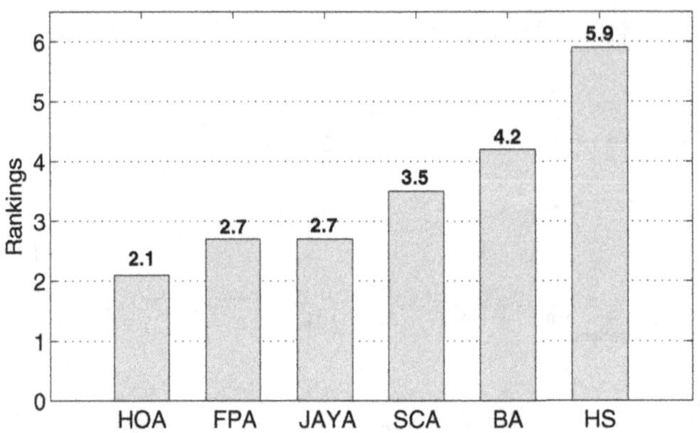

FIG. 4

Average rankings of the comparative algorithms using Friedman's statistical test.

Table 4 Holm's results between the control method (HOA-MLP) and other comparative methods.

Rank	Algorithm	*P*-value	α/Rank	Hypothesis
5	HS-MLP	2.66E-08	0.01	Reject
4	BA-MLP	0.00179	0.0125	Reject
3	SCA-MLP	0.03179	0.01667	Not reject
2	FPA-MLP	0.37978	0.025	Not reject
1	JAYA-MLP	0.37978	0.05	Not reject

However, it shows that there is no significant difference between the proposed algorithm and three comparative methods (i.e., SCA-MLP, FPA-MLP, and JAYA-MLP). The statistical analysis results prove the proposed HOA-MLP's ability in outperforming other comparative methods significantly.

5. Conclusion and future work

In this chapter, the training mechanism of MLP was improved through a recent swarm intelligence method called HOA. The HOA as an MLP trainer avoids the chronic problems (i.e., slow convergence and local optima trap) of the original gradient-descent algorithm through its publicity behaviors. The ultimate goal of the proposed HOA-MLP is to find the optimal configurations of MLP parameters (i.e., weights and biases) that minimize MSE values.

To measure the performance of the proposed HOA-MLP, we used 15 normalized classification datasets with various sizes and complexity. The ratio of the training-testing is 7:3 of the entire dataset collection. The training-testing process is achieved using a *stratified method* for splitting the dataset collection into a balanced shape. Since each classification data has a different number of class labels, the MLP is adjusted to have different numbers of input, hidden, and output nodes.

We compared our proposed algorithm with BA [31], HS [33], FPA, SCA [36], and JAYA algorithm [45]. Interestingly, the proposed HOA-MLP outperformed the others in 5 out of 15 datasets. Furthermore, the HOA-MLP achieved high-quality results for others. We also evaluated the convergence behavior of HOA-MLP against the comparative methods, and results show that HOA-MLP achieved better convergence. We also used Friedman's statistical test to prove the viability of HOA-MLP.

As the proposed HOA-MLP shows very efficient training behavior for MLP, other real-world classification data with more difficult characteristics can be investigated in the future, as can modified or hybridized versions of the algorithm.

References

[1] W.S. McCulloch, W. Pitts, A logical calculus of the ideas immanent in nervous activity, Bull. Math. Biophys. 5 (4) (1943) 115–133.

[2] A. Verikas, M. Bacauskiene, Feature selection with neural networks, Pattern Recogn. Lett. 23 (11) (2002) 1323–1335.

[3] F.H. She, L.X. Kong, S. Nahavandi, A.Z. Kouzani, Intelligent animal fiber classification with artificial neural networks, Text. Res. J. 72 (7) (2002) 594–600.

[4] S.G. Meshram, M.A. Ghorbani, S. Shamshirband, V. Karimi, C. Meshram, River flow prediction using hybrid PSOGSA algorithm based on feed-forward neural network, Soft Comput. 23 (20) (2019) 10429–10438.

[5] G. Bebis, M. Georgiopoulos, Feed-forward neural networks, IEEE Potentials 13 (4) (1994) 27–31.

[6] R. Hecht-Nielsen, Theory of the backpropagation neural network, in: Neural Networks for Perception, Elsevier, 1992, pp. 65–93.

[7] S. Lawrence, C.L. Giles, A.C. Tsoi, A.D. Back, Face recognition: a convolutional neural-network approach, IEEE Trans. Neural Netw. 8 (1) (1997) 98–113.

[8] S. Mirjalili, How effective is the Grey Wolf optimizer in training multi-layer perceptrons, Appl. Intell. 43 (1) (2015) 150–161.

[9] D.R. Hush, B.G. Horne, Progress in supervised neural networks, IEEE Signal Process. Mag. 10 (1) (1993) 8–39.

[10] M.T. Hagan, M.B. Menhaj, Training feedforward networks with the Marquardt algorithm, IEEE Trans. Neural Netw. 5 (6) (1994) 989–993.

[11] W.A.H.M. Ghanem, A. Jantan, A new approach for intrusion detection system based on training multilayer perceptron by using enhanced Bat algorithm, Neural Comput. Appl. 32 (2020) 11665–11698, https://doi.org/10.1007/s00521-019-04655-2.

[12] F. Han, J. Jiang, Q.-H. Ling, B.-Y. Su, A survey on metaheuristic optimization for random single-hidden layer feedforward neural network, Neurocomputing 335 (2019) 261–273.

[13] A.M. Hemeida, S.A. Hassan, A.-A.A. Mohamed, S. Alkhalaf, M.M. Mahmoud, T. Senjyu, A,B. El-Din, Nature-inspired algorithms for feed-forward neural network classifiers: a survey of one decade of research, Ain Shams Eng. J. 11 (3) (2020) 659–675.

[14] S. Ding, C. Su, J. Yu, An optimizing BP neural network algorithm based on genetic algorithm, Artif. Intell. Rev. 36 (2) (2011) 153–162.

[15] A. Slowik, M. Bialko, Training of artificial neural networks using differential evolution algorithm, in: 2008 Conference on Human System Interactions, IEEE, 2008, pp. 60–65.

[16] J. Ilonen, J.-K. Kamarainen, J. Lampinen, Differential evolution training algorithm for feed-forward neural networks, Neural Process. Lett. 17 (1) (2003) 93–105.

[17] Y. Zhang, P. Phillips, S. Wang, G. Ji, J. Yang, J. Wu, Fruit classification by biogeography-based optimization and feedforward neural network, Expert Syst. 33 (3) (2016) 239–253.

[18] H. Wu, Y. Zhou, Q. Luo, M.A. Basset, Training feedforward neural networks using symbiotic organisms search algorithm, Comput. Intell. Neurosci. 2016 (2016) 1–14, https://doi.org/10.1155/2016/9063065.

[19] G. Das, P.K. Pattnaik, S.K. Padhy, Artificial neural network trained by particle swarm optimization for non-linear channel equalization, Expert Syst. Appl. 41 (7) (2014) 3491–3496.

[20] M. Braik, H. Al-Zoubi, H. Al-Hiary, Artificial neural networks training via bio-inspired optimisation algorithms: modelling industrial winding process, case study, Soft Comput. 25 (6) (2021) 4545–4569.

[21] H. Faris, S. Mirjalili, I. Aljarah, Automatic selection of hidden neurons and weights in neural networks using grey wolf optimizer based on a hybrid encoding scheme, Int. J. Mach. Learn. Cybern. 10 (10) (2019) 2901–2920.

[22] D.A. Alboaneen, H. Tianfield, Y. Zhang, Glowworm swarm optimisation for training multi-layer perceptrons, in: Proceedings of the Fourth IEEE/ACM International Conference on Big Data Computing, Applications and Technologies, 2017, pp. 131–138.

[23] D. Bairathi, D. Gopalani, Salp swarm algorithm (SSA) for training feed-forward neural networks, in: Soft Computing for Problem Solving, Springer, 2019, pp. 521–534.

[24] W.A.H.M. Ghanem, A. Jantan, A cognitively inspired hybridization of artificial bee colony and dragonfly algorithms for training multi-layer perceptrons, Cogn. Comput. 10 (6) (2018) 1096–1134.

[25] A. Sheta, M. Braik, H. Al-Hiary, Modeling the Tennessee Eastman chemical process reactor using bio-inspired feedforward neural network (BI-FF-NN), Int. J. Adv. Manuf. Technol. 103 (2019) 1359–1380, https://doi.org/10.1007/s00170-019-03621-5.

[26] H. Moayedi, H. Nguyen, L.K. Foong, Nonlinear evolutionary swarm intelligence of grasshopper optimization algorithm and gray wolf optimization for weight adjustment of neural network, Eng. Comput. 37 (2021) 1265–1275, https://doi.org/10.1007/s00366-019-00882-2.

[27] A.A. Heidari, H. Faris, I. Aljarah, S. Mirjalili, An efficient hybrid multilayer perceptron neural network with grasshopper optimization, Soft Comput. 23 (17) (2019) 7941–7958.

[28] M. Khishe, M.R. Mosavi, Classification of underwater acoustical dataset using neural network trained by chimp optimization algorithm, Appl. Acoust. 157 (2020) 107005.

[29] H. Faris, I. Aljarah, S. Mirjalili, Improved monarch butterfly optimization for unconstrained global search and neural network training, Appl. Intell. 48 (2) (2018) 445–464.

[30] K. Socha, C. Blum, An ant colony optimization algorithm for continuous optimization: application to feed-forward neural network training, Neural Comput. Appl. 16 (3) (2007) 235–247.

[31] N.S. Jaddi, S. Abdullah, A.R. Hamdan, Multi-population cooperative bat algorithm-based optimization of artificial neural network model, Inform. Sci. 294 (2015) 628–644.

[32] S.M.J. Jalali, S. Ahmadian, P.M. Kebria, A. Khosravi, C.P. Lim, S. Nahavandi, Evolving artificial neural networks using butterfly optimization algorithm for data classification, in: International Conference on Neural Information Processing, Springer, 2019, pp. 596–607.

[33] S. Kulluk, L. Ozbakir, A. Baykasoglu, Training neural networks with harmony search algorithms for classification problems, Eng. Appl. Artif. Intel. 25 (1) (2012) 11–19.

[34] A.L. Bolaji, A.A. Ahmad, P.B. Shola, Training of neural network for pattern classification using fireworks algorithm, Int. J. Syst. Assur. Eng. Manag. 9 (1) (2018) 208–215.

[35] H. Faris, I. Aljarah, S. Mirjalili, Training feedforward neural networks using multi-verse optimizer for binary classification problems, Appl. Intell. 45 (2) (2016) 322–332.

[36] A.T. Sahlol, A.A. Ewees, A.M. Hemdan, A.E. Hassanien, Training feedforward neural networks using Sine-Cosine algorithm to improve the prediction of liver enzymes on fish farmed on nano-selenite, in: 2016 12th International Computer Engineering Conference (ICENCO), IEEE, 2016, pp. 35–40.

[37] H. Faris, I. Aljarah, N. Al-Madi, S. Mirjalili, Optimizing the learning process of feedforward neural networks using lightning search algorithm, Int. J. Artif. Intell. Tools 25 (6) (2016) 1650033.

[38] S. Mirjalili, S.Z.M. Hashim, H.M. Sardroudi, Training feedforward neural networks using hybrid particle swarm optimization and gravitational search algorithm, Appl. Math. Comput. 218 (22) (2012) 11125–11137.

[39] D.H. Wolpert, W.G. Macready, No free lunch theorems for optimization, IEEE Trans. Evol. Comput. 1 (1) (1997) 67–82.

[40] F. MiarNaeimi, G. Azizyan, M. Rashki, Horse herd optimization algorithm: a nature-inspired algorithm for high-dimensional optimization problems, Knowl.-Based Syst. 213 (2021) 106711.

[41] N. Mehrabi, E. Pashaei, Application of horse herd optimization algorithm for medical problems, in: 2021 International Conference on Innovations in Intelligent Systems and Applications (INISTA), IEEE, 2021, pp. 1–6.

[42] S. Basu, M. Basu, Horse herd optimization algorithm for fuel constrained day-ahead scheduling of isolated nanogrid, Appl. Artif. Intell. 35 (15) (2021) 1250–1270, https://doi.org/10.1080/08839514.2021.1975392.

[43] M.A. Awadallah, A.I. Hammouri, M.A. Al-Betar, M.S. Braik, M. Abd Elaziz, Binary horse herd optimization algorithm with crossover operators for feature selection, Comput. Biol. Med. 141 (2022) 105152, https://doi.org/10.1016/j.compbiomed.2021.105152.

[44] X.-S. Yang, Flower pollination algorithm for global optimization, in: International Conference on Unconventional Computing and Natural Computation, Springer, 2012, pp. 240–249.

[45] E. Uzlu, Application of Jaya algorithm-trained artificial neural networks for prediction of energy use in the nation of Turkey, Energy Sources Part B Econ. Plan. Policy 14 (5) (2019) 183–200.

[46] K. Sun, S.-H. Huang, D.S.-H. Wong, S.-S. Jang, Design and application of a variable selection method for multilayer perceptron neural network with LASSO, IEEE Trans. Neural Netw. Learn. Syst. 28 (6) (2016) 1386–1396.

[47] S.M. McDonnell, The Equid Ethogram: A Practical Field Guide to Horse Behavior, Eclipse Press, 2003.

[48] A.S.I. Wdaa, A. Sttar, Differential Evolution for Neural Networks Learning Enhancement (Ph.D. thesis), Universiti Teknologi Malaysia Johor Bahru, 2008.

[49] S. Mirjalili, S.M. Mirjalili, A. Lewis, Let a biogeography-based optimizer train your multi-layer perceptron, Inform. Sci. 269 (2014) 188–209.

[50] J.-R. Cano, S. García, F. Herrera, Subgroup discover in large size data sets preprocessed using stratified instance selection for increasing the presence of minority classes, Pattern Recogn. Lett. 29 (16) (2008) 2156–2164.

CHAPTER

20 Metaheuristics for clustering problems

Farhad Soleimanian Gharehchopogh[a], Benyamin Abdollahzadeh[a], Nima Khodadadi[b], and Seyedali Mirjalili[c,d]

[a]*Department of Computer Engineering, Urmia Branch, Islamic Azad University, Urmia, Iran,* [b]*Department of Civil and Environmental Engineering, Florida International University, Miami, FL, United States,* [c]*Centre for Artificial Intelligence Research and Optimisation, Torrens University Australia, Fortitude Valley, Brisbane, QLD, Australia,* [d]*University Research and Innovation Center, Obuda University, Budapest, Hungary*

1. Introduction

Data clustering is one of the most important and common problems of data analysis In the process of data clustering, we attempt each cluster to contain similar objects and, at the same time, with objects that are different from others [1,2]. In data clustering, the similarity is based on the distance between objects. In such a way, similar objects have less distance together and can be placed in a cluster. Data clustering has been used for financial and commercial applications as well as in marketing, data retrieval, E-commerce, information filtering, scientific observation, signal processing, web engineering, text mining, image processing, biology, stock prediction, and other fields of sciences and engineering [3–5].

Algorithms for solving data clustering problems are categorized into two groups: hierarchical and partitioned algorithms. Hierarchical algorithms are subcategorized into two types: divisive and collective clusters. In the divisive method, each observation starts in its cluster; then, using recursive algorithms, the lesser similarities in data are split into individual clusters. In collective algorithms, each object is considered a free cluster and the closest clusters in properties are merged until the individual clusters are formed. In partitioned algorithms, each cluster has its individual centers; then, according to similarities between objects and each data center, all objects are classified into one relevant cluster. These algorithms aim to maximize the similarities in one cluster, while connected to different clusters decreases. One of the most widely used algorithms in this area is the *k*-means algorithm, but it suffers from degraded performance in case of noises and outliers. This algorithm is also not able to detect clusters with nonconvex shapes. To alleviate these drawbacks, scientists use metaheuristics, which are typically nature-inspired optimization algorithms [6]. Such algorithms can find out general ways instead of technical methods in data clustering, which present the successful linear-complexity results [7,8]. Metaheuristic methods are the most successful and promising methods for resolving data clustering problems.

Van der Merwe and Engelbrecht proposed two new methods using Particle Swarm Optimization (PSO) [9] in data clustering problems [10]. They evaluated the approaches on six datasets and the results showed that the proposed methods achieved better convergence than *k*-means clustering. Kao et al.

Comprehensive Metaheuristics. https://doi.org/10.1016/B978-0-323-91781-0.00020-X

presented a combined approach for data clustering problems [11]. In this study, the authors combined the k-means algorithm with Nelder-Mead simplex search and PSO to create K-NM-PSO. The results of the simulation showed that the proposed method is both healthy and appropriate for data clustering.

Das et al. presented a modified Differential Evolution (DE) algorithm for automated data clustering [12]. Senthilnath et al. used the Firefly Algorithm (FFA) [13] to solve the data clustering problem. Alia et al. [14] applied the Harmony Search (HS) algorithm to data clustering. In this study, the clustering algorithm's main challenges were the physical sensitivity of Hard C-means (HCM) and Fuzzy C-means (FCM) to adjust the initial cluster centers. The HS algorithm [15] was employed to solve it. The proposed approach was implemented in two steps. First, the HS algorithm searches and finds the right cluster centers. Then, the best cluster centers are used as initial cluster centers for C-means algorithms. The approach was implemented on 10 datasets and the results showed that the proposed algorithm could be considered an appropriate data clustering problem. Huang et al. presented a compound method for data clustering by hybridizing continuous Ant Colony Optimization (ACO) and PSO [16]. The authors investigated four types of combinations: (1) sequence approach, (2) parallel approach, (3) sequence approach with an enlarged pheromone-particle table, and (4) global best exchange.

Kaushik et al. [17] presented a new compound method by hybridizing a modified Genetic Algorithm (GA) [18] and FFA. In this study, the primary population was selected by FFA. Jadhav and Gomath [19] created a new compound method using Gray Wolf Optimization (GWO) [20] and Whale Optimization Algorithm (WOA) [21] for data clustering. In another study, Kumar et al. [22] investigated combining the k-means algorithm and the Artificial Bee Colony (ABC) algorithm [23] for data clustering and found it to be successful. Ashish et al. [24] proposed a Parallel Bat Algorithm (PBA) for data clustering using MapReduce. Study results showed that PBA achieved more acceptable results than PSO and exhibited a significant increase in speed along with an increasing number of nodes.

The preceding paragraphs provided a brief literature review of metaheuristics in clustering, which demonstrated the merits of such techniques in this problem area. Recently, many metaheuristics have been designed and developed as black-box optimizers to solve a wide range of optimization problems. The No Free Lunch (NFL) theorem has been a motivator that logically proves that there is no optimization efficient enough to solve all optimization problems. Therefore, researchers constantly endeavor to develop new algorithms or change existing algorithms to solve problems. This motivated our attempts to provide a comprehensive comparative study of 12 algorithms on several clustering problems. The algorithms used are GA [18], PSO [9], HS [15], ABC [23], Biogeography-based Optimization (BBO) [25], DE [26], GWO, Symbiotic Organism Search (SOS) [27], WOA [21], FFA [28], African Vulture Optimization Algorithm (AVOA) [29], and Artificial Gorilla Troops Optimization (AGTO) [30]. The rest of this book chapter is organized as follows:

Section 2 provides the details of the data clustering problem and formulates it as an optimization problem. Section 3 describes the process of solving clustering problems using metaheuristics. Section 4 presents and discusses the results and findings. Finally, Section 5 concludes the chapter and suggests future directions.

2. Data clustering problem

Clustering of data can be considered as the process of dividing a dataset (with n samples and d variables) into k partitions. This is done using a clustering algorithm that can find similar samples in a d-dimensional space. The similarity of samples is usually quantified and evaluated using distances such as Euclidean distance. If we assume $S = (x_1, x_2, x_3, \ldots, x_n)$ with N samples, each sample in this set has d dimension that can be expressed as $(x_{i1}, x_{i2}, x_{i3}, \ldots, x_{id})$. If we want to cluster this set into k clusters, it is expressed as follows:

$$C = (C_1, C_2, C_3, \ldots, c_k) \tag{1}$$

$$\begin{aligned} & C_j \neq \varnothing, \forall j \in \{1, 2, 3, \ldots, k\} \\ & \bigcup_{j=1}^{k} C_j = S \\ & C_j \cap C_i = \varnothing, \forall j \neq i \text{ and } j, i \in \{1, 2, 3, \ldots, k\} \end{aligned} \tag{2}$$

Eqs. (1) and (2) divide the dataset into k clusters, in which none of the clusters are empty, the collection of all clusters will be equivalent to S, and there are no joint samples in the clusters.

Using the preceding elements, clustering can be expressed as an optimization problem as in Eq. (3).

$$f(S, C) = \sum_{j=1}^{n} min\left\{ \|x_j - C_i\|^2 \right\} \; i = \{1, 2, 3, \ldots, k\} \tag{3}$$

Eq. (3) shows that each sample in the dataset is assigned to a cluster with the least distance. As discussed, the distance can be calculated using Eq. (4).

$$d(x_i, x_j) = \sqrt{\sum_{m=1}^{d} |x_{im} - x_{jm}|^2} \tag{4}$$

The objective here is to minimize the distance in Eq. (4).

Now that we formulated clustering as an optimization problem, we can employ optimization algorithms to determine its global optimum. The next section discusses this process when using metaheuristics.

3. Data clustering using metaheuristic algorithms

Metaheuristic algorithms are used to solve optimization problems to find quality solutions. These algorithms have a good ability to obtain quality solutions in a short period of time by producing a variety of solutions. Due to their unique advantages, metaheuristic algorithms are used in solving various problems in many fields and are good options for solving problems in various search spaces, including continuous and discrete. Metaheuristic algorithms have other useful aspects such as relatively simple concepts and easy implementation and no need for information derived from the objective function. In addition, they have powerful and useful capabilities to generate various solutions that lead to local optimal escape. Given all the advantages mentioned, they can be a great option for solving the problem of data clustering.

The first step in solving data clustering problems using metaheuristics is to define the decision variables. In fact, the decision variable vector (candidate solution) should be formulated to be evaluated by the objective function. As per Eq. (4), if the dataset has n samples with d dimension each, for clustering the samples into k clusters, the decision variables should be of K^* size for metaheuristics. For example,

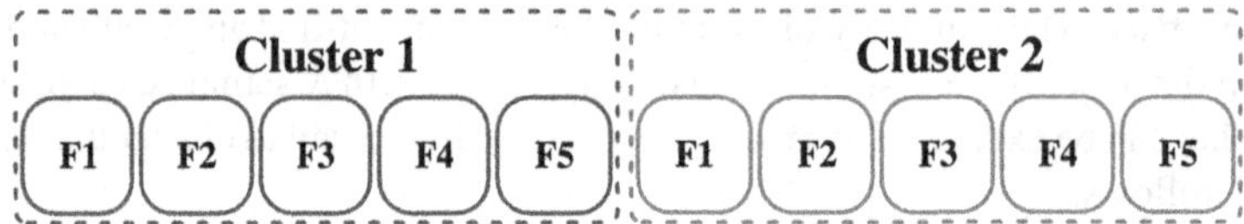

FIG. 1

Decision variable vector for a problem with five features and two clusters.

Fig. 1 shows the decision variable vector if the sample has five features, each of which can be assigned with values in the range of 0–10, and the goal is to cluster them into two clusters.

The representation of the decision variable vector allows a metaheuristic to define clusters using the samples and features. These solutions are then compared with the objective discussed in Eq. (3).

As discussed in the introduction, metaheuristics are stochastic algorithms that consider a problem as a black box. Despite evidence that conventional optimization algorithms show degraded performance, they provide different results on a given optimization problem. This means that we need to comparatively study several algorithms to identify an efficient algorithm from a pool of techniques and ensure that the solution obtained is reasonably good for the problem given the computational resources on hand. In the next section, we employ a set of 12 metaheuristics to solve a number of clustering problems.

4. Results and discussion

In this section, we use GA, DE, BBO, HS, PSO, GWO, AVOA, FFA, ABC, SOS, WOA, and AGTO metaheuristics to cluster different datasets. This set of algorithms is a mix of first-generation metaheuristics (e.g., GA, PSO, DE) and more recent ones (AGTO, AVOA, FFA, etc.). Table 1 lists the values used for the control parameters of these algorithms. All implementations and simulations were undertaken in MATLAB version 9.2 (R2017R) software installed on Windows 10 Enterprise 64-bit with a Core i7-4510U 2.6GHz processor and an 8GB RAM.

We used 10 datasets to test the algorithms. Table 2 lists the details of these datasets. As shown, these datasets have diverse numbers of samples, classes, and features.

To compare algorithms, we used 30 candidate solutions across 100 iterations over 30 independent runs. We also employed the ANOVA test as a quality assurance measure to ensure the accuracy and superiority of metaheuristics due to their stochastic nature.

4.1 Results of conventional metaheuristics

In this section, we compare the conventional metaheuristics GA, PSO, HS, DE, BBO, and ABC. Table 3 presents the results of this comparison. As shown, the best results are provided by GA and BBO. It is interesting that both algorithms are evolutionary and typically benefit from higher exploratory behavior. GA outperforms the other algorithms on four datasets and BBO is superior in three case studies.

Table 1 Parameter settings of optimization algorithms for evaluation and comparison.

Algorithm	Parameter	Value
AVOA	L_1	0.8
	L_2	0.2
	w	2.5
	P_1	0.6
	P_2	0.4
	P_3	0.6
GWO	Convergence constant a	[2,0]
DE	Crossover probability	0.5
	Scaling factor	0.5
PSO	Inertia factor	0.3
	c_1	1
	c_2	1
BBO	Immigration probability limits	[0,1]
	Habitat modification probability	1
	Max emigration (E) and Max immigration (I)	1
	Step size	1
	Mutation probability	0.005
FA	K value	2
	α	0.6
	β	0.4
	Q	0.7
	W	1
WOA	Spiral factor b	1
	Convergence constant a	[2,0]
AGTO	β	3
	W	0.8
	p	0.03
ABC	Number of Onlooker Bees	$1 \times nPop$
	Limit	100
SOS	Max fit eval	1.0
GA	Crossover Percentage	0.7
	Number of Offspring	$2 \times round(pc \times nPop/2)$
	Number of Mutants	$round(pm \times nPop)$
	Mutation Percentage	0.3
	Mutation Rat	0.1
HS	Pitch Adjustment Rate	0.1
	Harmony Memory Consideration Rate	0.8

Table 2 Case studies used in this chapter.

Data set	Instance number	Number of classes	Number of features
Iris	150	3	4
Wine	178	3	13
CMC	1473	3	9
Glass	214	6	9
Vowel	871	6	3
Cancer	683	2	9
Zoo	101	2	16
Spect	267	2	22
Ionosphere	351	2	34
Blood	748	2	4

Fig. 2 shows the convergence curves. As shown, BBO, DE, and GA show consistent improvement during the optimization process. The convergence rates of PSO and ABC are similar in all case studies.

Fig. 3 presents the results of the ANOVA test for ABC, PSO, DE, HS, BBO, and GA. As shown, GA provides the best results considering all independent runs and, as such, this algorithm can be regarded as statistically superior compared to the others. The BBO and DE algorithms also exhibited reasonably good results.

4.2 Results of recent metaheuristics

In this section, we apply and compare the more recently developed algorithms of GTO, FFA, WOA, GWO, SOS, and AVOA. Despite their being recent, these algorithms have been previously proven to be effective for a wide range of optimization problems. Table 4 and Fig. 5 show the statistical results of these algorithms on the clustering datasets.

The results demonstrate that AVOA is the best algorithm in four case studies and FFA is the second best. AGTO achieved the best results for two datasets.

The convergence curves in Fig. 4 show that the best rates are given by FFA, AGTO, and AVOA. All algorithms provide a better result than the metaheuristic optimization algorithms used in the preceding section. This shows that these algorithms might be better for clustering problems.

Fig. 5 provides the results of the ANOVA test. As shown, FFA, AGTO, and AVOA provide similar performance overall on impendent runs. A similar pattern can be observed for GWO, which indicates the merits of this algorithm in clustering problems. Taken together, all the experiments demonstrated that GA provides the best results for clustering problems among the conventional algorithms used in this work. Mechanisms such as cross-over in such evolutionary algorithms, abruptly change and mix solutions, resulting in having high exploratory behavior for an algorithm. This is even more important for large-scale clustering problems. Swarm-based algorithms such as FFA and AVOA provide the best results. Both algorithms benefit from a wide range of stochastic mechanisms that help them with the exploration of the data clustering problem.

Table 3 Comparison of traditional optimization algorithms.

Dataset	Metric	ABC	PSO	DE	HS	BBO	GA
Iris	Best	108.39	96.71	97.06	97.27	**96.66**	**96.66**
	Worst	124.84	138.95	105.46	100.43	**96.66**	**96.66**
	Mean	117.91	114.99	100.76	98.58	**96.66**	**96.66**
	STD	4	13.38	2.48	0.93	**0**	**0**
Wine	Best	16,326.55	16,336.43	16,304.25	16,302.38	16,292.18	**16,292.24**
	Worst	16,483.92	16,764.96	16,359.65	16,317.42	17,092.24	**16,295.39**
	Mean	16,392.27	16,435.32	16,319.23	16,308.36	16,476.43	**16,293.81**
	STD	40.98	96.26	11.63	3.96	246.4	**1.01**
CMC	Best	5755.36	5655.68	5552.86	5602.94	**5532.18**	5532.19
	Worst	6089.93	6838.36	5686.61	5686.02	**5532.19**	5534.91
	Mean	5909.63	6009.26	5603.11	5632.87	**5532.19**	5532.59
	STD	90.03	263.97	29.41	19.24	**0**	0.64
Cancer	Best	3091.54	3159.5	2964.71	2996.34	**2964.39**	**2964.39**
	Worst	3383.26	4330.73	3081.58	3062.59	**2964.39**	**2964.39**
	Mean	3224.58	3658.03	3006.47	3031.55	**2964.39**	**2964.39**
	STD	70.92	325.63	33.43	15.16	**0**	**0**
Glass	Best	295.41	277.04	263.21	285.14	210.44	**210.47**
	Worst	346.06	418.89	307.17	309.58	257.94	**251.76**
	Mean	318.66	346.93	280.7	294.67	229.74	**228.48**
	STD	12.96	41.23	10.14	5.61	15.92	**12.76**
Vowel	Best	173,019.93	153,637.04	156,903.89	158,594.67	153,328.98	**148,967.48**
	Worst	199,486.85	189,177.15	172,293.26	177,287.06	188,181.14	**150,875.57**
	Mean	188,896.49	170,179.78	164,529.1	165,431.28	165,061.52	**149,675.72**
	STD	6484.7	9682.67	3780.47	3657.83	7850.21	**620.87**
Zoo	Best	184.97	183.56	181.48	184.54	**181.24**	**181.24**
	Worst	194.7	198.54	187.73	188.56	**181.24**	**181.24**
	Mean	189.19	191.06	183.24	186.24	**181.24**	**181.24**
	STD	2.31	4.32	1.6	0.84	**0**	**0**
Spect	Best	514.26	513.61	498.16	512.72	**495.25**	495.25
	Worst	530.76	551.22	512.15	523.37	**496.15**	496.15
	Mean	521.13	537.06	503.41	517.48	**495.54**	495.55
	STD	4.42	9.95	3.43	2.34	**0.26**	0.24
Ionosphere	Best	1004.79	1016.36	835.08	989.86	**793.75**	793.74
	Worst	1083.66	1139.15	951.86	1046.05	**793.79**	793.88
	Mean	1037.21	1069.73	880.65	1016.61	**793.77**	793.79
	STD	19.64	34.72	33.62	14.5	**0.01**	0.04
Boold	Best	409,010.55	407,731.73	407,714.24	407,714.82	407,757.66	**407,714.23**
	Worst	411,640.97	419,363.96	408,024.19	409,639.67	447,239.95	**407,721.35**
	Mean	410,141.31	410,131.02	407,736.64	407,994.17	416,945.1	**407,714.53**
	STD	672.2	2895.75	59.96	363.86	10,154.06	**1.3**

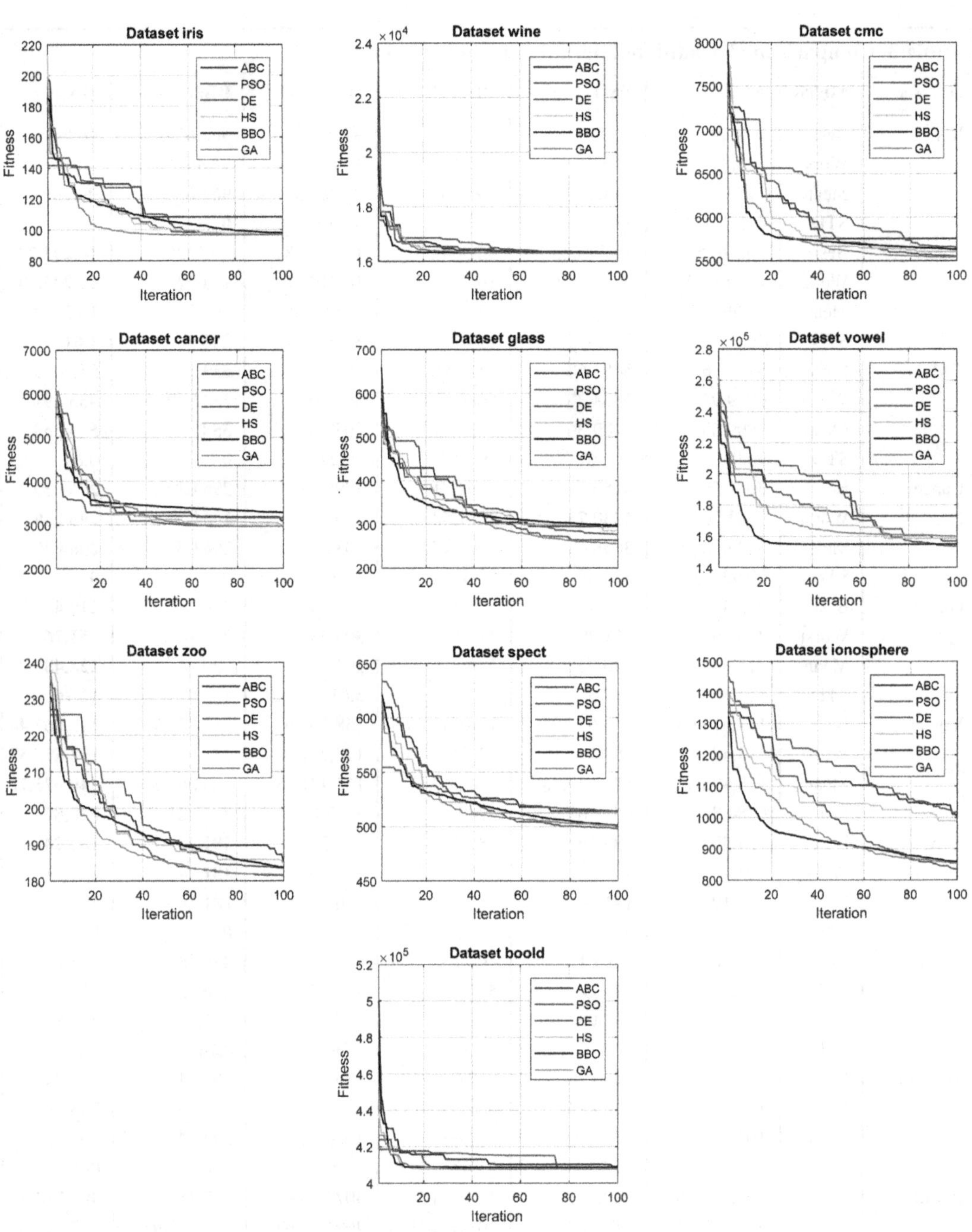

FIG. 2

Convergence curves of traditional optimization algorithms.

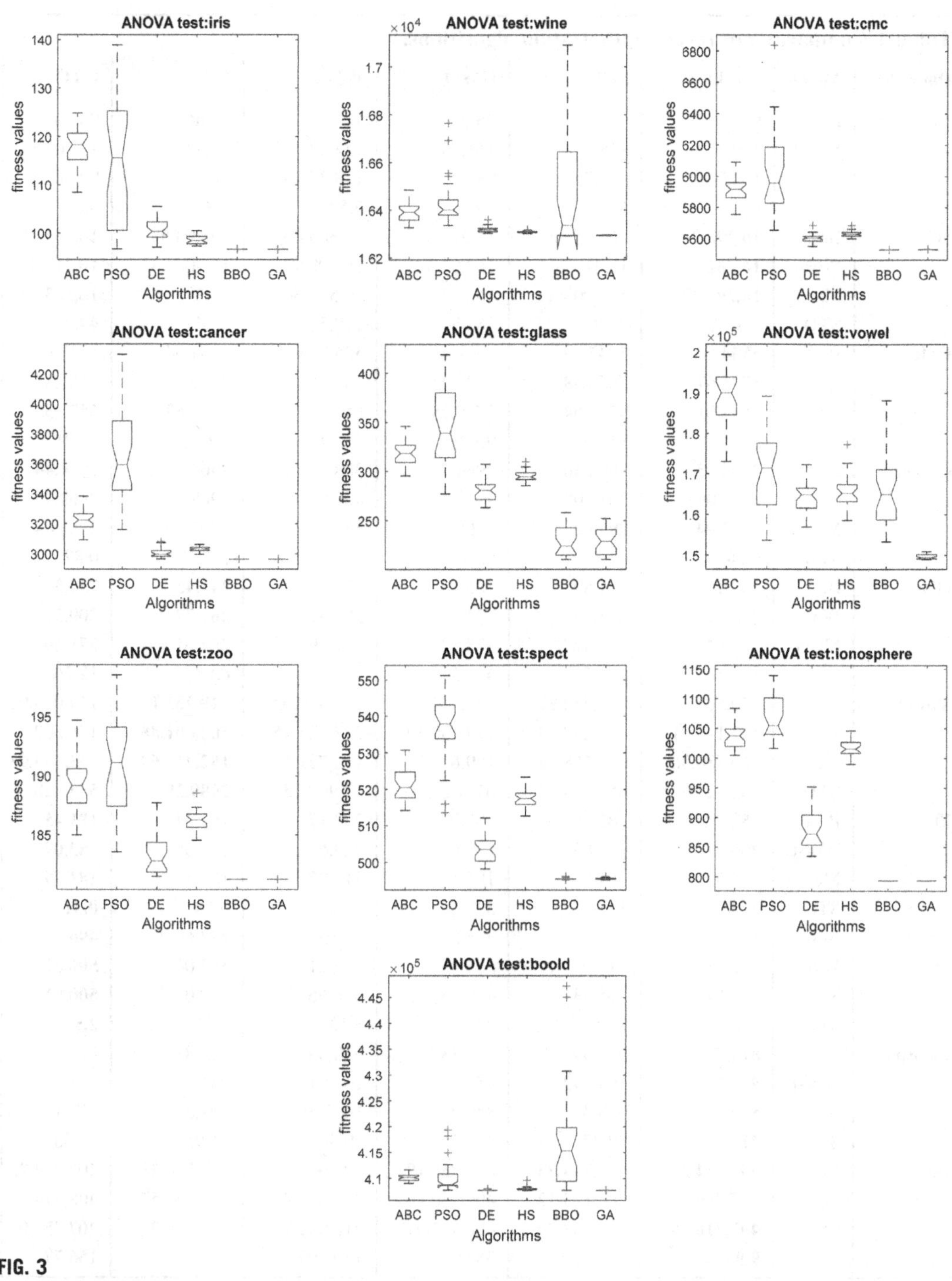

FIG. 3

ANOVA test result of traditional optimization algorithms.

Table 4 Comparison of recent optimization algorithms.

Dataset	Metric	AVOA	SOS	GWO	WOA	FFA	GTO
Iris	Best	96.66	96.71	96.9	109.01	**96.66**	96.66
	Worst	127.67	118.21	111.73	142.67	**97.63**	127.67
	Mean	100.78	99.67	98.52	128.33	**96.9**	99.58
	STD	9.63	5.17	2.88	9.52	**0.35**	8.8
Wine	Best	**16,293.96**	16,294.96	16,328.68	16,365.64	16,293.53	16,295.76
	Worst	**16,301.37**	16,310.26	16,504.3	16,808.92	16,308.24	16,317.49
	Mean	**16,296.27**	16,300.48	16,372.36	16,527.36	16,297.15	16,303.17
	STD	**1.6**	3.48	36.33	119.56	2.84	4.68
CMC	Best	5534.86	5547.19	5629.41	5855.45	**5534.64**	5537.6
	Worst	5775.67	5666.98	5992.07	7649.86	**5559.3**	5661.85
	Mean	5598.01	5591.64	5789.28	6504.3	**5541.63**	5571.55
	STD	52.16	31.49	96.88	423.44	**5.16**	26.21
Cancer	Best	**2964.45**	2966.46	2964.77	3082.51	2964.46	2964.44
	Worst	**2967.06**	3015.09	2968.11	4092.75	2968.59	2968.47
	Mean	**2965.04**	2978.2	2965.56	3508.09	2965.19	2965.04
	STD	**0.56**	11.3	0.69	291.71	0.87	0.87
Glass	Best	239.16	241.89	326.86	356.13	**241.43**	254.53
	Worst	310.07	313.76	574.52	505.31	**267.77**	309.57
	Mean	273.94	272.42	378.82	421.49	**252.01**	271.69
	STD	17.61	14.77	45.45	44.15	**6.69**	12.76
Vowel	Best	150,568.19	151,652.97	150,574.36	165,540.94	**149,233.11**	149,063.92
	Worst	192,150.41	179,232.03	179,130.19	222,723.45	**160,091.88**	176,562.52
	Mean	160,580.86	160,668.02	159,675.57	192,721.13	**152,219.64**	153,383.93
	STD	10,232.06	8543.64	6731.63	11,992.73	**2689.29**	5358.28
Zoo	Best	182.06	182.96	182.09	203.17	181.34	**181.45**
	Worst	200.57	186.83	199.03	223.68	183.65	**183.07**
	Mean	185.32	184.53	185.55	213.97	182.17	**181.79**
	STD	4.29	1.09	3.9	5.34	0.58	**0.31**
Spect	Best	497.99	501.61	496.21	520.53	498.69	**496.4**
	Worst	522.41	511.94	505.46	556.21	502.01	**506.81**
	Mean	505.54	505.33	499.75	536.66	500.02	**500.01**
	STD	6.4	2.55	2.9	8.13	0.81	**2.5**
Ionosphere	Best	**819.7**	925.49	805.73	920.33	855.81	823.71
	Worst	**923.68**	1046.42	952.47	1064.13	951.28	933.17
	Mean	**846.22**	978.44	849.2	1002.36	897.84	857.16
	STD	**21.99**	27.87	35.21	29.96	21.95	21.88
Boold	Best	**407,714.24**	407,714.23	407,756.18	407,980.91	407,714.23	407,714.23
	Worst	**407,769.21**	408,125.32	409,230.36	413,890.75	407,752.57	408,574.85
	Mean	**407,716.76**	407,788.29	408,427.91	410,042.93	407,717.27	407,759.05
	STD	**9.95**	138.31	384.77	1493.99	9.42	156.77

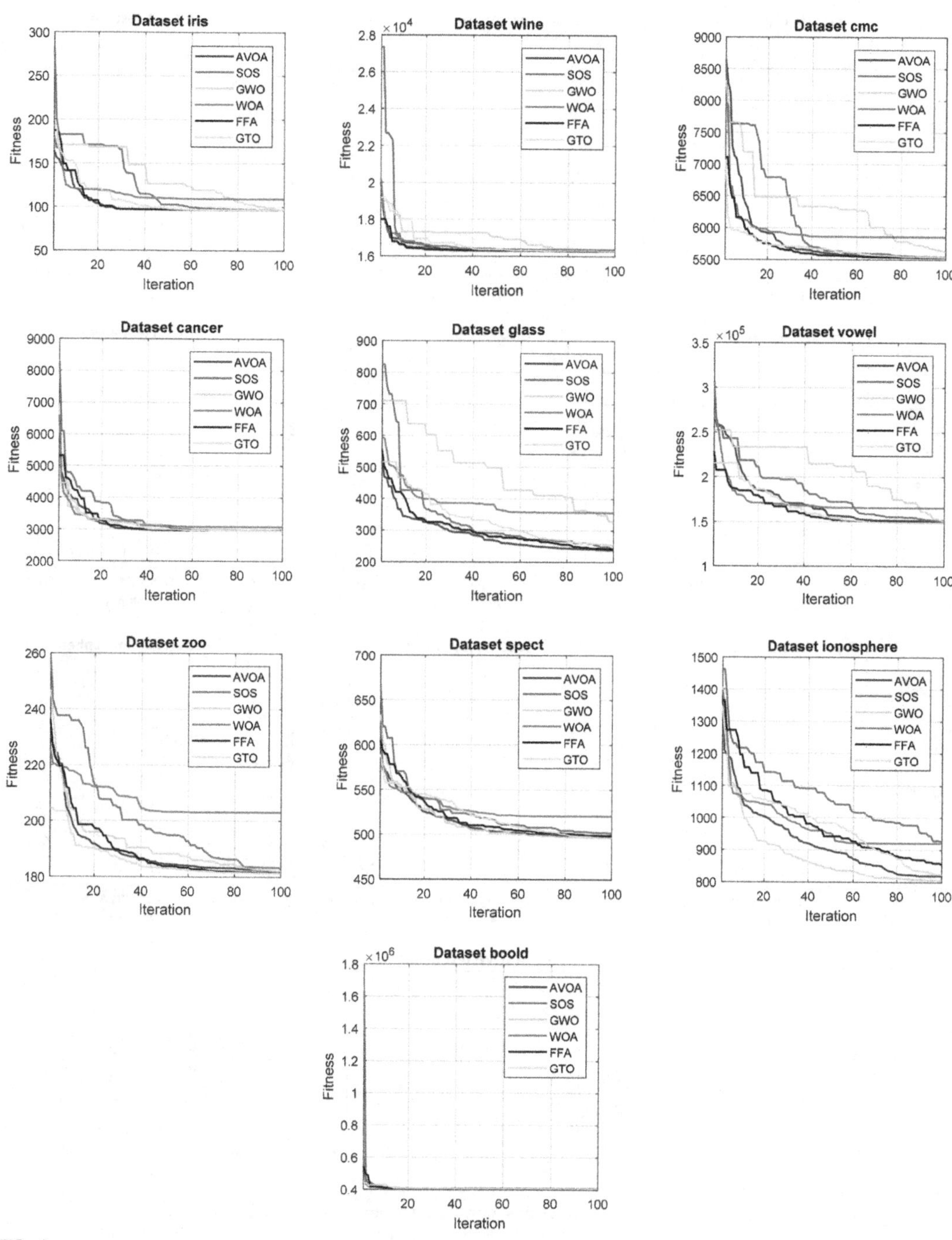

FIG. 4

Convergence curves of recent optimization algorithms.

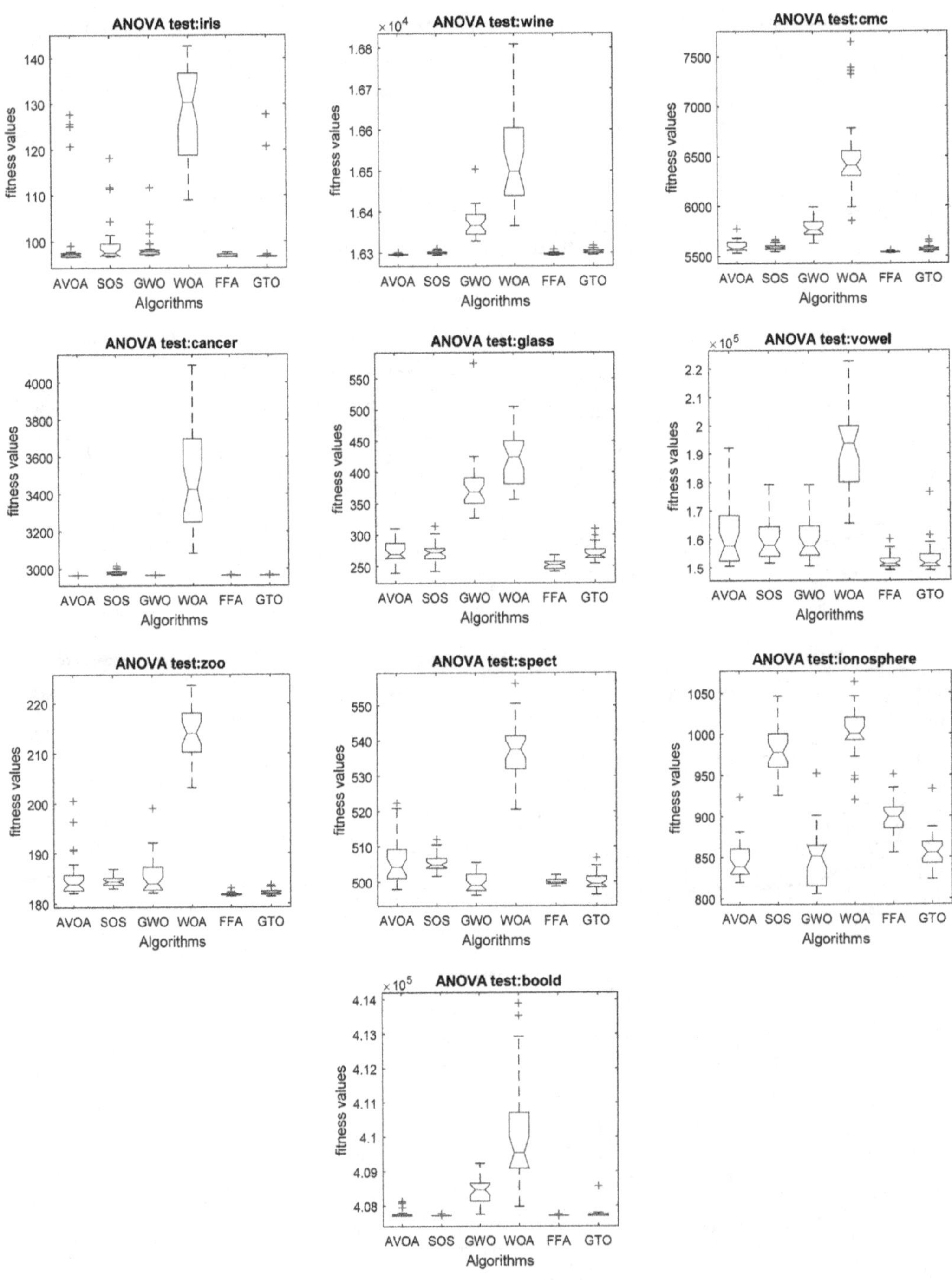

FIG. 5

ANOVA test results for recent optimization algorithms.

5. Conclusion and future works

Data clustering is one of the most important and popular data analysis techniques. It involves classifying unlabeled datasets into clusters. Each cluster's members have the minimum distance intra-cluster and are different and relatively far from the members of other clusters. Depending on the number of samples in a dataset and their dimension (features), the difficulty of this problem can be increased exponentially. This leads to the inefficiency of most conventional clustering algorithms. Metaheuristics algorithms are proving to be quite efficient in such cases, which motivated our attempt to apply 12 metaheuristics to 10 datasets. The comparative results provided some insights into the behavior of conventional, recent, evolutionary, and swarm-based algorithms.

It can be concluded that metaheuristic algorithms are good options for solving the data clustering problem because they can achieve acceptable and quality results in a reasonable time. Also, implementing metaheuristic algorithms is simple and easy. However, each algorithm has unique features and uses different procedures for optimization operations. For future studies, strengths and weaknesses can be examined in detail and various mechanisms and operators can be used to improve the performance of metaheuristic algorithms.

References

[1] N. Rahnema, F.S. Gharehchopogh, An improved artificial bee colony algorithm based on whale optimization algorithm for data clustering, Multimed. Tools Appl. 79 (43) (2020) 32169–32194.

[2] F.S. Gharehchopogh, Advances in tree seed algorithm: a comprehensive survey, Arch. Comput. Methods Eng. (2022) 1–24.

[3] F.S. Gharehchopogh, H. Shayanfar, H. Gholizadeh, A comprehensive survey on symbiotic organisms search algorithms, Artif. Intell. Rev. (2019) 1–48.

[4] F.S. Gharehchopogh, H. Gholizadeh, A comprehensive survey: Whale Optimization Algorithm and its applications, Swarm Evol. Comput. 48 (2019) 1–24.

[5] A. Kaveh, S. Talatahari, N. Khodadadi, Stochastic paint optimizer: theory and application in civil engineering, Eng. Comput. 38 (2022) 1921–1952.

[6] G. Zhang, C. Zhang, H. Zhang, Improved K-means algorithm based on density canopy, Knowl.-Based Syst. 145 (2018) 289–297.

[7] S.I. Boushaki, N. Kamel, O. Bendjeghaba, A new quantum chaotic cuckoo search algorithm for data clustering, Expert Syst. Appl. 96 (2018) 358–372.

[8] E.R. Hruschka, R.J. Campello, A.A. Freitas, A survey of evolutionary algorithms for clustering, IEEE Trans. Syst. Man Cybern. Part C Appl. Rev. 39 (2) (2009) 133–155.

[9] J. Kennedy, R. Eberhart, Particle swarm optimization, in: Proceedings of ICNN'95—International Conference on Neural Networks, IEEE, 1995.

[10] D. Van der Merwe, A.P. Engelbrecht, Data clustering using particle swarm optimization, in: Evolutionary Computation, 2003. CEC'03. The 2003 Congress on, IEEE, 2003.

[11] Y.-T. Kao, E. Zahara, I.-W. Kao, A hybridized approach to data clustering, Expert Syst. Appl. 34 (3) (2008) 1754–1762.

[12] S. Das, A. Abraham, A. Konar, Automatic clustering using an improved differential evolution algorithm, IEEE Trans. Syst. Man Cybern. Part A Syst. Hum. 38 (1) (2008) 218–237.

[13] J. Senthilnath, S. Omkar, V. Mani, Clustering using firefly algorithm: performance study, Swarm Evol. Comput. 1 (3) (2011) 164–171.

[14] O.M. Alia, M.A. Al-Betar, R. Mandava, A.T. Khader, Data clustering using harmony search algorithm, International Conference on Swarm, Evolutionary, and Memetic Computing, Springer, 2011, pp. 79–88.
[15] Z.W. Geem, J.H. Kim, G.V. Loganathan, A new heuristic optimization algorithm: harmony search, Simulation 76 (2) (2001) 60–68.
[16] C.-L. Huang, et al., Hybridization strategies for continuous ant colony optimization and particle swarm optimization applied to data clustering, Appl. Soft Comput. 13 (9) (2013) 3864–3872.
[17] K. Kaushik, V. Arora, A hybrid data clustering using firefly algorithm based improved genetic algorithm, Procedia Comput. Sci. 58 (2015) 249–256.
[18] D.E. Goldberg, J.H. Holland, Genetic Algorithms and Machine Learning, Springer Nature, 1988.
[19] A.N. Jadhav, N. Gomathi, WGC: hybridization of exponential grey wolf optimizer with whale optimization for data clustering, Alex. Eng. J. 57 (3) (2018) 1569–1584.
[20] S. Mirjalili, S.M. Mirjalili, A. Lewis, Grey wolf optimizer, Adv. Eng. Softw. 69 (2014) 46–61.
[21] S. Mirjalili, A. Lewis, The whale optimization algorithm, Adv. Eng. Softw. 95 (2016) 51–67.
[22] A. Kumar, D. Kumar, S. Jarial, A novel hybrid K-means and artificial bee colony algorithm approach for data clustering, Decision Sci. Lett. 7 (1) (2018) 65–76.
[23] D. Karaboga, B. Basturk, A powerful and efficient algorithm for numerical function optimization: artificial bee colony (ABC) algorithm, J. Glob. Optim. 39 (3) (2007) 459–471.
[24] T. Ashish, S. Kapil, B. Manju, Parallel bat algorithm-based clustering using MapReduce, in: Networking Communication and Data Knowledge Engineering, Springer, 2018, pp. 73–82.
[25] D. Simon, Biogeography-based optimization, IEEE Trans. Evol. Comput. 12 (6) (2008) 702–713.
[26] A.K. Qin, V.L. Huang, P.N. Suganthan, Differential evolution algorithm with strategy adaptation for global numerical optimization, IEEE Trans. Evol. Comput. 13 (2) (2008) 398–417.
[27] M.-Y. Cheng, D. Prayogo, Symbiotic organisms search: a new metaheuristic optimization algorithm, Comput. Struct. 139 (2014) 98–112.
[28] H. Shayanfar, F.S. Gharehchopogh, Farmland fertility: a new metaheuristic algorithm for solving continuous optimization problems, Appl. Soft Comput. 71 (2018) 728–746.
[29] B. Abdollahzadeh, F.S. Gharehchopogh, S. Mirjalili, African vultures optimization algorithm: a new nature-inspired metaheuristic algorithm for global optimization problems, Comput. Ind. Eng. 158 (2021) 107408.
[30] B. Abdollahzadeh, F. Soleimanian Gharehchopogh, S. Mirjalili, Artificial gorilla troops optimizer: a new nature-inspired metaheuristic algorithm for global optimization problems, Int. J. Intell. Syst. 36 (10) (2021) 5887–5958.

CHAPTER

Employment of bio-inspired algorithms in the field of antenna array optimization: A review

21

Krishanu Kundu[a] and Narendra Nath Pathak[b]

[a]*Department of Electronics and Communication Engineering, G.L. Bajaj Institute of Technology & Management, Greater Noida, Uttar Pradesh, India,* [b]*Department of Electronics and Communication Engineering, Dr. B.C. Roy Engineering College, Durgapur, West Bengal, India*

1. Introduction

Numerous variants of Evolutionary Algorithms (EAs) have been developed in the past few decades. The concept of EAs was first presented by Alan Turing in 1948. EAs are based on the characteristics and evolution of several biological species. Much research has been carried on Genetic Algorithms (GAs) [1,2] as well as Evolutionary Programming (EP) approaches [3,4]. EAs can provide effective solutions for single as well as multiobjective optimization problems. The GA was invented by John Holland in 1962. This algorithm is based on genetics and the concept of evolution. Genetics is based on inheritance and deviation of biological properties. According to GA, a set of chosen individuals evolves toward an optimal solution with respect to the selective pressure of the fitness function. The Tabu Search (TS) algorithm was developed by Fred Glover in 1986 [5]. The tendency of local search to get stuck in the region consisting of numerous equal-fit solutions is GA's greatest disadvantage. TS eliminates this drawback. Ant Colony Optimization (ACO) was developed by Marco Dorigo in 1992 [6]. It is inspired by the foraging behavior of ants and their usage of pheromones to exchange information. Ants deposit pheromones on the ground to mark favorable paths. The idea is that the rest of the colony members will follow these demarcated paths to reach the destination (food source) more conveniently. A similar mechanism is used in ACO to solve optimization problems. Significant progress in the history of EAs occurred with the invention of Particle Swarm Optimization (PSO) by James Kennedy in 1995 [7]. This algorithm is inspired by the swarm intelligence of birds as well as fish. Another vector-based EA known as Differential Evolution (DE) was industrialized in 1997 [8]. The Harmony Search (HS) algorithm was introduced by Geem et al. in 2001 [9]. This algorithm is based on music, as finding harmony is the same as finding optimality in an optimization process. For optimization of Internet-hosting centers, Nakrani and Tovey suggested a new algorithm called the Honey Bee Algorithm (HBA) in 2004 [10]. Yang invented the Virtual Bee Algorithm (VBA) in 2005 [11]. In the same year, Karaboga developed Artificial Bee Colony (ABC) [12]. In 2006, Chu and Tsai discovered Cat Swarm Optimization (CSO), which is inspired by the seeking and tracing behavior of cats [13]. In 2008 [14], Yang proposed the Firefly Algorithm (FA), taking inspiration from the flashing behavior

Comprehensive Metaheuristics. https://doi.org/10.1016/B978-0-323-91781-0.00021-1

of fireflies. Fireflies use flashing light to attract other fireflies. The brightness of the flashing light controls the strength of attraction, meaning less luminous fireflies are attracted to brighter fireflies. However, brightness is inversely proportionate to mutual distances between fireflies. A given firefly will carry on moving randomly until it finds a brighter firefly. The objective function must be associated with brightness property. The Gravitation Search Algorithm (GSA), which is built on the law of gravity plus mass interactions, was developed by Rashedi et al. in 2009 [15]. Yang et al. created the Cuckoo Search Algorithm (CSA) in 2009 [16]. CSA is inspired by obligate brood parasitism of cuckoo species. After discovering third-party eggs, host birds either throw away those eggs or settle in a new nest. The breeding ability of female parasitic cuckoos mimics shade and outline of eggs of little selected host species. In 2010 [17], Yang utilized the echolocation characteristics of bats to implement a new algorithm called the Bat Algorithm (BA). In 2012 [18], Yang developed the Flower Pollination Algorithm (FPA). The Gray Wolf Optimizer (GWO) [19] by Mirjalili et al. is based on the hunting behaviors of gray wolves. The Spider Monkey Optimization (SPO) algorithm was developed by Bansal et al. in 2014 [20]. The Crow Search Algorithm (CSA) was proposed by Askarzadeh in 2016 [21]. CSA is population based and impersonates the behavior of crows and their social interaction. The Whale Optimization Algorithm (WOA) is another of metaheuristic algorithm proposed by Mirjalili et al. in 2016 [22]. Based on the behavior of grasshopper swarms and their social interaction, the Grasshopper Optimization Algorithm (GOA) was proposed by Mirjalili et al. in 2017. The Butterfly Optimization Algorithm (BOA) is a population-based, nature-inspired algorithm proposed by Arora et al. in 2018. The BOA mimics the foraging and social behavior of butteries to solve global optimization problems. Additional evolutionary practices subjected to antenna pattern synthesizing problems include Wind-Driven Optimization (WDO) [23], Invasive Weed Optimization (IWO) [24–28], EP [29,30], and the Covariance Matrix Adaptation Evolution Strategy (CMA-ES) [31,32]. Wolpert et al. [33] experimented with certain optimization problems and presented several "no free lunch" (NFL) theorems that established that for any algorithm, any elevated performance over one class of problems is offset by performance over another class. Similarly, in Ref. [34] Ho et al. concluded that theoretically it is hard to find the best optimization tactic because the only way one strategy can outperform another is if it is specialized to the structure of the specific problem under consideration. A wide variety of optimization-related problems can occur in the antenna array domain. As such, we discuss several algorithms for solving these problems. Section 2 reviews FPA. Section 3 focuses on CSO and Section 4 discusses GSA. Section 5 concludes.

2. Flower pollination algorithm

Pollination of flowers is the base for FPA. The basic parameters of this algorithm are population strength (n) and switching probability (p). The latter is influenced by local pollination procedures, as neighboring flowers are far more likely to be pollinated than remote flowers. During FPA implementation, a random integer can be produced to be compared against switching probability. If the resultant integer is less than p, case global pollination takes place.

Global pollination:

$$x_i^{t+1} = x_i^t + L(x_i^t - g^*) \quad (1)$$

Local pollination:

$$x_i^{t+1} = x_i^t + \varepsilon\,(x_i^t - x_k^t) \quad (2)$$

L describes step size elected out of Lévy distribution. Here, Lévy steps confirm the following approximation:

$$L \sim (s+\beta)^{-1} \tag{3}$$

Here, β defines the Lévy exponent. The simplest way of engendering Lévy flights is to employ normal distributions (u, v), where x_j^t as well as x_k^t represent pollen belonging to diverse flowers with respect to the same parent plant (Fig. 1).

$$s = u/\left(|v|^{1+\beta}\right) \tag{4}$$

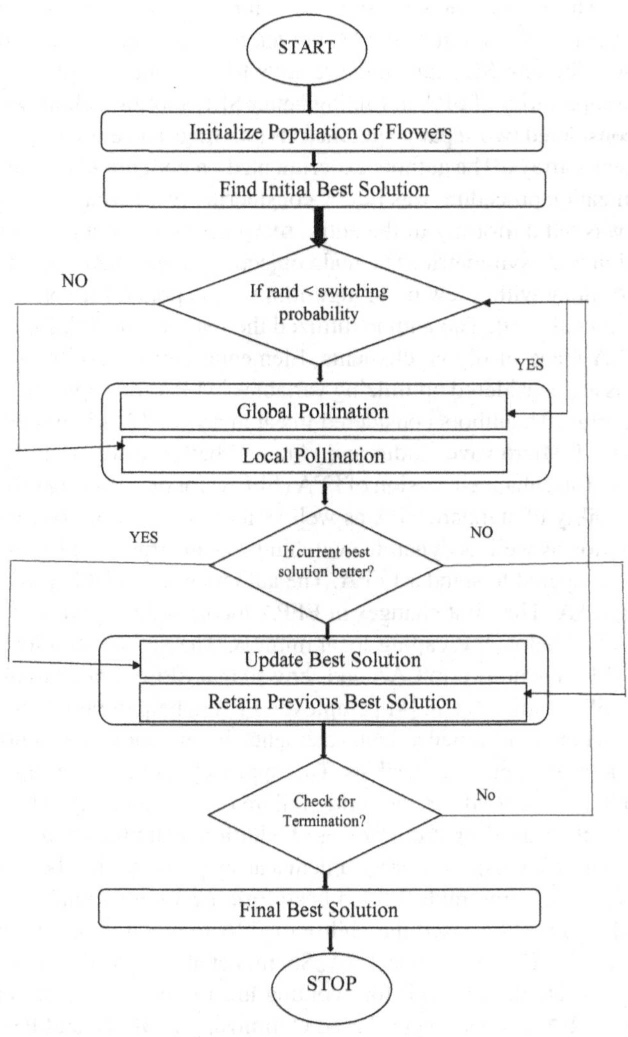

FIG. 1

Flowchart of Flower Pollination Algorithm (FPA).

The Scopus database showcases 223 conference papers, 398 journal papers, and 20 book chapters related to FPA in the period 2014–19. Furthermore, a search in same platform using the keywords "FPAs" and "Antennas" reveals 24 papers (both journal and conference). Patidar et al. employed FPA for synthesizing flat-top beam radiation pattern with respect to linear antenna arrays (LAAs) [35]. During the synthesis process, restrictions were made on the peak side lobe level (SLL), percentage of ripple in the flat-top portion, and maximum Voltage Standing Wave Ratio comprising two individual nulls. Binary phase shifters of 0 degrees and 180 degrees helped in the production of current amplitudes suitable for the array elements. Kundu et al. utilized FPA for optimizing the high SLL problem with respect to circular antenna array (CAA) [36]. A hybridized form of FPA with BA called BFP has the advantage of escaping local minima. Salgotra et al. [37] utilized this hybridized version of FPA for synthesizing unequally spaced LAA. The authors showcased the superiority of BFP over well-known algorithms for unequally spaced LAA synthesis. Singh et al. [38] proved the capability of FPA for synthesizing array pattern by maintaining minimum SLL plus nulls relative to the routes of interferences. Experimental results also proved the superiority of FPA in finding better SLL abolition along with proper null placement. Das et al. [39] considered two separate geometries (six rings and eight rings) based on concentric regular hexagonal antenna arrays. The authors experimented on both broadside and scanned array configurations. The optimization procedure was run by considering two variants of current distribution. In one type, the current was fed uniformly to the entire set of elements on a particular ring. In the other type, current distribution was asymmetric. The main objective of optimization was to obtain minimum SLL interference. FPA along with a few other algorithms was applied for solving the pattern optimization problem by Ram et al. [40]. The authors utilized the features of FPA for synthesizing the radiation pattern of an LLA made of dipole elements. Elemental current excitations as well as spacing between array elements are considered optimizing variables in FPA for improving the radiation pattern related to the antenna array. Th authors considered linear arrays of 12, 16, and 20 dipole elements for optimization through CST-Microwave studio and obtained better results in terms of SLL reduction. Singh et al. [41] proposed an enhanced version of FPA (EFPA) for overcoming problems such as underrated exploitation capability of standard FPA as well as its low speed of convergence. In EFPA, the theory of Cauchy mutation as well as dynamic switching was incorporated for improving exploration and exploitation rates compared to standard FPA. The authors utilized EFPA to synthesize the output pattern of nonuniform LAA. The vital changes in EFPA incorporate superior search capability compared to standard FPA and helps in escaping local minima. The authors employed the superiority of EFPA in minimizing SLL in nonuniform LAA and showed that EPFA tended to obtain results in fewer iterations compared to FPA. Patidar et al. [42] employed FPA to help in failure correction with respect to LAA by optimizing geometry of failed antenna elements. In this study, the authors considered length plus spacing between failed elements as variables. This approach proved to be superior to approaches in which excitation amplitudes are used for correcting failure of antenna array. The cost effectiveness of the method is proven by the canceling requirement of additional attenuators as well as phase shifters. Computational complexity with respect to large antenna array problems has been also reduced. Sallam et al. [43] used FPA for computing multifaceted beamforming weights with respect to phased array antenna. Experimental results showcased the ability of FPA to attain optimum Wiener weights very few iterations as compared to PSO as well as CSA. Sharma et al. [44] designed an LAA radiation pattern with minimal SLL by employing BFP for avoiding local minima. The performance of BFP was compared with standard FPA, Biogeography-based Optimization (BBO), and PSO. Saxena et al. [45] applied FPA for antenna array optimization. Here, FPA was employed on a LAA for attaining radiation pattern consisting of minimal SLL plus implantation of deeper nulls subjected to coveted directions.

3. Cat Swarm Optimization

CSO is a result of the close observation of the demeanor of cats in nature. The two most important qualities in cats are intense curiosity in moving objects and very good chasing skills. Cats are mostly found to be resting or in slow movement, but whenever they observe a prey, they spend profuse energy hunting for it. The emblematic behavior of cats is classified into searching mode and tracking mode. The cats in these two modes are distributed based on a ratio known as mixture ratio (MR). During seeking mode cats spend most of the time resting with curiosity as well as with slow motion. In tracking mode, as soon as they detect prey, cats move at a very high speed by spending maximum energy. To utilize this behavior, a mathematical model has been prepared for solving complex problems related to optimization.

3.1 Searching mode

According to searching mode, let C numbers of cats explore and look for the upcoming position to proceed. Essential factors include searching memory space (SMS), searching range of preferred proportion (SRP), Tally dimension for variation (TDV), and individual position consideration (IPC). SMS not only denotes the size of searching memory but also specifies the points analyzed by an individual cat. This very parameter is distinct for every cat. SRP asserts mutation ratio related to specified dimensions. TDV mimics the scope of dimensions to be varied. IPC represents a Boolean flag. Searching mode may be expressed accordingly.

Step 1: In this step, j numbers of copies are created based on value of IPC. For IPC equals to 1, $(C-1)$ numbers of copies of cat's current position are created using SMS. However, for IPC equals to 0, C copies of cat's current position are created based on SMS.

Step 2: For each copy of cat, according to TDV dimensions, SRP percentage of the current values is arbitrarily added or deducted; consequently, the old values are changed.

Step 3: Fitness values (FV) of every point indicated by cat are evaluated.

Step 4: If cost functions for each cat under consideration have exactly similar values, the candidate with equal probability to each is selected, else use following equation to find probability.

$$P'i = \frac{FV_i - FV_f}{FV_{max} - FV_{min}} \tag{5}$$

where FV_i represents fitness of ith cat and $FV_f = FV_{max}$, For maximal result, $FV_f = FV_{min}$.

Step 5: Utilize roulette wheel function to select the point to travel and accordingly modify the present position of selected candidate.

3.2 Tracking mode

In tracking mode, P number of cats attempt to track targets. During this the next move of individual cat is being decided by the cat's current velocity. The best position of the cats is calculated by the intensity of the swarm's members. Layout of this mode can be abstracted in the following three steps.

Step 1: Revise the velocities (v) relative to each dimension using Eq. (6)

$$(x_{best,d} - x_{k,d}),\ d = 1,\ldots,M v_{k,d} = v_{k,d} + r_1 c_1 \tag{6}$$

where $x_{best,d}$ represents cats with the best fitness value and $x_{k,d}$ is the current position of kth cat in dth dimension. Constant quantity, denoted as c_1, is between [0, 2], and r_1 is a random value between (0,1).

Step 2: Analyze whether the velocities stand inside the velocity limits. If the new velocity exceeds the tenacity of velocity, it is fixed within limits.

Step3: Revise the cat's position using Eq. (7)

$$x_{k,d} = x_{k,d} + v_{k,d} \tag{7}$$

Resulting position as well as velocity is examined for their validity for the range conditions.

According to the Scopus database, there are 114 conference papers, 163 journal papers, and 6 book chapters on CSO. A search using the keywords "CSO" and "Antennas" identified 14 papers (both journal and conference).

The CSO algorithm was introduced to array synthesis by Pappula et al. [46]. It was applied for optimizing the distinct elemental positions to attain minimum SLL as well as for channelizing of nulls in anticipated directions. Multiobjective optimization in antenna array was also carried out by Pappula et al. [47] by employing multiobjective binary CSO (MOBCSO). The authors presented a platform from which all targeted solutions were chosen for fulfilling all requirements needed to demonstrate supremacy of MOBCSO over multiobjective binary PSO (MOBPSO). The developed MOBCSO model proved to provide better PSLL as well as first null beam width (FNBW) in comparison with MOBPSO. It also proved capable of controlling the shape of array pattern. Ram et al. [48] utilized CSO for optimizing the multiple ring concentric CAA. CSO outperformed PSO, GA, SA, BBO, and FFA. Ram et al. [49] utilized CSO for diminishing SLL and enhancing directivity of a 9-ring concentric CAA (279 elements) pattern. Results showed increased sideband frequency and decreased sideband levels. Li et al. [50] introduced the two basic performance modes related to binary cat swarm algorithm (tracking mode and search mode) and presented a diverse location-updating scheme for the mentioned modes. Introduction of search pattern has ignited the search for the local area, which in turn helped in improving algorithm's accuracy. Introduction of the tracking mode helps in antenna selection to approach an optimal solution, which helps the BCSO algorithm avoid "premature" convergence. Pappula et al. [51] also proposed a multiobjective version of CSO for synthesizing dispersed planar-type antenna arrays. Ram et al. [52] applied CSO for optimizing control parameters related to radiation pattern formed by linear-type antenna arrays comprised of microstrip patch antenna. Resultant patterns were obtained by synthesizing the current excitation weight related to every single element. The antenna arrays were designed on the MATLAB platform and validated. Simulation results showed that the SLLs decreased more in comparison to the same configuration with negligible alteration in FNBW. Singh et al. [53] presented an hybridized version of CSO called CGSO by combining features of GA with CSO to optimize bandwidth as well as return loss of antenna. CGSO proved to be a cost-effective approach for solving multiobjective functions. CGSO not only reduced SLL but also improved directivity for shrunk multiband antenna. In experimental design configurations CGSO was able to improve null depth by approximately −80dB. Banerjee et al. [54] used PSO and CSO to reduce SLL of a steerable isotropic CAA. Here, CSO was also utilized to steer the main beam in a particular direction. CSO was successfully implemented for optimizing amplitude excitations with respect to steerable circular arrays. For 30- as well as 45-element steerable circular arrays, CSO was able to attain maximum SLL of −12.32dB and −11.45dB, respectively.

4. Gravitational Search Algorithm

Three specific parameters of inertia mass, location of mass with respect to dth dimension, gravitational mass (active and passive) control the working of agents in GSA. The problem solution can be represented by location of mass for a particular agent with respect to specific dimension, whereas inertia mass for any agent showcases its resistivity against its fast movement. The agent's velocity is controlled by gravitational mass and inertial mass. The agent's locations with respect to the specific dimensions are mostly revised through every iteration count, thus best fitness can be recorded. Closure condition can be well defined by a predefined iteration count. Post termination, the recorded best fitness with respect to the ultimate iteration resembles the global fitness against the specific problem. Location of mass for a particular agent with respect to specific dimension signifies global solution. The algorithm is as follows:

Step 1: Agent Initialization:

The first step involves random initialization of positions for N agents in the provided search interval in the following manner:

$$X_i = \left(x_i^1, \ldots, x_i^d, \ldots, x_i^n\right), \text{considering } i = 1,2,\ldots,N. \tag{8}$$

where x_i^d signifies the position for ith agent with respect to dth dimension, whereas n represents the space dimension.

Step 2: Fitness evolution and settlement of best fitness relative to each agent:

Execute fitness evolution for each agent at every iteration. Then, figure out the best as well as worst fitness relative to each iteration.

$$best(t) = \min_{j \in \{1,\ldots,N\}} fit_j(t) \tag{9}$$

$$worst(t) = \max_{j \in \{1,\ldots,N\}} fit_j(t) \tag{10}$$

where $fit_j(t)$ defines fitness for jth agent with respect to iteration t.

Step 3: Estimate gravitational constant:

Estimate gravitational constant denoted by G with respect to iteration t using Eq. (11).

$$G(t) = G_0 e^{\left(-\alpha^t / T\right)} \tag{11}$$

where $G_0 = 100$, $\alpha = 20$, and T denotes overall iteration count.

Step 4: Determine mass relative to agents:

Compute gravitational as well as inertia masses relative to every agent with respect to iteration t employing the following equations:

$$M_{ai} = M_{pi} = M_{ii} = M_i, \; i = 1,2,\ldots,N. \tag{12}$$

$$m_i(t) = \frac{fit_i(t) - worst_i(t)}{best(t) - worst(t)} \tag{13}$$

$$M_i(t) = \frac{m_i(t)}{\sum_{j=1}^{N} m_j(t)} \tag{14}$$

where M_{ai} and M_{pi} represent active as well as passive gravitational mass with respect to ith agent, and M_{ii} represents inertia mass for that agent.

Step 5: Compute acceleration:

Acceleration for ith agent relative to iteration t:

$$a_i^d(t) = \frac{F_i^d(t)}{M_{ii}(t)} \tag{15}$$

where $F_i^d(t)$ is total force exerting on ith agent:

$$F_i^d(t) = \sum_{j \in Kbest,\, j \neq i} \text{rand}_j F_{ij}^d(t) \tag{16}$$

where *Kbest* represents a set of first K agents having best fitness value and heaviest mass. $F_{ij}^d(t)$ is force exerted on agent "i" by agent "j" with respect to dth dimension and tth iteration:

$$F_{ij}^d(t) = G(t)\frac{M_{pi}(t) \times M_{aj}(t)}{R_{ij}(t) + \varepsilon}\left(x_j^d(t) - x_i^d(t)\right) \tag{17}$$

where $R_{ij}(t)$ is Euclidian distance between two agents ("i", "j") with respect to iteration t.

Step 6: Revise agent velocity and respective position:

Calculate velocity plus position relative to the agents for subsequent iteration $(t+1)$ utilizing:

$$v_i^d(t+1) = \text{rand}_i \times v_i^d(t) + a_i^d(t) \tag{18}$$

$$x_i^d(t+1) = x_i^d(t) + v_i^d(t+1) \tag{19}$$

Step 7: Redo steps 2–6 until maximum iteration count is attained. Best fitness obtained from final iteration is designated as global fitness, whereas positions with respect to corresponding agent subjected to specified dimensions are considered as the global solution against that particular problem (Fig. 2).

The Scopus database includes 132 conference papers, 197 journal papers, and 5 book chapters related to GSA. A search of same databased using the keywords "FPAs" and "Antennas" identified seven papers (both journal and conference). Darzi et al. [55] improved the weight vectors of the linear constraint minimum variance (LCMV) technique by incorporating GSA into existing LCMV. Better known for its adaptive beamforming capabilities, LCMV is mostly used for canceling signal interference and producing a strong beam through its weighted vectors. Experimental results prove that the integration of LCMV with GSA results in minimization of interference as well as increased signal-to-noise ratio. This proposed GSA-LCMV also proved to be applicable for synthesis of radiation pattern related to smart antenna due its better accuracy as well as its freedom from complex mathematical functions. Ram et al. [56] presented a hybridized form of GSA by blending it with perception of opposition-based methodology for improving the radiation pattern of single-ring and multiple-ring nonuniform CAA. OGSA delivers optimal radiation pattern through abatement of maximal SLL as well

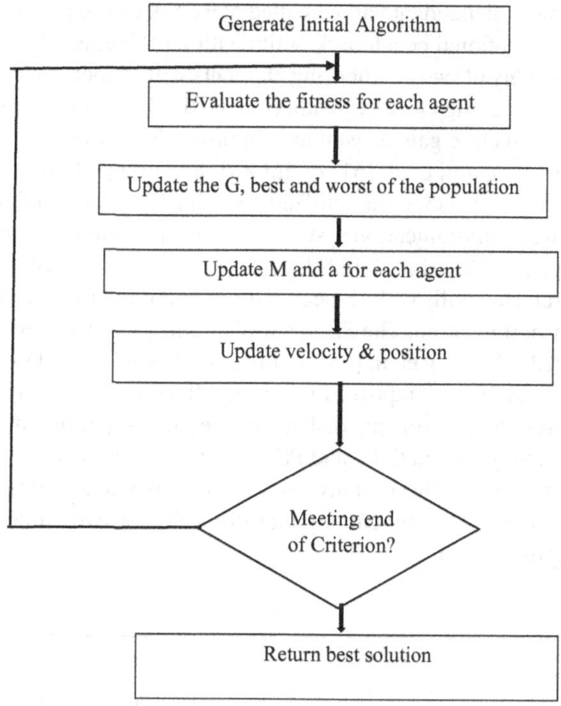

FIG. 2

Flowchart of Gravitation Search Algorithm (GSA).

as FNBW. This in turn improves directivity for CAA as well as enhances maximal reduction of SLL with respect to CAA. The authors also highlighted the superiority of OGSA in handling composite nonlinear problems.

Sun et al. [57] applied a hybrid version of GSA called PSOGSA for synthesizing radiation pattern and choosing activated elements applicable to conformal array. Simulation of PSOGSA is basically controlled by bench function. With consideration of uncoupled as well as coupled elemental pattern, a hybrid PSOGSA algorithm was applied in case of conformal pattern synthesis along with full-wave calculation. The outcome of simulation along with full-wave calculation validated the ability of PSOGSA to search for a global optimum with efficient selection of active elements.

Standard GSA employs the best agents devoid of any sort of randomization; therefore, it tends to converge at suboptimal outcomes. To enhance global search capability and convergence rate, Darzi et al. [58] presented a hybridized version called Stochastic Leader GSA (SL-GSA), which randomly selects k agents from an entire population to enrich exploration rate at an early stage. In SL-GSA, the entire set is progressively diminished by rejecting the agents showing poorest performances, which helps to attain rapid convergence. Six different benchmark functions were analyzed using SL-GSA. The algorithm was also applied to the minimum variance distortionless response (MVDR) beamforming approach to prove its compatibility to deal with various real-time optimization complications. To increase the beamforming performance by prevention of loss with respect to optimal trajectory, Darzi et al. [59] presented a memory-dependent variant called MBGSA whose performance was validated by finding solutions for 14 conventional benchmark optimization problems. The elemental length as well as spacing between an assembly of two neighboring elements with respect to linear-type dipole antenna array (LDAA) were optimized using GSA by Swain et al. [60]. The performance of GSA was validated by achievement of greater directive gain as well as minimized SLL level with respect to the radiation pattern formed by LDAA. Mangaraj et al. [61] optimized radiation pattern of a 12-element linear antenna array by using GSA. Here, GSA was utilized to obtain low value of half power beam width (HPBW) relevant to wireless communication system (WCS) application operating at extreme high frequency. GSA was employed by Hesari et al. [62] for pattern synthesis related to linear but nonuniform-type planar arrays with preferred nulls with respect to direction of interference as well as minimal SLL by means of position-only optimization. The SLL as well as null depth attained using GSA in designing planar array were enhanced. Mohanty et al. [63] optimized a 12-element LDAA with the help of GSA and found the directivity was further improved by proper placement of a huge reflecting surface adjacent to one side (90 degrees to the array axis). This resulted in a superdirective LDAA. To cope with the issues of premature convergence in GSA and PSO, the two methods were merged. Bera et al. [64] considered GSA-PSO for pattern synthesis of diverse sets of CCAA geometries. Results shows percentage of thinning was greater than 40% by employing GSA-PSO for experimental array sets, which in turn reduced mutual coupling effect.

5. Case study

As per the literature, FPA is widely used for optimizing uniform linear and circular antenna arrays. In this current work, to prove efficiency of FPA, we optimized a nonuniform CAA. FPA delivered better SLL (−25.11 dB) after synthesizing the CAA with optimized spacing (Table 1 and Fig. 3).

Table 1 Excitation amplitude distribution of 28 elements of circular antenna array with optimized spacing using FPA.

Algorithm	Amplitude distribution/inter element spacing	Max SLL
FPA	**Amplitude**: 0.9995 0.0530 0.3196 0.0001 0.1648 0 0.4624 0.2273 0 0.3781 0.8773 1.0000 1.0000 0.7472 0.9739 0.9598 1.0000 0.9286 0 0.4488 0.5248 0.0022 0.0023 0.0004 0.3853 0.8957 0.9518 1.0000 **Spacing**: 0.4975 0.4212 0.4228 0.5874 0.5615 0.4205 0.5472 0.5887 0.4205 0.4880 0.4226 0.4981 0.4320 0.4273 0.4457 0.5719 0.5887 0.5764 0.5149 0.5470 0.5887 0.4361 0.5887 0.5074 0.4642 0.5887 0.4205 0.4262	−25.11 dB

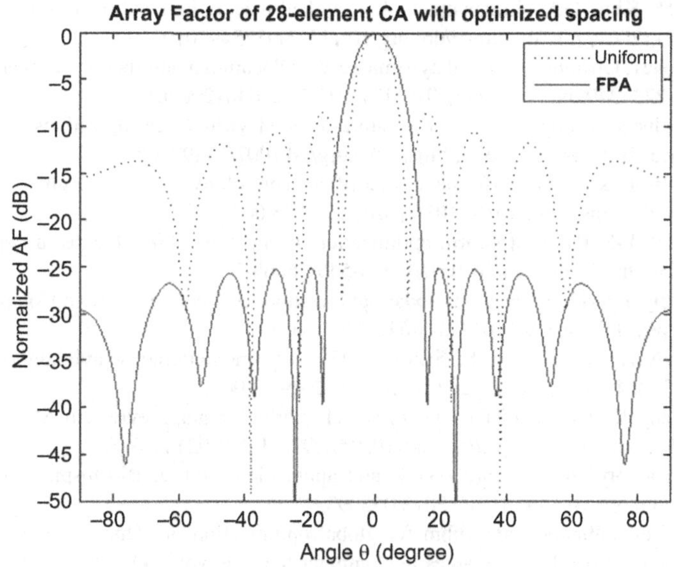

FIG. 3

Array pattern obtained after optimization using FPA considering circular array with optimized spacing.

6. Conclusion

In this chapter, we analyzed FPA, CSO, and GSA and their applications in antenna array optimization. Results show that these bio-inspired algorithms were effective in finding solutions for specific antenna array design problems.

References

[1] D.E. Goldberg, Genetic Algorithms in Search, Optimization and Machine Learning, Addison-Wesley, New York, NY, 1989.

[2] J.H. Holland, Adaptation in Natural and Artificial Systems, The University of Michigan Press, Ann Arbor, MI, 1975.

[3] D.B. Fogel, Evolutionary Computation: Toward a New Philosophy of Machine Intelligence, IEEE Press, Piscataway, NJ, 1995.

[4] H.-G. Beyer, H.-P. Schwefel, Evolution strategies—a comprehensive introduction, Nat. Comput. 1 (1) (2002) 3–52.

[5] F. Glover, M. Laguna, Tabu search, in: Handbook of Combinatorial Optimization, 1998, pp. 2093–2229, https://doi.org/10.1007/978-1-4613-0303-9_33.

[6] M. Dorigo, Ant colony optimization, Scholarpedia 2 (3) (2007) 1461.

[7] J. Kennedy, R. Eberhart, Particle swarm optimization, in: Proceedings of ICNN'95—International Conference on Neural Networks, vol. 4, IEEE, 1995, pp. 1942–1948, https://doi.org/10.1109/icnn.1995.488968.

[8] R. Storn, K. Price, Differential evolution—a simple and efficient heuristic for global optimization over continuous spaces, J. Glob. Optim. 11 (1997) 341, https://doi.org/10.1023/A:1008202821328.

[9] Z.W. Geem, J.H. Kim, G.V. Loganathan, A new heuristic optimization algorithm: harmony search, Simulation 76 (2) (2001) 60–68, https://doi.org/10.1177/003754970107600201.

[10] S. Nakrani, C. Tovey, On honey bees and dynamic server allocation in internet hosting centers, Adapt. Behav. 12 (3–4) (2004) 223–240, https://doi.org/10.1177/105971230401200308.

[11] X.-S. Yang, Engineering optimizations via nature-inspired virtual bee algorithms, in: Lecture Notes in Computer Science, 2005, pp. 317–323, https://doi.org/10.1007/11499305_33.

[12] D. Karaboga, B. Basturk, On the performance of artificial bee colony (ABC) algorithm, Appl. Soft Comput. 8 (1) (2008) 687–697, https://doi.org/10.1016/j.asoc.2007.05.007.

[13] S.-C. Chu, P. Tsai, J.-S. Pan, Cat swarm optimization, in: PRICAI 2006: Trends in Artificial Intelligence, 2006, pp. 854–858, https://doi.org/10.1007/978-3-540-36668-3_94.

[14] X.-S. Yang, Firefly algorithms for multimodal optimization, in: Lecture Notes in Computer Science, 2009, pp. 169–178, https://doi.org/10.1007/978-3-642-04944-6_14.

[15] E. Rashedi, H. Nezamabadi-pour, S. Saryazdi, GSA: a gravitational search algorithm, Inf. Sci. 179 (13) (2009) 2232–2248, https://doi.org/10.1016/j.ins.2009.03.004.

[16] I. Fister, X.-S. Yang, D. Fister, I. Fister, Cuckoo search: a brief literature review, in: Studies in Computational Intelligence, 2013, pp. 49–62, https://doi.org/10.1007/978-3-319-02141-6_3.

[17] X.-S. Yang, Bat algorithm: literature review and applications, Int. J. Bio-Inspir. Comput. 5 (3) (2013) 141–149, https://doi.org/10.1504/IJBIC.2013.055093.

[18] X.-S. Yang, Flower pollination algorithm for global optimization, in: Unconventional Computation and Natural Computation 2012, Lecture Notes in Computer Science, vol. 7445, 2012, pp. 240–249.

[19] S. Mirjalili, S.M. Mirjalili, A. Lewis, Grey wolf optimizer, Adv. Eng. Softw. 69 (2014) 46–61, https://doi.org/10.1016/j.advengsoft.2013.12.007.

[20] J.C. Bansal, H. Sharma, S.S. Jadon, M. Clerc, Spider monkey optimization algorithm for numerical optimization, Memet. Comput. 6 (1) (2014) 31–47, https://doi.org/10.1007/s12293-013-0128-0.

[21] A. Askarzadeh, A novel metaheuristic method for solving constrained engineering optimization problems: crow search algorithm, Comput. Struct. 169 (2016) 1–12, https://doi.org/10.1016/j.compstruc.2016.03.001.

[22] S. Mirjalili, A. Lewis, The whale optimization algorithm, Adv. Eng. Softw. 95 (2016) 51–67, https://doi.org/10.1016/j.advengsoft.2016.01.008.

[23] Z. Bayraktar, M. Komurcu, J.A. Bossard, D.H. Werner, The wind driven optimization technique and its application in electromagnetics, IEEE Trans. Antennas Propag. 61 (5) (2013) 2745–2757.

[24] Z.D. Zaharis, C. Skeberis, T.D. Xenos, P.I. Lazaridis, J. Cosmas, Design of a novel antenna array beamformerusing neural networks trained by modified adaptive dispersion invasive weed optimization based data, IEEE Trans. Broadcast. 59 (3) (2013) 455–460.

[25] Z.D. Zaharis, P.I. Lazaridis, J. Cosmas, C. Skeberis, T.D. Xenos, Synthesis of a near-optimal high-gain antennaarray with main lobe tilting and null filling using taguchi initialized invasive weed optimization, IEEE Trans. Broadcast. 60 (1) (2014) 120–127.

[26] Y.-Y. Bai, S. Xiao, C. Liu, B.-Z. Wang, A hybrid IWO/PSO algorithm for pattern synthesis of conformal phased arrays, IEEE Trans. Antennas Propag. 61 (4) (2013) 2328–2332.

[27] G.G. Roy, S. Das, P. Chakraborty, P.N. Suganthan, Design of non-uniform circular antenna arrays using a modified invasive weed optimization algorithm, IEEE Trans. Antennas Propag. 59 (1) (2011) 110–118.

[28] S. Karimkashi, A.A. Kishk, Invasive weed optimization and its features in electromagnetics, IEEE Trans. Antennas Propag. 58 (4) (2010) 1269–1278.

[29] A. Hoorfar, Evolutionary programming in electromagnetic optimization: a review, IEEE Trans. Antennas Propag. 55 (3) (2007) 523–537.

[30] A. Hoorfar, J. Zhu, S. Nelatury, Electromagnetic optimization using a mixed-parameter self-adaptive evolutionaryalgorithm, Microw. Opt. Technol. Lett. 39 (4) (2003) 267–271.

[31] E. Boudaher, A. Hoorfar, Electromagnetic design optimization using mixed-parameter and multiobjective CMA-ES, in: Proceedings of the IEEE Antennas and Propagation Society International Symposium (APSURSI '13), IEEE, Orlando, FL, July 2013, pp. 406–407.

[32] E. BouDaher, A. Hoorfar, Electromagnetic optimization using mixed-parameter and multiobjective covariance matrix adaptation evolution strategy, IEEE Trans. Antennas Propag. 63 (4) (2015) 1712–1724.

[33] D.H. Wolpert, W.G. Macready, No free lunch theorems for optimization, IEEE Trans. Evol. Comput. 1 (1) (1997) 67–82.

[34] Y.C. Ho, D.L. Pepyne, Simple explanation of the no-freelunch theorem and its implications, J. Optim. Theory Appl. 115 (3) (2002) 549–570.

[35] H. Patidar, G.K. Mahanti, R. Muralidharan, Synthesis of flat-top beam pattern of linear antenna arrays with restricted side lobe level, VSWR and independent nulls using Flower Pollination algorithm, Int. J. Electron. (2019) 1–14, https://doi.org/10.1080/00207217.2019.1636297.

[36] K. Kundu, N.N. Pathak, Circular antenna array optimization using flower pollination algorithm, in: U. Biswas, A. Banerjee, S. Pal, A. Biswas, D. Sarkar, S. Haldar (Eds.), Advances in Computer, Communication and Control, Lecture Notes in Networks and Systems, vol. 41, Springer, Singapore, 2019, https://doi.org/10.1007/978-981-13-3122-0_40.

[37] R. Salgotra, U. Singh, A novel bat flower pollination algorithm for synthesis of linear antenna arrays, Neural Comput. Applic. 30 (7) (2018) 2269–2282, https://doi.org/10.1007/s00521-016-2833-3.

[38] U. Singh, R. Salgotra, Synthesis of linear antenna array using flower pollination algorithm, Neural Comput. & Applic. 29 (2) (2018) 435–445, https://doi.org/10.1007/s00521-016-2457-7.

[39] S. Das, R. Bera, D. Mandal, S. Prasad Ghoshal, R. Kar, Evolutionary algorithms based synthesis of low sidelobe hexagonal arrays, Swarm Evol. Comput. 38 (2018) 139–157, https://doi.org/10.1016/j.swevo.2017.07.003.

[40] G. Ram, R. Kar, D. Mandal, S.P. Ghoshal, Optimal design of linear antenna arrays of dipole elements using flower pollination algorithm, IETE J. Res. 65 (5) (2019) 694–701, https://doi.org/10.1080/03772063.2018.1452639.

[41] U. Singh, R. Salgotra, Pattern synthesis of linear antenna arrays using enhanced flower pollination algorithm, Int. J. Antennas Propag. 2017 (2017), 7158752. 11 pages https://doi.org/10.1155/2017/7158752.

[42] H. Patidar, G.K. Mahanti, Failure correction of linear antenna array by changing length and spacing of failed elements, Prog. Electromagn. Res. M 61 (2017) 75–84, https://doi.org/10.2528/PIERM17072002.

[43] T. Sallam, A. Abdel-Rahman, M. Alghoniemy, Z. Kawasaki, Flower pollination algorithm for adaptive beamforming of phased array antennas, J. Mach. Intell. 2 (2) (2017) 1–5, https://doi.org/10.21174/jomi.v2i2.71.

[44] S.K. Sharma, N. Mittal, R. Salgotra, U. Singh, Linear antenna array synthesis using bat flower pollinator, in: 2017 International Conference on Innovations in Information, Embedded and Communication Systems (ICIIECS), Coimbatore, 2017, pp. 1–4, https://doi.org/10.1109/ICIIECS.2017.8276119.

[45] P. Saxena, A. Kothari, Linear antenna array optimization using flower pollination algorithm, SpringerPlus 5 (2016) 306, https://doi.org/10.1186/s40064-016-1961-7.

[46] L. Pappula, D. Ghosh, Linear antenna array synthesis using cat swarm optimization, AEU Int. J. Electron. Commun. 68 (6) (2014) 540–549, https://doi.org/10.1016/j.aeue.2013.12.012.

[47] L. Pappula, D. Ghosh, Planar thinned antenna array synthesis using multi-objective binary cat swarm optimization, in: 2015 IEEE International Symposium on Antennas and Propagation & USNC/URSI National Radio Science Meeting, 2015, https://doi.org/10.1109/aps.2015.7305620.

[48] G. Ram, D. Mandal, R. Kar, S.P. Ghoshal, Circular and concentric circular antenna array synthesis using cat swarm optimization, IETE Tech. Rev. 32 (3) (2015) 204–217, https://doi.org/10.1080/02564602.2014.1002543.

[49] G. Ram, D. Mandal, R. Kar, S.P. Ghoshal, Cat swarm optimization as applied to time-modulated concentric circular antenna array: analysis and comparison with other stochastic optimization methods, IEEE Trans. Antennas Propag. 63 (9) (2015) 4180–4183, https://doi.org/10.1109/tap.2015.2444439.

[50] S. Li, J. Ma, M. Zhuang, Y. Chen, Research of antenna selection based on binary cat swarm optimization, in: 2016 10th IEEE International Conference on Anti-Counterfeiting, Security, and Identification (ASID), 2016, https://doi.org/10.1109/icasid.2016.7873933.

[51] L. Pappula, D. Ghosh, Synthesis of thinned planar antenna array using multiobjective normal mutated binary cat swarm optimization, Appl. Comput. Intell. Soft Comput. 2016 (2016), 4102156. 9 pages https://doi.org/10.1155/2016/4102156.

[52] G. Ram, D. Mandal, S.P. Ghoshal, R. Kar, Optimal array factor radiation pattern synthesis for linear antenna array using cat swarm optimization: validation by an electromagnetic simulator, Front. Inf. Technol. Electron. Eng. 18 (4) (2017) 570–577, https://doi.org/10.1631/fitee.1500371.

[53] A. Singh, R. Mehra, V. Pandey, A hybrid approach for antenna optimization using cat swarm based genetic optimization, Adv. Electromagn. 7 (3) (2018) 23–34, https://doi.org/10.7716/aem.v7i3.624.

[54] S. Banerjee, D. Mandal, Array pattern optimization for steerable circular isotropic antenna array using cat swarm optimization algorithm, Wirel. Pers. Commun. 99 (3) (2018) 1169–1194, https://doi.org/10.1007/s11277-017-5171-6.

[55] S. Darzi, T.S. Kiong, M.T. Islam, M. Ismail, S. Kibria, B. Salem, Null steering of adaptive beamforming using linear constraint minimum variance assisted by particle swarm optimization, dynamic mutated artificial immune system, and gravitational search algorithm, Sci. World J. 2014 (2014), 724639. 10 pages https://doi.org/10.1155/2014/724639.

[56] G. Ram, D. Mandal, R. Kar, S.P. Ghoshal, Opposition based gravitational search algorithm for synthesis circular and concentric circular antenna arrays, Sci. Iran. 22 (6) (2015) 2457–2471.

[57] B. Sun, C. Liu, L. Yang, X. Wu, Y. Li, X. Wang, Conformal array pattern synthesis and activated elements selection strategy based on PSOGSA algorithm, Int. J. Antennas Propag. 2015 (2015), 858357. 11 pages https://doi.org/10.1155/2015/858357.

[58] S. Darzi, M.T. Islam, S.K. Tiong, S. Kibria, M. Singh, Stochastic leader gravitational search algorithm for enhanced adaptive beamforming technique, PLoS One 10 (11) (2015) e0140526, https://doi.org/10.1371/journal.pone.0140526.

[59] S. Darzi, T. Sieh Kiong, M. Tariqul Islam, H. Rezai Soleymanpour, S. Kibria, A memory-based gravitational search algorithm for enhancing minimum variance distortionless response beamforming, Appl. Soft Comput. 47 (2016) 103–118, https://doi.org/10.1016/j.asoc.2016.05.045.

[60] P. Swain, S.K. Mohanty, B.B. Mangaraj, Linear dipole antenna array design and optimization using gravitational search algorithm, in: 2016 2nd International Conference on Advances in Electrical, Electronics, Information, Communication and Bio-Informatics (AEEICB), 2016, https://doi.org/10.1109/aeeicb.2016.7538343.

[61] B.B. Mangaraj, P. Swain, An optimal LAA subsystem designed using gravitational search algorithm, Eng. Sci. Technol. Int. J. 20 (2) (2017) 494–501, https://doi.org/10.1016/j.jestch.2017.01.002.

[62] M. Hesari, A. Ebrahimzadeh, Introducing deeper nulls and reduction of side-lobe level in linear and non-uniform planar antenna arrays using gravitational search algorithm, Prog. Electromagn. Res. B 73 (2017) 131–145.

[63] S.K. Mohanty, B.B. Mangaraj, An optimal design of super-directive dipole linear antenna array using gravitational search algorithm and large perfect reflecting surface, Recent Adv. Electr. Electron. Eng. 11 (2) (2018) 227–238, https://doi.org/10.2174/2352096511666180124152013.

[64] R. Bera, K. Kundu, N.N. Pathak, Optimal pattern synthesis of thinned and non-uniformly excited concentric circular array antennas using hybrid GSA-PSO technique, Radioengineering 27 (2) (2019) 369–385, https://doi.org/10.13164/re.2019.0369.

CHAPTER

Foundations of combinatorial optimization, heuristics, and metaheuristics

22

Bochra Rabbouch[a], Hana Rabbouch[b], Foued Saâdaoui[c], and Rafaa Mraihi[d]
[a]*Higher Institute of Applied Sciences and Technology of Sousse, University of Sousse, Sousse, Tunisia,* [b]*Higher Institute of Management of Tunis, University of Tunis, Tunis, Tunisia,* [c]*Department of Statistics, Faculty of Sciences, King Abdulaziz University, Jeddah, Saudi Arabia,* [d]*Ecole Supérieure de Commerce de Tunis, Campus Universitaire de Manouba, Manouba, Tunisia*

1. Introduction

Combinatorial optimization (CO) problems are an important class of problems where the number of possible solutions grows combinatorially with the problem size. These kinds of problems have attracted the attention of researchers in computer sciences, operational research, and artificial intelligence. The fundamental objective of CO is to find the optimal or near-optimal solution for a complex problem using different optimization techniques to minimize costs and maximize profits and performances.

This chapter provides background on different materials, notations, and algorithms in CO. The chapter is organized as follows: Sections 3 and 4 discuss the analysis and complexity of algorithm theories. Section 5 discusses modeling CO problems to simplify realistic complications and reduce problem difficulties. Modeling consists of illustrating the graph theory concepts, the mathematical programming, and the constraint programming paradigms. Finally, Section 6 presents solution methods, including exact methods, heuristics, and metaheuristics.

2. Combinatorial optimization problems

A CO problem is an optimization problem where the number of possible solutions is finite and grows combinatorially with the problem size. It aims to look for the perfect solution from a very huge solution space and allows an excellent usage of limited resources in order to attain a fundamental objective within a running time bounded by a polynomial in the input size. The quality of optimization relies on how quickly it is possible to find the optimal solution. The CO problem has emerged in industrial, production, logistic environments, and computer systems.

A CO problem can be modeled by a set of variables to find a satisfying solution respecting a set of constraints while optimizing an objective function. The problem P can be written as:

Comprehensive Metaheuristics. https://doi.org/10.1016/B978-0-323-91781-0.00022-3

$$P = \langle X, D, C, f \rangle,$$

where

- $X = \{x_1, \ldots, x_n\}$ is the set of variables;
- $D = \{D(x_1), \ldots, D(x_n)\}$ is the set of domains of variables, $D(x_n)$ is the domain of x_n;
- $C = C_1, \ldots, C_n$ is the set of constraints over variables; and
- f is the objective function to be optimized.

It is possible to write the objective function to optimize as:

$$\text{optimize } \{f(F) : F \in \mathcal{F}\},$$

where *optimize* is replaced by either *minimize* or *maximize* and including the following settings:

- A finite set, $E = \{e_1, \ldots, e_n\}$.
- A weight function, $w : E \rightarrow \mathbb{Z}$, $w(e_i)$ is the weight of e_i.
- A finite family, $\mathcal{F} = \{F_1, \ldots, F_m\}$, $F_i \subseteq E$ are the feasible solutions.
- A cost function, $f : \mathcal{F} \rightarrow \mathbb{Z}, f(F) = \sum_{e \in F} w(e)$ (additive cost function).

For the most part, similar problems are able to be defined as integer programs with binary variables to verify if every member of the collection is a part of the subset or not. Furthermore, because optimization problems are of minimization or maximization type, so for maximization problems for example, note that if the function to maximize is well defined, then minimizing the negation of this will maximize the original function.

CO is the most popular type of optimization and it is often linked with graph theory and routing and it includes the vehicle routing problem [1, 2], the traveling salesman problem [3], the knapsack problem [4], the bin packing and cutting stock problem [5], and the bus scheduling problem [6]. Some well-known combinatorial games and puzzles include Chess [7], Sudoku [8], and Go [9].

3. Analysis of algorithms

An algorithm consists of an ordering set of constructions for solving a problem. In computer science, the analysis of algorithms is the determination of the whole quantity of resources (such as time and storage) necessary to execute them. This helps scientists to compare algorithms in terms of their efficiency, speed, and resource consumption by specifying an estimate number of operations without regard to the specific implementation or input used. Then, it presents a regular measure of algorithm complexity regardless of the platform or the problem cases solved by the algorithm.

Generally, when analyzing algorithms, the fact or most used to measure performance is the time spent by an algorithm to solve the problem. Time is expressed in terms of number of elementary operations such as comparisons or branching instructions.

A complete analysis of the running time of an algorithm requires the following steps:

- implement the whole algorithm completely;
- determine the time needed for each basic operation;
- identify unknown quantities that can be used to describe the frequency of execution of the basic operations;

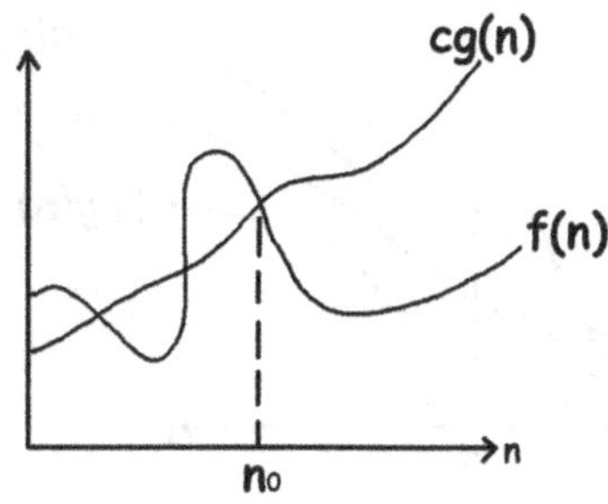

FIG. 1

$g(n)$ is an asymptotic upper bound for $f(n)$.

- establish a realistic model for the input to the program;
- analyze the unknown quantities, supposing the modeled input; and
- calculate the total running time by multiplying the time by the frequency for each operation, then adding all the products.

In a related context, the asymptotic efficiency of an algorithm reveals that the running time of an algorithm increases when the size of an input tends to infinity.

To specify an asymptotic upper bound [10] in algorithm analysis, we generally use the O notation and we say that such algorithm is of an Order n ($O(n)$) where n is the size of the problem (the total number of operations executed by the algorithm is at most a constant time n). The O notation is used to describe the complexity of an algorithm in a worst-case scenario. So, let f and g be functions from $\mathbb{N}$ to $\mathbb{N}$, $f(n)$ is of $O(g(n))$ if there exist positive constants c and n_0 and for all $n \geq n_0$, $0 \leq f(n) \leq cg(n)$ and that means informally that f grows as g or slower (Fig. 1).

If the opposite happens, that is, if $g(n) = O(f(n))$, we specify an asymptotic lower bound [10]; we generally use the Ω notation. More specifically, $f(n)$ is of $\Omega(g(n))$ if there exist positive constants c and n_0 such that $0 \leq cg(n) \leq f(n)$, for all $n \geq n_0$ (Fig. 2).

Finally, to specify an asymptotically tight bound [10], we use the θ notation and we say that $f(n)$ is of $\theta g(n)$ if there exist positive constants c_1, c_2, and n_0 and for all $n \geq n_0$, $0 \leq c_1 g(n) \leq f(n) \leq c_2 g(n)$. The latter means that f and g have precisely the same rate of growth (Fig. 3).

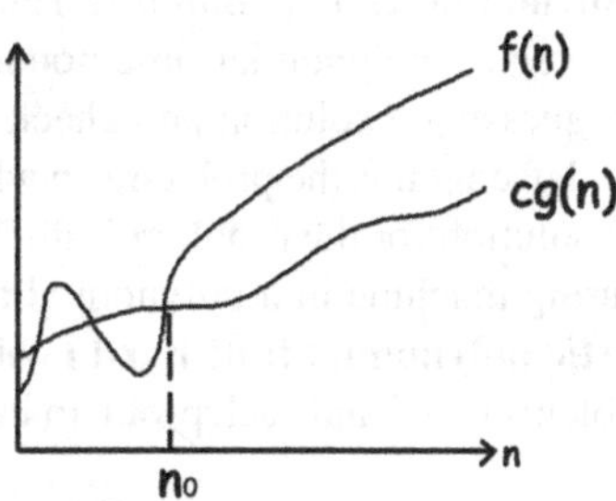

FIG. 2

$g(n)$ is an asymptotic lower bound for $f(n)$.

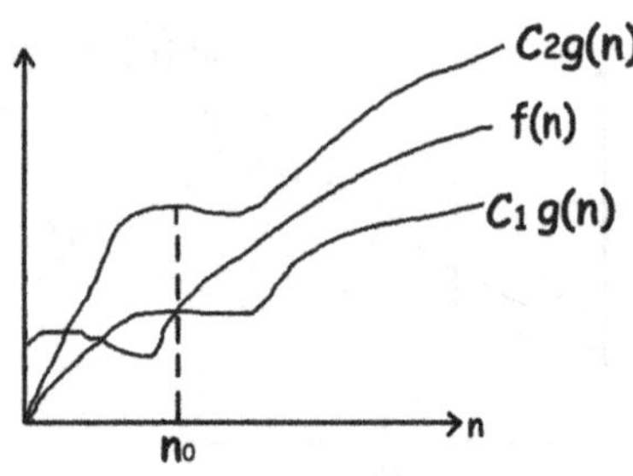

FIG. 3

g(*n*) is an asymptotically tight bound for *f*(*n*).

4. Complexity of algorithms

Computational complexity theory is a field in theoretical computer science and applied mathematics, used to analyze the efficiency of algorithms and help solve a problem with handling with resources required. Most complexity problems ask a question and expect an answer. The specific kind of question asked characterizes the problem, which requires either a "yes" or a "no" answer. Such problems are called decision problems. In complexity theory, we usually find it convenient to consider only decision problems, instead of problems requiring all sorts of different answers. For example, any maximization or minimization problem where the aim is to find a solution with the maximum possible profit or the minimum possible cost can be automatically transformed to a decision problem formulated as: "is there a real number *X* that represents the solution to the problem?" Then, the complexity theory can be applied to optimization problems.

An important issue that comes up when considering CO problems is the classification of problems. In this context, those remarkable classes of problems are generally described as presented in the sections that follow.

- **Class *P*: Polynomial time problems:** This class of problem consists of all decision problems for which exists a polynomial time deterministic algorithm to solve them efficiently. This class of decision problems includes tractable problems that can be solved sufficiently by a deterministic Turing machine in a polynomial time.
- **Class *NP*: Nondeterministic polynomial time problems:** This class of problems consists of all decision problems that can be solved by polynomial time nondeterministic algorithms. We can obtain the solution algorithm by guessing a solution and check whether the guessed solution is correct or not. Furthermore, Class NP contains the problems in which "verifying" the solution of the problem is quick, but finding the solution for the problem is difficult. Problems of this class can be solved by a nondeterministic Turing machine in a polynomial time.
- **Class *NP Hard*: Nondeterministic polynomial time hard problems:** The *NP*-hard problems are at least as hard as the hardest problems in *NP* and each problem in *NP* can be solved by reducing it to different polynomial problems.
- **Class *NP Complete*: Nondeterministic polynomial time complete problems:** Those problems are the most difficult problems in their class. A problem is classified as *NP* complete if it satisfies two

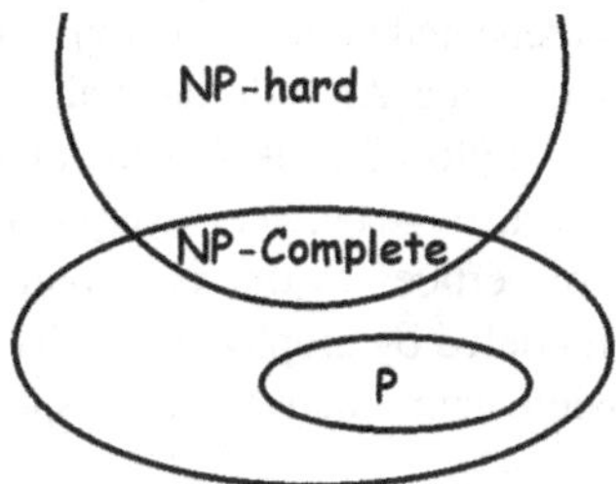

FIG. 4

Complexity classes.

conditions: it should be in the set of *NP* class and at the same time it should be *NP hard*. This class contains the hardest problems in computer science.

Fig. 4 illustrates how the different complexity classes are related to each other.

5. Modeling a CO problem

Solving an optimization problem
Problem⇒Model⇒solution(s)

Modeling an optimization problem helps to simplify the reality complications and reduces the difficulties of the problem. The design of a "good" model projects expert knowledge of the domain, variables, and constraints, as well as computer-based resolution methods. In addition, models can capture and organize the understanding of the system, permit early exploration of alternatives, increase the decomposition and modularization of the system, facilitate the system's evolution and maintenance, and finally, facilitate the reuse of parts of the system in new projects.

5.1 Graph theory concepts

A graph is a data structure that can be used to model relations, processes, and paths in practical systems and realistic problems. Graph theory was developed in 1735 when Leonhard Euler solved negatively (proved that the problem has no solution) the historically notable problem in mathematics called the Seven Bridges of Königsberg.

A graph can be defined as an ordered pair $G=(V, A)$ that links a set of V vertices by a set of A arcs or edges (each edge is generally associated with two vertices) where V and A are usually finite. If there is more than one edge between two vertices, the graph is a multigraph. If edges have no orientation, the graph is undirected and there is no difference between an arc (x, y) and an arc (y, x), but when edges have orientations, the graph is directed. In addition, a mixed graph is a graph in which some edges have orientations and others have no orientations, and a weighted graph is a graph in which there is a number or a weight associated to each edge (cost, distance, etc.).

A walk is defined as a sequence of adjacent nodes (neighbors) and is denoted by $w=[v_1, v_2, \ldots, v_n]$. The walk is closed (circuit) if $v_1 = v_k (\forall k > 1)$. If there are no repeated nodes in a walk, it is called path. A walk $w = [v_1, v_2, \ldots, v_k]$ with no repeated nodes except $v_1 = v_k$ is a cycle. A Hamiltonian cycle is a cycle that includes every vertex exactly once. An Eulerian walk is a walk in which each edge appears exactly once, and an Eulerian circuit is a closed walk in which each edge appears exactly once.

A graph is called complete if there is an edge between every pair of vertices and is called connected if there is a path between every pair of vertices. A tree is a connected graph with no cycles.

Many practical problems can be modeled by graphs especially in telecommunications, mathematics, computer sciences, data transmission, neural networks, transportation, logistics, and even in chemistry, biology, and linguistics.

5.2 Mathematical optimization model

A mathematical optimization model is an abstraction that allows to simplify and summarize a system. Once a mathematical model is made, a list of inputs, outputs, decision variables, and constraints is prepared and hence the problem is decorticated and the system behavior can be predicted.

To model an optimization problem, mathematical programming formulations represent the main goal as objective function, which is the output (a single score) to maximize or minimize. In addition, mathematical programs represent problem choices as decision variables that influence the objective function with consideration to the constraints on variable values expressing the limits on how big or small possible decision variables can get. Decision variables may be discrete or continuous:

- A variable is discrete if its set of values is fixed above or is countable (only have integer values, for example).
- A variable is continuous if it may take any variable in a specified interval.

Different families of optimization models are used in practice to formulate and solve decision-making problems. In general, the most successful models are based on mathematical programming and constraint programming. To classify mathematical programming formulations, five categories can be founded:

- Linear programming (LP): A class of optimization problems where objective function and constraints are expressed by linear functions.
- Integer programming (IP): Linear problems that restrict the variables to be integers.
- Mixed-integer linear programming (MILP): Linear problems in which decision variables are both discrete and continuous.
- Nonlinear programming (NLP): Objective function and constraints are expressed by nonlinear functions.
- Mixed-integer nonlinear programming: Nonlinear problems in which some decision variables are integers.

Mathematical programs and their terminology are due to George B. Dantzig, the inventor of the simplex algorithm for solving the LP problems in 1947 [11]. Since then, LP has been widely used and has attained a strong position in practical optimization. An interesting survey paper providing references to papers and reports whose purpose is to give overviews of linear, integer, and mixed-integer optimization, including solution techniques, solvers, languages, applications, and textbooks, is presented by Newman and Weiss [12].

5.2.1 Linear programming

An LP model is an optimization method to find the optimal solution to some special case mathematical optimization model where the objective and the constraints are required to be linear. The LP also contains the decision variables, which are the unknown quantities or decisions that are to be optimized.

An LP model can be considered as:

$$\text{Minimize } F_{LP} = c_1x_1 + c_2x_2 + \cdots + c_jx_j + \cdots + c_nx_n. \tag{1}$$

Subject to:

$$a_{11}x_1 + a_{12}x_2 + \cdots + a_{1j}x_j + \cdots + a_{1n}x_n(\leq, =, \text{or} \geq)b_1, \tag{2}$$

$$a_{i1}x_1 + a_{i2}x_2 + \cdots + a_{ij}x_j + \cdots + a_{in}x_n(\leq, =, \text{or} \geq)b_i, \tag{3}$$

$$a_{m1}x_1 + a_{m2}x_2 + \cdots + a_{mj}x_j + \cdots + a_{mn}x_n(\leq, =, \text{or} \geq)b_m, \tag{4}$$

$$x_1, \ldots, x_j, \ldots, x_n \geq 0, \tag{5}$$

where

- $x_1, \ldots, x_j, \ldots, x_n$ are the decision variables that are constrained to be nonnegative.
- $c_1, \ldots, c_j, \ldots, c_n$ are the coefficients in the objective function for the optimization problem.
- $b_1, \ldots, b_i, \ldots, b_m$ are the right-hand side values in the constraints sets.
- $a_{11}, \ldots, a_{ij}, \ldots, a_{mn}$ are the variables coefficients in the constraints set.

Eq. (1) presents the objective function that assesses the quality of the solution and which aims to minimize (or maximize) the value of F_{LP}.

The inequalities (2, 3, 4, 5) are the constraints that a valid solution (called a feasible solution) to the problem must satisfy. Each constraint requires that a linear function of the decision variables is either equal to, not less than, or not more than a scalar value. Besides, a common condition simply states that each decision variable must be nonnegative.

A solution that satisfies all constraints is called a feasible solution. The feasible solution with the best objective value (either the lowest in the case of minimization problem or the highest in the case of maximization problem) is called the optimal solution.

The development of LP is considered one of the most important scientific advances of the mid-20th century. A very efficient solution procedure for solving linear programs is the simplex algorithm. The simplex method proceeds by enumerating feasible solutions and improving the value of the objective function at each step. The termination step is reached after a finite number of different transitions. A strength of this method is its robustness: it allows solving any linear program even for problems with one or more optimal solutions (can provide all optimal solutions, if they exist), finding redundant constraints in the formulation, discovering instances when the objective value is unbounded over the feasible region, and generating feasible solutions to start the procedure or to prove that the problem has no feasible solution. For a deep explanation of LP and the simplex method, the reader is referred to the textbook [13].

5.2.2 Integer linear programming

Integer linear programs are extension of linear programs where decision variables are restricted to be integer values. They generally concern model problems counting some values (e.g., commodities) and enforce representing them by integer decision variables. A special case of integer linear programs is when modeling problems with yes/no decisions then the decision variables are initiated as binary values and are restricted to 0/1 values only.

5.2.3 Mixed-integer linear programming

Mixed-integer optimization problems appear naturally in many contexts of real-world problems and overcome several new technical challenges that do not appear in the nonmixed counterparts. In addition, MILPs can optimize decisions that take place in complex systems by including integer, binary, and real variables to form logical statements.

Mixed-integer programs are linear programs composed by linear inequalities (constraints) and linear objective function to optimize with the added restriction that some, but not necessarily all, of the variables must be integer-valued. The term "integer" can be replaced by the term "binary" if the integer decision variables are restricted to take on either 0 or 1 values. A standard mixed-integer linear problem is of the form:

$$z^* = Min\ c^T x,\ \ x \in X. \tag{6}$$

Subject to:

$$Ax(\leq, =, \text{or} \geq)b, \tag{7}$$

$$l \leq x \leq u, \tag{8}$$

where Eq. (7) refers to the linear constraints and Eq. (8) refers to the bound constraints. In addition, in an MILP, there are three main categories of decision variables:

- positive real variables ($x_j \geq 0, \forall x_j \in \mathbb{R}_+$);
- positive integer variables ($x_j \geq 0, \forall x_j \in \mathbb{Z}_+$); and
- binary variables ($x_j \in \{0, 1\}$).

In MILPs, the best objective feasible solution is called the optimal solution. The notation $x^* \in X$ designs the optimal solution to find and $\overline{x} \in X$ designs any feasible solution to the problem. In general, solving an MILP aims to generate not only a decreasing sequence of upper bounds $u_1 > \cdots > u_k \geq z^*$ but also an increasing sequence of lower bounds $l_1 < \cdots < l_k \leq z^*$. For minimization problems, any feasible solution $\overline{x} \in X$ yields an upper bound $\overline{u} = c\overline{x}$ on the optimal value, namely $\overline{u} \geq z^*$. In some cases, since it is possible to achieve function values without converging to the optimal values, MILPs then have no optimal solution but only feasible ones; those problems are called unbounded. In some other cases, no solution exists to an MILP and thus the problem is called infeasible.

To obtain lower bounds for minimization problems, we must have recourse to a relaxation of the problem.

Given (*P*) a minimization problem and a problem (*RP*) where

$$(P) \quad z^* = \min\{c(x) : x \in X \subseteq \mathbb{R}^n\}$$

and

$$(RP) \quad \tilde{z} = \min\{\tilde{c}(x) : x \in \tilde{X} \subseteq \mathbb{R}^n\},$$

(*RP*) is a relaxation of (*P*) if $X \subseteq \tilde{X}$ and $\tilde{c}(x) \leq c(x)\ \forall x \in X$.

Proposition 1. If *RP* is a relaxation of *P*, $\tilde{z} \leq z^*$.

Proof. Let x^* be an optimal solution of P, then $x^* \in X \subseteq \widetilde{X}$ and $\widetilde{c}(x^*) \leq c(x^*) = z^*$. Since $x^* \in \widetilde{X}$, we have $\widetilde{z} \leq \widetilde{c}(x^*)$.

Proposition 2. *Let x^*_{RP} be an optimal solution of* RP. *If x^*_{RP} is feasible for $P(x^*_{RP} \in X)$ and $\widetilde{c}(x^*_{RP}) = c(x^*_{RP})$, then x^*_{RP} is also optimal for* P.

The different types of relaxations that can be applied to an MILP problem are:

- Linear relaxation: This consists of removing the integrality constraint of a given decision variable. The resulting relaxation is a linear program. This relaxation technique transforms an NP-hard optimization problem (integer programming) into a related problem that is solvable in polynomial time (LP) and the solution to the relaxed linear program can be used to gain information about the solution to the original integer program.
- Relaxation by elimination: This relaxation technique is the most used, especially to easily compute a lower bound for a minimization MILP problem. This straightforward method consists of deleting one or more constraints.
- Lagrangian relaxation: This is a relaxation method that approximates a difficult problem of optimization by a simpler problem. The idea consists of omitting the "difficult" constraints and penalizing the violation of those constraints by adding a Lagrange multiplier for each one of them, which imposes a cost on violations for all feasible solutions. The resulted relaxed problem can often be solved more easily than the original problem.

Modeling optimization problems as MILPs can be done by different ways respecting some arrangements and by bringing the problem characteristics on constraints, decision variables, and objective functions in these different modeling approaches. Some models may be narrower than others (in terms of the number of constraints and variables required) but may be more difficult to solve than larger models. Improving the efficiency of MILP models requires a deep understanding on how solvers work. Then, the model can be solved quickly and practically.

5.2.4 LP solvers

Linear programs were efficiently solved using the simplex method, developed by George Dantzig in 1947, or interior point methods (also called barrier methods) introduced by Khachiyan in 1979 [12]. The simplex algorithm has an exponential worst-case complexity. Interior point methods were first only a proof that linear programs can be solved in polynomial time. Simply because the theoretical complexity of the simplex method is worse than that of interior point algorithms does not mean that the simplex method always performs worse in practice. The implementations of the two types of algorithms are important for run-time performance, as is the type of linear program being solved. Improved versions of both methods present the base for most powerful computer programs for solving LP problems.

During the 1950s, even small LP problems were not tractable. Several decades later, large-scale MILPs involving huge numbers of variables, inequalities, and realistically sized integer programming instances become tractable. In addition to algorithmic features, especially heuristic algorithms and cutting plane methods, progress is due to both hardware capabilities and software effectiveness (LP solvers). Nowadays, experts can easily solve an optimization model through a direct use of LP solvers.

The most competitive programming solvers [12] are CPLEX [14], GuRoBi [15], and Xpress [16]. However, CPLEX (ILOG, Inc.) is the most powerful, well-known commercial LP/CP solver and is considered an efficient popular solver due to its computational performance features. The actual computational performance is the result of a combination of different types of improvements such as primal simplex, dual simplex, and barrier methods; branch-and-bound, branch-and-cut, and cutting-planes techniques; heuristics and Relaxation-Induced Neighborhood Search (RINS); parallelization; and so on. An interesting explanation that helps understanding the key principles in CPLEX including methods and algorithms is presented in [17].

Solving an optimization model through an LP solver requires generating vectors containing objective function coefficients, constants, and matrices of data in accepting forms; then, the specialized solution technique can be implemented. Programming languages such as C++ have been used to convert parameters and to write scripts and code them in the required form by the optimization solvers. However, modeling languages present a real advance in the ease of use of mathematical programming by making it simple to express very large and complex models. The modeling languages allow using special constructs similar to the way in which they appear when written mathematically, such as the expression of objective functions and inequalities. Thus, the development of models became more rapid and more flexible in debugging and replacing or adding constraints.

Furthermore, different algebraic modeling languages used for programming problems appeared, such as AMPL [18], GAMS [19], Mosel [16], and OPL [14]. AMPL and GAMS modeling languages can be used on a variety of solvers such as CPLEX, GuRoBi, and Xpress, and other solvers including nonlinear ones. However, Mosel is compatible with a limited number of solvers but has the advantage that it is a compiled language, which is faster to read into the solver for large models than AMPL and GAMS are.

The OPL is an algebraic programming language that makes coding easier and shorter than it is with a general-purpose programming language. It is part of the CPLEX software package and therefore tailored for the IBM ILOG CPLEX and IBM ILOG CPLEX CP Optimizers (can also resolve the constraint programming approaches). In recent years, integer programming and constraint programming approaches have been complementary strengths to solve optimization problems. The OPL programming language supports this complementarity by backing the traditional algebraic mathematical notations and integrating the modern language of constraint programming, involving the ability to choose search procedures to benefit from both technologies. It includes [20]:

- OPLScript: a script language where models are objects to make easy the solving of a sequence of models, multiple instances of the same model, or decomposition schemes such as column generation and Benders decomposition algorithms.
- The OPL component library: to integrate an OPL model directly in an application by generating C++ code or by using a COM component.
- A development environment.

5.3 Constraint programming

CO problems can be expressed on a declarative or on a procedural formulation. The declarative formulation directly expresses constraints and the objective function then attempts to find the solution without the distraction of algorithmic features. The procedural formulation defines the manner to solve the problem by providing an algorithm. The artificial intelligence community invented ways to twist declarative and procedural formulations to work together. This idea brought about the introduction of logic programming, constraint logic programming, constraint handling rules, and constraint programming.

Constraint programming is a new paradigm evolved within operations research, computer sciences, algorithms, and artificial intelligence. As its name reveals, constraint programming is a two-level architecture including a constraint and a programming component. It is considered an emerging declarative programming language for describing feasibility problems. It consists of parameters, sets, decision variables (like mathematical programming), and an objective function and deals with discrete decision variables (binary or integer).

A constraint is a logical relation among several variables, each taking a value in a given domain. Its purpose is to restrict the possible values that variables can take. Thus, constraint programming is a logic-based method to model and solve large CO problems based on constraints. The idea of constraint programming is to solve problems by stating constraints that must be satisfied by the solution. Considering its roots in computer science, constraint programming's basic concept is to state the problem constraints as invoking procedures; then, constraints are viewed as relations or procedures that operate on the solution space. Constraints are stored in the *constraint store* and each constraint contributes some mechanisms (enumeration, labeling, domain filtering) to reduce the search space by discarding infeasible solutions using constraint satisfiability arguments. A constraint solver is used to solve the problem dealing with a *constraint satisfaction problem* (CSP), which is a manner to describe the problem declaratively by using constraints over variables. It can be seen as a set of objects that must satisfy the given constraints and hold among the given decision variables. It can be seen as a set of objects that must satisfy the given constraints and hold among the given decision variables. Therefore, it introduces the domain of possible values of finite decision variables and presents relations between those variables, then it tries to solve them by constraint satisfaction methods.

A CSP is defined as a triple (X, D, C) where

- $X = \{x_1, x_2, \ldots, x_n\}$ is the set of variables.
- $D = \{d_1 \times d_2 \times \cdots \times d_n\}$ is the domain (sets of possible values) of the variables d_i is a finite set of potential values for x_i.
- $C = \{c_1, c_2, \ldots, c_m\}$ is a set of constraints. Each constraint is a relation on the subset of domains $C_i \subseteq d_{i1} \times \cdots \times d_{il}$ which defines the possible values of the variables $x_{i1}, \ldots, x_{il}$ acceptable by the constraint C_i.

The solution of the CSP is a tuple $\langle v_1, \ldots, v_n \rangle$ assigned to the variables $x_1, \ldots, x_n$ where each variable x_i has a possible value from its domain set d_i and all constraints C are satisfied.

5.3.1 Constraint propagation

Solving a constraint programming is ensured by combining a systematic search process with inference techniques. The systematic search process consists of enumerating all possible variable-value combinations and construing a search tree where the tree's root represents the problem to be solved. At each node in the search tree, domain filtering and constraint propagation techniques are used to perform inference. Each constraint is related by a domain filtering algorithm to eliminate inconsistent values (cannot be part of any solution) from the variable domains. The domain filtering algorithm take turns until no one is able to prune a domain any more or one domain becomes empty (i.e., the problem has no solution). Practically, the most effective filtering algorithms are those associated with global constraints. The constraint propagation process is iteratively applying the domain filtering algorithm. It is handled then by communicating the updated domains (domains where the unpromising values are discarded) to all of the constraints that are stated over this variable. Note that the aim of constraint propagation is to bring down the original problem and to know whether the problem is solved, may be feasible, or may be infeasible.

5.3.2 Global constraints

An important aspect of constraint programming is global constraints. A global constraint is a way to strengthen the constraint propagation by performing the pruning process. It consists of choosing one constraint (so-called global constraint) rather than several simple constraints. This concept may remove inconsistent solutions efficiently, eases the task of modeling the problem as CSP, and allows saving complexity and computational time.

5.3.3 Systematic search

The fundamental algorithmic step in solving a CSP is search. The search technique can be complete or not. A complete systematic search algorithm guarantees finding the solutions if they exist, then deducing the optimal solutions or proving that the problem is insoluble. The main disadvantage of these algorithms is that they take a very long computational time. An example of systematic complete algorithms is the backtracking search algorithm. It is a technique that performs a depth-first traversal of a search tree. The algorithm incrementally maintains a partial assignment that specifies consistent values with the constraints whose variables are all assigned. The beginning is with an empty assignment, and at each step, the process chooses a variable and a value in its domain. Whenever it detects a partial assignment that violates any of the constraints and cannot be extended to a solution, backtracking is performed and it returns to the most recently instantiated decision that still had alternatives available.

An example of systematic incomplete search algorithms is the local search for solving CSP. Whereas the nodes in the search tree in backtracking search are presented as partial sets of assignments to the variables, they represent complete assignments in local search. Each node is associated with a cost value resulted from a cost function and a neighborhood function is applied to detect adjacent nodes. The search progresses by iteratively selecting neighbors and applying the algorithm to look for neighbors of lowest cost because a standard cost function is the measure of the number of constraints that are not satisfied.

For further information about constraint programming, the reader is referred to [20–25].

6. Solution methods

Solving NP-hard problems has been a challenging topic for many researchers almost since the beginning of computer history (long before the concept of NP-hardness was discovered). Those problems can be solved by exact algorithms, but the best approach is to search for good approximation algorithms, heuristics, and metaheuristics. In this section, we review some of the methods used to solve the vehicle routing problem mentioned in the next chapter. However, because exact algorithms do not present practical solutions to our problem in this thesis, discussion of the techniques will be brief compared to techniques of approximation algorithms.

6.1 Exact algorithms

An exact algorithm is an algorithm that solves a problem to optimality. In most cases, this kind of algorithm generates optimal solutions (if they exist) by reducing the solution space and the number of different alternatives needed to be analyzed or by evaluating implicitly every possible solution to obtain the best one. Those methods perform well for small problems but are not very suitable to solve NP-hard problems, especially when considering their limitations for large-scale problems. In addition, they are time-consuming algorithms even for small instances where they require an exponential time.

In what follows, we review two exact approaches to solve CO problems: branching methods (Branch & Bound, Branch & Cut, Branch & Price, Branch & Infer) and dynamic programming, both of which divide a problem in different subproblems. Branching algorithms split problems into independent subproblems, solve the subproblems, and output as the optimal solution to the original problem the best feasible solution found along the search. Alternatively, a dynamic programming algorithm is appropriate for a smaller set of optimization problems. It breaks down the initial problem into not independent subproblems in a recursive manner and then recursively finds the optimal solution to the subproblems.

6.1.1 Branching algorithms

The basic branching algorithm determines a repeated dived-and-conquer strategy to find the optimal solution to a given hard problem P_0. A branching step is presented by creating a list of subproblems $P_1,\ldots,P_n$ of P_0 and each P_i is solved if it is possible. The obtained solution becomes a candidate solution of P_0 and the best candidate solution is called the *incumbent solution*. Else, if a P_i is too hard, it is branched again (treated as P_0) and so on, recursively, until all the subproblems are treated, and then the incumbent solution is the optimal solution required *Opt* (Algorithm 1).

In literature, there are many popular algorithms that exploit the branching strategy. For example, the *Branch & Bound algorithm*, which combines the branching strategy and the bounding procedure, the

Algorithm 1 An elementary branching algorithm for a CO problem.

```
Begin
   Opt←∞and P←{P0}
   WHILE P is nonempty DO
   Define subproblems Pi=P1,...,Pn of P0, add them to P and remove P0
   Choose a subproblem Pi, remove it from P and solve it
   IF infeasible THEN
     define subproblem Pi1=P11,...,Pn1 of Pi, add them to P and remove Pi
   ELSE
     set x is the candidate solution of Pi
     set Opt←optimize{x,Opt}
   END
   RETURN Opt the optimal solution of the problem P
End
```

Branch & Cut algorithm using the branching strategy with the cutting steps, and the *Branch & Infer algorithm*, which designs the branching method with inference.

- **Branch & Bound**: The Branch & Bound algorithm is the most popular branching algorithm that can greatly reduce the solution space. It is a complete tree-based search used in both constraint programming and integer programming, but in rather different forms. In CP, the branching is combined with inference aimed at reducing the amount of choices needed to explore. In IP, the branching is linked with a relaxation (to provide the bounds), which eliminates the exploration of nonpromising nodes for which the relaxation is infeasible or worse than the best solution found so far.

  This algorithm is based on:
  - **Branching**: This step consists of dividing recursively the feasible set of a problem P_0 into smaller subproblems, organized in a tree structure, where each node refers to a subproblem P_i that only looks for the subset ("leaves" of the tree) at that node.
  - **Bounding**: By solving the equivalent relaxed problem (some constraints are ignored or replaced with less strict constraints), a lower bound and/or an upper bound are calculated for the cost of the solutions within each subset.
  - **Pruning**: Pruning by optimality where a solution cannot be improved by decomposing more of the tree, pruning by bound where the current solution in that node is worse than the best known bound, or pruning by infeasibility when the solution is infeasible. Then, there are no reasons to apply the branching for this node and the subset is discarded.

  Otherwise, the subset will be further branched and bounded and the algorithm ends when all solutions are pruned or fixed.

  The search strategy refers to how to explore the tree and how to select the next node to process: either in depth or in progress. A depth-first search strategy is most economical in memory: the last constructed node is split and the exploration is done deep down leaves of the tree until a node is pruned and discarded. In this case, the strategy returns to the last unexplored node and goes back in depth, and so on. Only one branch at a time is explored and is stored in a stack.

  In a progressive strategy (frontier search), the separate node is the one of weakest evaluation. The advantage is often to more quickly improve the best solution and thus accelerate the search since it will be more easily possible to prune nodes. This strategy consumes more memory (it can require a lot of memory to store the candidate list) and generally is much faster than the depth-first strategy. Besides, in order to accelerate the method as much as possible, it is elementary to have the best possible bounds.

  The Branch & Bound method can reduce the search step, but it is not always easily applicable because sometimes it is very hard to calculate effective lower and upper bounds. In addition, it may not be sufficient for large-scale problems where finding an optimal solution requires a long computation time.
- **Branch & Cut**: The Branch & Cut algorithm is an extension of the Branch & Bound algorithm, with the addition of cutting-plane methods to calculate the lower bound at each node of the tree to define a small feasible set of solutions. This method is exploited especially in large integer programming problems. A CO problem is formulated as an integer programming problem and is solved with exponentially many constraints where those constraints are generated on demand. The solution is determined by adding a number of valid linear inequalities (cuts) to the formulation to improve the

relaxation bound and then solving the sequence of LP relaxations resulting from the cutting-plane methods.

Cuts can be used all over the tree and are called globally valid or used only in the resultant node, in which case they are called locally valid.

The algorithm is based on:
- adding the valid inequalities to the initial formulation: this step consists of creating a formulation with better LP relaxation;
- solving current LP relaxation by cutting off the solution with valid inequalities;
- cutting and branching: only with the initial LP relaxation (root node); and
- branching and cutting: at all nodes in the tree.

The Branch & Cut approach is the heart of all modern mixed-integer programming solvers. It succeeds in reducing the number of nodes to explore and define feasible regions, which helps in easily finding the optimal solutions.

- **Branch & Price**: The Branch & Price algorithm is an extension of the Branch & Bound algorithm. It is a useful method for solving large integer linear programming (ILP) and MILP where the task is to generate strong lower and upper bounds. It is useful also for solving a list of LP relaxations of the integer LP. This method combines the Branch & Bound algorithm and column generation method. The column generation search method solves an LP with exponentially many variables where variables represent complex objects. Then, the Branch & Price method solves a mixed LP with exponentially many variables branching over column generation where at each node of the search tree, columns (or variables) can be added to the LP relaxation to improve it. At the approach's beginning, some of those variables (columns) are excluded from the LP relaxation of the large MIP in order to reduce memory and are then added back to the LP relaxation as needed where most of them may be assigned to zero in the optimal solution. Then, the major part of the columns are irrelevant for solving the problem.

The algorithm involves:
- developing an appropriate formulation to form the master problem, which contains many variables;
- generating a restricted master problem to be solved, which is a modified version of the master problem and contains only some columns;
- solving a subproblem called a pricing problem to obtain the column with a negative reduced cost; and
- adding the column with the negative reduced cost to the restricted master problem and thus optimizing relaxation.

If the solution to the relaxation is not integral, the branching occurs and resolves a pricing problem.

If cutting planes are used to resolve LP relaxations within a Branch & Price algorithm, the method is known as Branch & Price & Cut.

- **Branch & Infer**: The Branch & Infer framework is a promising approach that combines the branching methods with inference and unifies the classical Branch & Cut approach from integer LP with the usual operational semantics of finite domain constraint programming [23]. This method is used for integer and finite domain constraint programming. For an ILP, and considering the constraint language of ILP with the primitive and nonprimitive constraints, we can obtain the relaxed version of a combinatorial problem by the primitive constraints. Then, inferring a new primitive constraint corresponds to the generation of a cutting plane that cuts off some part of this relaxation. But the basic inference principle in finite domain constraint programming is domain reduction (reduction of nonprimitive constraints to primitive constraints) and for each nonprimitive constraint, a propagation algorithm plans to remove inconsistent values from the domain of the variables occurring in the constraint. Note that arithmetic constraints are primitive in ILP, whereas in CP they are nonprimitive. That is why, each arithmetic constraint is handled individually in CP.

In addition, because the primitive constraints are not expressive enough or because the complete reduction is computationally not feasible, the branching methods are applied to split the problem into subproblems and process each subproblem independently.

The Branch & Infer method is very useful not only for resolving the CO problem but also for solving CSPs. The rule system of the Branch & Infer consists of the rules *bi_infer*, *bi_branch*, and *bi_clash* together with the basic three alternatives:

- *bi_sol* (satisfiability)
- *bi_climb* (branch and bound): using lower bounding constraints to find an optimal solution
- *bi_bound, bi_opt* (branch and relax): In contrast to branch and bound, which uses only lower bounds, branch and relax works with two bounds: global lower bound *glb* and local upper bound *lub* (computed for each subproblem). The local upper bounds may introduce a new rule to prune the search tree. If a local upper bound is smaller than the best known global lower bound, then the corresponding subproblem cannot lead to a better solution and therefore can be discarded

6.1.2 Dynamic programming

A dynamic programming algorithm (whose study began in 1957 [26]) is a class of backtracking algorithm where subproblems are repeatedly solved. It is based on simplifying complicated problems by splitting them, in a recursive manner, into a finite number of related but simpler problems, resolving those subproblems, and then deducing the global solution for the original problem.

The first step is to generalize the original problem by creating an array of simpler subproblems and giving a recurrence relating some subproblems to other subproblems in the array. The second step is to find the optimal value searched for each subproblem and compute the best values (solutions) for the more complicated subproblems by exploiting values already computed for the easier problems, which is guaranteed through the above recurrence feature. The last step to use the array of values calculated to choose the best solution for the original problem.

6.2 Heuristics

To solve CO problems, exact methods are not very suitable considering their limitations for large-scale problems. Hence, in the last decades, there has been extensive research and efforts allotted to the development of approximate algorithms (heuristics and metaheuristics) that are often applied to produce good-quality solutions for optimization problems in a reasonable computation time (polynomial time).

The term heuristic is inspired from the Greek word "Heuriskein," which means to find or to discover. Heuristics are approaches applied after discovering, learning, and looking for the good-quality solutions to solve problems. Heuristics can be considered as search procedures that iteratively generate and evaluate possible solutions. However, there is no guarantee that the solution obtained will be a satisfactory solution, or that the same solution quality will be retrieved every time the algorithm is run. But generally, the key to a heuristic algorithm's success relies on its ability to adapt to the specifications of the problem, to deal with the behavior of its basic structure, and especially to stay away from local optima.

6.2.1 The constructive heuristics

A constructive heuristic is a heuristic method that starts with an empty set of solutions and repeatedly constructs a feasible solution, according to some constructive rules, until a general complete solution is built. It varies from local search heuristics, which began from a complete solution and tried to improve it using local moves. Constructive heuristics are efficient techniques and have great interest as they generally succeed in finding satisfactory solutions in a very short time. Such solutions have the ability to be applied to identify the initial solution for metaheuristics.

The construction of the solution is made through some construction rules. Such rules are generally related to various aspects of the problem and the objective looked for. In this section, we describe constructive heuristics applied for routing problems [27, 28] where the objective is principally to minimize the travel distance while servicing all customers [29]. In this context, constructive heuristics are called *Route construction heuristics* and aim to select nodes or arcs sequentially based on cost minimization criteria until creating a feasible solution [30].

Based on cost minimization criterion, we can distinguish:

- **Nearest neighbor heuristic**: The nearest neighbor heuristic was originally initialized by Tyagi [31] in 1968. This heuristic builds routes by adding un-routed customers that respect the near neighbor criterion. At each iteration, the heuristic looks for the closest customer to the depot and then to the last customer added to the route. A new route is started unless the search fails adding customers in the appropriate positions or no more customers are waiting.
- **Savings heuristic**: The savings heuristic was originally proposed by Clarke and Wright [1] in 1964. It builds one route at a time by adding un-routed customers to the current route, with satisfying the savings criterion. The saving criterion is a measure of cost formula by combining given weights α_1, α_2, and α_3.
- **Insertion heuristic**: The insertion heuristic was originally conceived by Mole and Jameson [32] in 1976. It constructs routes by inserting un-routed customers in appropriate positions. The first step consists of initializing each route using an initialization criterion. After initializing the current route, and at each iteration, the heuristic uses two other formulations to insert a new un-routed customer into the selected route between two adjacent routed customers.
- **Sweep heuristic**: The sweep heuristic was originally implemented by Gillet and Miller [33] in 1974. It can be considered a two-phase heuristic as it includes clustering and routing steps. It begins by computing the polar angle of un-routed customers and ranking them in an ascending order. The first customer in the list is called "the reference line." Hence, un-routed customers are added to the current route with consideration to their polar angle and with respect to the depot. Physically, this process is similar to a counterclockwise sweep movement considering the depot as central point, and starting from and ending in the reference line.

Constructive heuristics are able to build feasible solutions quickly, which is an important feature because many real-world optimization problems need fast response time. Whereas, the solutions generated are for low-quality until those methods cannot look more than one iteration step ahead.

6.2.2 The improvement heuristics

In contrast to constructive heuristics, which start from an empty solution and build it iteratively, improvement heuristics begin with a complete solution and aim to improve it using simple local modifications called moves. Improvement heuristics are considered local search heuristics that only accept possibilities that enhance the solution's quality and improve its objective function.

From the beginning, the heuristic deals with a complete solution and attempts to yield a good solution by iteratively applying modification metrics. Once the alterations improve it, the current solution is replaced by the new improved solution. However, if it causes a worsening objective function, the process will be repeated and the heuristic performs acceptable operations until no more improvements are found. This improvement metric can be considered a solution intensification procedure or a guided local search.

Improvement operators proposed for this purpose are diversified. In the context of route improvement heuristics, operators can manage to move a customer from one route to another, it is the case of *inter route operators* modify multiple routes at the same time. In addition, operators can manage to exchange the position of customers in the same route, it works on a single route and is called *intraroute operators*.

Heuristic techniques are related to some vocabulary in the context of exploring promising solutions. The *search space* describes the space covering all the feasible solutions for the problem. The *local search* designs the process that improve the solution by moving from a candidate solution to another, which is generated through the *neighborhood* aspect. We say that the new improved solution is the *neighbor* of the previous one. In the search space, the neighbor of a candidate solution generated by the local search process to improve the objective function can be a local optimum or can be the best searched solution and is called the global optimum. Fig. 5 illustrates a search space including local and global optimums.

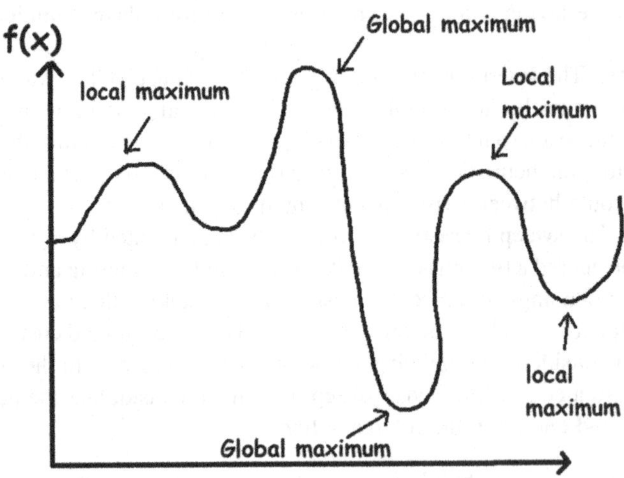

FIG. 5

Local and global optimums in a search space.

Formally, an optimization problem (minimization problem of the form $\min_{x \in S} f(x)$) may be expressed by the couple (S, f) where S represents the set of finite feasible solutions and $f: S \rightarrow \mathbb{R}$ is the objective function to optimize. The objective function assigns to each solution $x \in S$ of the search space a real number aiming to evaluate x and determining its cost. A solution $x^* \in S$ is a global optimum (global minimum) if it has a better objective function than all solutions of the search space, that is, $\forall x \in S, f(x^*) \leq f(x)$. Hence, the main goal in solving an optimization problem is to find a global optimal solution x^*.

A neighbor solution x' is an altered solution by a set of elementary moves to improve the objective function. The set of all solutions that can be reached from the current solution with respect to the move set is called the neighborhood of the current solution and is notated as $N(x) \subset S$. Relative to the neighborhood function, a solution $x \in N(x)$ is a local optimum if it has a better quality than all its neighbors, that is, $\forall x' \in N(x), f(x) \leq f(x')$.

The straightness of a local search heuristic can appear if it has the ability to escape from the local optima solutions and finds then the global optimum. In addition, improvement heuristics are powerful concepts helping conceiving and guiding the search under metaheuristics, especially when used to formulate metaheuristic initial populations.

6.3 Metaheuristics

Unlike exact algorithms, metaheuristics deliver satisfactory solutions in a reasonable computation time even when tackling large-size problems. This characteristic shows their effectiveness and efficiency to solve large problems in many areas. The term *metaheuristic* was invented by Glover in 1986 [34]. Many classification criteria may be used for metaheuristics [35]:

- **Nature inspired versus nonnature inspired**: A large number of metaheuristics are inspired from natural processes. Evolutionary algorithms are inspired by biology, ant and bee colony algorithms are inspired by social species and social sciences, and simulated annealing is inspired by physical processes.
- **Memory usage versus memoryless methods**: A metaheuristic is called memoryless if there is no information extracted dynamically to use during the search, like as in Greedy Randomized Adaptive Search Procedure (GRASP) and simulated annealing metaheuristics. However, some other metaheuristics can use some information extracted during the search to explore more promising regions and then intensify the search, like the case of short-term and long-term memories in tabu search.
- **Deterministic versus stochastic**: A deterministic metaheuristic solves an optimization problem by making deterministic decisions (e.g., tabu search), whereas stochastic metaheuristics apply some random rules during the search (e.g., simulated annealing, evolutionary algorithms). In deterministic algorithms, using the same initial solution will lead to the same final solution, whereas in stochastic metaheuristics, different final solutions may be obtained from the same initial solution.
- **Population-based search versus single-solution-based search**: Single-solution-based algorithms (e.g., tabu search, simulated annealing) manipulate and transform a single solution during the search, whereas population-based algorithms (e.g., ant colony optimization, evolutionary algorithms) develop a whole population of solutions. These two families have complementary characteristics: single-solution-based metaheuristics are exploitation oriented; they have the power

to intensify the search in local regions. Population-based metaheuristics are exploration oriented; they allow a better diversification in the whole search space.

- **Iterative versus greedy**: Most of the metaheuristics are iterative algorithms that start with a population of solutions (a complete solution) and alter it at each iteration using some search operators. Alternatively, greedy algorithms start from an empty solution, and at each step a decision variable of the problem is assigned until a complete solution is constructed (e.g., GRASP metaheuristic).

In the following, we present some important metaheuristic techniques.

6.3.1 Simulated annealing

This is one of the most popular local search-based metaheuristics. It is inspired from the physical annealing process for crystalline solids, where a solid placed in a heat bath and heated up by elevating temperature. When it melts, it is slowly cooled by gradually lowering the temperature, with the objective of relaxing toward a low-energy state and obtaining a strong crystalline structure. The fundamental aspect to a successful annealing system is related to the rate of cooling metals, which must begin with a sufficiently high temperature to not obtain flaws (metastable states) and lowering it gradually until it converges to an equilibrium state [36].

The simulated annealing search algorithm was introduced in 1982 by Kirkpatrick et al. [37] who applied the Metropolis algorithm [38] from statistical mechanics. This algorithm simulates a thermodynamical system by creating a sequence of states at a given temperature. Then, in each iteration, an atom is randomly displacing, where E is the change in energy resulting from the displacement. If the displacement causes a decrease in the system energy, that is, the energy difference ΔE between the two configurations is negative: ($\Delta E < 0$), then the new configuration is accepted and is used as a starting point for the next step. Either, if $\Delta E > 0$, the new configuration is accepted with the probability:

$$P(\Delta E) = \exp^{(-\Delta E/bT)},$$

where T is the current temperature and b is a constant called a Boltzmann constant. Depending on the Boltzmann distribution, the displacement is accepted or the old state is preserved. By repeating this process for a long time, an equilibrium state called thermal equilibrium would be reached.

Table 1 illustrates the analogy between the physical system and the optimization problem.

The simulated annealing method is a memoryless stochastic algorithm. There is no exploitation of information gathered during the search process and it includes randomization aspects to the selection process. The algorithm method starts from a feasible solution x (considering a problem of cost

Table 1 Simulated annealing analogy.

Physical system	Optimization problem
System state	Feasible solution
Energy	Objective function
Ground state	Global optimum solution
Metastable state	Local optimum
Careful annealing	Simulated annealing

Algorithm 2 The simulated annealing algorithm.

```
Begin
   Generate an initial solution x
   T←Tmax/*initialize temperature*/
   Repeat
   For(i=0;i<numIterations;i++) Do
     Generate a new solution x' within the neighborhood of x(x' ε N(x))
     ΔE ← f(x')-f(x)
     If ΔE<0 Then
       x ← x'
     Else
       p= Random (0,1)/*generate a random number in the interval (0,1)*/
       If p < exp^(-ΔE/bT) Then
         x ← x'
   T ← α * T/*reduce current temperature*/
  Until a stopping criterion has reached /*T<Tmin*/
  Return the best solution x
End
```

minimization $f(x)$) and by analogy to the annealing process, a random neighbor x' is generated where only improvement moves are accepted and that if it satisfies the equation $\Delta E = f(x') - f(x) < 0$. Furthermore, the algorithm begins with a high temperature T to allow a better exploration of the search space and this temperature decreases gradually during the search process. The temperature function is updated using a constant variable α on $T(t+1) = \alpha T(t)$, where t is the current iteration and the typical values of α vary between 0.8 and 0.99. These values can provide a very small diminution of the temperature. For each temperature, a number of moves according to the Metropolis algorithm is executed to simulate the thermal equilibrium. Finally, the algorithm will be stopped when a stopping criterion has reached (Algorithm 2).

To summarize, simulated annealing is a robust and generic probabilistic technique for locating good approximations to the global optimum in numerous search spaces to resolve CO problems [39, 40] and especially the vehicle routing problem [41–49].

6.3.2 The tabu search

The tabu search is a famous search technique coined by Glover [34] in 1986. The metaheuristic name was inspired from the word "taboo" which means forbidden. Moreover, in the tabu search approach, some possible solutions, with consideration to a short-term memory, are tabu and are then merged to a tabu list that stores nonpromising recently applied moves or solutions. Based on this process, tabu search allows a smart exploring of the search space, which helps to avoid the trap of local optima.

The tabu list is a central element of the tabu search, which aims to record the recent search moves to discard neighbors recently visited and then intensify the search in unexplored regions to encourage looking for optimal solutions. The tabu list avoids cycles and prevents visiting already visited solutions, as it memorizes previous search trajectory through a so-called short-term memory. The short-term memory is updated at each iteration and allows the storage of the attributes of moves in the tabu list. The size of the tabu list is generally fixed and contains a constant number of tabu moves. When it is filled, some old tabu moves are eliminated to allow recording new moves and the duration that a move is maintained as tabu is called the tabu tenure.

However, when necessary, the tabu search process selects the best possible neighboring solution even if it is already in the tabu list if it would lead to a better solution than the current one. Such improvements are applied using algorithmic tools and are called aspiration criterion, which consists of accepting a tabu move if the current solution improves best till now.

The main steps of a basic tabu search algorithm for a cost minimization problem can be described in Algorithm 3.

The tabu search algorithm deals also with advanced lists called medium-term memory and long-term memory that improves the intensification and the diversification facts in the search process. The intensification is maintained through the medium-term memory where recording attributes related to the best solutions found ever (the elite solutions) to exploit them by extracting the common features to guide the search in promising areas of the search space and then to intensify the search around good solutions. The long-term memory has been proposed in tabu search to stimulate the diversification of

Algorithm 3 The tabu search algorithm.

```
Begin
    Initialize an empty tabu list T
    Generate an initial solution x
    Let x*←x/ * x* is the best so far solution */
    Repeat
     Generate a subset S of solutions in N(x)/ * N(x) is the current neighborhood
     of x */
     Select the best neighborhood move x' ε S, where f (x')<f (x)
     If f(x')<f(x*) Then
       x*←x'/* aspiration condition: if current solution improves best now, accept
       it even if it is in the tabu list */
       x←x'/* update current solution */
        T←T+x'/* update tabu list */
    Else
     If (x' ε N(x)/T) Then
       x←x'/* update current solution if the new solution is not tabu */
       If (f(x')<f(x*)) Then
         x*←x'/* update the best till now solution if the new solution is better in
         quality */
       T←T+x'/* update tabu list */
    Until a stopping condition has reached
    Return the best solution x*
End
```

Algorithm 4 The tabu search pattern.

```
Begin
   Initialize an empty list; empty medium term and long term memories
   Generate an initial solution x
   Let x*←x/* x* is the best so far solution */
   Repeat
    Generate a subset S of solutions in N(x)/* N(x) is the current neighborhood of
    x */
    Select the best neighborhood move x' ∈ S./* nontabu or aspiration criterion
    holds */
      x*←x'
      Update current solution, tabulist, aspiration conditions, medium - and long
      - term memories
      If intensification criterion is maintained, Then intensification
      If diversification criterion is maintained, Then diversification
   Until a stopping condition has reached
   Return solution x*
End
```

the search. The information on explored regions along the search are stored in the long-term memory to investigate the unexplored regions not examined yet and then to enhance exploring the unvisited areas of the solution space.

The tabu search advanced mechanisms presented as a general tabu search pattern designed in Algorithm 4.

The tabu search is a powerful algorithmic approach. It has been widely applied with success to tackle numerous CO problems and especially the vehicle routing problem [47, 50–54] considering its flexibility and its competence to deal with complicated constraints that are typical for real-life problems.

6.3.3 Greedy randomized adaptive search procedure (GRASP)

The GRASP metaheuristic is a multistart, iterative, greedy, memoryless metaheuristic that was introduced by Feo and Resende [55] in 1989 to solve CO problems. The basic algorithm contains two main steps: construction and local search, which are called in each iteration. The construction phase consists of building a solution through a randomized greedy algorithm. Once a feasible solution is reached, its neighborhood is investigated during the local search step, which is applied to improve the constructed solution. This procedure is repeated until a fixed number of iterations is attained and the best overall solution is kept as the result (Algorithm 5).

A greedy algorithm aims to construct solutions progressively by including new elements into a partial solution until a complete feasible solution is obtained. An ordered list containing decreasing values of all candidate elements that can be included into a complete solution (that do not destroy its feasibility) is introduced and a greedy evaluation function is applied to evaluate the list and to select the next

Algorithm 5 The greedy randomized adaptive search procedure pattern.

```
Begin
    Initialize the number of iteration
    Repeat
     x=The_gready_randomized_algorithm;
     x'=The_local_search_procedure (x)
    Until A given number of iterations
    Return Best solution found
End
```

elements. This function aims to compute the function cost after including the element to be incorporated into the partial solution under construction and evaluate it. If this element brings the smallest incremental increase to the function cost, it is selected and then removed from the candidate set of solutions (Algorithm 6).

Solutions generated through greedy algorithms are not usually optimal solutions. Hence the idea to apply random steps to greedy algorithms. Randomization can diversify the search space, escape from the local traps, and generate various solutions. Greedy randomized algorithms have the same principle as the greedy procedure illustrated earlier but make use of the randomization process. The procedure starts as in simple greedy algorithm by a candidate set of elements that may incorporated in a complete solution. The evaluation of those elements is presented through the greedy evaluation function that leads to a second list including only the best elements with the smallest incremental costs, called the restricted candidate list (RCL). The element to be incorporated in the current solution is selected randomly from this list and then the set of candidate elements is updated and the incremental costs are reevaluated (Algorithm 7).

Algorithm 6 The greedy algorithm.

```
Begin
    x={}/* initial solution is null */
    Initialize the candidate set of solutions C ∊ S
    Evaluate the incremental cost: c(e) ∀e ∊ C
    Repeat
     Select an element e' ∊ C with the smallest incremental cost c(e')
     Incorporate e' into the current solution x←x ∪{e'}
     Update candidate set C
     Reevaluate the incremental cost c(e)∀e ∊ C
    Until C=∅
    Return solution x
End
```

Algorithm 7 The greedy randomized algorithm.

```
Begin
  x={}/* initial solution is null */
  Initialize the candidate set of solutions C ∊ S
  Evaluate the incremental cost: c(e) ∀e ∊ C
  Repeat
   Construct a list including the best elements with the smallest incremental
   costs
   Select randomly an element e′ from the restricted candidate list
   Incorporate e′ into the current solution x←x∪{e′}
   Update candidate set C
   Reevaluate the incremental cost c(e)∀e ∊ C
  Until C=∅
  Return solution x
End
```

Even by using the greedy randomized algorithms, solutions generated are not guaranteed to be optimal. The neighborhood of the best solution constructed is investigated and a local search technique is applied to replace the current solution by the best solution in its neighborhood. The local search procedure is stopped when no more improvements are found in the neighborhood of the current solution (Algorithm 8).

The GRASP metaheuristic is a successful method to solve CO problems, especially vehicle routing problems [56] since it benefits from good initial solutions that usually lead to promising final solutions due to randomization. In addition to the local search procedures, it ensures the intensification and diversification of the search space, allowing it to find the global optimum. Moreover, it is easy to implement in context of algorithm development and coding, which is an especially interesting characteristic of the GRASP metaheuristic.

Algorithm 8 The local search procedure (*x*).

```
Begin
  x′←x /* The initial solution is the solution generated by the greedy
   randomized algorithm */
  Repeat
   Find x″ ∊ N(x′) with f(x″)< f(x′)
   x′←x″
  Until No more improvement found
  Return solution x′
End
```

6.3.4 Ant colony optimization (ACO)

The ACO algorithm is a probabilistic population-based metaheuristic for solving CO problems that was developed by Dorigo et al. [57] in 1991. The basic idea in ACO algorithms is to mimic the collective behaviors of real ants when they are looking for paths from the nest to food locations that are usually the shortest paths. The system can be considered a multiagent system where a highly structured swarm of ants cooperates to find food sources employing coordinated interactions and indirectly communicating through chemical substances known as pheromones, which is left during their trip on the ground to mark the trails to and from the food source.

The mechanism of the foraging behavior of ants is described as follows: new ants move randomly until detecting a pheromone trail and, with a high probability, they will follow this route and the pheromone is enhanced by the increasing number of ants. Generally, an ant colony is able to find the shortest path between two points due to the greater pheromone concentration. Since every ant goes back to the nest after having visited the food source, it conceives a pheromone trail in both directions. Ants taking a short path return quickly, which leads to a more rapid increase of the pheromone concentration than would occur on a longer path. Thus, concurrency ants tend to follow this path due to the fast accumulation of the pheromone, which is known as positive feedback or autocatalysis. Hence, most of the ants are directed to use the shortest path based on the experience of previous ants.

The pheromone is a volatile substance having a decreasing effect over time (evaporation process). It will vanish into the air if it is not refreshed (reinforcement process). The quantity of pheromone left depends on the remaining quantity of food, and when that amount has finished, ants will stop putting pheromones onto the trail.

Based on the ants' behavior and the concepts illustrated, the ACO algorithm relies on artificial ants to solve hard CO problems. Artificial ants will randomly construct the solution by adding some solution components iteratively to the partial solution with regard to the concentration of pheromone until a complete solution is initialized. The randomization fact is used to allow the construction of a variety of different solutions. The pheromone trails save problem-specific information about promising constructed solutions in a common memory to guide generating other solutions. It is considered the memory of the whole ant search system.

The pheromone trails are altered dynamically during the search and the pheromone is updated. Hence, given two nodes i and j, we indicate the edge in between as $(i;j)$ and the associated pheromone value as τ_{ij}. When passing an edge $(i;j)$, an ant modifies the pheromone value τ_{ij} by simply increasing it using a constant value and then applying the reinforcement phase:

$$\tau_{ij} \leftarrow \tau_{ij} + \Delta\tau.$$

The evaporation phase when pheromone values decreases over the time is similar to the evaporation of real pheromone to avoid being trapped in local optima, for example. The decrease of the pheromone concentration usually occurs in an exponential way, using an evaporation parameter $\rho \in]0, 1]$:

$$\tau_{ij} \leftarrow (1-\rho)\tau_{ij}.$$

Due to the evaporation phase, solutions of bad quality are eliminated and new areas of the search space can be discovered that enhance the diversification process to not be trapped in local optimum solutions.

The steps of ACO are described in Algorithm 9.

Algorithm 9 The ant colony optimization algorithm.

```
Begin
   Determine number of ants n_a
   Initialize pheromone trails
   Repeat
    For (k=1;k≤n_a;K++) Do /* For each ant Do */
     Construct solution x_k
     Update pheromone trails
    End
    Apply pheromone evaporation
    Apply pheromone reinforcement
   Until a stopping condition has reached
   Return solution x*
End
```

ACO algorithms have become a very popular tool to solve CO problems and they have achieved widespread success in solving different optimization problems such as real-world vehicle routing problems [6, 58–60].

6.3.5 Genetic algorithms

The baselines of heredity from parents to offsprings and the theory of evolution have inspired computer scientists since the 1950s to simulate them in designing evolutionary algorithms [61]. The most known evolutionary algorithm is the genetic algorithm developed in 1975 by John Holland [62]. Genetic algorithms are stochastic population-based metaheuristics that have been successfully applied to solve widespread complex CO problems. The main idea of genetic algorithms is to iteratively mimic the evolution of species and the survival of the fittest theories. Table 2 illustrates the analogy between the evolutionary system and the optimization problem.

Basic genetic algorithms start from a randomly generated population (Algorithm 10). Every individual of the population is composed of a set of chromosomes and is evaluated by a fitness value to determine its ability to survive. The fittest two individuals are selected to reproduce (the crossover step) and generate a super-fit offspring. The offspring is altered with a small probability (the mutation step)

Table 2 Genetic algorithm analogy.

Metaphor	Optimization problem
Individual	Feasible solution
Population	The solution set
Chromosomes	The encoded solution
Genes	Elements of the encoded solutions
Fitness	Objective function
Offspring	Generated solution

Algorithm 10 The basic genetic algorithm.

```
Begin
   1:Generate the initial population
   2:Representation
   3:Repeat
   4: Fitness evaluation
   5: Selection
   6: Crossover
   7: Mutation
   8: Replacement
  10:Until a stopping condition has reached
  11:Best individual or best population found
End
```

and the new generated offspring take place in the new population (the replacement step) until a termination criterion is reached and the best individual or population is found.

In the first step, an initial population of solutions is generated, which will be improved in the next steps of the genetic algorithm. Basically, the initial population is produced randomly, but it can be generated using constructive heuristics or some single solution-based heuristics such as tabu search, GRASP, and so on. Next, the representation step consists of initializing and encoding the initial population and split the individuals composing it on a set of chromosomes including the genes. Traditionally, the binary representation was the most used method to encode the initial population, but nowadays various representations are used and are generally problem-specific representations such as real-valued vectors, permutations, and real discrete vectors.

The fitness function refers to the objective function that associates a cost to each feasible solution to describe its quality. The fitness evaluation in a genetic algorithm consists of judging the ability of individuals to survive through a fitness value to compare them at each iteration. Then, based on the fitness evaluation, the fittest individuals are selected. The selection step is one of the fundamental components of a genetic algorithm that may lead to encouraging solutions. Individuals are chosen for reproducing according to their fitness by mean of selection strategies [35], such as:

- Roulette wheel selection: This is the most applied selection strategy. It consists of accrediting to each individual a selection probability that is proportional to its fitness value. The probability P_i of an individual i having a fitness f_i is:

$$P_i = f_i / \left(\sum_{j=1}^{n} f_j \right).$$

- Tournament selection: In tournament selection, the size k of the tournament group is fixed from the beginning and the strategy involves randomly selecting k individuals from the population. Then a tournament is applied to them and the best one is chosen.

- Rank-based selection: A high rank is associated to individuals with good fitness and the rank is scaled linearly using the following formula:

$$P(i) = 2 - s/\mu + 2 \cdot r(i)(s-1)/\mu(\mu-1)$$

where s is the selection pressure ($1.0 < s < 2.0$), μ is the size of population, and $r(i)$ is the rank associated with the individual i.

The selection strategy leads to choosing two parents to reproduce. A crossover method is applied by combining the genes of two individuals, which produces a new offspring. The traditional genetic algorithm used strings of bits to present chromosomes. Three classical crossover operators have been widely implemented:

- One-point crossover: A point from each chromosome is selected randomly and the two chromosomes are cut at the corresponding points and the blocks are exchanged. The one-point crossover is the basic crossover operator.
- Two-point crossover: In each chromosome, two points are chosen randomly and the blocks between the two points are exchanged by the two chromosomes.
- Uniform crossover: Two individuals are recombined independently of their size. Each gene in the offspring is selected randomly from one of the parents. Each parent has equal probability to generate the offspring.

In addition to these three crossover operators, various crossover operators have been produced for real-valued representations in addition to others that are problem-specific crossover operators that enhance the chance to develop the fittest offsprings [35, 63].

Unlike the crossover that is considered a binary operator, the mutation is a unary operator acting on a single individual. It aims to alter with small probability the generated offspring. The most popular mutation operator is the inversion mutation where a random gene changes its position and there exist other mutation techniques to bring small alterations to the individual and thus guarantee the diversification in the search space. The whole process will be repeated until a stopping criterion is reached, which can be a fixed number of iterations, a lapse of time that passed, or when there are no more improvements in the generated results. The genetic algorithm usually finds promising results until it deals with both intensification and diversification on the search space. For these reasons, it has been widely applied to tackle different optimization problems, especially vehicle routing problems [63–71].

7. Conclusion

In this chapter, we illustrated a technical background for different materials and notations related to the CO problem. First, we introduced CO problems and discussed the analysis and complexity of algorithms. Then, as solving a problem requires modeling it before hand, we described concepts of modeling CO problems, including graph theory, mathematical models and programming, and constraint programming techniques. Finally, we highlighted some solution methods for solving those problems, including exact algorithms, heuristics, and metaheuristics.

References

[1] G. Clarke, J.W. Wright, Scheduling of vehicles from central depot to number of delivery points, Oper. Res. 12 (4) (1964) 568–581.
[2] G. Dantzig, J. Ramser, The truck dispatching problem, Manag. Sci. 6 (1) (1959) 80–91.
[3] M. Flood, The traveling-salesman problem, Oper. Res. 4 (1) (1956) 61–75.
[4] H. Kellerer, U. Pferschy, D. Pisinger, Knapsack Problems, Springer, Berlin, Germany, 2004, https://doi.org/10.1007/978-3-540-24777-7.
[5] J.M.V. De Carvalho, LP models for bin packing and cutting stock problems, Eur. J. Oper. Res. 141 (2002) 253–273.
[6] J. Euchi, R. Mraihi, The urban bus routing problem in the Tunisian case by the hybrid artificial ant colony algorithm, Swarm Evol. Comput. 2 (2012) 15–24.
[7] J. Watkins, Across the Board: The Mathematics of Chessboard Problems, Princeton University Press, 2004.
[8] R. Lewis, Metaheuristics can solve Sudoku puzzles, J. Heuristics 13 (2007) 387–401.
[9] B. Bouzy, T. Cazenave, Computer go: an AI oriented survey, Artif. Intell. 132 (2001) 39–103.
[10] T.H. Cormen, C.E. Leiserson, R.L. Rivest, C. Stein, Introduction to Algorithms, second ed., MIT Press, 2001.
[11] M.S. Bazaraa, J.J. Jarvis, H.D. Sherali, Linear Programming and Network Flows, second ed., John Wiley & Sons, Inc., New York, NY, 1990.
[12] A.M. Newman, M. Weiss, A survey of linear and mixed-integer optimization tutorials, INFORMS Trans. Educ. 14 (1) (2013) 26–38.
[13] G.B. Dantzig, Linear Programming and Extensions, Princeton University Press, 1998.
[14] IBM, ILOG CPLEX IBM, International Business Machines Corporation, Incline Village, NV, 2009.
[15] GuRoBi, GuRoBi Optimizer, GuRoBi Optimization Inc., Houston, 2009.
[16] FICO, Xpress-MP Optimization Suite, 2008. http://www.fico.com/en/Products/OMTools/Pages/FICO-Xpress-Optimization-Suite.aspx,Minneapolis.
[17] R. Lima, IBM ILOG CPLEX What is Inside of the Box?, EWO Seminar Carnegie Mellon University, 2010.
[18] R. Fourer, D. Gay, B. Kernighan, AMPLA Modelling Language for Mathematical Programming, Thomson Brooks/Cole, Pacific Grove, CA, 2003.
[19] GAMS, GAMS Distribution 23.9.1, GAMS, Washington, DC, 2012.
[20] P. Van Hentenryck, Constraint and integer programming in OPL, INFORMS J. Comput. 14 (4) (2002) 345–372.
[21] P. Laborie, J. Rogerie, P. Shaw, P. Vilam, IBM ILOG CP optimizer for scheduling, Constraints 23 (2) (2018) 210–250, https://doi.org/10.1007/s10601-018-9281-x.
[22] K.R. Apt, Principles of Constraint Programming, Cambridge University Press, 2003.
[23] A. Bockmayr, T. Kasper, Branch and Infer: a unifying framework for integer and finite domain constraint programming, INFORMS J. Comput. 10 (3) (1998) 287–300.
[24] B. De Backer, V. Furnon, P. Shaw, P. Kilby, P. Prosser, Solving vehicle routing problems using constraint programming and metaheuristic, J. Heuristics 6 (4) (2000) 501–523.
[25] P. Shaw, Using constraint programming and local search methods to solve vehicle routing problems, in: Proceedings of the CP-98, 1998, pp. 417–431.
[26] R. Bellman, Dynamic Programming, Princeton University Press, 1957.
[27] O. Bräysy, M. Gendreau, Vehicle routing problem with time windows, part I: route construction and local search algorithms, Transp. Sci. 39 (1) (2005) 104–118.
[28] O. Bräysy, M. Gendreau, Vehicle routing problem with time windows, part II: metaheuristics, Transp. Sci. 39 (1) (2005) 119–139.
[29] P. Toth, D. Vigo, An overview of vehicle routing problems, in: The Vehicle Routing Problem, Society for Industrial and Applied Mathematics, Philadelphia, PA, 2001, pp. 1–26 (Chapter 1).

[30] M.M. Solomon, Algorithms for the vehicle routing and scheduling problems with time window constraints, Oper. Res. 35 (2) (1987) 166–324.

[31] M. Tyagi, A practical method for the truck dispatching problem, J. Oper. Res. Soc. Jpn 10 (1968) 76–92.

[32] R. Mole, S. Jameson, A sequential route-building algorithm employing a generalised savings criterion, Oper. Res. Q. 27 (1976) 503–511.

[33] B. Gillet, L. Miller, A heuristic algorithm for the vehicle-dispatch problem, Oper. Res. 22 (1974) 340–349.

[34] F. Glover, Future paths for integer programming and links to artificial intelligence, Comput. Oper. Res. 13 (5) (1986) 533–549.

[35] E.G. Talbi, Metaheuristics: From Design to Implementation, John Wiley & Sons, 2009.

[36] B. Rabbouch, F. Saâdaoui, R. Mraihi, Efficient implementation of the genetic algorithm to solve rich vehicle routing problems, Oper. Res. 21 (2021) 1763–1791.

[37] S. Kirkpatrick, C.D. Gelatt, M.P. Vecchi, Optimization by simulated annealing, Science 220 (4598) (1983) 671–680.

[38] N. Metropolis, A. Rosenbluth, M. Rosenbluth, A. Teller, E. Teller, Equation of state calculations by fast computing machines, J. Chem. Phys. 21 (6) (1953) 1087–1092.

[39] P.J.M. Van Laarhoven, E.H.L. Aarts, Simulated Annealing: Theory and Applications, D. Reidel/Kluwer Academic Publishers, Dordrecht, Boston, Norwell, Massachusetts, 1987.

[40] R.W. Eglese, Simulated annealing: a tool for operational research, Eur. J. Oper. Res. 46 (3) (1990) 271–281.

[41] S. Afifi, D. Dang, A. Moukrim, A simulated annealing algorithm for the vehicle routing problem with time windows and synchronization constraints, in: 7th International Conference, Learning and Intelligent Optimization (LION7), Catania, Italy, 2013, pp. 259–265.

[42] S. Birim, Vehicle routing problem with cross docking: a simulated annealing approach, Procedia Soc. Behav. Sci. 235 (2016) 149–158.

[43] A.V. Breedam, Improvement heuristics for the vehicle routing problem based on simulated annealing, Eur. J. Oper. Res. 86 (3) (1995) 480–490.

[44] W.-C. Chiang, R.A. Russell, Simulated annealing metaheuristics for the vehicle routing problem with time windows, Ann. Oper. Res. 63 (1) (1996) 3–27.

[45] S.C.H. Leung, J. Zheng, D. Zhang, X. Zhou, Simulated annealing for the vehicle routing problem with two-dimensional loading constraints, Flex. Serv. Manuf. J. 22 (1) (2010) 61–82.

[46] S.-W. Lin, V.F. Yu, S.-Y. Chou, Solving the truck and trailer routing problem based on a simulated annealing heuristic, Comput. Oper. Res. 36 (2009) 1683–1692.

[47] I.H. Osman, Metastrategy simulated annealing and tabu search algorithms for the vehicle routing problem, Ann. Oper. Res. 41 (1993) 421–451.

[48] R. Tavakkoli-Moghaddam, N. Safaei, Y. Gholipour, A hybrid simulated annealing for capacitated vehicle routing problems with the independent route length, Appl. Math. Comput. 176 (2) (2006) 445–454.

[49] S. Yu, C. Ding, K. Zhu, A hybrid GA-TS algorithm for open vehicle routing optimization of coal mines material, Expert Syst. Appl. 38 (8) (2011) 10568–10573.

[50] J.F. Cordeau, G. Laporte, A. Mercier, A unified tabu search heuristic for vehicle routing problems with time windows, J. Oper. Res. Soc. 52 (8) (2001) 928–936.

[51] J.F. Cordeau, G. Laporte, A. Mercier, Improved tabu search algorithm for the handling of route duration constraints in vehicle routing problem with time windows, J. Oper. Res. 55 (5) (2004) 542–546.

[52] S. Faiz, S. Krichen, W. Inoubli, A DSS based on GIS and tabu search for solving the CVRP: the Tunisian case, Egypt. J. Remote Sens. Space Sci. 17 (1) (2014) 105–110.

[53] S. Krichen, S. Faiz, T. Tlili, K. Tej, Tabu-based GIS for solving the vehicle routing problem, Expert Syst. Appl. 41 (14) (2014) 6483–6493.

[54] J.A. Sicilia, C. Quemada, B. Royo, D. Escuin, An optimization algorithm for solving the rich vehicle routing problem based on variable neighborhood search and tabu search metaheuristics, J. Comput. Appl. Math. 291 (2016) 468–477.

[55] T.A. Feo, M.G.C. Resende, A probabilistic heuristic for a computationally difficult set covering problem, Oper. Res. Lett. 8 (1989) 67–71.
[56] H. Yahyaoui, S. Krichen, A. Dekdouk, A decision model based on a GRASP genetic algorithm for solving the vehicle routing problem, Int. J. Appl. Metaheuristic Comput. 9 (2) (2018) 72–90.
[57] M. Dorigo, V. Maniezzo, A. Colorni, The ant system: an autocatalytic optimization process, Dept. of Electronics, Politecnico di Milano, Italy, 1991. Tech. Rep.
[58] Y. Li, H. Soleimani, M. Zohal, An improved ant colony optimization algorithm for the multi-depot green vehicle routing problem with multiple objectives, J. Clean. Prod. 227 (2019) 1161–1172.
[59] X. Wang, T.M. Choi, H. Liu, X. Yue, A novel hybrid ant colony optimization algorithm for emergency transportation problems during post-disaster scenarios, IEEE Trans. Syst. Man Cybern. Syst. 48 (4) (2018) 545–556.
[60] T. Yalian, An improved ant colony optimization for multi-depot vehicle routing problem, Int. J. Eng. Technol. 8 (5) (2016) 385–388.
[61] A. Fraser, Simulation of genetic systems by automatic digital computers I. Introduction, Aust. J. Biol. Sci. 10 (1957) 484–491.
[62] J.H. Holland, Adaptation in Natural and Artificial Systems, MIT Press, Cambridge, MA, 1975.
[63] J.Y. Potvin, S. Bengio, The vehicle routing problem with time windows—part II: genetic search, INFORMS J. Comput. 8 (1996) 165–172.
[64] B.M. Baker, M.A. Ayechew, A genetic algorithm for the vehicle routing problem, Comput. Oper. Res. 30 (5) (2003) 787–800.
[65] T.D. Berov, A vehicle routing planning system for goods distribution in urban areas using Google maps and genetic algorithm, Int. J. Traffic Transp. Eng. 6 (2) (2016) 159–167.
[66] P.L.N.U. Cooray, T.D. Rupasinghe, Machine learning-based parameter tuned genetic algorithm for energy minimizing vehicle routing problem, J. Ind. Eng. 2017 (2017). 3019523.
[67] S. Karakatic, V. Podgorelec, A survey of genetic algorithms for solving multi depot vehicle routing problem, Appl. Soft Comput. 27 (2015) 519–532.
[68] M.A. Mohammed, M.K.A. Ghani, R.I. Hamed, S.A. Mostafa, M.S. Ahmad, D.A. Ibrahim, Solving vehicle routing problem by using improved genetic algorithm for optimal solution, J. Comput. Sci. 21 (2017) 255–262.
[69] P.R.O. da Costa, S. Mauceri, P. Carroll, F. Pallonetto, A genetic algorithm for a green vehicle routing problem, Electron Notes Discrete Math. 64 (2018) 65–74.
[70] A. Ramalingam, K. Vivekanandan, Genetic algorithm based solution model for multi-depot vehicle routing problem with time windows, Int. J. Adv. Res. Comput. Commun. Eng. 3 (11) (2014) 8433–8439.
[71] T. Vidal, T.G. Crainic, M. Gendreau, N. Lahrichi, W. Rei, A hybrid genetic algorithm for multidepot and periodic vehicle routing problems, Oper. Res. 60 (3) (2012) 611–624.

Index

Note: Page numbers followed by *f* indicate figures, *t* indicate tables, and *b* indicate boxes.

G

H

R

S

T

Printed in the United States
by Baker & Taylor Publisher Services